Molecular Medical Parasitology

Molecular Medical Parasitology

Edited by

J. Joseph Marr, M.D.
Estes Park, Colorado 80517, USA

Timothy W. Nilsen, Ph.D.
Case Western Reserve University School of Medicine,
Cleveland, Ohio 43606, USA

Richard W. Komuniecki, Ph.D.
Department of Biology, University of Toledo,
Toledo, Ohio 43606, USA

ACADEMIC PRESS

An imprint of Elsevier Science

Amsterdam • Boston • London • New York • Oxford • Paris
San Diego • San Francisco • Singapore • Sydney • Tokyo

This book is printed on acid-free paper.

Academic Press
An Imprint of Elsevier Science
84 Theobald's Road, London WC1X 8RR, UK
http://www.academicpress.com

Academic Press
An Imprint of Elsevier Science
525 B Street, Suite 1900, San Diego, California 92101-4495, USA
http://www.academicpress.com

ISBN 0–12–473346–8

Library of Congress Catalog Number: 2002111022

A catalogue record for this book is available from the British Library

Cover figure: Adapted from *The International Journal for Parasitology*, Vol. 31, Issue 12 (October 2001), 1343–1353, Figure 1A, courtesy of Dr K. A. Joiner and The Australian Society for Parasitology Inc.

Typeset by Charon Tec Pvt. Ltd, Chennai, India
Printed and bound in Great Britain by MPG Books, Bodmin, Cornwall
02 03 04 05 06 07 MP 9 8 7 6 5 4 3 2 1

Contents

List of contributors

Mark Blaxter, Ph.D.
Institute of Cell, Animal and Population Biology
Darwin Building
King's Buildings
University of Edinburgh
West Mains Road
Edinburgh EH9 3JT
UK

Stephen M. Beverley, Ph.D.
Professor and Chairman
760B McDonnell Science Building
Box 8230
Department of Molecular Microbiology
Washington University School of Medicine
660 South Euclid Avenue
St. Louis, MO 63110-1093
USA

Nicola S. Carter, Ph.D.
Assistant Professor
Department of Biochemistry and Molecular Biology
Oregon Health and Sciences University
3181 S. W. Sam Jackson Park Road
Portland, OR 97239
USA

Juan José Cazzulo, Ph.D.
Instituto de Investigaciones Biotecnológicas
Universidad Nacional de General San Martín
Av. General Paz y Albarellos
INTI, Edificio 24, Casilla de Correo 30
(1650) San Martín, Provincia de Buenos Aires
Argentina

Debopam Chakrabarti, Ph.D.
Associate Professor
Department of Molecular Biology & Microbiology, University of Central Florida
12722 Research Parkway
Orlando, Fl 32826
USA

Christine Clayton, B.A., Ph.D.
Zentrum für Molekulare Biologie
Im Neuenheimer Feld 282
D-69120 Heidelberg
Germany

Michael J. Crawford, Ph.D.
Department of Biology
University of Pennsylvania
415 South University Avenue
Philadelphia, PA 19104-6018
USA

George A.M. Cross, Ph.D.
Andre and Bella Meyer Professor
Laboratory of Molecular Parasitology
(Box 185)
The Rockefeller University
1230 York Avenue
New York, NY 10021-6399
USA

Tim Day, Ph.D.
Assistant Professor
Biomedical Sciences
College of Veterinary Medicine
Christensen Drive
Iowa State University
Ames, IA 50011
USA

Martin J. Fraunholz, Ph.D.
Department of Biology
University of Pennsylvania
415 South University Avenue
Philadelphia, PA 19104-6018
USA

Timothy G. Geary, Ph.D.
Discovery Research
Pharmacia Corp.
301 Henrietta St.
Mailstop 7923-25-423
Kalamazoo, MI 49006
USA

Jonatha M. Gott, Ph.D.
Center for RNA Molecular Biology
Case Western Reserve University
School of Medicine
10900 Euclid Avenue
Cleveland, OH 44106-4960
USA

Arthur Günzl, Ph.D.
Associate Professor of Genetics and
Developmental Biology
Center for Microbial Pathogenesis
University of Connecticut Health Center
263 Farmington Avenue
Farmington, CT 06030-8130
USA

Erik L. Hewlett, M.D.
Professor of Medicine and Pharmacology
University of Virginia School of Medicine
Box 800419 School of Medicine
University of Virginia
Charlottesville, VA 22908
USA

John M. Kelly, Ph.D.
Pathogen Molecular Biology and
Biochemistry Unit
Department of Infectious and Tropical Diseases
London School of Hygiene and Tropical
Medicine
Keppel Street
London WC1E 7HT
UK

Richard W. Komuniecki, Ph.D.
Distinguished University Professor of
Biological Sciences
Department of Biological Sciences
University of Toledo
2801 West Bancroft St.
Toledo, OH 43606
USA

J. Joseph Marr, M.D.
President, BioMed, LLC.
General Partner, Pacific Rim Ventures
180 Centennial Dr.
Estes Park, CO 80517
USA

Richard. J. Martin, Ph.D.
Professor and Chair
Biomedical Sciences
College of Veterinary Medicine
Christensen Drive
Iowa State University
Ames, IA 50011-1250
USA

Paul A.M. Michels, Ph.D.
Research Unit for Tropical Diseases
Christian de Duve Institute of Cellular Pathology
Catholic University of Louvain
Avenue Hippocrate 74–75
B-1200 Brussels
Belgium

Geoffrey Ian McFadden, Ph.D.
ARC Professorial Fellow
Plant Cell Biology Research Centre
School of Botany
University of Melbourne
Victoria 3010
Australia

Miklós Müller, M.D.
Laboratory of Biochemical Parasitology
The Rockefeller University
1230 York Avenue
New York, NY 10021-6399
USA

Timothy W. Nilsen, Ph.D.
Professor and Director
Center for RNA Molecular Biology
Case Western Reserve University
School of Medicine
10900 Euclid Avenue
Cleveland, OH 44106-4960
USA

Fred R. Opperdoes, Ph.D.
Research Unit for Tropical Diseases and Laboratory of Biochemistry
Christian de Duve Institute of Cellular Pathology
Catholic University of Louvain
Avenue Hippocrate 74–75
B-1200 Brussels
Belgium

Marc Ouellette, Ph.D.
MRC Scientist
Burroughs Wellcome Fund Scholar
Centre de Recherche en Infectiologie
CHUQ, pavillon CHUL
2705 Boul. Laurier
Québec, QuéG1V 4G2
Canada

Richard D. Pearson, M.D.
Professor of Medicine and Pathology
Division of Infectious Diseases and International Health, Box 801378
Departments of Internal Medicine and Pathology
University of Virginia
School of Medicine
University of Virginia Health System
Charlottesville, VA 22908
USA

William A. Petri, Jr., M.D., Ph.D.
Professor of Medicine, Microbiology, and Pathology
University of Virginia Health System
MR4 Building, Room 2115
P.O. Box 801340
Charlottesville, VA 22908-1340
USA

Jenny Purcell, Ph.D.
Department of Preclinical
Veterinary Sciences R. (D.) S. V. S., Summerhall
University of Edinburgh
Edinburgh EH91QH
UK

Nicolle Rager, B.A.
Senior Research Assistant
Department of Biochemistry and Molecular
Biology
Oregon Health and Sciences University
3181 S. W. Sam Jackson Park Road
Portland, OR 97239
USA

Alan P. Robertson, Ph.D.
Adjunct Assistant Professor
Biomedical Sciences
College of Veterinary Medicine
Christensen Drive
Iowa State University
Ames, IA 50011
USA

David S. Roos, Ph.D.
Merriam Professor of Biology
Director, Genomics Institute
University of Pennsylvania
415 South University Avenue
Philadelphia, PA 19104-6018
USA

Larry Ruben, Ph.D.
Professor and Chairman
Department of Biological Sciences
Southern Methodist University
6501 Airline
Dallas, TX 75275
USA

David P. Thompson, Ph.D.
Discovery Research
Pharmacia Corp.
301 Henrietta St.
Mailstop 7923-25-410
Kalamazoo, MI 49007
USA

Aloysius G.M. Tielens, Ph.D.
Department Biochemistry and Cell Biology
Faculty of Veterinary Medicine
Utrecht University
P.O. Box 80176
3508 TD Utrecht
The Netherlands

Salvatore J. Turco, Ph.D.
Anthony S. Turco Professor of Biochemistry
Department of Biochemistry
University of Kentucky Medical Center
Lexington, KY 40536
USA

Buddy Ullman, Ph.D.
Department of Biochemistry and
Molecular Biology
Oregon Health and Sciences University
3181 S. W. Sam Jackson Park Road
Portland, OR 97239
USA

Steve Ward
Walter Myers Professor of Parasitology
Liverpool School of Tropical Medicine
Pembroke Place
Liverpool L35 5QA
UK

Preface

Parasitology was born as the tropical stepchild of medicine but has become a well recognized scientific and medical discipline in its own right in our increasingly globally conscious world. It began as a descriptive medical curiosity but the remarkable adaptive mechanisms evinced by these astoundingly versatile organisms have stimulated significant research. Many advances in basic science have come from the study of this increasingly fascinating, phylogenetically diverse group of organisms. Parasitology, in the past decade, has undergone another consequential metamorphosis. The entry of molecular biology with its elucidation of the genetics, genomics, and proteomics of these organisms has provided increasingly sophisticated explanations of their capacities to persist under intense ecological and physiological pressures.

Molecular Medical Parasitology had its inception in an earlier volume entitled *Biochemistry and Molecular Biology of Parasites*. This earlier work has been subsumed in the present text. *Molecular Medical Parasitology* presents parasitology in the context of current molecular biology, biochemistry and cell biology. Throughout the text, emphasis has been placed on the commonality of biochemical and cellular biological processes among these varied organisms. In some discussions, traditional taxonomy, which grouped certain organisms according to similarities in morphology or disease processes, has not been adhered to rigorously. This has been done judiciously in order to emphasize the universality of biochemical and molecular biological mechanisms. Wherever appropriate, information from one chapter has been cross-referenced to another in order to strengthen the important molecular relationships among groups.

The first section, entitled Molecular Biology, opens with a chapter on genomics that is the stage on which the next five chapters play. These chapters include RNA editing and processing, transcription, and post-transcriptional events and describe the interplay of molecular biology and physiology that is manifest in such specific topics as antigenic variation of African trypanosomes and the genetics of virulence.

The second section encompasses the biochemistry and cell biology of the protozoa and then the helminths. Energy metabolism, probably the most thoroughly studied aspect of the biochemistry of these organisms, is presented first in each part. In the sub-section on protozoa, chapters on amino acid and nucleic acid metabolism are followed by specific topics of special interest including surface antigens, intracellular signaling, and intracellular organelles, each with a special emphasis on the commonalities and notable differences in the genomics of the organisms involved. In the sub-section on helminths, the chapter on energy metabolism is followed by an important chapter on neurotransmitters and their receptors. These are critical to the parasite in maintaining its niche in the host and, from a medical perspective, are major therapeutic targets. This section concludes with a chapter on the structure and function of helminth surfaces with emphasis the anatomy and physiology of these critical interfaces that protect the parasites from most host defenses.

Throughout the volume, the authors and editors have emphasized the actual or potential medical importance of major biochemical or molecular biological advances. These considerations are expanded in the third section. The first chapter is on drug resistance, which, in fact, is a medical manifestation of molecular biology and biochemistry bringing about alterations in the cell biology as a result of environmental pressures. It has become a significant medical problem in recent years. The chapter on therapy discusses the implications of the basic science presented in the earlier sections as well as specifics of treatment.

This is the first parasitology text that integrates current molecular biology, biochemistry, and cell biology with the control of these heterogeneous organisms. The authors are among the best in their respective fields and the knowledgeable scientist will recognize their contributions. They have written clearly, comprehensively, and well. Presentations by these seasoned investigators should be of interest to the experienced scientist, the graduate student, and the physician.

We must list first among the acknowledgements, our authors. Much credit, however, must go to Ms. Claire Minto, an extraordinary editor, who has been an exceptional resource in the preparation of this book.

J. Joseph Marr, M.D.
Richard W. Komuniecki, Ph.D.
Timothy W. Nilsen, Ph.D.

S E C T I O N

I

MOLECULAR BIOLOGY

C H A P T E R

1

Parasite genomics

Mark Blaxter
Institute of Cell, Animal and Population Biology,
University of Edinburgh, Edinburgh, UK

INTRODUCTION

Genomics, like parasitology, is a research field that thrives on the intersection between different disciplines. Parasitologists study a phylogenetically disparate assemblage of organisms chosen from global diversity on the basis of their trophic relationships to other 'host' organisms, and use the tools and paradigms of biochemistry, molecular biology, physiology and behaviour (amongst others) to illuminate the biology of these important taxa. Genomics uses data arising from karyotypic analysis, genetic and physical mapping of traits and anonymous markers, DNA sequencing and bioinformatic prediction of function-structure relationships. The meld of parasitology and genomics is thus necessarily and productively hybrid.

Genomics research in parasitology can be divided, pragmatically, into two sectors. One is a drive to generate resources: clone banks, sequence, annotated genes, functional genomics platforms. The other is a hypothesis-driven search for pattern and process in the structure, expression and evolution of genomes: how does the organism self-assemble given this set of genotypic data? These two sectors overlap, as resource generation necessarily underlies the testing of hypotheses of genome-wide function. While the methodologies used to analyse the genomes of protozoan, nematode and platyhelminth genomes may differ because of the ways the genomes of these organisms are organized, the aims of programs on individual species are in general the same:

1. The determination of the complete sequence of the chromosomal (and plastid) genome of the organism
2. The identification of the coding genes (both protein and RNA) on the sequence (also termed 'gene discovery')
3. The prediction of function of each of the genes, and the prediction of function of operator/promoter/control regions in the non-coding DNA
4. The integration of functional, sequence and architectural information into biological models of the structure of the chromosomes

Molecular Medical Parasitology
ISBN 0-12-473346-8

and of the interaction between the expressed parts of the genome

5. Investigating natural variation in the genome in the context of the host, population structure, drug treatment and other selective forces.

Along this difficult path additional goals can be found, such as the identification of candidate sequences, genes, or gene products that may be of utility in diagnosis, surveillance, drug targeting or vaccine component development.

Genomics and genome sequencing is still a young field. The first genomes sequenced were those of parasites: viruses infecting bacteria (phiX174 and lambda phage are landmarks). Progress to whole genome sequencing of self-reproducing organisms had as stepping-stones the determination of the complete genome sequence of the human mitochondrion (again relevant to parasitology as mitochondria arise from an ancient symbiotic event). The first genome sequence determined for a self-reproducing organism was that of *Helicobacter pylori*, an important human-pathogenic bacterium. In the field of bacterial genomics, the focus has remained on pathogenic species, and most of the over 100 sequenced genomes are from human pathogens. For parasitology, these genomes give insight into the differences at the level of the genome between free-living bacteria (such as *Escherichia coli*) and endoparasitic bacteria (such as the Chlamydiae and Rickettsiales). Importantly, it is now technically feasible to sequence the genomes of eukaryotes with large genomes (>20 Mb) and thus several parasite genome projects are underway. As with the sequencing of the nematode *Caenorhabditis elegans*, the fly *Drosophila melanogaster* and the human genomes, this will in turn bring about a revolution in the way parasite biology research is done.

THE SIZE OF THE PROBLEM

Bacterial genomes are relatively small (0.6 to 15 Mb) compared to those of eukaryotes (10 Mb to >10 000 Mb). Parasitic eukaryote genomes range from ~9 Mb (*Theileria*) to 5000 Mb (*Ascaris*) and above (Table 1.1). The number of genes encoded by a genome is roughly proportional to its size, but is modified by the presence of intronic DNA and of junk, or non-coding repetitive DNA. For example, while the genome of the nematode *Caenorhabditis elegans* is 100 Mb, and contains 20 000 protein coding genes, the human genome is 3000 Mb (30-fold larger) but encodes only 30 000–40 000 genes. The average gene density in *C. elegans* is thus about one gene per 5 kb, while in humans it is one gene per >70 kb. Protozoa have relatively small genomes that are often rich in non-coding repeats, and are likely to have in the region of 6000 to 15 000 protein coding genes. Parasitic nematodes have genomes of a similar size to *C. elegans* in the main, but several species have much larger DNA contents per haploid genome. In *Ascaris* and related taxa, the genome is both highly repetitive and much larger than that of *C. elegans*. Overall, parasitic nematodes are likely to have similar gene complements to *C. elegans* (20 000). The genomes of platyhelminths are much less well known, but *Schistosoma* species have genomes of ~270 Mb that are rich in repetitive sequences. Again the gene count is likely to be in the 20 000 range. Arthropod parasites have larger genomes than the model arthropod, *Drosophila melanogaster*, which has 15 000 genes in a 160 Mb genome. For example, *Anopheles* has a genome of 280 Mb but is expected to have a gene count similar to *D. melanogaster*.

The multitude and phylogenetic diversity of parasites means that genomic approaches to parasite biology and control need to be carefully

TABLE 1.1 Parasite genomes: genome sizes, karyotype, gene number and genome project status of selected parasites

Species	Genome size	Karyotype (2n)	Genome survey sequences in dbGSS	Genome sequencing status	Methods used
Nematode parasites					
Brugia malayi	100 Mb	12	18 000	Selected genome segments	
Ascaris suum	5000 Mb	48		–	
Haemonchus contortus	100 Mb	12		–	
Platyhelminth parasites					
Schistosoma mansoni	270 Mb	14	42 000		Physical map based
Apicomplexan parasites					
Plasmodium falciparum	35 Mb	14	(also 18 000 GSS from other plasmodial species)	Completed	Whole genome and chromosome-by-chromosome shotgun
Theileria annulata	10 Mb	4		Genome sequencing initiated	
Trypanosomatid parasites					
Trypanosoma cruzi	35 Mb Haploid	35	21 000	Near completion	Chromosome-by-chromosome and physical map based
Trypanosoma brucei	35 Mb	11	90 000	Near completion	Chromosome-by-chromosome shotgun and physical map based
Leishmania major	33.6 Mb	36	15 000	Near completion	Chromosome-by-chromosome shotgun and physical map based
Other protozoan parasites					
Entamoeba histolytica	<20 Mb	14	80 000	Near completion	Whole genome shotgun
Giardia intestinalis	12 Mb	5		Near completion	Whole genome shotgun
Vectors of parasites and arthropod parasites					
Anopheles gambiae	280 Mb	3	60 000	Near completion	Whole genome shotgun

tailored to each target organism. The World Health Organisation in collaboration with the national funding agencies of both endemic and developed countries have therefore sponsored genome projects on target organisms representing the major human and animal parasitic diseases. Each project has used tools based on the peculiarities of their system and the knowledge/skill base present in the interested community. The parasite genome projects are models of north–south, endemic–developed cooperation, and, in this open spirit, most of the data produced is freely available through the internet to interested researchers.

GENERATING GENOMICS DATA

Genomics uses data from many sources. The parasite genome projects use layers of related data types to build first physical and genetic maps of the target genomes, followed by finer detail sequence and expression maps, ultimately yielding an annotated genome. Most of the projects are still in the midst of the data generation part of the process (see http://www.ebi.ac.uk/parasites/parasite-genome.html for the latest news on the various parasite genome projects), and no simple summary will adequately cover all the projects. The field is also changing extremely rapidly, and a summary given today may be rendered obsolete with tomorrow's database release.

Genetic maps

Genetic maps are available for many parasitic organisms. The maps are built by examining the genotype of recombinant cross progeny of marked parents. The markers can be phenotypic (eye color, resistance to filarial nematode infection) or anonymous genetic markers (microsatellite sequence tagged sites or restriction fragment length polymorphisms for example). The result is a linkage map showing the association of the markers and their relative order. This map is of utility in verifying the correctness of related genome maps made at the physical (DNA) level, as markers placed adjacent by genetics should also be adjacent in any physical map. Genetic mapping is necessarily restricted to organisms that reproduce sexually, and operationally is further restricted by considerations of practicality (is it possible to carry out controlled crosses and score progeny in the laboratory?).

To overcome this need for sex, a method for genetic mapping without sexual recombination has been developed, called HAPPY mapping. HAPPY mapping is based on the observation that in a population of large DNA fragments generated by random shearing of a complete genome, the chance that two sequence tagged markers will be on the same individual molecule is proportional to their separation on the genome. This mapping procedure uses PCR-based genotyping assays to screen sub-haploid quantities of sheared DNA for association between markers, and the association is then used to build a 'genetic' map as one would with real genetic data. The benefit of the HAPPY map is that the markers are cloned and sequenced at the outset, allowing rapid progression to complete physical mapping (see below).

Karyotyping

Chromosomes are the units of genome organization. Mapping of genes or other molecular markers to physical chromosomes is a useful and often central step in genomics. At a gross level, chromosomes can be separated by morphology (for example the filarial nematode sex determining X and Y chromosomes) and by differential banding staining with intercalary dyes. In the protozoa, the chromosomes are often too small to be resolved usefully by microscopy, but are within the range that is resolved by pulsed field gel electrophoresis (PFGE). PFGE karyotypes are available for all of the major parasitic protozoan species, and comparative karyotyping of strains and related species has yielded valuable information on conservation of linkage, and patterns of genome evolution. Fluorescence *in situ* hybridization (FISH) involves the 'painting' of a chromosomal copy of a gene with a fluor-labelled probe in a preparation of metaphase cells. It is useful in confirming linkage of cloned markers, and in joining otherwise unlinked segments of a physical map. For chromosomes separable by PFGE, Southern hybridization can be used to similar effect.

For many organisms, including the nematodes, the chromosomes are too large (>10 Mb) to be separated by PFGE and too small to be useful for FISH and banding studies. It is possible to separate these chromosomes using a fluorescence-activated cell sorting instrument, though this technique has not been used yet in parasite genomics.

Physical maps

It is often useful to have a genomic copy of a gene of interest cloned. Large-insert genomic DNA clones can be constructed in a number of different vector–host systems. These range from lambda bacteriophage (maximal insert capacity ~21 kb of foreign DNA), through cosmids (~35 kb), bacterial artificial chromosomes (BAC, ~200 kb) and yeast artificial chromosomes (YAC, ~3000 kb). Each vector–host combination also differs in copy number within the host cell: in general vectors maintained at low copy number tend to be more stable against recombination, rearrangement and deletion. Yeast host cells are often more tolerant of skewed base-composition insert DNA, such as that from *Plasmodium*, and of repeat-rich insert DNA.

The inserts of large-insert clone libraries can be compared to each other and the overlap data used to build a map of the cloned genome, a physical map. Overlap between clones can be predicted in two ways. One is derived from restriction enzyme fingerprinting of each clone. A fingerprint is the pattern of bands observed when the clone is cut with one or two enzymes. Clones containing DNA from the same genomic region will share more fingerprint bands than would be expected to occur by chance, and can be overlapped on the basis of shared fragments. The other method of building a physical map is by sequence-tagged site mapping, where the library is screened with probes by hybridization or clones are identified using STS-based PCR. The two methods (fingerprinting and STS mapping) can be, and usually are, combined in the production of a map. Maps have been produced or are in production for many parasites. FISH hybridization to spread chromosomal segments can also be used to build maps, and for smaller genomes it is possible to construct restriction fragment-based maps using stretched chromosomes cut *in situ* on the slide.

Physical map construction is compromised by the sheer volume of data that must be produced and analysed, and the known sorts of confounding errors that can occur. In fingerprinting, there is (usually known) error in band size estimation, and two bands can be scored as matching by size despite being different in sequence. The method is very sensitive to the number of shared bands required to score a real overlap, as too high a score requirement will result in failure to link overlapping clones, whereas too low a score will result in multiple, incompatible overlaps being accepted. In STS mapping, errors can arise from the presence of repetitive sequence (either duplicated genes or non-coding repeats) that will wrongly join two distinct genomic regions. The number of clones required to map a genome depends on the genome size, the mean size of the inserts (and the distribution around the mean), and the representativeness of the library. Some genomic regions clone poorly (if at all), and often multiple cloning systems must be used to obtain closure.

The experience of the *C. elegans* project is of relevance here, as it was one of the first to build a 'complete' physical map. A map built from 17 000 cosmid fingerprints yielded ~3000 contigs of overlapping clones. When a larger-insert YAC (yeast artificial chromosome) library was added to the map by hybridization, the number of contigs dropped to ~600. Much additional work, involving constructing and screening libraries in multiple vector–host systems, was necessary to achieve the final 98% coverage. In *P. falciparum*, YACs have also been used to construct a physical map, but in this case the process was facilitated by using hybridization to PFGE blots, and hybridization of PFGE-separated chromosomes to YAC libraries, to assign cloned YACs to chromosomes. Similarly, in *S. mansoni*, FISH is being used to assign clone contigs to the chromosomes.

Genome sequencing

Expressed sequence tags

The genome sequence of a parasitic organism can be obtained in a number of convergent ways. The choice of experimental route is dependent on the resources available, the genomic biology of the organism and the needs of the researchers. For some parasites, genomics effort has focused on gene discovery, and rapid and cost effective methods have been used to obtain sequence tags on many of the genes of the organisms. The coding portion of a genome (the portion that is transcribed as RNA, and is translated to give protein) is typically less than 50% for eukaryotic organisms. For eukaryotes with introns this proportion drops further still. A method that sampled and sequenced only the expressed portions of a genome would thus be an efficient gene discovery tool.

The expressed sequence tag (EST) strategy is one such method. To generate ESTs, a cDNA library, representative of the genes expressed in a particular stage, sex or tissue of the parasite, is sampled at random. From each random clone, a single-pass sequence is generated. This sequence serves both to tag the transcript from which it derived, and also offers sequence data that can be used to perform informatic analyses to identify the function of the encoded protein.

For some parasitic genome projects, ESTs are the main or only mode of genomics data production, while in others they play a minor role: the balance is based on the needs and opportunities available for each target species. The dbEST division of the public databases contains over 10 million ESTs, from over 390 organisms (Table 1.2). Of these, only ~2% (200 000) are from parasitic organisms and their vector hosts, but parasites make up ~15% of the different species represented. The overrepresentation by species is due to the generally smaller size of the parasitic datasets than those from humans and model organisms. EST acquisition is relatively cheap, and is a 'low tech' genomics option for laboratories and communities without funding and infrastructure for larger programs. The yield, in terms of 'interesting new genes' per unit of effort, is very high, and EST projects can substitute for more 'hypothesis-driven' gene cloning efforts where the aim is to define the transcriptional features of a particular stage or tissue of the parasite of interest.

The diversity of genes represented in an EST dataset reflects not only the size of the dataset,

TABLE 1.2 EST datasets from parasitic organisms (December 2001)

Species	Number of ESTs	Expected number of genes per genome	Species	Number of ESTs	Expected number of genes per genome
Nematode parasites	104 222	20 000	Apicomplexan parasites	39 138	7 000
Brugia malayi	22 439		*Plasmodium yoelii yoelii*	12 471	
Onchocerca volvulus	14 922		*Eimeria tenella*	11 438	
Strongyloides stercoralis	11 392		*Plasmodium falciparum*	6 769	
Ascaris suum	7 410		*Plasmodium berghei*	5 345	
Ancylostoma caninum	7 259		*Plasmodium yoelii*	3 091	
Strongyloides ratti	6 562		*Theileria parva* and *T. annulata*	24	
Meloidogyne javanica	5 600				
Haemonchus contortus	4 843		Trypanosomatid parasites	17 479	10 000
Parastrongyloides trichosuri	4 541		*Trypanosoma cruzi*	10 133	
Heterodera glycines	4 327		*Trypanosoma brucei*, *T. b. brucei* and *T. b. rhodesiense*	5 133	
Trichinella spiralis	4 238				
Meloidogyne arenaria	3 334				
Trichuris muris	1 388		*Leishmania major*, *L. infantum* and *L. mexicana*	2 213	
Globodera pallida	1 246				
Ancylostoma ceylanicum	1 110				
Necator americanus	961		Other protozoan parasites	1 070	6 000+
Globodera rostochiensis	894		*Entamoeba histolytica*	463	
Toxocara canis	519		*Acanthamoeba healyi*	377	
Ostertagia ostertagi	450		*Entamoeba dispar*	139	
Teladorsagia circumcincta	315		*Giardia intestinalis*	91	
Litomosoides sigmodontis	198		Vectors of parasites and arthropod parasites	9 587	15 000+
Wuchereria bancrofti	131				
(seven additional species with less than100 ESTs each)			*Anopheles gambiae*	6 037	
			Aedes aegypti	1 518	
Platyhelminth parasites	19 709	20 000	*Biomphalaria glabrata*	1 426	
Schistosoma mansoni	16 813		*Sarcoptes scabiei*	396	
Schistosoma japonicum and *S. haematobium*	2 097		*Boophilus microplus*	143	
			Culex pipiens pallens and *Anopheles stephensi*	64	
Echinococcus granulosus	799				

but also the representativeness of the library, and the faithfulness of the cDNA cloning procedure. In general, the abundance of ESTs corresponding to one gene transcript will reflect the steady state mRNA levels of the transcript in the organism, but smaller mRNAs are reverse transcribed and cloned more easily than larger ones, and thus some bias can arise. To gain access to low-expression-level transcripts, a large number of ESTs must be sequenced from a given library. To improve the efficiency of new gene discovery in EST libraries, subtraction or normalization procedures can be carried out. Normalization aims to make the levels of each transcript in the library approximately equivalent by selecting against highly expressed genes. Subtraction aims to eliminate from the library sequences that derive from transcripts also present in another stage or tissue. Normalization and subtraction can be carried out previous to the cloning stage, or on libraries by hybridization-elimination of unwanted clones.

The EST strategy brings with it problems. One is that the EST dataset will only be as good as the library from which it derives. Another is that it becomes increasingly more difficult to identify new genes as sequences are accumulated from a species: the yield of new genes per sequence can drop to less than one per ten ESTs quite rapidly. If libraries are not available from all life-cycle stages, it will not be possible to sample all the genes of the organism, as many will have close stage-specific regulation. Finding rarely expressed genes will be a stochastic process. For example, even in 'mature' genome projects, EST analysis typically yields only ~50% of the genes later discovered by genome sequencing. Normalization and subtraction can improve this, but in general the returns for effort fall off rapidly. The bioinformatic analysis of ESTs is discussed below.

Genome survey sequencing

ESTs can only sample a gene when it is expressed in the tissue or stage from which the cDNA library is made. In addition, many genes may be expressed at such low levels (for example in one neuron of a metazoan parasite), or in very particular environmental circumstances (for example during the process of entering a new host cell) that it is very unlikely that they will be identified by ESTs. While genes are represented in cDNA libraries in proportion to their level of expression, in genomic DNA libraries they are present in proportion to their representation in the genome. For relatively unbiased gene discovery the genomic equivalent of ESTs, genome survey sequences (GSS), are often useful.

GSS surveys of eukaryotic genomes yield interagenic, non-coding, intronic and coding sequences. This feature makes them ideal for establishing an overview of the patterns of sequence present in a genome. For organisms with few or no introns, and thus relatively gene-dense genomes, GSS surveys can be as efficient as EST surveys for gene discovery. In *Trypanosoma brucei* for example, coding sequence makes up ~50% of the genome, and thus GSS surveying yields many open reading frames. For metazoan genomes, which typically contain less than 25% coding sequence, GSS are less efficient at gene discovery, but are still a valuable adjunct to EST-based analyses.

In addition to coding genes, GSS can help define repeat sequences useful for diagnostic or population genetics programs, and also reveal features such as transposons and retrotransposons. As outlined below, the GSS concept can be taken further to provide shotgun sequencing for most of a genome. GSS are usually determined from large insert libraries in cosmid or BAC vectors, using primers that yield reads of each end of the cloned insert (and thus called, for example, 'BAC end sequences').

Map-based genome sequencing

For genomes where it is valid and possible to proceed to whole genome sequencing, two main approaches are taken. One is to use the resource of a physical map of the genome, and to sequence it 'clone by clone'. The other is to perform a shotgun sequencing project on the whole genome, or PFG separated individual chromosomes. Both approaches require the ability to assemble large (usually >35 kb) genomic segments of DNA from a large number of individual reads of ~500 bp each. To robustly and credibly assemble a sequence, it is usual to first make one (or several) small insert sublibraries (inserts of ~2 kb) from the target fragment, and to sequence a number of these selected at random to give a 6–10-fold sequence coverage of the fragment. This is called shotgun sequencing. The redundancy of 6–10-fold is required both to ensure the correct sequence is determined free of errors (by verifying the nature of any particular base using independent sequence reads) and to attempt to cover the whole fragment. If the whole of the fragment were equally clonable, then ~6-fold coverage would be required to ensure that all regions of the fragment are sampled. In practice, not all regions are equally clonable, and assembly of shotgun reads is usually followed by a 'finishing' phase where missing regions are sequenced by more directed methods (such as primer walking) and ambiguities in the sequence are clarified.

Sequencing a genome using a physical map thus involves a large number of small shotgun-assembly-finishing projects based on a set of clones that overlap minimally and cover the whole (or most of) the genome. The regions of overlap between the clones serve to add confirmation to the determined sequence and to the mapping process.

Rather than build a map first, a map-as-you-go strategy has been proposed, utilizing extensive GSS data from large insert clones. The BAC or other library is first completely end sequenced to yield one GSS every 5–10 kb of the genome on average. A BAC clone is selected (at random) and shotgun sequenced. Using the finished sequence, a new clone that minimally overlaps is selected on the basis of end sequence comparison. In this manner, the project can 'walk' from each seed clone into the flanking genome.

Shotgun sequencing of whole chromosomes and whole genomes

Initially, the complexity of genomic DNA, in particular the presence of local and disseminated repeats, suggested that it might not be possible to assemble and finish fragments of DNA larger than ~100 kb. The limits were set by the efficiency of the computer algorithms used for assembly, and the overall sequencing strategy. However, with advances in sequencing, algorithms and computing power, it is now possible to assemble even the largest genomes in one go, from tens of millions of individual reads (Figure 1.1).

The success of the clone-by-clone strategy is based in part on its breaking down of the assembly problem in to a set of smaller, manageable ones. For many protozoan parasites, chromosomes can be separated by PFGE. These chromosomes, ranging from <100 kb to over 7 Mb, can also be shotgun sequenced and thus the whole-genome assembly is reduced to a set of smaller projects. Chromosome shotgun sequencing is now a mainstay of many protozoan genome sequencing projects. For organisms with larger genomes, where chromosomal separations are not possible, the success of the human (3000 Mb) and *Drosophila melanogaster* (160 Mb) whole genome shotgun assemblies suggests that this method should also be applicable to metazoan parasites.

For whole genome and whole chromosome shotgun projects, the large numbers of reads from small insert (2 kb) clones are usually supplemented with reads from shotgun libraries with larger mean insert size (10 kb and above; BAC end GSS are also used). These longer clones serve as a scaffold that is used to orientate and affirm the assembly made with the 500 bp reads from the 2 kb clones. If a sequence contig suggests that the two ends of a large insert clone are too close to each other, it is likely to be in error. Similarly, large-insert sequences can serve to link contigs derived from shotgun sequencing.

The whole genome shotgun method requires a large amount of data to be effective. For a 100 Mb genome (such as is found in many nematode parasites) a one-fold shotgun is

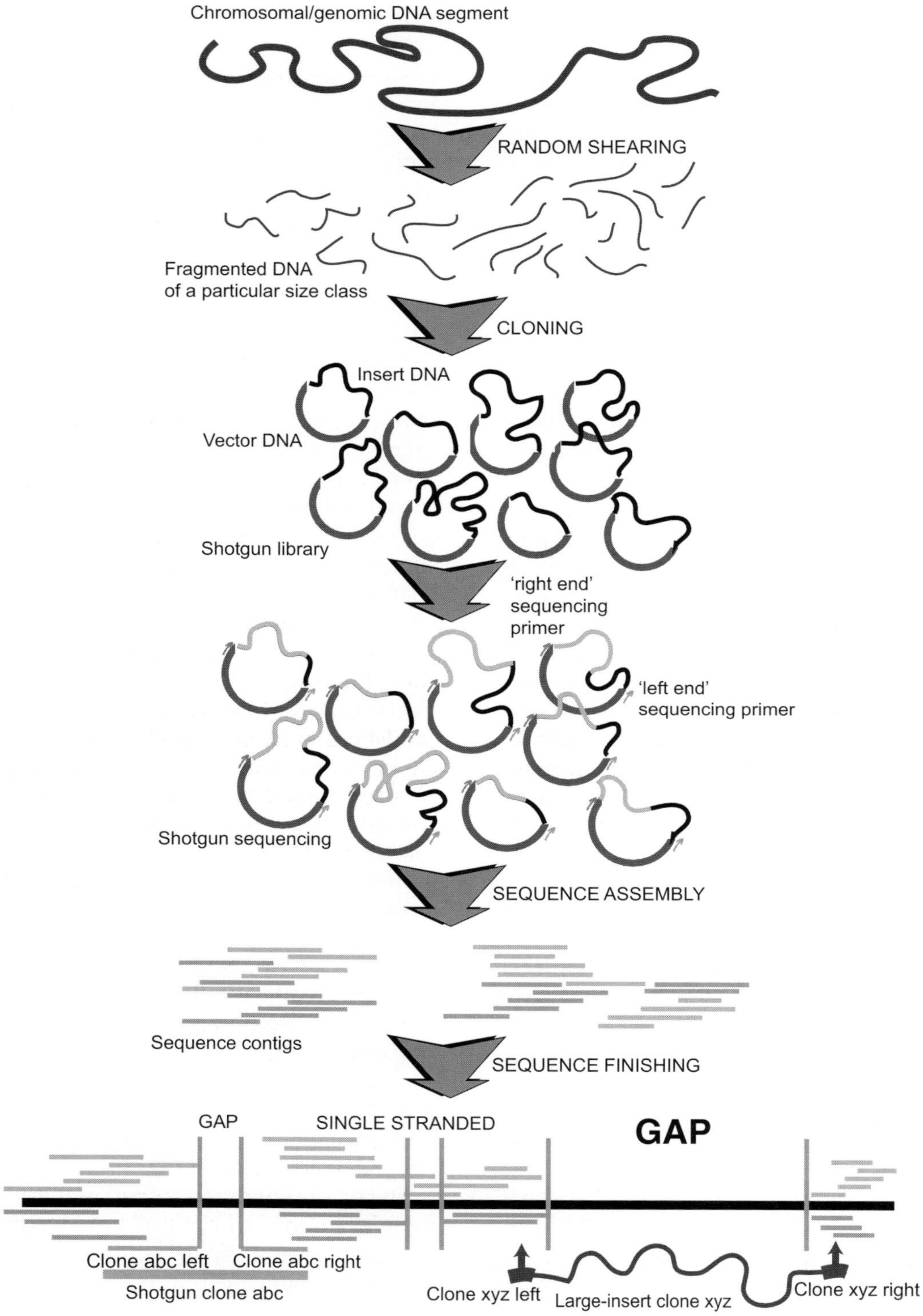
Chromosomal/genomic DNA segment
RANDOM SHEARING
Fragmented DNA
of a particular size class
CLONING
Insert DNA
Vector DNA
Shotgun library
'right end'
sequencing
primer
'left end'
sequencing primer
Shotgun sequencing
SEQUENCE ASSEMBLY
Sequence contigs
SEQUENCE FINISHING
GAP
SINGLE STRANDED
GAP
Clone abc left
Clone abc right
Shotgun clone abc
Clone xyz left
Large-insert clone xyz
Clone xyz right

~200 000 sequences of 500 bp. A ten-fold shotgun is thus ~2 million sequences. For each shotgun project, the reads have to be assembled with resolution of ambiguities resulting from the presence of repetitive DNA. The finishing process is necessarily more protracted, and utilizes data inherent in the scaffold of larger insert clones, and often also long-range restriction mapping and PFG southern blotting.

A genome sequence is a hypothesis

The result of a shotgun sequencing project, be it of a clone, a chromosome or a genome, is a DNA sequence that has been verified to the best ability of the sequencers. The error rate in most sequencing projects is estimated to be one miscalled base in 10 000. This is a maximal estimate of the error, and often independent resequencing surveys reveal much lower actual error rates. The final public sequence must therefore be regarded as a hypothesis, and used and interpreted with reference to the strength of the supporting evidence. In particular it is technically difficult to resolve the sequence of tandem short repeats and regions of low complexity or biased base composition. The genome sequencers will strive to resolve all conflicts in the data, but rely on the user communities to communicate to them any errors found.

During the shotgun and finishing phases of genome sequencing projects, many genome sequencing centres will release preliminary assemblies of the data. These are works in progress, and users should be aware that contiguated sequences present in one day's preliminary data release might be absent from the next due to the discovery of errors in assembly. Even the published sequence will change as errors are corrected, and care should be taken to use the latest data release in analyses.

The problems and benefits of the reference strain

Sequencing a genome, even the relatively small genomes of bacteria, is a major undertaking, and resources are unlikely to be available for multiple genome sequences of disease organisms. For model organisms, the choice of strain for full genome sequencing is often obvious: the history of genetic research will have defined an isolate or strain as being the 'wild type', and this will be the most appropriate for sequencing. Thus, for *C. elegans* the N2 strain was sequenced, and for *Arabidopsis thaliana*, the Landsberg ecotype was chosen. For parasites the choice is rarely as easy. Parasitologists are often interested in the diversity of their target organisms, and research focuses on between-strain and between-species variation in virulence and other important traits. The best studied, best 'domesticated' strains used in many laboratories may have lost key traits during adaptation to laboratory hosts, or due to inbreeding or other genetic selection. The best known strain may include in its phenotypes the

FIGURE 1.1 (See also Color Plate 1) Sequencing a genome. This figure illustrates the steps involved in determining the sequence of a genome (or genomic segment, such as a chromosome or large-insert clone). The DNA is first sheared and cloned to make a shotgun library, which is then sequenced using universal 'left' and 'right' primers to generate a shotgun sequence set for the segment. This shotgun sequence set is then assembled into contigs of sequence reads that overlap each other. These contigs are turned into a finished product through the use of linking clone data: either matching left and right reads from individual clones within the shotgun sequence dataset, or end-sequences from larger-insert clones. It is usual to finish the sequence by obtaining sequence from both strands, resolving all ambiguities in the predicted consensus, and linking all contigs.

loss of many of the most critical phenotypes of the parasite as experienced in the wild. In addition, much research is carried out on model parasites, often natural parasites of model laboratory hosts. A conflict can then arise between the desire to know the genotype of the disease organism, and the utility of knowing the genotype of the well tested model. Parasite genome projects have struggled with this conflict and different outcomes have been based on balancing competing needs. For filarial nematodes, the choice of human parasite was simple: only one can readily be maintained in laboratory culture, and one 'strain' of this species is almost universally used in research: thus the TRS strain of *B. malayi* was chosen. For leishmaniasis, the many species causing different disease syndromes suggested that it might be necessary to examine multiple genomes, and even multiple strains within each: a compromise was reached once it became evident that *Leishmania* species share extensive synteny, and thus a well studied *L. major* strain was chosen. Users of genome data should be aware of this necessary simplification in considering and using genome data: the genotype determined may not reflect the highly variable genotypes of real-world populations.

BIOINFORMATICS AND THE ANALYSIS OF GENOME DATASETS

The completion of a genome sequence is not the final product. The wealth of data encoded in the millions of contiguous bases must be interpreted and linked to biology. The size of genome datasets has required the development of a new way of analysing data in biology, generally grouped under the term bioinformatics. The linear sequence of the genome DNA encodes (in response to the environment) the four-dimensional organism. Bioinformatics aims to 'compute' the organism given the DNA sequence.

All of the parasite genome projects include a dedicated bioinformatic component, and learning the concepts and skills of bioinformatics is essential if parasitologists are to exploit the data in their research. The informatics of interpretation of linear strings of data, in terms of pattern recognition and 'emergent' higher order properties, has been well developed outside biology, but biology-driven informatics is now a dynamic and fruitful field. The genome sequence can be likened to a book, where each character has been painstakingly determined, but where we have little idea of the language, syntax and grammar in which it was written. Bioinformatics aims to derive language, grammar and syntax from the data. The project starts from some universals, such as the genetic code, and a general view of structural features such as exons and introns, but the rest has to be derived from the data. For example, given a conflicting set of exon predictions (open reading frame segments flanked by valid splice sites), is it possible to predict with accuracy the correct gene model, and thus the correct encoded protein sequence? Questions such as these are yielding to ever improving gene-finding and predictive algorithms. Features such as overall sequence complexity, the presence of local or global repeats, the pattern of di- or tri-nucleotide sequences, the relationship between repeats and predicted exons, and the pattern of exonic versus intronic or extragenic DNA can now be computed and used to annotate a genome.

Many of the tools used are based on probabilistic methods, and thus need a training set of 'known' genes from the organism of interest in order to start extracting information. As a genome project advances, the training set can be enlarged to include the predictions achieved, and thus the algorithms can bootstrap to afford higher levels of accuracy in prediction. It is important for a genome project to reanalyse and reannotate sequence segments 'finished' in the early phase with the tools developed for annotation as the project advances: new genes can be found and old definitions refined on the basis of larger datasets. Probabilistic methods can also be used to define RNA-coding genes such as tRNAs and the new class of microRNAs that play roles in gene regulation.

One powerful method of annotation of genome data is by comparison to genes from other organisms. Depending on the closeness of the phylogenetic relationship between the organisms, this comparative genomics approach can yield data on promoter and other regulatory, non-protein-coding regions, on the pattern of exons of otherwise anonymous genes, and on the presence of potential open reading frames encoding conserved peptide segments. The simplest comparative method is the use of database similarity search algorithms such as BLAST that find segments of significant similarity between sequences. If the query sequence is a genome dataset, and the database searched is a protein database (the BLASTX option for example) then segments with high similarity scores will probably identify coding exons in the DNA. These BLAST similarity matches can then be used to confirm gene predictions made on the basis of DNA-intrinsic signals such as open reading frames and splice sites.

Bioinformatics annotation tools also provide information to help with the finishing phase of genomic analysis: for example, a predicted DNA segment that had similarity to a known protein split by a frameshift might suggest that the segment contained a sequencing error. This recursive interaction between annotation and finishing aids both in providing a higher quality product. Annotation tools are also used in parallel. Prediction of a feature using two independent tools suggests that it is much more likely to be real.

Analysis of EST sequences

EST sequences define gene transcripts. Because they are short, and single-pass, they do not necessarily yield information about the complete and correct sequence of the original mRNA. To achieve this sort of data quality, ESTs are first grouped together into clusters. Clusters of ESTs are built using sequence identity information (identical sequence segments in two ESTs suggest that they derive from mRNAs derived from the same gene) and also clone-of-origin information (many EST projects derive two reads for each clone, one from either end of the insert). These clusters of ESTs can then be used as a substrate on which to perform additional analyses. Importantly, the clustered dataset will be less complex (in terms of numbers of individual objects requiring analysis) than the original EST dataset: the usual mean compression ratio for low-number (<20 000) EST projects is between 1.6 and 4 ESTs per cluster. The clusters can be aligned to each other and consensus sequences predicted, which should be more reliable than the single pass ESTs. For datasets generated from non-normalized libraries, the relative abundance of ESTs per cluster will reflect the abundance of the original mRNAs in the organism, and thus some expression pattern information can be derived from the clusters. Clustered datasets are also much more useful for downstream 'functional genomics' programs, as they allow one (or a few) clones to be selected for each putative gene. ESTs are also important for genome annotation, as they confirm that a segment of DNA is indeed expressed as mRNA.

Several integrated tool sets are available for analysis of EST datasets. These usually include clustering by identity, and some BLAST-based analysis. More complete analyses require an underlying genomic database, often customized to deal with the 'fragmented' genome data resulting from ESTs.

Analysis of genome sequence

Genome annotation is an art that is evolving into a science. The basis of genome annotation is a reference sequence housed in a database. This database can then be filled with annotation 'objects' that refer to coordinates of the sequence (and to each other). Thus each parasite genome project, as it has matured, has developed a genome database. These range from simple, hypertext mark-up language (html) websites to full relational databases with interactive graphical viewers.

Annotations can be thought of as 'hard' and 'soft'. Hard annotation, given a completed sequence, refers to objects that rely only on the primary DNA sequence such as possible open reading frames, splice sites, and local repeats. Soft annotation refers to comments on the sequence and its encoded proteins based on their relative similarity to other objects annotated in other genomes. These soft objects include such things as functional identification of peptides based on BLAST similarity to a protein of known function, or decoration of DNA sequence with promoter motifs derived from probabilistic searches.

With complete genomes to hand, whole genome analyses can be carried out to infer the presence of repetitive DNA segments of various types, and to perform within-genome classification of encoded peptides into protein families. These sorts of analyses are important in defining the overall structure of the genome, and in identifying novel features of its encoded genes.

The annotation snowball

The annotation of a sequence as having a particular property should be qualified by the relative goodness-of-fit of that sequence to the features of the property. For assigning function, sequence similarity to a protein of known function is often used. There is a hidden, but known, danger in this process: what if the functional assignment of the protein of 'known' function was also based on similarity to another protein? And what if the chain of functional assignment is long (many steps before the original biologically verified functional assignment is accessed), or one of the assignations is wrong? For example, many proteins have been functionally characterized through their roles in model organisms defined by genetic studies. If the parasite lacks the structures or biochemistry displayed by the model organism, the 'function' may not exist in the parasite, even though its proteins are annotated with the function. One classic example, of relevance to parasites, is that of the DoxA2 'phenoloxidase' of *Drosophila melanogaster*. This gene was identified by mutation as controlling tanning of the insect cuticle, and knockouts lacked active phenoloxidase. Such phenoloxidase activity could underlie the crosslinking and tanning of nematode cuticles and eggshells, and of platyhelminth eggshells. The DoxA2 phenoloxidase has homologues in many other genomes, including nematodes. However, there are also homologues in protozoan parasites (such as *P. falciparum* and *T. brucei*) not expected to encode such activity. In addition, plant homologues appear to locate to a perinuclear site. DoxA2 is in fact a regulatory component of the 26S proteasome, which in *D. melanogaster* is involved in targeting the pre-pro-phenoloxidase protein for processing: mutants in DoxA2 are indeed negative for phenoloxidase, but not because the phenoloxidase gene is disrupted.

Analysis of the parasite genes under the misapprehension that they are tanning enzymes would be a mistake.

This possibility of promulgation of low-quality (or simply wrong) annotation through databases by association is termed the 'annotation snowball' and users of annotated genome data would be advised to check the functional assignments made for their chosen proteins, and to communicate any revised assignments to the genome database curators. The two-way flow of annotation information between actively curated genome databases and research users is essential if the genome databases are to keep up with research findings and the researchers are to be supplied with the highest quality of annotation.

THE POST-GENOMICS ERA, AND THE OTHER 'OMES'

With a finished and annotated genome, or significant EST datasets, to hand, parasite biologists have access to the genotype of their chosen organism. The next step in genome-based analysis of parasite biology is the use of these data in assays that screen the whole genome rather than small select segments. Whole genome assays can be based on DNA (screening genotypes), RNA (screening expressed genotypes) or protein (screening the translated genotype). Each of these types of assay has been christened with a new 'omics' name, and the array of 'omes' can be bewildering. The genome yields, on transcription, the transcriptome, which in turn yields the proteome on translation.

Short segments of genomic DNA, or oligonucleotides derived from and representative of the whole genome sequence, can be arrayed on a macro- (spots of DNA/oligonucleotide on nylon membranes) or micro- (thousands of spots of DNA/oligonucleotide per square centimetre) format. These arrays can then be screened with genomic DNA from parasite isolates to define segments of DNA altered or missing in the new isolate compared to the reference genome. This sort of screen may be useful in defining components of virulence, if a virulent strain is arrayed, and avirulent strains derived from loss-of-function mutations are tested. The system has been proven for yeast strain comparison. One problem with this sort of assay is that it cannot identify DNA segments present in the new strain but missing from the reference.

Similar arrays can be constructed from the mRNA-encoding parts of the genome, by identifying coding exons and specifically PCR-amplifying and arraying these 'ORFeome' (all the open reading frames of an organism) DNA fragments. EST clone sets are very useful for this sort of array, as they provide 'prespliced' gene sets in a readily accessible format. The ORFeome arrays can be screened with mRNA populations derived from different strains (to identify changes associated with different phenotypes in the strains, such as virulence), from the reference strain after treatment with different conditions (such as drugs, or immune cells) or from different stages of the parasite (to identify stage-specifically expressed genes). Transcriptome analysis is very powerful, and generates datasets with often bewildering complexity. The statistical analysis of these datasets is an emerging and difficult (bio) informatics problem, and standards have yet to emerge. An alternative approach to transcriptomics is through the use of serial analysis of gene expression methods (SAGE). SAGE is a high-throughput method that relies on the identification of unique, short sequence tags in the 3′ end of mRNAs. These tags are isolated by a reverse-transcriptase-PCR-cloning method using type II restriction enzymes, can be sequenced simply in a multiplex format, and yield expression level data for the corresponding genes. Typically, tens of thousands of SAGE tags are generated and analysed to decipher which genes they derive from, and the number of tags per gene indicates relative expression level.

The ORFeome sets can also be used for protein-based genomics assays. By expressing the ORF segments in a yeast two-hybrid vector system, the 'interactome' can be defined. Yeast two-hybrid analyses require that two proteins interact (the bait and the prey) in order to rescue a yeast mutant strain. The interaction between two parasite proteins can therefore be assayed, and high-throughput screens allow the all-versus-all analysis of whole proteomes. Methods are being developed that permit immobilization of proteins in active conformations on high-density arrays, and these will allow more sensitive assays of protein–protein interactions.

For parasitology, interest often focuses on how best to kill the target organism. ORFeome clone sets can be used to develop medium throughput vaccine candidacy tests, and to test drugs *in vitro* in model transgenic systems (such as yeast or bacteria). High throughput screens with nematode parasite proteins in the model nematode *C. elegans* are also (theoretically) possible. To be useful, a drug target must be essential to the survival of the parasite. While some parasite systems permit specific knockout of chosen genes, and thus genetic assays for their roles, it is hard to prove that a gene is essential in organisms with no genetics. For these parasites (such as most nematodes, platyhelminths) an alternative knockout system may be available in the phenomenon of double-stranded RNA-mediated gene knockout (RNAi, or RNA interference), where endogenous mRNA is specifically degraded in response to exposure of the organism to double-stranded RNAs corresponding to the gene of interest. This process has been shown to function in all the major eukaryote lineages, and should be adaptable for most parasites. It is already a reality in trypanosomatids, and positive results have been reported for nematode parasites (A. Aboobaker and M. Blaxter, unpublished observations).

THE PARASITES AND THEIR GENOMES

The field of parasite genomics is rapidly changing. The speed with which genome data can be acquired will make any simple review of the current state of parasite genomics date rapidly. Therefore, below I have indicated the status of the genome projects, the sorts of data available, and highlighted some of the striking research findings that have already emerged. Each parasite genome project has developed databases to display and analyse its data. A core centre for access to parasite genome data has been established at the European Bioinformatics Institute, with funding from the WHO TDR. This www site (http://www.ebi.ac.uk/parasites/parasite-genome.html; see Table 1.3) provides access to genome databases as well as other genomics related sites for parasites. An excellent collection of review and primary data papers on parasite genomics was recently published as a special issue of the *International Journal for Parasitology* (volume **30**, issue number 4, 2000) and is especially recommended for readers looking for more detail on specific genomes.

Parasite genome size and organization

Parasite genomes vary widely in size (from <10 Mb to >5000 Mb) and organization (from one chromosome to hundreds). There are, however, several themes that derive from the phylogenetic disparity of parasitic organisms. Protozoans tend to have smaller genomes

TABLE 1.3 Parasite genomics internet resources*

Parasite genome central resources	
Parasite-Genome WWW site	http://www.ebi.ac.uk/parasites/parasite-genome.html
Parasite-Genome BLAST server	http://www.ebi.ac.uk/blast2/parasites.html
Parasite-Genome Proteome Analyses	http://www.ebi.ac.uk/parasites/proteomes.html
Parasite Genome email network	parasite-genome@jiscmail.ac.uk
Parasite Genome contact	Martin Aslett aslett@ebi.ac.uk
Structural Genomics of Pathogenic Protozoa	http://www.sgpp.org
The Molecular Parasitology Network	http://www.rna.ucla.edu/par/molpar/index.html
Malaria (*Plasmodium*) genome project	
PlasmoDB	http://PlasmoDB.org
Malaria gene identification program	http://parasite.vetmed.ufl.edu/
Malaria genome sequencing at the Sanger Institute	http://www.sanger.ac.uk/Projects/P_falciparum
Malaria genome sequencing at TIGR	http://www.tigr.org/tdb/edb/pfdb/pfdb.html
Malaria genome sequencing at Stanford	http://sequence-www.stanford.edu/group/malaria/index.html
NCBI Malaria Genetics and Genomics	http://www.ncbi.nlm.nih.gov/Malaria/
Toxoplasma gondii genome project	
ToxoDB and *T. gondii* genome project WWW site	http://www.ebi.ac.uk/parasites/toxo/toxpage.html
Toxoplasma EST program	http://genome.wustl.edu/est/toxo_esthmpg.html
Cryptosporidium parvum genome project	
C. parvum genome sequencing project	http://www.parvum.mic.vcu.edu/
C. parvum EST project	http://medsfgh.ucsf.edu/id/CpTags/home.html
Entamoeba histolytica genome project	
TIGR *E. histolytica* genome sequencing	http://www.tigr.org/tdb/e2k1/eha1/
African trypanosome (*T. brucei*) genome project	
T. brucei genome project WWW site	http://parsun1.path.cam.ac.uk
Sanger Institute *T. brucei* genome sequencing	http://www.sanger.ac.uk/Projects/T_brucei/
TIGR *T. brucei* genome sequencing	http://www.tigr.org/tdb/mdb/tbdb/index.html
Trypanosoma cruzi genome project	
TIGR *T. cruzi* genome sequencing	http://www.tigr.org/tdb/mdb/tcdb/
Swedish *T. cruzi* genome sequencing	http://cruzi.genpat.uu.se/
Swedish *T. cruzi* EST sequencing	http://www.genpat.uu.se/tryp/tryp.html
Leishmania genome project	
Leishmania genome project WWW site	http://www.ebi.ac.uk/parasites/leish.html
Sanger Institute *L. major* genome sequencing	http://www.sanger.ac.uk/Projects/L_major/
Brugia malayi genome project	
Genome project WWW site	http://nema.cap.ed.ac.uk/fgn/filgen.html
Onchocerciasis WWW site	http://math.smith.edu/OnchoNet/OnchoNet.html
EST projects on parasitic nematodes	
Edinburgh parasitic nematodes EST project	http://www.nematodes.org/
GSC-WUSTL parasitic nematode EST Project	http://www.nematode.net
Schistosoma genome project	
Schistosome genome project WWW site	http://www.nhm.ac.uk/hosted_sites/schisto/

(*Continued*)

TABLE 1.3 (*Continued*)

TIGR *S. mansoni* genome sequencing	http://www.tigr.org/tdb/e2k1/sma1/
S. mansoni EST project	http://verjo18.iq.usp.br/schisto/
Vectors of parasites	
Mosquito genomics WWW and genome databases	http://klab.agsci.colostate.edu/ http://konops.imbb.forth.gr/AnoDB/

*This table is a shortened form of the original (by M. Aslett, A. Ivens and M. Blaxter) available on the world wide web (WWW) at http://www.ebi.ac.uk/parasites/parasite-genome.html.
The genome sequencing centre and informatics centre acronyms used above are:

- TIGR: The Institute for Genomic Research, Gaithersburg, MMD, USA
- Sanger Institute: The Pathogen Sequencing Unit, Wellcome Trust Sanger Institute, Hinxton, Cambridge UK
- GSC-WUSTL: The Genome Sequencing Center, Washington University, St. Louis, USA
- Edinburgh: The Nematode Genomics Lab, ICAPB, University of Edinburgh, UK
- Stanford: The Stanford University Genome Sequencing Center, Stanford, CA, USA
- NCBI: The National Center for Biotechnology Information, NLM, NIH, Bethesda, MD, USA.

(though the genome of *Trypanosoma cruzi*, at 70 Mb, is larger than that of the smallest nematode) organized as many small (0.1 to 5 Mb) chromosomes. Nematodes in general have genomes of ~100 Mb, though smaller (*Meloidogyne* estimated to be ~50 Mb) and much larger (*Ascaris*, estimated to be 5000 Mb) genomes are known. The karyotype of nematodes tends also to be conservative with between 5 and 20 chromosomes being normal, though several species have up to 100 chromosomes. Some ascaridids undergo somatic chromosome diminution, where the genome of the somatic cells has many thousands of chromosome fragments that result from the fragmentation of larger germline chromosomes. Platyhelminth karyotypes are also generally conservative with <50 chromosomes, but genome sizes range up to ~300 Mb. Parasitic arthropods and arthropod vectors of parasites have genomes similar to that of *Drosophila melanogaster* (genome size $2\text{–}8 \times 10^8$).

Trypanosomatid parasite genomes

The genomes of trypanosomatid parasites are the subjects of advanced genome projects. The small chromosome 1 of *Leishmania donovani* was the first full chromosome of any protozoan to be sequenced. The chromosome had a simple structure, with repetitive telomeric regions flanking less repetitive subtelomeric sequences, and the telomeres bracketing a gene dense chromosomal core. The genes on chromosome 1 were organized as one transcription unit: all are on one strand and appear to be transcribed from a single promoter. This pattern has also been found in other chromosomes, though larger chromosomes may have more than one transcription unit. Homologous chromosomes can differ significantly in length, but this is mainly due to variations in the subtelomeric repeats. Within species, there is much karyotypic variation, but this genome 'instability' appears not to affect the protein-coding regions of the genome, only the telomeres. Comparison between species of *Leishmania* reveals that despite stark differences in karyotype as measured by PFGE, all species have very conserved linkage groups, and show a high degree of conservation of synteny. The genome of *Leishmania major* Friedland is being sequenced by a chromosome-by-chromosome strategy utilizing both physical maps and whole chromosome shotgun projects. As the genes of *L. major* are being defined, functional genomics efforts are underway to examine stage-specific expression, involvement in virulence and potential vaccine candidacy through microarray, genetic and mass-vaccine approaches. A small but significant EST dataset accompanies this: trypanosomatid mRNAs all carry the same 5′ trans-spliced leader sequence, permitting facile construction of full length cDNA libraries which have been sampled to define both stage-specific transcripts and housekeeping genes. The genes of *Leishmania* species revealed by EST and genomic sequencing display relatively limited overall similarity to genes of other organisms, with a relatively high proportion being classified as 'unknown' or 'novel', as would be expected from their phylogenetic distance from other sequenced eukaryotes.

The general patterns of genome organization found in *Leishmania* are also observed in other trypanosomatids, but with taxon-specific variations. In *T. cruzi*, homologous chromosomes can differ in size by over 50% and there are suggestions of aneuploidy, and thus genome sequencing of this species must rely on a clone-by-clone scaffold informing chromosome shotguns. The *T. cruzi* genome is also more replete with multigene families, and appears to be more variable (in gene content) between isolates, again posing problems in analysis and application. The *T. cruzi* genome is being sequenced from the reference strain CL Brener TC3 by a combination of physical map-based and whole chromosome strategies. Many *T. cruzi* ESTs have been generated.

The genome of *T. brucei* is also more repetitive overall than *Leishmania*, but these repeats are in general restricted in distribution to particular segments of chromosomes. One repetitive gene family, the variable surface glycoprotein (vsg) family, is distributed between chromosome ends (where it resides in 'expression sites', only one of which is active in each cell), chromosome-internal sites, and a population of very small minichromosomes. Active transposition and gene conversion serve to move genes from the silent sites (chromosome internal, minichromosome, and most telomeric sites) to the active expression site. This plasticity in the genome for vsg genes does not appear to be true for other genes. The genome sequence of the reference strain of *T. brucei*, TREU 927/4, is being determined by a combination of whole-chromosome and physical mapping strategies assisted by GSS and EST sequencing.

Apicomplexan parasite genomes

The apicomplexan parasites (*Plasmodium*, *Toxoplasma*, *Eimeria* and related genera) have relatively simple genomes, but genomic analysis is significantly complicated by the extreme sequence composition bias of their DNA. For example, the human malaria parasite *P. falciparum* has an AT content of over 80% overall, with genes (protein-coding regions) having a slightly lower overall % AT than intergenic and other non-coding segments. This AT bias makes sequencing very difficult, as *Plasmodium* DNA does not clone well in bacterial systems (and those clones that are recovered are often derived from only a subsection of the genome) and sequencing chemistries do not yield very long reads from AT-biased DNA . Despite these problems, the *P. falciparum* genome is near completion, and there are significant genomics datasets available for *Toxoplasma*, *Babesia*, *Eimeria* and *Cryptosporidium*.

The first malaria chromosomes to be sequenced were from the 3D7 strain of *P. falciparum*, and reveal the overall composition and structure of apicomplexan chromosomes. As in trypanosomatids, relatively repetitive telomeric and subtelomeric regions flank a gene-rich core. In the malaria chromosomes, there is an obvious centromere, central to the core region, which has a remarkable >95% AT composition. Malaria genes often contain introns, making gene model building more difficult for this group than in the trypanosomatids. The subtelomeric repetitive DNA includes several important gene families, such as the *var* and *rifin* genes, implicated in antigenic variation in malaria. This congruent localization of variant genes to chromosome ends in *T. brucei* and malaria is likely to be a convergent solution to the need to evade the immune response of the host, rather than an ancestral feature of these genomes. Apicomplexans appear to have arisen through a symbiotic merger of two eukaryote genomes, as they have a plastid genome as well as a mitochondrial extranuclear genome, and these two genomes have discordant origins. This merger event has had its effects on the nuclear genome, which includes both 'plant-like' and 'protozoan' components: the plant-like components in particular are being investigated as possible new routes to drug development.

As part of the malaria genome program, genomics data (low-depth whole-genome shotgun data in the main) are also being generated from other *Plasmodium* species (particularly the human parasites *P. vivax* and *P. malariae*, and the murine model *P. chabaudi*). Due to the high AT content of the genome, the construction of cDNA libraries from malarial RNA is technically difficult. The many A- or T-rich segments in the genome result in frequent mispriming from mRNAs, rRNAs, and contaminating genomic DNA. A large number of ESTs have been generated, however, and have provided significant help in annotation of the sequenced genomic segments. The AT bias of the genome is more extreme in the extragenic segments, and it has thus proved possible to generate gene libraries from malarial parasite DNA by using a nuclease (derived from mung beans) that degrades single stranded 'bubbles' of high-AT sequence in partially denatured DNA. These 'mung bean nuclease libraries' yield DNA segments that usually derive from coding regions, and can be sequenced as if they were cDNAs. These libraries are useful because they are normalized: they will include genome segments relative to their concentration in the genome not their expression level.

Other protozoan genomes

The taxon 'Protozoa' is probably not monophyletic. For this reason alone, the genomes of 'protozoa' should not be expected to show any greater sharing of features than would, say, plants and animals. The genomes of several divergent protozoans are being sequenced, and these data are yielding new information on the possibilities for genome organization in eukaryotes, and on the origin of eukaryotes.

The 12 Mb genome of the diplomonad *Giardia lamblia* is nearing completion, and reveals that this 'amitochondriate', basal protozoan has many genes that appear to derive from the alphaproteobacterial mitochondrial ancestor. This finding is most simply explained by the amitochondriate condition being a secondary loss phenomenon rather than a frozen snapshot of the origin of eukaryotes. *Entamoeba histolytica* has a genome of ~15 Mb arranged as 14 chromosomes, and the complete sequence of isolate HM1:IMSS is being sought. One intriguing feature of the *E. histolytica* genome is the presence of multiple copies of an extrachromosomal circular replicon encoding the ribosomal RNA genes. This circle (and variants thereof) has a variable but high copy number and may be a mechanism for increasing expression of ribosomal RNA: there are no chromosomal copies.

Nematode genomes

As described above, the nematode *C. elegans* was the first metazoan to have its genome sequenced. This program was directed mainly at the development of genomic technologies for sequencing larger genomes, such as humans, and at providing a resource for exploitation by the community of researchers using *C. elegans* as a model for investigating medically relevant phenotypes. It is, however, obvious that *C. elegans* also should be a good model for parasitic nematodes in both genomics and biology. The phylum Nematoda is also very diverse, and parasitism of animals and plants has arisen several independent times. However, nematodes are relatively conservative morphologically, and, in contrast to the platyhelminths, conservative in life history characteristics. This conservatism may also extend to genome biology.

The genomes of nematodes are usually ~100 Mb, but range from 50 Mb to 5000 Mb. The largest genomes are those of the ascaridids, which undergo somatic chromatin diminution. This process, first noted in the early years of the discovery of chromosomes and central to the development of the chromosome theory of heredity, is perhaps at its apogee in *Parascaris univalens*. *P. univalens* has but a single chromosome (of ~5000 Mb) in each of its germ cells, but has several thousand chromosomes in each somatic cell. Analysis of *A. suum* showed that the chromosomal breakage was a carefully orchestrated event that resulted in breakage in specific regions of the chromosome and generation of new telomeres in the somatic cells. Some genomic segments are lost during the process, and these can contain genes expressed only in the germ line cells. Chromatin diminution is unknown from other nematode orders, even those most closely related to the Ascaridida.

The *C. elegans* genome encodes about 20 000 protein-coding genes, of which about 17 000 have been confirmed by cDNA and EST sequencing. EST-based genomics programs are underway for a dozen parasitic nematode species. For the filarial nematodes *Brugia malayi* and *Onchocerca volvulus* there are 23 000 and 12 000 EST sequences respectively, and these have been analysed to reveal unique gene content and relative similarity to each other, other nematodes and other phyla. The *B. malayi* dataset defines about 8800 different transcripts, and thus comprises 40% of the expected gene complement of 20 000. Of these 8800 genes, about 40% do not have a recognizable homologue in *C. elegans*, or other non-filarial nematodes. This divergence reflects the deep phylogenetic separation between *C. elegans* and the filarial species. Comparison of the closely related *O. volvulus* and *B. malayi* reveals, as would be expected, much closer conservation of gene sequences, but there are still marked differences between these two human parasites. Genes expressed at high levels in one species are absent or rarely found in the EST dataset from the other, and there are significant numbers of putative transcripts unique to each dataset. Whether this is due to the stochastic nature of clone selection for sequencing, or rapid divergence under selection for parasitic phenotypes remains unclear. The EST datasets for other nematodes are generally smaller (100–10 000 sequences) but confirm the general findings of the filarial–*C. elegans* comparison. Thus parasites such as hookworms (*Necator americanus*) and other strongylid nematodes closely related to *C. elegans* show high conservation, while genes from the trichocephalid species *Trichinella* and *Trichuris* often show higher pairwise similarity to genes from species in other phyla compared to *C. elegans*. Other nematodes, like *C. elegans*, use *trans*-splicing to provide a 5′ cap to their mRNAs and all species tested use the same sequence (spliced leader 1, SL1). This feature makes it easy to generate full length cDNA libraries for any nematode species even when material is limited. About 80% of the mRNA in

Ascaris and *C. elegans* is SL1-*trans*-spliced, and about 60% of *C. elegans* genes are *trans*-spliced.

The cross comparison between the free-living *C. elegans* and the parasitic nematodes can reveal genomic features associated with parasitism. For example, *C. elegans* expresses a member of the nematode polyprotein allergen (NPA) family of genes. In *Ascaris* and *B. malayi* the different internal repetitive blocks of the NPA protein are very similar, while in *C. elegans* each repeat unit is strikingly different. As the NPA protein binds lipids, this may reflect the different requirements and environments of the nematodes: *C. elegans* exploiting a varied soil bacterial flora, and the parasites exploiting a less variable host environment. Several genes have apparently undergone expansion to form a gene family in one branch of the Nematoda, while remaining unique in others. Filarial nematodes have but one FAR protein gene (another lipid-binding protein) while *C. elegans* has seven; conversely *B. malayi* has eight ALT (abundant larval protein) genes while *C. elegans* has one (B. Gregory and M. Blaxter, unpublished observations). Emerging microarray data for *C. elegans* defines sets of genes involved in various life-cycle transitions and tissue development. These data can be used to cross compare with parasite EST datasets to identify conservation of involvement of genes in basic processes such as gonadogenesis or development of the third stage dauer larva/infective larva. These EST datasets also revealed one new set of problems in dealing with these large metazoan parasites, those deriving from uncloned starting organisms. The populations used for the different libraries were not all from the same strain, and comparison of EST sequences has identified numerous single base changes that are likely to have arisen through allelic variation (rather than experimental error). For parasites where genetics is difficult or even absent, and where the standard laboratory strains have not been specifically selected to be inbred, such variation within and between strains has to be carefully considered. Similar problems are also seen in the schistosome genome project (see below).

To date only limited genome sequence data are available from nematodes other then *C. elegans* and the congeneric *C. briggsae*. Analysis is most advanced in *B. malayi*. A striking feature of *C. elegans* gene organization is that approximately 20% of the genes are arranged as operons, where two or more genes are cotranscribed from a single promoter. The polycistronic pre-mRNA is then resolved into individual mature mRNAs by linked polyadenylation of the upstream transcript and *trans*-splicing of a novel family of spliced leader exons (the SL2 family) to the 5′ end of the downstream transcript. This polycistronic organization has also been demonstrated in other closely related free-living nematode species, and also in strongylid, filarid and strongyloidoid parasites (D. Guiliano and M. Blaxter, unpublished observations). The functional significance of polycistrons remains unclear, but may relate to the shared use of strong promoters. Gene density on the chromosomes is relatively high in *C. elegans*, at approximately one per 5 kb. The regions of the *B. malayi* genome that have been sequenced (about 100 kb) have a density of one gene per 6 kb (D. Guiliano and M. Blaxter, unpublished observations). Particularly striking is the finding that introns in *B. malayi* are usually much longer than those in *C. elegans*.

The *C. elegans* genome is organized into five autosomes and one sex chromosome: sex determination is by an XX (female)–XO (male) system. While *B. malayi* also has five autosomes, sex determination is by an XX (female)–XY (male) system. Sequence analysis of the Y chromosome (which appears heterochromatic in stained chromosome preparations) revealed that it has many retrotransposon elements, suggesting that like mammalian and fly sex chromosomes it may have lost much coding function. As genetics is impractical in filaria, no comparative genetic maps have been constructed. Likewise, FISH has not yet been developed sufficiently to allow comparison of linkage of conserved genes. However, comparisons between the sequenced segments of the *B. malayi* genome and that of *C. elegans* have revealed that there is long-range conservation of linkage and local conservation of synteny between these two species (D. Guiliano and M. Blaxter, unpublished observations). As *B. malayi* and *C. elegans* are estimated to have diverged several hundred million years ago, this in turn suggests that other nematodes may also display conserved genomic organizations.

Mitochondrial genomics of nematodes was again started in *C. elegans* but has also included parasitic species. Nematode mitochondria display a remarkable plasticity of organization compared to other metazoan phyla. The protein and rRNA genes of the *C. elegans* and *Ascaris* mitochondrial genomes are collinear, but in a very different order from that of *O. volvulus*. None of these three mitochondrial genomes displays any significant collinearity with those from other phyla. The mitochondrial genome of *Trichinella spiralis* is different again, but can be compared to other phyla, suggesting that the major rearrangements observed have occurred within the Nematoda. The mitochondrial genome of the plant parasitic *Globodera pallida* is even stranger. While the normal mitochondrial genome is a single circular (sometimes linear) element of 14 kb (in nematodes) to 30 kb, the *G. pallida* mitochondrion contains multiple subgenomic circles of 8–9 kb, each of which carries a different subset of mitochondrial genes.

Platyhelminth genomes

The platyhelminth parasites span a huge phylogenetic diversity, yet only a few are major parasitic diseases of humans and our animals. The genus *Schistosoma* has been the subject of a coordinated genomics programme for many years. Starting with EST analysis of stage-specific libraries derived from both vector and definitive host stages of the human parasite *Schistosoma mansoni*, the project has also generated EST data for the other major schistosome species of humans (*S. haematobium* and *S. japonicum*). The EST dataset has been complemented by 'ORESTES' tags – ESTs derived from the central portions of mRNAs that help in tying together 5′ and 3′ EST sequences.

The *S. mansoni* genome is estimated to be ~270 Mb, and is organized as eight chromosome pairs, including a ZW sex determination pair. The chromosomes condense well during mitosis, and have a well defined heterochromatic banding pattern. The karyotype of *S. mansoni* has been defined, and FISH can be used to assign DNA markers to specific chromosomes. The banding pattern probably derives from the presence of large segments of the genome made up of non-coding simple repeats. The *S. mansoni* genome is indeed rich in repetitive DNA, and many classes of simple, satellite and transposon-derived repeats have been defined. Some of these locate to the heterochromatic regions of the chromosomes by FISH, while others have a more dispersed distribution. Large insert DNA libraries have been constructed for *S. mansoni* and are being used to initiate physical mapping and sequencing of the genome. Over 40 000 end sequences are available from these libraries in dbGSS.

The other *Schistosoma* species are also being studied as a comparison to the major human parasite. Comparative mitochondrial genomics suggests that the schistosomes have undergone rapid and divergent evolution in the recent (geological) past, and that the human parasites arose from parasites of domesticated animals.

Genome-based research on other platyhelminths is in its infancy, with only a small number of ESTs from *Echinococcus* spp. available (Table 1.2). Most if not all platyhelminths use a 5′ *trans*-spliced leader sequence, which differs between species, to cap some of their mRNAs, and platyhelminths may also utilize an operonic gene organization.

Parasite vector genomes

The sequence of the host genome (human) for the major human parasites is available in draft, allowing parasitologists to investigate possible interactions between parasite and host. Many parasites also pass through intermediate or vector hosts, and thus the genomes of these vector species are also being investigated. For some arthropod vectors, additional urgency is afforded by their being vectors also for viral and protozoal diseases. The mosquitoes *Aedes aegypti* and *Anopheles gambiae* have genomes significantly larger than that of the model dipteran, *D. melanogaster*, but much of this increase in genome size appears to be due to a higher proportional content of repetitive and retrotransposon DNA. Genetic maps are well advanced for mosquitoes, particularly *A. gambiae*, and loci conferring relative resistance to vectoring filarial nematodes have been mapped. Genomic sequencing has been initiated based on physical maps spanning these and other important vector traits. Similarly, the *Schistosoma mansoni* snail host, *Biomphalaria glabrata*, has been analysed using EST techniques to investigate the schistosome–snail interaction, and the basis of differential vectoring capacity in snail populations. Analysis of the *B. glabrata* genome is hampered by the lack of genomics data from any closely related comparators. The tick vector for *Babesia*, *Boophilus microplus*, has been the subject of a small EST project aimed at discovery of protein products involved in feeding, for vaccine development (see Table 1.2).

Genomics analyses have yet to be applied to other metazoan parasites in any major way. Small sequence datasets are available for the medicinal leech *Hirudo medicinalis*, and for scab and scabies mites (*Psoroptes ovis* and *Sarcoptes scabei*) (see Table 1.2).

THE FUTURE

The past ten years have seen the emergence of genomics as a field, and the application of genomics analyses to problems of parasite biology. The future of this mode of analysis is bright. As genome datasets reach completion (for example, the *P. falciparum* genome is 'essentially finished' and other protozoa are not far behind) the opportunities for whole-system analyses will grow. What metabolic pathways are present in the chosen parasite? Which are absent? Do these metabolic data suggest essential bottleneck enzymes that might be good drug targets? What potential immunomodulators does the parasite genome encode, and how might these function in the context of the host or vector genome? Can the evolutionary history of the origins of parasitism be read from comparison between free-living and parasitic species? Does this information suggest potential routes to effective intervention? Which genes are behaving anomalously compared to their genomic environment? Are these possible virulence factors, evolving rapidly in a Red Queen race with host defences? Can the genes involved in susceptibility and resistance to drugs be identified by a combination of genetics and map-based cloning, or genome and cDNA microarray screening? When a parasite is treated with a drug, which genes are switched on or off, and are these potential second-site targets for prolonging the effective clinical life of a drug that is being rendered useless through the evolution of genetic resistance? Many parasites evade adaptive immunity by antigenic switching: how is this achieved?

As all facets of genomic and functional genomic analysis become cheaper (per base and per assay) the river of data will turn to a flood, and the current trickle-down will truly benefit all areas of parasitology.

ACKNOWLEDGEMENTS

I would like to thank my colleagues in the Nematode Genomics program in Edinburgh, and the community of WHO-sponsored parasite genome projects, particularly Martin Aslett, the WHO parasite-genome informatics specialist at the European Bioinformatics Institute, and Dr. Al Ivens of the Sanger Institute (both Hinxton, UK) for inspiration and support.

FURTHER READING

The Parasite Genomes issue of the *International Journal for Parasitology* (2000, volume **30**, issue number 4) and the Parasite Genomics special issue of *Parasitology* (1999, supplement to volume **118**) are replete with articles on general genomics and bioinformatics, as well as species-centred reviews of progress-to-date.

General parasite genomics

Degrave, W.M., Melville, S., Ivens, A. and Aslett, M. (2001). *Int. J. Parasitol.* **31**, 532–536.

Johnston, D.A., Blaxter, M.L., Degrave, W.M., Foster, J., Ivens, A.C. and Melville, S.E. (1999). *Bioessays* **21**, 131–147.

Lawson, D. (1999). *Parasitology* **118**, S15–S18.

Plasmodium genomics

Bowman, S., Lawson, D., Basham, D. *et al.* (1999). *Nature* **400**, 532–538.

Gardner, M.J., Tettelin, H., Carucci, D.J. *et al.* (1998). *Science* **282**, 1126–1132.

Trypanosomatid genomics

Ivens, A.C. and Blackwell, J.M. (1999). *Parasitol. Today* **15**, 225–231.

Myler, P.J., Audleman, L., deVos, T. *et al.* (1999). *Proc. Natl. Acad. Sci. USA* **96**, 2902–2906.

Nematode and schistosome genomics

Blaxter, M., Daub, J., Guiliano, D., Parkinson, J., Whitton, C. and The Filarial Genome Project (2001). *Trans. Roy. Soc. Hygiene Trop. Med.* **96**, 1–11.

Blaxter, M.L. (1998). *Science* **282**, 2041–2046.

Blaxter, M.L., Aslett, M., Daub, J., Guiliano, D. and The Filarial Genome Project (1999). *Parasitology* **118**, S39–S51.

The *C. elegans* Genome Sequencing Consortium (1998). *Science* **282**, 2012–2018.

Williams, S.A. and Johnston, D.A. (1999). *Parasitology* **118**, S19–S38.

CHAPTER

2

RNA processing in parasitic organisms: *trans*-splicing and RNA editing

Jonatha M. Gott and Timothy W. Nilsen
Center for RNA Molecular Biology, Case Western Reserve University School of Medicine, Cleveland, OH, USA

INTRODUCTION

The flow of genetic information from DNA to protein requires a mRNA intermediate. In all eukaryotes, mature mRNAs (i.e. mRNA that is translated into protein by ribosomes) are not simple copies of the DNA sequence. Rather, mRNAs are transcribed by RNA polymerase as precursor molecules which must be processed prior to translation. mRNA processing is complex and includes 5′ end-maturation (capping), 3′ end-maturation (usually cleavage and polyadenylation) and removal of non-coding sequences via splicing. Perhaps not surprisingly, these fundamental steps of RNA processing are shared between parasitic organisms and their hosts. Remarkably, however, the analysis of parasite gene expression has revealed novel and unusual types of mRNA processing reactions. The two best understood examples of such reactions are *trans*-splicing (the joining of two separate RNAs by splicing) and RNA editing (the process whereby the sequence of mRNA is changed after transcription). The purpose of this chapter is to provide an in depth discussion of our current understanding of these two mRNA maturation pathways.

TRANS-SPLICING

Discovery of *trans*-splicing

The phenomenon of antigenic variation in African trypanosomes involves the expression of proteins that comprise the external 'coat' of the parasite. These proteins known collectively

Molecular Medical Parasitology
ISBN 0-12-473346-8

as Variant Surface Glycoproteins (or VSGs), are the main target of the host immune response. In order to understand how VSG expression was controlled and how one VSG was 'switched' for another during the course of infection (see Chapter 5), an early priority was to obtain cDNA and genomic clones encoding VSGs.

Comparison of such clones revealed that the cDNA and genomic sequences were collinear except that the cDNA contained a 39 nucleotide (nt) sequence at its 5′ end not found in the genomic clone. Interestingly, the genomic sequence at the point of divergence from the cDNA corresponded to a potential splice acceptor site (3′ splice site). It was well established that most genes in higher eukaryotes contain intervening sequences (introns) that must be removed by splicing to create functional mRNAs. Accordingly, it seemed likely that the 39 nucleotide sequence found at the 5′ end of VSG cDNA clones was derived by splicing. However, extensive analysis of chromosomal DNA sequences upstream of the VSG gene did not reveal the 39 nucleotide sequence. Where then did this sequence come from? Surprisingly, it was shown that the sequence was present in the genome in multiple copies, organized within a tandem repeat. Furthermore, RNA analysis revealed that these repeated units were transcribed to yield a small non-polyadenylated RNA (~140 nt) with the 39 nt sequence at its 5′ end. Strikingly, the 39 nt sequence was immediately upstream of a consensus splice donor site (5′ splice site). Collectively, all of these observations were consistent with the possibility that the mature VSG mRNA was generated by splicing of the small RNA to the pre-VSG mRNA. The joining of two separate molecules by splicing (*trans*-splicing) was unprecedented. However, definitive proof that *trans*-splicing indeed occurs in trypanosomatids came with the demonstration that intermediates predicted from such a reaction actually existed (Figure 2.1).

It is now clear that *trans*-splicing is not confined to VSG mRNAs. Indeed, available evidence indicates that all mRNAs in trypanosomes receive the 39 nt sequence (now known as the spliced leader, SL) from the small RNA (SL RNA) via *trans*-splicing. The biological roles and mechanism of this unusual RNA processing reaction are discussed below.

Phylogenetic distribution and evolutionary origin of *trans*-splicing

For some time after its discovery, *trans*-splicing was thought to be idiosyncratic to trypanosomes. This notion was dispelled when a careful analysis of actin genes and their mRNAs was performed in the free-living soil nematode, *C. elegans*. Here, it was observed that mRNAs derived from three of four actin genes contained a common 22 nt sequence at their 5′ ends. Similar studies to those described above in trypanosomes showed that the 22 nt sequence was acquired via *trans*-splicing from a small SL RNA. Subsequently, *trans*-splicing was shown to be a common feature of gene expression in all nematodes, both parasitic and free-living. In these organisms, it is now evident that most (but not all) pre-mRNAs receive the SL sequence. The fact that some mRNAs in nematodes are not subject to *trans*-splicing contrasts with the situation in trypanosomatids (see above). A second difference between nematode and trypanosomatids is the prevalence of conventional (*cis*-) introns; such introns are extremely rare in trypanosome genes but quite abundant in nematodes. The significance (if any) of this difference is not yet known.

Trypanosomatids branched extremely early in eukaryotic evolution. Nematodes, although

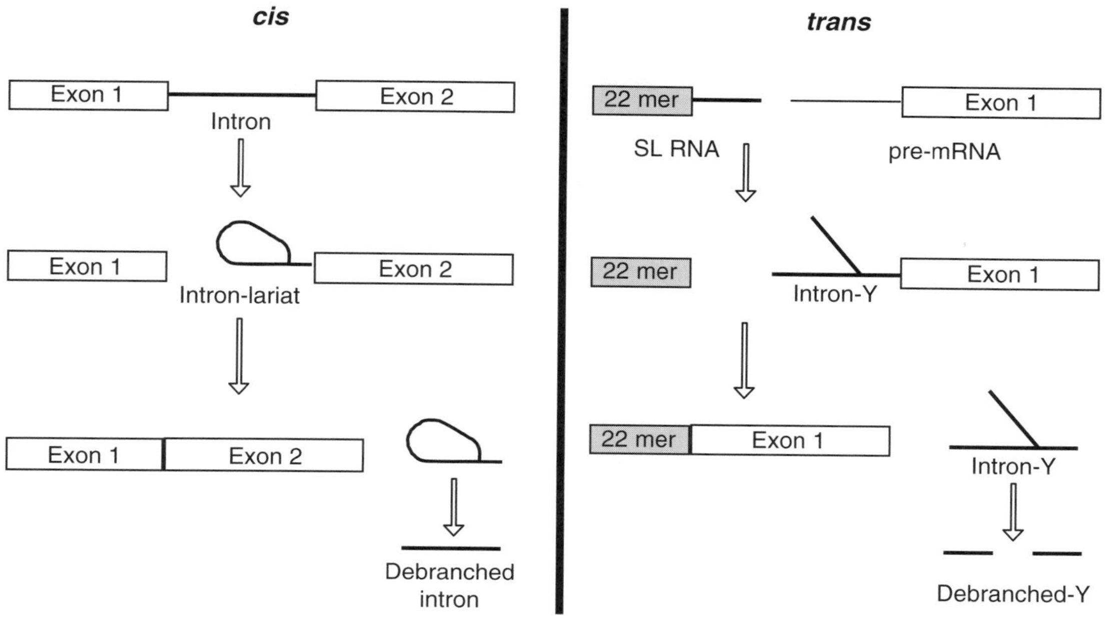

FIGURE 2.1 *Cis* and *trans*-splicing proceed through identical reaction pathways. Identification and characterization of the Y-branched intermediate (analogous to the lariat generated in *cis*-splicing) provided definitive evidence for the existence of *trans*-splicing.

among the most primitive of metazoans, branched much later. Without information on the occurrence of *trans*-splicing in intervening lineages, it was not possible to determine if *trans*-splicing in kinetoplasts and nematodes was evolutionarily related. However, during the past decade, genes and their transcripts have been examined in a large (but not comprehensive) number of diverse organisms. Intriguingly, mRNA maturation via *trans*-splicing has been 'rediscovered' several times. It is quite clear that *trans*-splicing is present in euglenoid protists, some parasitic flatworms, cnidarians such as *Hydra*, and tunicates (primitive chordates). *Trans*-splicing is unambiguously absent in budding and fission yeast, higher plants, insects and mammals (including humans) (Figure 2.2).

These observations indicate that *trans*-splicing has a much wider phylogenetic distribution than originally suspected, but its evolutionary origin(s) is still unclear. The available evidence is compatible with either multiple origins or a single origin accompanied by multiple losses. Unfortunately, there is no direct or easy way to determine whether *trans*-splicing does or does not occur in a particular organism. Accordingly, a definitive answer to the evolutionary origin question awaits a detailed characterization of gene expression in more organisms. Nevertheless, it is evident that *trans*-splicing (although essential for viability in a wide variety of parasitic organisms) is neither restricted to parasites nor is it an adaptation to parasitism. However, because *trans*-splicing is not used by the human host, it

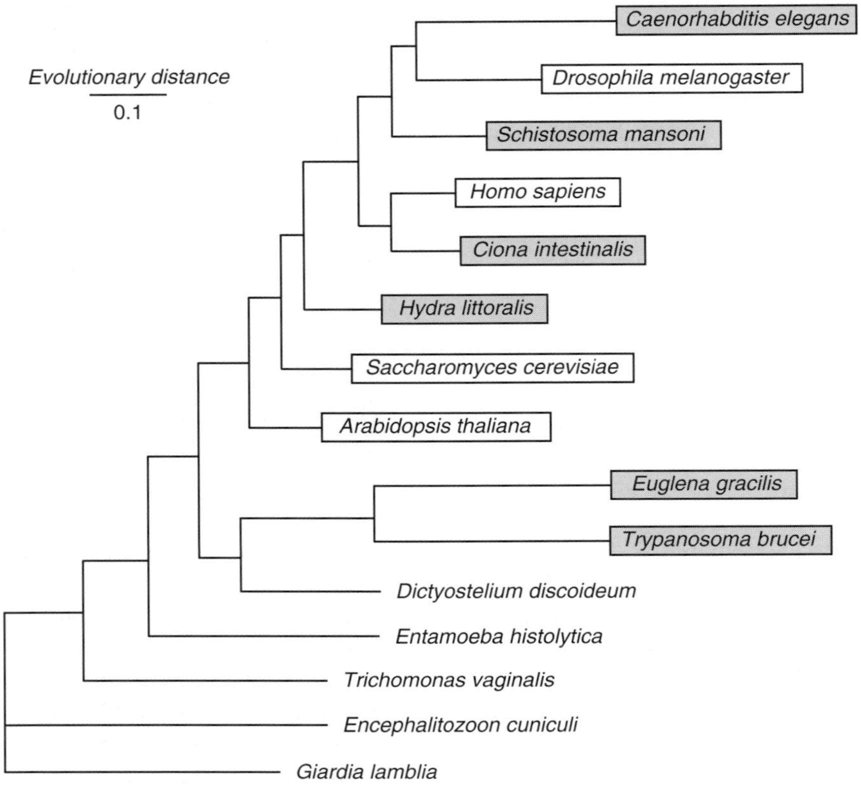

FIGURE 2.2 Phylogenetic distribution of SL *trans*-splicing. Representative species of phyla known to carry out *trans*-splicing are boxed in gray. Representative species of phyla in which *trans*-splicing is absent are in clear boxes. For those organisms that are unboxed, there is insufficient information to determine the presence or absence of *trans*-splicing.

remains an attractive potential target for therapeutic intervention (see below).

Mechanism of *trans*-splicing

Prior to considering the biochemical mechanism of *trans*-splicing, the process of *cis*-splicing will be reviewed briefly since the mechanisms are related. As noted above, the vast majority of genes in most eukaryotes are transcribed as long precursor RNAs which must be processed to yield translatable mature mRNAs. Required processing steps include 5′ end capping, 3′ end cleavage and polyadenylation, and splicing (the removal of internal non-coding regions). Extensive biochemical and genetic analyses in a wide variety of systems have shown that splicing is catalyzed in a massive ribonucleoprotein complex known as the spliceosome. Essential components of the *cis*-spliceosome include five small RNAs (U1, U2, U4, U5 and U6) and more than fifty proteins. Because the small RNAs function as RNA–protein complexes, they are known as snRNPs (small nuclear ribonucleoproteins). Thus, the spliceosome rivals the ribosome in complexity.

Unlike the ribosome however, the spliceosome does not exist as a preformed entity, but rather forms anew on each intron. Spliceosome assembly involves the ordered recognition of three conserved sequence elements within the intron, the 5′ splice site (splice donor), the branch point region, and the 3′ splice site (splice acceptor): First, U1 snRNP recognizes the 5′ splice site via a base-pairing interaction; U1 snRNP then promotes U2 snRNP binding to the branch point region, an interaction that also involves base pairing. Subsequent to engagement of the substrate by U1 and U2 snRNPs, U4, U5 and U6 snRNPs join to form a complete spliceosome. After assembly, but prior to catalysis, a complex series of RNA/RNA rearrangements occurs, resulting in destabilization of U1 and U4 snRNPs. Following spliceosome maturation, catalysis commences. The actual removal of intervening sequences occurs through two consecutive transesterification reactions (the replacement of one phosphodiester with another). In the first transesterification, the 2′ hydroxyl of the branch point adenosine attacks the 5′ splice site. This reaction liberates the 5′ exon and creates the so-called lariat intermediate. The second transesterification then occurs; here the free 3′ hydroxyl of the 5′ exon attacks the 3′ splice site resulting in ligated exons and release of intron in the form of a lariat (Figure 2.1). Several lines of evidence suggest (but have not definitively proven) that the catalytic moieties of the spliceosome are its RNA constituents. It therefore seems likely that the spliceosome (like the ribosome) is a ribozyme.

As noted above, the sequence elements defining the 5′ and 3′ splice sites used in *trans*-splicing conform to consensus *cis* splice sites. Furthermore, the intermediates and products of a *trans*-splicing reaction are analogous to those produced by *cis*-splicing (Figure 2.1). Accordingly, it seemed likely that the fundamental mechanism of *trans*-splicing (i.e. consecutive *trans*-esterification reactions) would be identical to that of *cis*-splicing. Moreover, it also seemed likely that the catalytic machinery (the spliceosome) would be similar if not identical. These predictions were borne out in studies both in trypanosomes and nematodes (see below). Here, it should be noted that experimental systems useful for studying *trans*-splicing are quite limited. Most of our understanding of the *cis*-splicing mechanism has derived from biochemical studies in cell-free systems derived from either mammalian cells or yeast. In addition, yeast is genetically tractable and genetic approaches have been extremely valuable in dissecting the splicing mechanism in this organism.

The vast majority of organisms that carry out *trans*-splicing are not amenable to either genetic analysis or biochemical manipulation. However, certain *in vivo* approaches (e.g. transfection, gene inactivation and permeabilization) are available in trypanosomes, and biochemical approaches are possible in extracts derived from homogenized embryos of the parasitic nematode, *Ascaris*; these extracts faithfully and efficiently catalyze both *cis* and *trans*-splicing. Essentially all of our understanding of the mechanism of *trans*-splicing has been obtained in these two systems, and while it seems likely that results obtained in nematodes and trypanosomes will apply to other systems, this has not been proven. The following discussion makes the assumption that the mechanism of *trans*-splicing is similar in all organisms that carry out the reaction. It is now clear that *cis* and *trans*-splicing are identical in terms of the chemistry of catalysis. Furthermore, both reactions occur in large spliceosomal complexes that share most essential components in common. In this regard, U2, U4, U5 and U6 snRNPs are required for both types of splicing as are most, if not all, non-snRNP associated factors. The most striking difference between the two types of splicing is that U1 snRNP, an essential cofactor for *cis*-splicing, is not involved in *trans*-splicing. As noted above, the function of U1 snRNP is to recognize *cis* 5′ splice sites; it also helps U2 snRNP associate with the branch point region and it is of central importance in the joining of 5′ and 3′ splice sites prior to catalysis.

How these functions are performed in *trans*-splicing and how the SL RNA (which contains the 5′ splice site) efficiently associates with the pre-mRNA (which contains the 3′ splice site) have been the most intriguing questions with regard to the mechanism of *trans*-splicing. Recent biochemical analysis has begun to provide answers to these long-standing questions.

Early on, it was recognized that the SL RNA did not function as a naked RNA, but rather as a ribonucleoprotein particle that shares striking similarities with the small RNAs that are cofactors for *cis*-splicing. Specifically, a subset of those RNPs (U1, U2, U4 and U5) are known as Sm-snRNPs because they bind a common set of seven proteins known as Sm proteins (Sm simply stands for an antigenic determinant). All known SL RNPs are Sm snRNPs but unlike the *cis*-spliceosomal Sm snRNPs, which are stable cofactors, SL RNPs are consumed during the *trans*-splicing reaction.

We now know that assembly into an Sm-snRNP is a necessary prerequisite for SL RNP function in *trans*-splicing. Indeed, the fact that all SL RNAs known are Sm-snRNPs is one of the strongest arguments for a common evolutionary ancestor for *trans*-splicing (see above). We also know that assembly into an Sm-snRNP promotes the association of two additional SL RNP-specific proteins. These proteins in turn are essential for SL RNP function. First, they facilitate association of the SL RNP with U4, U5 and U6 snRNPs such that the SL RNA's 5′ splice site is recognized independent of U1 snRNP. Second, they provide the mechanism whereby the SL RNP and pre-mRNA associate. In this regard, one SL RNP-specific protein interacts specifically with a protein associated with the branch point region of the pre-mRNA, and this interaction is required for *trans*-splicing. The mechanisms of cross-intron bridging in *cis*-splicing and 'intron-intron' bridging in *trans*-splicing are illustrated schematically in Figure 2.3. The two functions of the SL RNP specific proteins adequately explain how *trans*-splicing can proceed without U1 snRNP. Perhaps more importantly, the mechanistic details of *trans*-splicing provide an object lesson in how the same fundamental problem (i.e. splice site recognition and juxtaposition) can be solved in different ways. Finally, because the SL RNP-specific proteins are unambiguously absent in organisms that do not carry out *trans*-splicing (e.g. human hosts) they provide attractive targets for chemotherapeutic intervention.

Prior to leaving a discussion of mechanism, it is necessary to consider how the 3′ splice site of the pre-mRNA is recognized. In *cis*-splicing (as noted above) 3′ splice site recognition is coupled to and facilitated by 5′ splice site recognition. Obviously, this cannot apply in *trans*-splicing, where splice sites are recognized independently. We now know that 3′ splice site identification in *trans*-splicing is determined by sequence elements present in the downstream exon. These elements, known as exonic splicing enhancers, serve as bindings sites for proteins known as SR splicing factors. These SR proteins (so called because they possess domains rich in serine arginine repeats) recruit specific proteins to the 3′ splice site

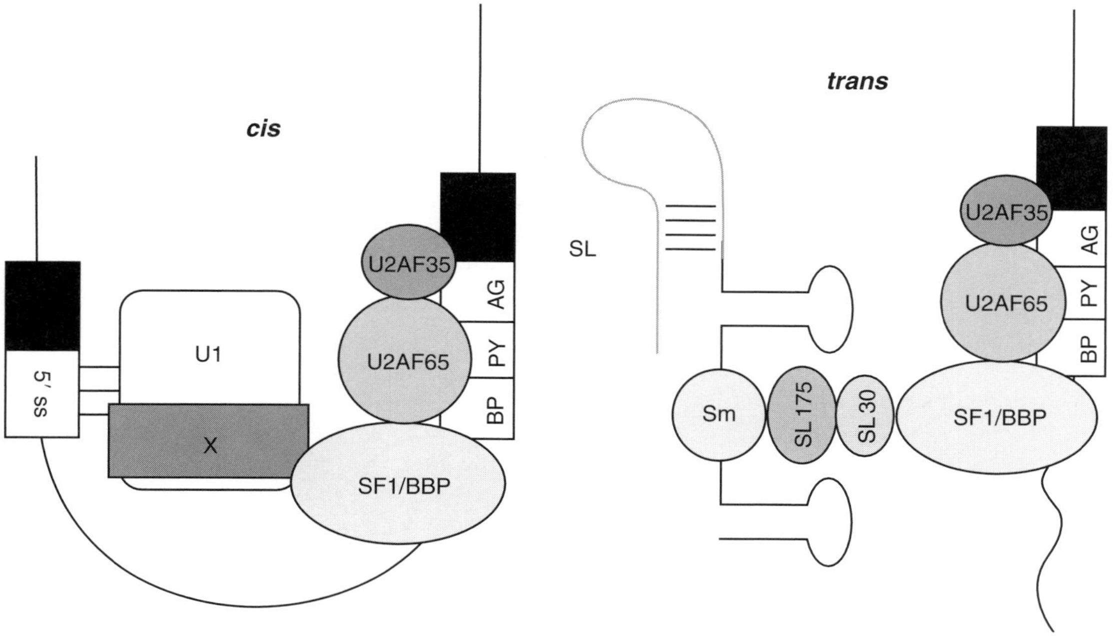

FIGURE 2.3 Splice site bridging complexes in *cis* and *trans*-splicing. In *cis*-splicing the 5′ splice site is recognized by U1 snRNP; in *trans*-splicing the 5′ splice site is present in the SL RNA. SF1/BBP and U2AF are splicing factors that recognize the branch point and 3′ splice site respectively. SL 175 and SL 30 are SL RNP specific proteins characterized in the nematode, *Ascaris*. The protein that interacts with U1 snRNP and SF1/BBP in *cis* splicing has not yet been definitively characterized.

which in turn recruit U2 snRNP to the branch point region (schematically illustrated in Figure 2.4).

In summary, the basic outline of the mechanism of *trans*-splicing is emerging. There exist clear similarities as well as intriguing differences between this process and *cis*-splicing. The challenge for the future in both types of splicing will be defining, in much greater detail, the protein/protein and protein/RNA interactions that are required for catalysis.

Biological function of *trans*-splicing

A fundamental and longstanding question has been why do certain organisms (Figure 2.2) use *trans*-splicing during mRNA maturation. Obviously, this type of RNA processing is not *a priori* necessary for gene expression in most eukaryotes. Unfortunately, very little information about either gene organization or expression is available in most of the organisms that perform *trans*-splicing. However, we do have a reasonable understanding of these topics in trypanosomes. In this regard, it is clear that most protein-coding genes in trypanosomes do not have their own promoters (Chapter 3). As a consequence, individual genes are synthesized as parts of long multicistronic (polygenic) transcription units. It is well established that eukaryotic ribosomes (unlike their prokaryotic counterparts) cannot translate coding regions

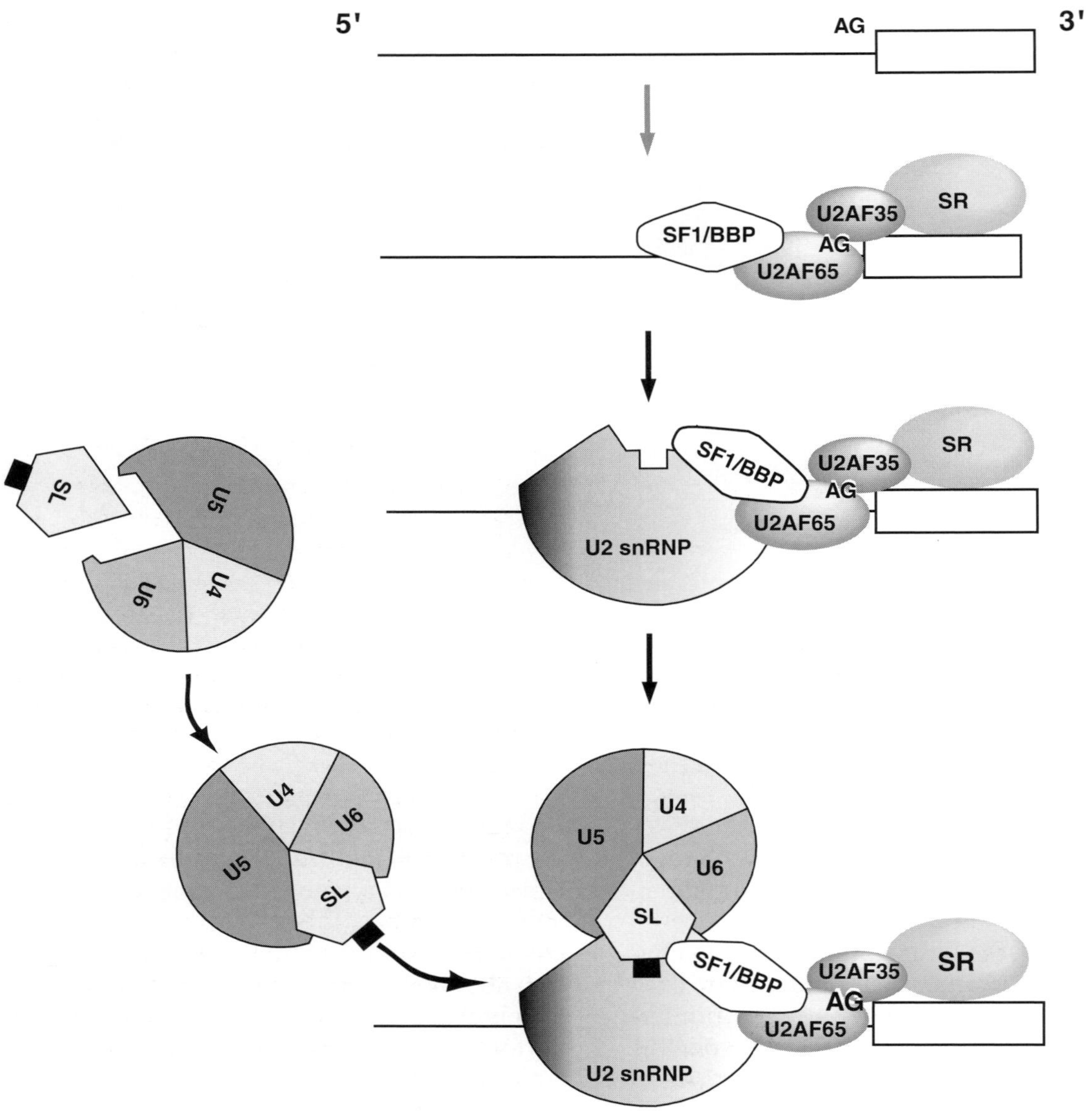

FIGURE 2.4 Schematic depiction of *trans*-spliceosome assembly.

that are internal in such transcripts. In this regard, the vast majority of functional mRNAs in eukaryotes are monocistronic; a typical mRNA contains a 5′ cap (which serves as a ribosome recognition signal), a 5′ untranslated sequence, the protein coding sequence, a 3′ untranslated sequence, and a poly-A tail. To create such an mRNA from an internal coding

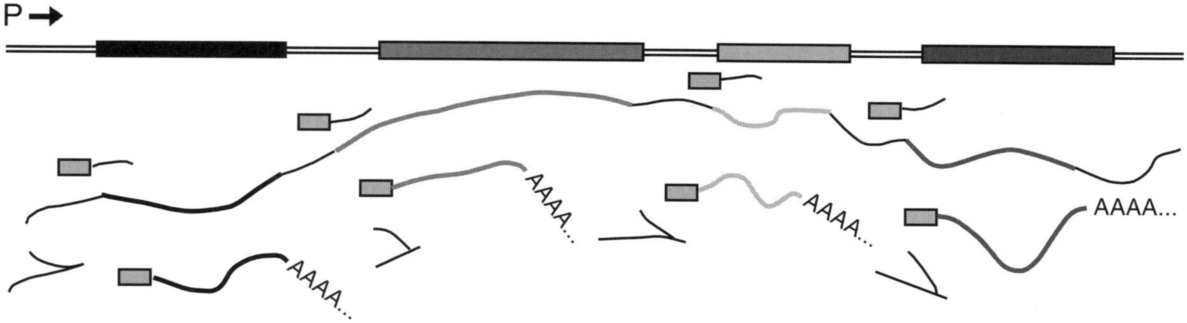

FIGURE 2.5 Processing of multicistronic transcription units in trypanosomatids. mRNA coding regions are boxed; 5′ end maturation involves *trans*-splicing, 3′ end maturation involves cleavage and polyadenylation.

region in a polycistronic transcript, it is necessary to provide a discrete 5′ end with a cap and a 3′ end. It is now clear that *trans*-splicing serves the 5′ end maturation function in trypanosomes, i.e. addition of the spliced leader provides both a cap and a 5′ end; 3′ end maturation is achieved via endonucleolytic cleavage and polyadenylation (Figure 2.5). Therefore, *trans*-splicing is required in trypanosomes because of the unusual mode by which genes are transcribed in these organisms. Although it is possible that *trans*-splicing serves additional functions (e.g. in mRNA transport from the nucleus, or stability), there is no experimental evidence for such roles.

What is the biological function of *trans*-splicing in organisms other than trypanosomes? We know that multicistronic transcription units exist in nematodes (*C. elegans* in particular) and that *trans*-splicing serves to process these units. Accordingly, it is tempting to speculate that this is the 'universal' function of *trans*-splicing. However, many more mRNAs are *trans*-spliced in nematodes than are present as part of polygenic transcripts. Therefore, it seems likely that *trans*-splicing is necessary for something else. What that is awaits further investigation. It is clear the *trans*-splicing is not used as a regulatory mechanism, nor do *trans*-spliced mRNAs fall into specific functional classes.

In summary, with the exception of trypanosomes, the biological function(s) of *trans*-splicing remain largely unclear. Elucidation of the role(s) of this unusual RNA processing reaction will undoubtedly provide insight into the evolutionary pressures which have caused it to be retained in multiple phylogenetic groups.

RNA EDITING

Discovery of RNA editing

The kinetoplast, the single mitochondrion of kinetoplastid protozoa, has a highly unusual gene organization, even by mitochondrial standards. Kinetoplastid DNA is composed of a concatenated network of roughly 20–50 large and 5000–10 000 small circular DNAs. The larger, 'maxicircle' DNAs, which range in size from 20–39 kb in different species, contain large regions of homology to other mitochondrial genomes, but initially no function could be ascribed to the 0.8–2.5 kb 'minicircles', even though both types of DNAs were known to be transcribed.

Curiously, although mitochondria usually encode the genes for the integral membrane proteins involved in electron transport and oxidative phosphorylation, homologs of some of these mitochondrial genes seemed to be missing entirely, while other maxicircle-encoded 'genes' did not contain contiguous open reading frames, appearing to require one or more frameshifts for expression. Noting that one such frameshift was conserved between species, Benne and colleagues characterized the *coxII* mRNA from both *Trypanosoma brucei* and *Crithidia fasciculata*, and found that kinetoplast *coxII* mRNAs contained four uridine (U) residues that were not encoded within maxicircle DNA. After an exhaustive search for additional 'intact' copies of the *coxII* gene, they concluded that the extra nucleotides were added at the RNA level via an unknown 'editing' process.

With acceptance of this unorthodox idea came insights into other previously unexplained mysteries of mitochondrial gene expression, such as the apparent use of an altered genetic code in plant mitochondria. Examination of plant mitochondrial (and chloroplast) mRNAs showed that they also differed in specific ways from the genes from which they were transcribed, but in this case the differences were due to cytidine (C) to uridine (and more rarely U to C) changes rather than alterations in overall mRNA length. Thus, this form of RNA editing results in the restoration of canonical codon usage patterns rather than causing shifts in the reading frame. Although subsequent work has demonstrated that editing is quite widespread and is not confined exclusively to organelles, kinetoplastid RNA (kRNA) editing still provides the most fantastic examples of this extraordinary form of gene expression.

Characterization of editing events in trypanosomatids

Since its initial discovery, many additional examples of kRNA editing have been described, involving both insertion of non-templated U residues and deletion of Us that are encoded within the mitochondrial genome. These range from cases requiring relatively modest changes to 'cryptogenes', genes in which a large percentage of the nucleotides present in kRNAs are not encoded in a traditional manner (Figure 2.6). Although not recognizable as genes at the DNA level, extensive editing throughout the

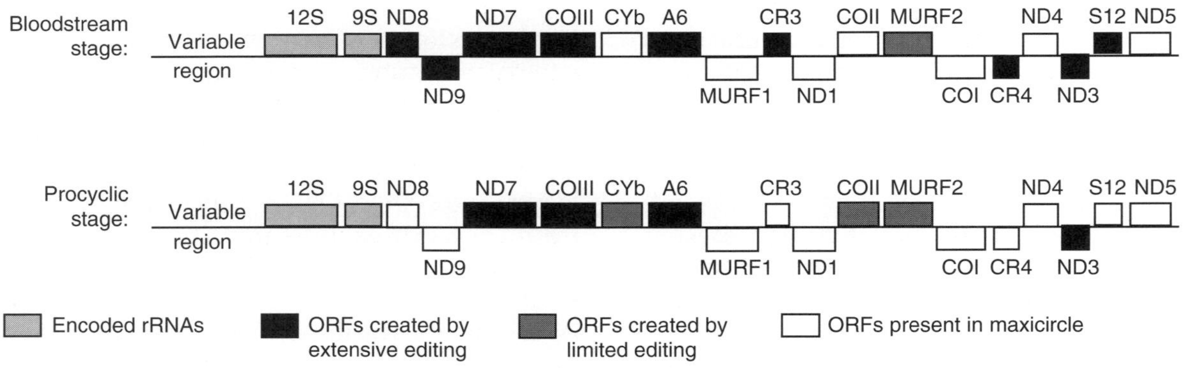

FIGURE 2.6 Editing of *T. brucei* genes at different developmental stages.

entire transcript results in the creation of long, translatable mRNAs. Perhaps the most spectacular example of this type of 'pan-editing' is provided by the 969 nt *T. brucei coxIII* mRNA, which contains 547 U insertions and 41 U deletions relative to the maxicircle genome. The end result of RNA editing is the creation of open reading frames predicted to encode polypeptides homologous to mitochondrial proteins found in other organisms. These changes, which involve the addition of 1–8 Us at insertion sites and the removal of 1–5 Us at deletion sites, often result in the creation (or removal) of initiation and termination codons.

The frequency of uridine insertion and deletion varies considerably among different groups of kinetoplastid protozoa, with more extensive editing generally found in early-diverging branches of the phylogenetic tree. For instance, 12 of 17 maxicircle genes in both *T. brucei* and *Leishmania tarentolae* are edited, but *T. brucei* editing involves the insertion of 3030 U residues and the deletion of 322 Us, whereas *L. tarentolae* kRNAs are subject to 'only' 348 U insertions and 56 U deletions. At first it was thought that editing might have arisen as an adaptation to a parasitic lifestyle, but this seems not to be the case, since bodonids, a free-living group of kinetoplastid protists, also edit their RNAs in a similar manner. Interestingly, many editing events are developmentally regulated (see Figure 2.6), but the extent to which these alterations are used to regulate gene expression is still unclear.

Potential clues as to mechanism

Despite the lack of obvious patterns or sequence contexts surrounding editing sites, mechanistic hints began to emerge upon characterization of the steady-state RNAs present in kinetoplasts. The identification of a substantial pool of partially edited RNAs that were processed only near their 3′ ends suggested that editing occurs with a 3′ to 5′ polarity, a feature that has since been confirmed (see below).

The key to the mechanism was the discovery of guide RNAs (gRNAs) by Blum, Bakalara, and Simpson. These small (55–70 nt) transcripts are perfectly complementary to short stretches of fully edited mRNAs if non-Watson–Crick G-U pairs are allowed, suggesting that gRNAs are the source of the 'missing' information. (Despite the fact that only A-U and G-C pairs are normally used during RNA synthesis, such G-U 'wobble' pairs are common in structured RNAs and are sometimes used in codon recognition by tRNAs during translation.) Further characterization of these small RNAs led to the recognition that they contain three functionally distinct regions: (i) an anchor region (5–12 nt at the 5′ end of the gRNA) that pairs with the mRNA just 3′ of an editing site, (ii) a 25–35 nt long guiding region that specifies 1–20 sites of U insertion/deletion using both standard A-U pairs and G-U wobble pairs, and (iii) a non-encoded 3′ oligoU tail of 5–24 nt (Figure 2.7). Many, but not all, gRNAs are encoded within minicircle sequences, with the number of gRNAs derived from each minicircle varying between species.

Editing models

The discovery of gRNAs, partially-edited potential intermediates, and 'mis-edited' RNAs containing odd junction sequences led to the elaboration of several editing models. Each required pairing of a specific gRNA with a region of the pre-edited mRNA just downstream of an editing site, cleavage of the RNA chain, addition (or deletion) of uridines based on the sequence of the guiding region of the gRNA, and reformation of the altered mRNA. These models differed, however, in the means by which these steps were accomplished. Most

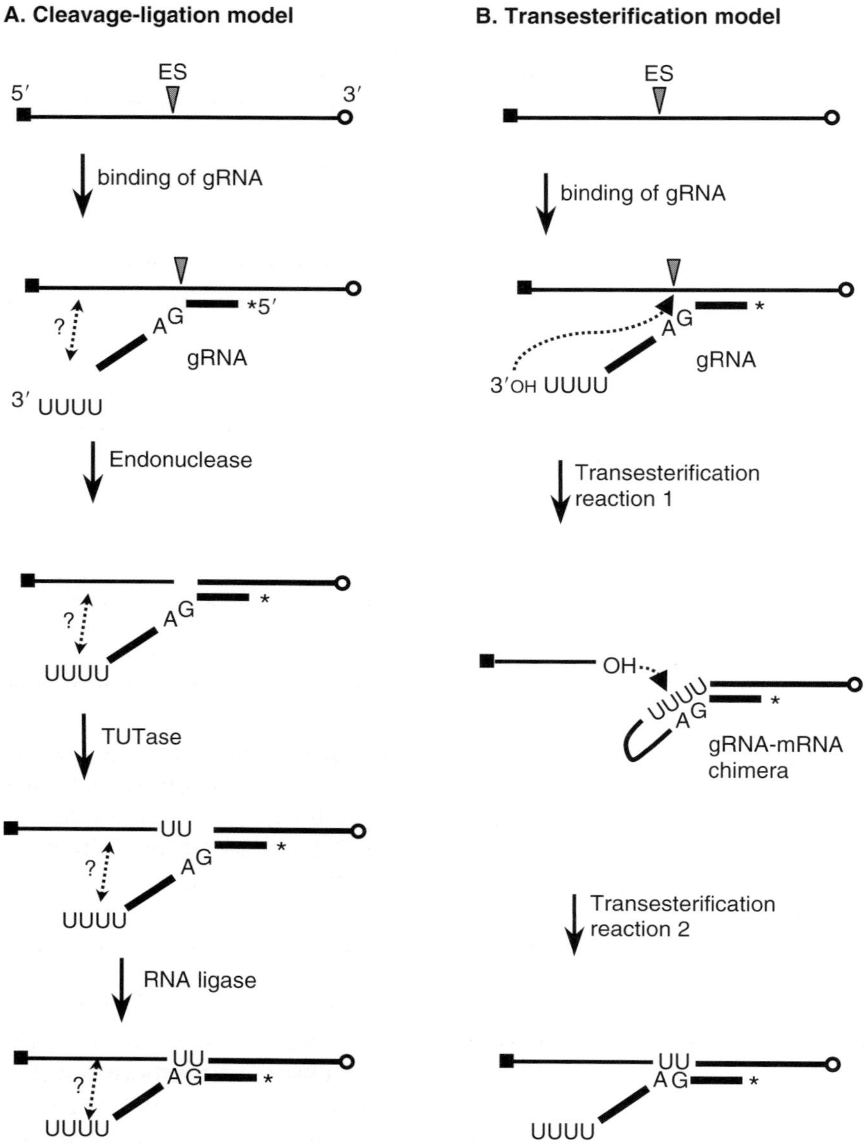

FIGURE 2.7 Schematic depiction of two models for kinetoplast mRNA editing.

editing models were variations of the two mechanisms depicted in Figure 2.7. The cleavage–ligation model (Figure 2.7A) was initially proposed by Simpson and colleagues based on the presence of enzymatic activities, such as terminal uridine transferase (TUTase), RNA ligase, and endonucleases, in *L. tarentolae* mitochondria. A second, transesterification model

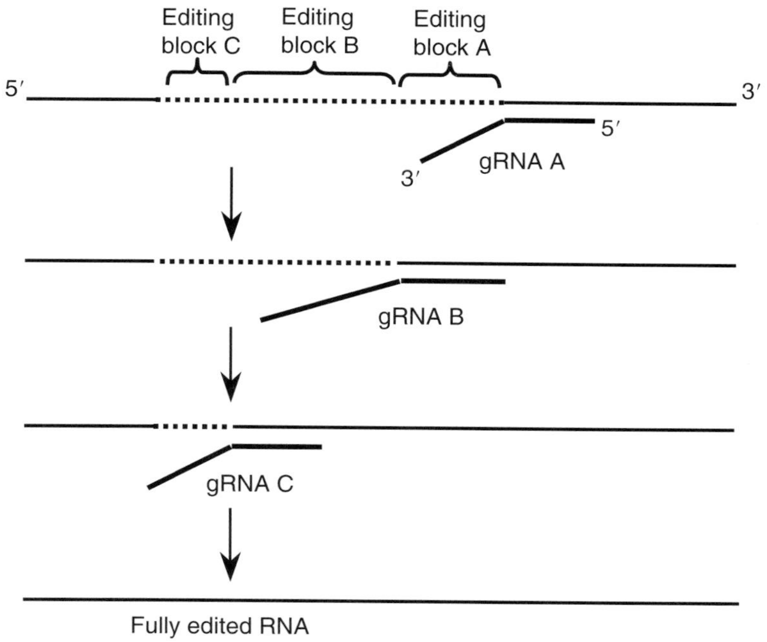

FIGURE 2.8 Sequential use of different guide RNAs accounts for the 3′ to 5′ polarity of editing.

(Figure 2.7B) was proposed shortly thereafter, based largely on the exciting developments in the fields of catalytic RNAs and RNA splicing. This model predicted the existence of hybrid RNAs consisting of 5′ gRNA sequences covalently attached to mRNA sequences at their 3′ ends. The presence of such 'chimeras' in kinetoplastid RNAs, albeit at very low abundance (<1 per cell), further intensified the debate regarding the relative merits of individual models.

Each of the proposed mechanisms served to explain many of the observations present in the literature. For instance, in all models of gRNA-directed editing the apparent 3′ to 5′ polarity could be explained by sequential utilization of gRNAs (Figure 2.8). In a pre-edited mRNA requiring the utilization of many overlapping gRNAs, only the most 3′ gRNA binding site would be present, with the anchor regions for the subsequent gRNAs being created by insertion and deletions of Us within the preceding editing block. Likewise, examples of mis-editing could be explained by 'mis-guiding', i.e. the use of the wrong gRNA for U insertion-deletion within a given region.

These models also proved extremely valuable in directing experimental approaches, since each made specific predictions regarding reaction intermediates (see below). The most obvious difference between the models was the role of the oligoU tail of gRNAs, which in the cleavage–ligation model serves as a means of stabilizing the editing complex, perhaps retaining the 5′ cleavage product after endonuclease cleavage, while in the transesterification model it provides the attacking group in the first transesterification reaction and serves as a reservoir for added or deleted uridine residues. However, in the absence of an assay system that could utilize artificial gRNAs and editing substrates, it proved difficult to distinguish between these potential roles.

Development of *in vitro* editing systems and the current model for kRNA editing

An important breakthrough in the field was provided by the development by Stuart and colleagues of *in vitro* systems capable of supporting editing of defined substrates using cognate gRNAs. This allowed confirmation of the hypothesis that gRNAs do indeed direct the insertion and deletion of uridines and set the stage for specific tests of each model. Initially, the efficiency of these systems was so low that intermediates and products could only be characterized by RT-PCR analysis, but eventually the efficiency reached levels allowing the use of isotopically labeled substrates and gRNAs. This was a critical advance, since it allowed the direct analysis of reaction intermediates and products.

The power of this approach can be illustrated by comparing the predicted outcomes of experiments utilizing end-labeled molecules for each of the models depicted in Figure 2.7. For example, if a 5′ end-labeled gRNA (* in Figure 2.7) is used, the transesterification model predicts that this short RNA would become incorporated into a larger gRNA-mRNA chimera during the course of the reaction, whereas in the cleavage–ligation model the size of the labeled gRNA would remain unchanged. Addition of a 3′ end-labeled pre-edited mRNA (○ in Figure 2.7) would also distinguish between these models, with the cleavage–ligation model predicting the production of a smaller 3′ cleavage fragment rather than the larger chimera anticipated by the transesterification model. Similarly, because the size of the 5′ cleavage intermediate is only expected to change in the cleavage–ligation model (upon addition or deletion of uridines), a 5′ end-labeled pre-edited mRNA (■ in Figure 2.7) could also be used to differentiate between these two potential mechanisms.

Data derived from such labeling experiments strongly support the original cleavage–ligation model for kRNA editing shown in Figure 2.7A. Short, 3′ cleavage intermediates are produced in these *in vitro* assays, and uridines are added to or removed from the 5′ cleavage intermediates. Although chimeric gRNA-mRNA molecules have been observed in these experiments, the kinetics of the appearance of these chimeras has led to the conclusion that these hybrid molecules are rare, non-productive side-products rather than authentic intermediates in the editing pathway.

Proteins involved in kRNA editing

The cleavage–ligation model of uridine insertion/deletion predicts the existence of a number of kinetoplastid proteins, including endonuclease(s), TUTase, an exonuclease (for U deletion from the 5′ cleavage product), and RNA ligase(s). These activities have been detected in, and in some cases purified from, mitochondrial lysates. Roles for additional proteins in regulation, assembly of the editing machinery, processivity, and gRNA unwinding have been postulated, and candidates for many of these activities have also been identified. A number of these enzymatic activities cosediment on glycerol gradients (fractionating between 10–40 S, depending on conditions and species), suggesting that they act as part of one or more large complexes. A combination of biochemical (fractionation, UV crosslinking, RNA binding, adenylation/deadenylation, and helicase assays), genetic (gene disruption, replacement, knockouts), and antibody (inhibition, immunoprecipitation, immunofluorescence microscopy) approaches are currently being used to define the proteins involved in the editing pathway. However, given the complexity of the system, the large number of components, and the potential overlap between activities responsible for U insertion and deletion, thus far it has been quite difficult to definitively assign a function (or functions) to most of the proteins under study.

Editing is an essential process

Because no 'correct' copies of most mitochondrial genes have been found, it has long been assumed that editing is required for expression of a number of kinetoplastid gene products. However, due to the highly hydrophobic nature of the proteins encoded by these edited mRNAs, direct evidence for translation of the products of editing has only lately been achieved. Very recently, gene knockout experiments have been used to demonstrate that the gene encoding an RNA ligase associated with editing complexes is essential for both RNA editing and survival of the bloodstream form of *T. brucei*, indicating that RNA editing is both a critical aspect of gene expression and an inviting target for intervention.

Outstanding questions

Although the general outlines of the editing mechanism are now clear, many important challenges still remain. Work is continuing on the identification and characterization of the components of the editing apparatus and the determination of their functional roles in the editing process. The current *in vitro* systems are somewhat limited in that they support only a single round of nucleotide insertion or deletion and that only a subset of substrates are efficiently edited. Examination of the roles of proteins, such as helicases, that may contribute to editing processivity will require the use of assay systems that are capable of carrying out editing at consecutive sites, and eventually, sequential use of multiple gRNAs.

The assembly and disassembly of the editing machinery is another issue that has yet to be seriously addressed. Although it is clear that targeting of editing sites is accomplished by annealing of the gRNA anchor to its cognate mRNA binding site, the steps involved in this initial recognition are unknown. It seems likely that gRNAs are utilized as ribonucleoprotein particles and it is possible that they are actively recruited to specific sites by assembly factors. Many kinetoplastid proteins have been shown to crosslink to gRNAs, with postulated roles including gRNA maturation, folding, stability, and turnover as well as more direct involvement in the editing process. Other related questions include (i) the point at which the rest of the editing machinery is added, (ii) whether the components of the editing apparatus are added as a pre-formed complex, as small subassemblies, or individually, (iii) whether or not enzymatic activities required for uridine insertion and deletion are present in the same complexes, and (iv) the extent of overlap between insertion and deletion activities (for example, are separate endonucleases and/or RNA ligases required for insertion and deletion?). Ultimately, *in vitro* reconstitution of a processive editing activity from purified components will be needed to definitively answer these difficult questions.

Finally, the role of editing in the regulation of kinetoplastid gene expression has not been fully explored. Mitochondrial function is drastically reduced in bloodstream parasites and many aspects of RNA metabolism are developmentally regulated. There is a differential abundance of individual mRNAs and gRNAs at various life cycle stages, and the extent of both editing and polyadenylation vary with developmental stage. However, since the abundance of gRNAs does not correlate well with the level of edited mRNAs, regulatory mechanisms other than simple transcript abundance are likely to affect editing efficiency. In addition, editing could possibly be linked to other processes, including RNA synthesis, maturation (polyadenylation, alternative processing), or message stability. And because it is currently unknown whether partially edited mRNAs are translated, the potential also exists for the production of different protein isoforms during development.

SUMMARY

There are a number of notable features of kRNA editing that make this form of gene expression unique. These include the findings that (i) some (pan-edited) mRNAs require many separate genetic segments to assemble a functional mRNA product, (ii) non-Watson–Crick (G-U) pairs are involved in the production of mRNA transcripts, and (iii) only part of the mRNA is copied directly from the DNA, the rest is templated by gRNA transcripts.

Despite significant differences in mechanism, there are a number of interesting parallels between kRNA editing and *trans*-splicing. First, each entails the expression and assembly of genetic information encoded at multiple locations within the genome. Secondly, in each case the ultimate end-product of the 'gene' is assembled at the RNA level using multiple substrates. Finally, each of these processes is apparently catalyzed by large ribonucleoprotein complexes potentially requiring *de novo* assembly at each splice site or editing block. The existence of these alternative modes of gene expression within different cellular compartments of a non-traditional 'model' organism emphasizes the importance of exploring all evolutionary niches as we seek to elucidate the underlying mechanisms of gene expression in all of its diverse forms.

FURTHER READING

Trans-splicing

Agabian, N. (1990). *Trans*-splicing of nuclear pre-mRNAs. *Cell* **61**, 1157–1160.

Blumenthal, T. and Steward, K. (1997). RNA processing and gene structure. In: Riddle, D.L., Blumenthal, T., Meyer, B.J. and Priess, J.R. (eds). *C. elegans II*, Cold Spring Harbor: Cold Spring Harbor Laboratory Press, pp. 117–146.

Bonen, L. (1993). *Trans*-splicing of pre-mRNA in plants, animals, and protists. *FASEB J.* **7**, 40–46.

Bruzik, J.P., van Doren, K., Hirsh, D. and Steitz, J.A. (1988). *Trans*-splicing involves a novel form of small ribonucleoprotein particles. *Nature* **335**, 559–562.

Davis, R.E. (1996). Spliced leader RNA *trans*-splicing in metazoa. *Parasitol. Today* **12**, 33–40.

Denker, J.A., Maroney, P.A., Yu, Y.-T., Kanost, R.A. and Nilsen, T.W. (1996). Multiple requirements for nematode spliced leader RNP function in *trans*-splicing. *RNA* **2**, 746–755.

Denker, J.A., Zuckerman, D.M., Maroney, P.A. and Nilsen, T.W. (2002). New components of the spliced leader RNP required for nematode *trans*-splicing. *Nature* **417**, 667–670.

Ferguson, K., Heid, P. and Rothman, J. (1996). The SL1 *trans*-spliced leader RNA performs an essential embryonic function in *Caenorhabditis elegans* that can also be supplied by SL2 RNA. *Genes Dev.* **10**, 1543–1556.

Hannon, G.J., Maroney, P.A., Denker, J.A. and Nilsen, T.W. (1990). *Trans*-splicing of nematode pre-messenger RNA *in vitro*. *Cell* **61**, 1247–1255.

Krause, M. and Hirsh, D. (1987). A *trans*-spliced leader sequence on actin mRNA in *C. elegans*. *Cell* **49**, 753–761.

Murphy, W.J., Watkins, K.P. and Agabian, N. (1986). Identification of a novel Y branch structure as an intermediate in trypanosome mRNA processing: evidence for *trans*-splicing. *Cell* **47**, 517–525.

Nilsen, T.W. (1994). RNA–RNA interactions in the spliceosome: Unraveling the ties that bind. *Cell* **78**, 1–4.

Nilsen, T.W. (1997). *Trans*-splicing. In: Krainer, A.R. (ed.). *Frontiers in Molecular Biology: Eukaryotic mRNA Processing*, Oxford: IRL Press, pp. 310–334.

Nilsen, T.W. (2001). Evolutionary origin(s) of SL-addition *trans*-splicing: still an enigma. *Trends Genet.* **17**, 678–680.

Sutton, R. and Boothroyd, J.C. (1986). Evidence for *trans*-splicing in trypanosomes. *Cell* **47**, 527–535.

Tschudi, C. and Ullu, E. (1990). Destruction of U2, U4, or U6 small nuclear RNAs blocks *trans*-splicing in trypanosome cells. *Cell* **61**, 459–466.

RNA editing

Benne, R., Van den Burg, J., Brakenhoff, J.P., Sloof, P., Van Boom, J.H. and Tromp, M.C. (1986). Major transcript of the frameshifted *cox II* gene from trypanosome mitochondria contains four nucleotides that are not encoded in the DNA. *Cell* **46**, 819–826.

Blum, B., Bakalara, N. and Simpson, L. (1990). A model for RNA editing in kinetoplastid mitochondria: 'guide' RNA molecules transcribed from maxicircle DNA provide the edited information. *Cell* **60**, 189–198.

Estevez, A.M. and Simpson, L. (1999). Uridine insertion/deletion RNA editing in trypanosome mitochondria – a review. *Gene* **240**, 247–260.

Hajduk, S.L. and Sabatini, R.S. (1998). Mitochondrial mRNA editing in kinetoplastid protozoa. In: Grosjean, H. and Benne, R. (eds). *Modification and Editing of RNA*, Washington: ASM Press, pp. 377–393.

Kable, M.L., Siewert, S.D., Heidmann, S. and Stuart, K. (1996). RNA editing: a mechanism of gRNA-specified uridylate insertion into precursor mRNA. *Science* **273**, 1189–1195.

Seiwert, S.D. and Stuart, K. (1994). RNA editing: transfer of genetic information from gRNA to precursor mRNA *in vitro*. *Science* **266**, 114–117.

Seiwert, S.D., Heidmann, S. and Stuart, K. (1996). Direct visualization of uridylate deletion in vitro suggests a mechanism for kinetoplastid RNA editing. *Cell* **84**, 831–841.

Simpson, L. (1999). RNA editing – an evolutionary perspective. In: Gesteland, R.F., Cech, T.R. and Atkins, J.F. (eds). *The RNA World* (second edition), Cold Spring Harbor: Cold Spring Harbor Laboratory Press, pp. 585–608.

Stuart, K., Allen, T.E., Heidmann, S. and Seiwert, S.D. (1997). RNA editing in kinetoplastid protozoa. *Microbiol. Mol. Biol. Rev.* **61**, 105–120.

Databases and informative web sites

Brewster, S., Aslett, M. and Barker, D.C. (1998). Kinetoplast DNA minicircle database. *Parasitol. Today* **14**, 437.
http://www.ebi.ac.uk/parasites/kDNA/Source.html

Hinz, S. and Goringer, H.U. (1999). The guide RNA database (3.0). *Nucl. Acids Res.* **27**, 168.
http://www.biochem.mpg.de/~goeringe/

Simpson, L., Wang, S.H., Thiemann, O.H., Alfonzo, J.D., Maslor, D.A. and Avila, H.A. (1998). U-insertion/deletion edited sequence database. *Nucl. Acids Res.* **26**, 170.
http://www.rna.ucla.edu/trypanosome/database.html

kRNA editing website
http://www.rna.ucla.edu/trypanosome/index.html

C H A P T E R

3

Transcription

Arthur Günzl
University of Connecticut Health Center,
Farmington, CT, USA

INTRODUCTION

Mechanisms of transcription have been analyzed in great detail in a few model organisms such as the budding yeast *Saccharomyces cerevisiae*, mouse and human. In contrast, our knowledge of the transcription machinery and the mechanisms of transcriptional regulation in parasites is rather limited. There is one notable exception: *Acanthamoeba castellani*, a facultative parasite causing keratitis and acute encephalitis in humans, has been an excellent subject for transcriptional studies and, in particular, RNA polymerase (pol) I-mediated transcription has been meticulously investigated in this organism. Details of this work can be found in several excellent reviews by M.R. Paule. In most other parasites, the lack of suitable assay systems has prevented the investigation of transcriptional processes. However, DNA transfection has now been established for several important parasites, and *in vitro* transcription systems have been developed for trypanosomatid and nematode species. This technology has enabled structural analysis of gene promoters. Moreover, specific promoter element/protein complexes have been identified, and a few proteins of the transcription machinery characterized. Thus far, the most detailed knowledge has been obtained in trypanosomatid species and, therefore, they are a focus in this chapter.

The general picture

The details of eukaryotic transcription are described in numerous books and reviews. Nevertheless, the following brief overview should be helpful to understand parasite-specific aspects of transcription presented in the remainder of the chapter.

Whereas prokaryotes have a single DNA-dependent RNA polymerase, eukaryotic organisms harbor three such enzymes in the nucleus, each serving a distinct function in RNA synthesis. RNA pol I transcribes exclusively the large ribosomal (r)RNA gene unit (rDNA), RNA pol II synthesizes mRNA plus several small RNAs, e.g. the U1–U5 small nuclear (sn)RNAs,

Molecular Medical Parasitology
ISBN 0–12–473346–8

and RNA pol III synthesizes other small RNAs, including tRNAs, 5S rRNA, and U6 snRNA. Eukaryotic RNA pols are multi-subunit enzymes consisting of 12 or more different polypeptides. The largest subunit (~160–220 kDa), the second largest subunit (~115–135 kDa) and a third polypeptide with a size of approximately 40 kDa represent the core-enzyme subunits; they contain highly conserved sequence motifs with homology to prokaryotic RNA pol domains and presumably form the active center of the enzyme. The other subunits are either specific for a particular RNA pol or common to two or to all three enzymes. Eukaryotic RNA pols are unable to recognize specific DNA sequence motifs and cannot accurately transcribe genes by themselves. For correct transcription initiation, auxiliary factors are needed which assemble on promoter sequences and, by recruiting the RNA pol, form a stable transcription initiation complex. Each RNA pol interacts with a specific set of transcription factors and, therefore, depends on characteristic promoter structures. Accordingly, eukaryotic genes are divided into class I, class II, or class III genes. Some factors take part in transcription of genes belonging to different classes. The most prominent example of a ubiquitous factor is the TATA-box binding protein (TBP), a single polypeptide that, as a component of various multi-subunit transcription factors, is important for transcription initiation of all three RNA pols. Auxiliary factors and RNA pol in conjunction with a functional promoter are able to direct a basal level of correctly initiated transcription and constitute the general or basal transcription machinery. In addition, several groups of transcription regulators and cofactors have been described which modulate transcription efficiency at the level of transcription initiation. In the nucleus, the DNA is organized as chromatin, and the nucleosomal structure formed by histones is a general transcription repressor because it obstructs the interaction of the basal transcription machinery with the promoter region. Therefore, chromatin remodeling is an essential step for efficient transcription initiation at chromatin templates. After transcription initiation, the RNA pol detaches from the auxiliary factors, escapes from the promoter, and transforms into a transcription elongation complex. Transcription then proceeds until a termination signal is encountered. Each RNA pol has its specific signal and termination mode. RNA pol III terminates at short runs of 4 to 7 T residues in the sense strand, independent of a *trans*-acting factor. In contrast, termination determinants of class I genes bind a protein factor which, by specifically interacting with RNA pol I, disrupts transcription. Termination of RNA pol II transcription is less well characterized, but it requires G-rich sequences and, in the case of mRNA synthesis, appears to be coupled to RNA polyadenylation.

UNUSUAL MODES OF TRANSCRIPTION IN TRYPANOSOMATIDS AND NEMATODES

Polycistronic transcription of protein coding genes

In eukaryotes, a protein coding gene typically constitutes a single transcription unit, i.e. it is transcribed monocistronically. Cotranscriptional capping of mRNA, a process in which a 7-methylguanosine triphosphate is fused in a 5′–5′ linkage to nascent pre-mRNAs, appears to be the main reason for this mode of transcription because the capping enzyme requires a free 5′ end. Nonetheless, several polycistronic transcripts have been identified in higher eukaryotes, but, characteristically, these molecules mature as polycistronic units.

Trans-splicing of an already capped Spliced Leader (SL) sequence, which comprises the 5′ terminal region of the SL RNA, to the 5′ end of a mRNA is an alternative, post-transcriptional mode of capping and, in combination with polyadenylation, allows the excision of individual mRNA molecules from polycistronic precursors. Accordingly, polycistronic transcription of protein coding genes has been found in those organisms harboring *trans*-splicing of SLs. In trypanosomatids, protein coding genes are tandemly linked, separated by short intergenic regions and, as has been demonstrated by various experimental procedures, are transcribed polycistronically. The organization of the *Leishmania major* Friedlin chromosome 1 sheds some light on the extent of trypanosomatid transcription units. The 300-kbp long chromosome encodes 79 genes. Of those, 50 genes are tandemly located on one strand and the remaining 29 genes sit in tandem on the other strand. Both gene arrays are arranged head-to-head and, most likely, each represents a single transcription unit. However, it is not clear where transcription starts or how RNA pol II is recruited to the DNA. In nematodes, as has been estimated in the free-living worm *Caenorhabditis elegans*, about 25% of all genes are organized in polycistronic transcription units. In contrast to trypanosomatid units, these units are defined and consist of 2–8 genes.

α-Amanitin-resistant transcription of genes encoding the major cell surface antigens in *T. brucei*

Cotranscriptional capping locks mRNA synthesis to RNA pol II-mediated transcription, because the capping enzyme directly and specifically interacts with the phosphorylated, carboxy-terminal domain of the RNA pol II largest subunit. Post-transcriptional *trans*-splicing of a capped Spliced Leader, however, uncouples this linkage and raises the possibility that other RNA pols are recruited for mRNA production. The ability of RNA pol I to efficiently synthesize functional mRNA in combination with *trans*-splicing was demonstrated in *T. brucei* by transient, rDNA promoter-directed reporter gene expression. Evidence that *T. brucei* utilizes RNA pol I for transcription of some endogenous protein coding genes was first obtained by nuclear run-on experiments in which transcription elongation on a variant surface glycoprotein (VSG) gene was resistant to α-amanitin. This mushroom toxin is a strong inhibitor of RNA pol II and a moderate inhibitor of RNA pol III, but does not affect RNA pol I transcription. VSG is the constituent of the cell surface coat in bloodstream form trypanosomes, and a VSG gene is expressed from a telomeric VSG gene expression site (VSG ES) which harbors a single VSG gene and several associated genes. VSG ESs are not the only units transcribed by an α-amanitin-resistant RNA pol in *T. brucei*. Procyclin gene expression sites (procyclin ES), encoding the major cell surface antigens of insect form trypanosomes (procyclics), are located at chromosome-internal positions and are transcribed by an analogous enzyme activity. Furthermore, metacyclic trypanosomes, in the salivary gland of the tsetse, express VSG genes in an α-amanitin-resistant manner from monocistronic transcription units called metacyclic (m)VSG ES. Transcription of other protein coding genes, however, is highly sensitive to α-amanitin and, therefore, mediated by RNA pol II.

α-Amanitin-resistant transcription of protein coding genes is not restricted to *T. brucei*, but has also been observed in the parasites *Trichomonas vaginalis* and *Entamoeba histolytica*. However, in these organisms resistance appears to be a general phenomenon and is not restricted to a few genes. α-Amanitin inhibits transcription elongation by binding to the RNA pol II largest subunit, and studies in other organisms have identified four invariant amino acid residues in this polypeptide which are essential for toxin binding. The *T. vaginalis* protein is divergent at three of the four amino acid residues, whereas in trypanosomatids three residues are identical and the fourth is a conservative change. Thus, in *T. vaginalis,* mRNA is synthesized by a divergent, α-amanitin-resistant RNA pol II, whereas VSG and procyclin ES transcription in *T. brucei* is most likely mediated by RNA pol I. The latter hypothesis is supported by several lines of indirect evidence. For example, rDNA, VSG ES and procyclin ES transcription is much more resistant to the detergent sarkosyl and, as measured in nuclear run-on assays, has a higher rate than transcription of protein-coding housekeeping genes. Furthermore, rDNA, VSG ES and procyclin ES promoters share structural features and interact with common transcription factors (see below, Class I transcription in trypanosomatids). Finally, depletion of RNA pol I from cell extracts specifically and strongly reduced rDNA, VSG ES and procyclin ES promoter transcription *in vitro*. Nevertheless, there are two findings which, apparently, are not consistent with RNA pol I-mediated transcription of VSG and procyclin ESs. Firstly, varying the divalent cation concentration in nuclear run-on experiments conducted with bloodstream-form trypanosomes differentially affected transcription elongation on a VSG gene and on rDNA. The pattern observed with the VSG gene resembled that of a tubulin gene, and it was proposed that a modified, α-amanitin-resistant RNA pol II mediates VSG ES transcription. However, since the *T. brucei* RNA pol II has an intact α-amanitin binding site and toxin resistance is associated with amino acid changes within this site, the nature of such a modification remains obscure. Moreover, rDNA and VSG ES promoter transcription in a procyclic cell extract were both sensitive to elevated manganese ion concentrations, in contrast to procyclin ES promoter transcription. Hence, the observed effects are rather explained by factors which modify the RNA pol I elongation complex in a life-cycle-specific way at VSG and procyclin ES promoters than by an α-amanitin-resistant RNA pol II. Secondly, RNA pol I-mediated transcription of rDNA takes place in the nucleolus and it might be expected that, in *T. brucei,* procyclin and VSG ESs are transcribed in the same compartment. However, at least for VSG ES transcription this is not the case. On the other hand, a nucleolus typically lacks RNA splicing factors and, accordingly, antibodies directed against *trans*-splicing factors did not stain the trypanosome nucleolus. The dependence of VSG mRNA maturation on the *trans*-splicing machinery may have forced *T. brucei* to evolve means for extranucleolar VSG ES transcription by RNA pol I.

In sum, there is a lot of evidence for but no convincing evidence against RNA pol I-mediated procyclin and VSG ES transcription, but a final proof for this hypothesis is still missing. Nonetheless, in this chapter, procyclin and VSG ESs are treated as class I gene units.

CLASS I TRANSCRIPTION IN TRYPANOSOMATIDS

As has been proven for *S. cerevisiae,* eukaryotes possess a single type of class I gene unit encoding the large ribosomal RNAs. Procyclin and (m)VSG ESs of *T. brucei* are the only additional class I gene units characterized in eukaryotes thus far. The three types of class I promoters in *T. brucei* share no obvious sequence homology, they exhibit structural differences, and there is substantial evidence that life-cycle-specific gene expression is in part regulated at the transcriptional level. Hence, it is possible that procyclin and (m)VSG ES promoter transcription depends on novel transcription factors or factor domains.

Transcription regulation

T. brucei has two diploid procyclin ESs which are exclusively expressed in procyclic cells. Procyclin ES promoters are active in both bloodstream and insect forms, but post-transcriptional processes prevent procyclin expression in the bloodstream. On the other hand, the procyclin ES promoter is up to tenfold more active in procyclics than in bloodstream-form trypanosomes. This differential activity was observed independently of the genomic context in which the procyclin ES promoter was integrated. Moreover, procyclin ES promoter transcription in a procyclic cell extract was in comparison to VSG ES and rDNA promoter transcription exceptional by its fourfold higher efficiency, a distinct lag phase, a high template DNA concentration optimum, and its tolerance to manganese cations. Taken together, these data suggest that procyclin ES transcription in procyclic cells is enhanced by a life cycle-specific component.

There are at least 27 mVSG ESs and about 20 bloodstream-form VSG ESs in a *T. brucei* cell. However, the VSG coat of metacyclic and bloodstream-form trypanosomes consists of identical molecules which are expressed from a single gene. Therefore, only one (m)VSG ES is maximally expressed, while the others are inactivated. Regulation occurs at the level of transcription, but it is currently under debate whether in bloodstream-form trypanosomes inactivation is predominantly due to inefficient transcription initiation or to transcription attenuation. Differential expression of VSG ESs does not depend on promoter sequences, because replacement of the VSG ES promoter by an rDNA promoter did not impair activation/inactivation control of VSG ESs. Furthermore, the detection of DNase I hypersensitive sites in both active and inactive VSG ES promoters indicated that auxiliary transcription factors are assembled on all VSG ESs. These findings suggest an epigenetic control of VSG expression in the bloodstream. In procyclic cells, VSG ES promoters generally direct a low level of transcription which is terminated within 700 bp of the transcription initiation site (TIS). Interestingly, rDNA and procyclin ES promoters integrated into a VSG ES are fully active and not repressed, suggesting that repression of VSG ES transcription is dependent on the promoter sequence. Finally, mVSG ES promoters in their chromosomal contexts are only active in the metacyclic stage and, according to nascent RNA analysis, are completely inactivated in both proyclic and bloodstream-form cells.

Promoter structures

A well characterized example of a class I gene promoter from lower eukaryotes is the *S. cerevisiae* rDNA promoter which comprises three sequence blocks denoted as domains I, II, and III (Figure 3.1). Domain I ranges from position −28 to position +8 (−28/+8) relative to the TIS and represents the core promoter, being defined as the minimal structure required for accurate transcription initiation. Typically, it is the only absolutely essential promoter element. Domain II (−76/−51) and domain III (−146/−91) constitute a bipartite upstream element which, unlike the core promoter, stably interacts with *trans*-acting factors. Finally, a sequence motif was identified at position

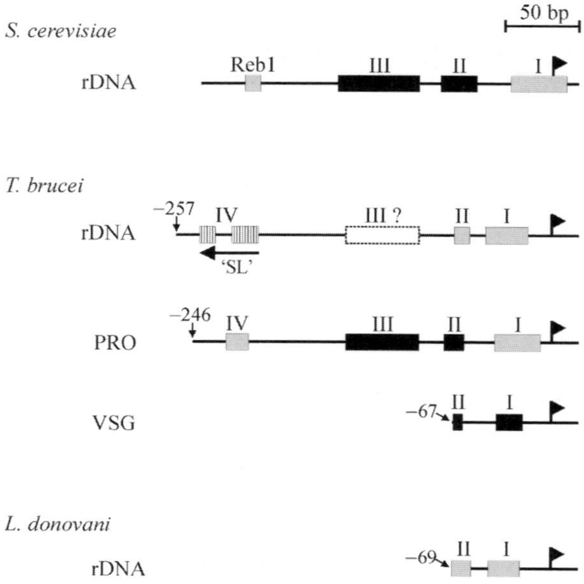

FIGURE 3.1 Class I promoters of *Saccharomyces cerevisiae, Trypanosoma brucei,* and *Leishmania donovani.* Schematic drawing to scale of rRNA gene (rDNA) promoters, and procyclin (PRO) and variant surface glycoprotein (VSG) gene expression site promoters. Promoter domains, as identified by linker-scanner or block substitution analyses (see text), are indicated by rectangles; they are numbered according to the yeast nomenclature. Promoter domains important for stable binding of *trans*-activating factors in transcription competition experiments are drawn in black. The two elements of the *T. brucei* rDNA promoter domain IV (striped rectangles) share sequence homologies with SL RNA gene promoter elements, are orientated in the opposite direction to that of the SL RNA gene (arrow) and, *in vitro*, bind a *trans*-acting factor essential for SL RNA gene transcription. The region of the putative domain III in the *T. brucei* rDNA promoter has not been analyzed. Domain IV of the *S. cerevisiae* promoter is the Reb1p binding domain (Reb1) which is involved in chromatin remodeling. For the *T. brucei* and *L. donovani* promoters, the drawings show minimal promoter regions required for full transcriptional activity. Positions of the 5′ ends are relative to the TISs which are indicated by flags.

−215 which, by binding the protein Reb1, is involved in chromatin remodeling and is important for transcription *in vivo* within the rDNA repeat array.

In comparison, the two procyclin ES promoters of *T. brucei* are nearly identical in sequence and are similarly structured to the yeast rDNA promoter (Figure 3.1). According to detailed mutational analyses *in vivo* and *in vitro*, the region between positions −246 and −7 is sufficient for full transcriptional activity and contains four distinct promoter domains. Domain I (−40/−7) is absolutely essential for transcription and presumably represents the core promoter. Alteration of domain II (−72/−57) reduced transcription efficiency dramatically but did not abolish it, and mutation of domain III (−143/−90) resulted in a moderate drop in transcriptional activity. Furthermore, changing the distance between domains I, II, and III strongly reduced transcription efficiency, demonstrating that the positional arrangement of these promoter elements is of crucial importance. In all studies, the results for domains I to III were in close agreement. In contrast, the most distal part of the promoter containing domain IV (−222/−207) proved to be important *in vivo* but not *in vitro*. Hence, size and location of the *T. brucei* procyclin ES promoter domains are very similar to those of the yeast rDNA promoter. Moreover, *in vitro* competition of procyclin ES promoter transcription revealed that, as in the yeast rDNA promoter, domain III in cooperation with domain II is essential for stable binding of *trans*-activating factors whereas domain I is not. The procyclin ES promoter domain IV is located at approximately the same position as the Reb1 binding domain of the yeast rDNA promoter (Figure 3.1) and, since it exerts its effect only *in vivo*, it may function in chromatin remodeling in the same way as its yeast counterpart.

The rDNA promoter of *T. brucei,* although less well characterized, has a similar organization (Figure 3.1). The essential region ranges from position −257 to −13 and most likely contains four domains like the procyclin ES promoter. While domain I (−42/−13) and domain II (−62/−53) were unambiguously identified by a block substitution analysis in the proximal promoter region, 5′ promoter deletions indicated the presence of domains III and IV in the distal part of the promoter. Subsequently, domain IV was characterized in more detail by *in vitro* transcription competition experiments, which uncovered a surprising piece of information: a linear DNA fragment containing the rDNA promoter competed transcription of the SL RNA gene, which served as a control template in the competition assays. This unexpected finding suggested that rDNA and SL RNA gene promoters, which recruit different RNA pols, bind a common *trans*-activating transcription factor. A closer look at the rDNA promoter revealed two sequence blocks in the most distal promoter region which, in opposite orientation, are homologous to two essential SL RNA gene promoter elements (Figure 3.1; and see below). When these sequences were mutated, the rDNA fragment was unable to compete SL RNA gene transcription *in vitro* and the rDNA promoter lost its ability to direct efficient gene expression *in vivo,* demonstrating the functional importance of these elements in the rDNA promoter. Domain IV is dispensable for efficient transcription *in vitro* indicating that it is functionally analogous to domain IV of the procyclin ES promoter. However, domains IV of rDNA and procyclin ES promoters are distinct elements because they share no sequence homology, and the procyclin ES promoter does not bind the activator of SL RNA gene transcription. Hence, domain IV of the procyclin ES promoter presumably interacts with a specific protein.

In addition to their study in *T. brucei,* rDNA promoters have been investigated by transient transfection assays in several *Leishmania* species. The most detailed study was conducted in *L. donovani,* and revealed a structure surprisingly different from the corresponding *T. brucei* promoter. In *L. donovani,* the 69-bp region upstream of the TIS is sufficient to direct full transcriptional activity, and it contains two distinct promoter domains (Figure 3.1). Interestingly, transcription efficiency was reduced when the sequence around the initiation site was changed. However, since mutation of the initiation nucleotide itself may have a dramatic effect on transcription efficiency, it is currently not clear whether the *L. donovani* rDNA promoter includes an element at this site essential for transcription complex assembly.

VSG ES promoters represent the third type of class I promoters in *T. brucei.* Several promoter sequences have been determined, and the few single nucleotide polymorphisms found appear to have no influence on transcription activity. Interestingly, the structure of the VSG ES promoter closely resembles that of the *L. donovani* rDNA promoter; it is of the same short size and it has the same two-domain structure (Figure 3.1). Hence, in *T. brucei,* the VSG ES promoter differs significantly from procyclin ES and rDNA promoters because it lacks domains III and IV. Moreover, transcription competition experiments *in vitro* showed that VSG ES promoter domains I and II, in contrast to their counterparts in the procyclin ES promoter, are essential for and cooperate in stable binding of transcription factors. Hence, it appears that domains I and II of these two promoters are not functionally analogous.

To date, it has not been clarified whether promoters of metacyclic VSG ES represent a fourth type of class I gene promoters in *T. brucei.* This may be the case because mVSG gene expression is regulated exclusively at the level of transcription and is different from that of bloodstream-form VSG genes. In addition, mVSG ES promoter sequences are less well conserved and deviate significantly from the consensus sequence of bloodstream-form VSG ES promoters. On the other hand, mVSG ES promoters are functional when truncated to the same short size as their counterparts in bloodstream-form VSG ESs and they appear to have the same two-domain structure. However, analysis of the *MVAT7* mVSG ES promoter suggested that the promoter domains are not located at the same positions as their counterparts in bloodstream forms. In summary, class I promoters of trypanosomatids can be divided into two groups: the *T. brucei* rDNA and procyclin ES promoters represent the first group which is characterized by a multi-domain structure resembling that of the yeast rDNA promoter. The second group comprises the *L. donovani* rDNA and *T. brucei* VSG ES promoters which are very short, not extending further upstream than position −70, and possess two distinct domains upstream of the TIS. A two-domain structure is reminiscent of vertebrate rDNA promoters, but in these organisms domain II is much larger and located approximately 90 bp further upstream. Therefore, the structure of the two trypanosomatid promoters in a strict sense does not resemble any known class I promoter.

Finally, as has been shown in other eukaryotic organisms, rDNA transcription is stimulated by enhancer elements which work relatively independently of distance and which are typically arranged as tandem repeats. In this respect, it is interesting to note that repetitive elements are present in trypanosomatid rRNA gene units. In several *Leishmania* species and in *Trypanosoma cruzi,* tandem repeats were identified close to the promoter region, and in the *T. brucei* rRNA gene unit, repetitive elements are located approximately 9 kb upstream of the promoter region. Furthermore, VSG ESs in *T. brucei* have a 50-bp repeat array approximately 1.4 kb upstream of the TIS whereas no repeats have yet been reported for mVSG and procyclin ESs. Interestingly, in *L. chagasi,* the repeats moderately enhanced transient rDNA promoter-directed expression of a reporter gene, whereas those of the *L. donovani* rDNA promoter and of the *T. brucei* VSG ES promoter exhibited no enhancer effect. It should be noted though that, in the yeast rDNA promoter, the enhancer repeats were only functional in a chromosomal context and not in an episomal vector construct. Hence, it still needs to be clarified whether repetitive elements in trypanosomatid class I gene units function as transcriptional enhancers.

DNA–protein interactions at promoter domains

In a first step to characterize the transcription machinery in *T. brucei,* the interaction of proteins with class I gene promoter elements has been investigated by gel retardation assays, and specific band-shifts were reported for all three promoter types. Interestingly, for both procyclin and VSG ES promoters, specific protein binding to single-stranded promoter sequences was observed. A polypeptide of 40 kDa was identified binding to the antisense and sense strands of VSG ES promoter domains I and II, respectively, due to an inverted sequence motif present in these domains. Furthermore, competition in DNA/protein complex formation suggested that the same protein binds to the antisense strand of domain I in the rDNA and procyclin ES promoters. The involvement of single-stranded DNA-binding proteins in transcription regulation is not unprecedented, and several cases have been reported in which such proteins activate or repress transcription. In *T. brucei*, however, the functional significance of single-stranded DNA–protein complexes for transcription remains to be determined. Basal transcription factors of other eukaryotes interact with double-stranded DNA. *T. brucei* may be no exception here because specific, double-stranded DNA–protein complexes were identified with probes spanning the procyclin ES promoter domain I or both domains of the VSG ES promoter.

Transcription termination

The mechanism of RNA pol I transcription termination is conserved between yeast and mammals, and requires a terminator consisting of a termination factor and its specific DNA binding site. Termination is caused by specific interactions of RNA pol I and the termination factor and depends on the right orientation of the terminator. Berberof and colleagues used transient transfection assays to search for termination determinants in a procyclin ES. By cloning putative terminator-containing sequences between the procyclin ES promoter and a reporter gene, they were able to identify three regions which efficiently reduced reporter gene expression, presumably by stopping RNA pol I. In accordance with the general model, these DNA fragments were functional in an orientation-dependent manner and did not inhibit RNA pol II-driven gene expression. Interestingly, in one of these fragments, an important sequence motif was identified which resembled domain II of the VSG ES promoter and which interacted with the same 40 kDa single-stranded DNA binding factor (see above, Promoter structures). While further experiments are necessary to assess the role of this single-stranded DNA–protein complex in transcription termination, the connection between terminator and promoter elements made in this study is typical for rDNA repeat units of yeast and higher eukaryotes. These gene units contain termination signals downstream of the ribosomal genes, as expected, but additionally possess a single terminator, denoted as proximal terminator or T_0, approximately 200 bp upstream of the TIS. The proximal terminator can be multifunctional, but in all cases analyzed it is involved in remodeling the chromatin structure of the promoter region. In yeast, the termination signal and proximal terminator is the Reb1p binding domain (Figure 3.1). Since domains IV of the *T. brucei* procyclin ES and rDNA promoters may be functionally analogous to the proximal terminator (see section on Promoter structures), they may also function in transcription termination.

RNA pol I transcription machinery

The conventional way to characterize a transcription component involves purification of the protein, polypeptide sequencing, and cloning of the respective gene. However, the very low abundance of most transcription factors makes them difficult to purify and, in trypanosomatids, no polypeptide involved in RNA pol I-mediated transcription has yet been discovered in this way, despite the characterization of specific band-shifts and the development of an *in vitro* transcription system.

Another strategy is data mining of genome databases for conserved sequence motifs to identify homologs of polypeptides already characterized in other eukaryotes. Thus far, this has been successful for *bona fide* RNA pol I subunits. Many eukaryotic RNA pol subunits are functionally conserved and homologs exist even in archaebacteria. By exploiting this high degree of conservation, the gene of the largest subunit of *T. brucei* RNA pol I was identified and characterized. It is a single-copy gene and encodes a 1781 amino acid-long polypeptide with a predicted molecular weight of 197 kDa. The deduced amino acid sequence harbors RNA pol I-specific regions as well as eight conserved domains present in all eukaryotic RNA pol largest subunits. Recently, we used a partial *T. brucei* sequence similar to the RNA pol I second largest subunit (gene bank accession no. W84090) to clone the complete cDNA of this polypeptide. The deduced amino acid sequence contains conserved domains of second largest subunits as well as sequences homologous to RNA pol I signatures defined in higher eukaryotes. The predicted molecular weight of 180 kDa of this polypeptide, which was confirmed by partial purification of RNA pol I, is unusually large due to an approximately 300 amino acid-long N-terminal domain which is not present in homologs of other eukaryotes. If this extra domain is of functional significance, it may serve as a target against the parasite.

While the script-based functional classification of sequences present in trypanosomatid genome databases identified several putative RNA pol subunit genes (see the Parasite Proteome Analysis web page at http://www.ebi.ac.uk/parasites/proteomes.html), no similarities to auxiliary factor genes have yet been detected. This may not be possible at all because in contrast to most RNA pol subunits, class I factors are not well conserved among eukaryotes. Exceptions to this rule are TBP which generally is part of the class I basal machinery and histones H3 and H4 which are subunits of the yeast upstream activation factor. In trypansomatids, a TBP homolog has not been identified thus far, and it remains to be determined whether histones play a role as transcription factor components in these organisms.

CLASS II TRANSCRIPTION OF PROTEIN CODING GENES

Trichomonas vaginalis and *Giardia lamblia*

T. vaginalis and *G. lamblia* belong to phylogenetic lineages which diverged very early from the main eukaryotic lineage. In *T. vaginalis*, the sequence surrounding the TIS of protein coding genes is conserved and transient reporter gene expression assays have shown that it is an essential promoter element directing accurate transcription. In addition, the sequence specifically binds a nuclear factor and, therefore, was identified as an initiator element. Initiator elements have been characterized predominantly in mammalian systems. They interact with the basal transcription machinery and are important for transcription initiation especially in TATA-less promoters. Interestingly, a mammalian initiator element could functionally substitute for the *Trichomonas* element, suggesting that initiator elements are highly conserved and evolved very early in eukaryotic evolution. In *G. lamblia*, the TIS of protein coding genes is surrounded by an A/T-rich sequence which, as has been shown for several genes, is essential for *in vivo* expression of a reporter gene and, by itself, is able to direct a low level of expression. In contrast to the *T. vaginalis* initiator element, the A/T-rich sequence is not well conserved and cannot be substituted by a mammalian initiator element. In both *G. lamblia* and *T. vaginalis* genes, promoter elements were identified in the region upstream of the TIS. In *G. lamblia*, a consensus TATA element has not been identified, but A/T-rich elements, located between positions −20 and −60, were shown to be important transcription determinants. *T. vaginalis* promoters apparently lack TATA boxes. Analysis of the hydrogenosomal α-succinyl CoA synthetase gene revealed in the upstream promoter region two important and closely spaced elements centered at positions −91 and −75, while mutation of sequences between −49 and the initiator element did not greatly influence transient reporter gene expression. Besides two 10 bp sequence blocks at positions −54 and −194 which exhibited moderate transcriptional effects, deletional analysis indicated the presence of additional positive and negative regulatory elements further upstream.

As mentioned above (α-Amanitin-resistant transcription), the RNA pol II largest subunit of *T. vaginalis* has been characterized and found to be divergent at the α-amanitin binding site. In addition, it has an unusual C-terminal domain (CTD) because it lacks the essential heptapeptide repeats of higher eukaryotes but is relatively rich in serine and proline residues. This unusual CTD structure is characteristic for lower protists and is also found in *G. lamblia* and trypanosomatids.

Trypanosomatidae

In trypanosomatids, gene expression is mainly controlled post-transcriptionally, because the polycistronic mode of transcription does not leave much scope for regulation at the level of transcription initiation. Interestingly, transcription initiation by RNA pol II at polycistronic transcription units appears to have a low specificity. In several trypansomatid species it has been observed that genes can be expressed from episomal, promoterless vectors as long as sequence determinants for RNA processing are present. Nevertheless, in *T. brucei,* two putative promoters have been reported upstream of the actin gene cluster and the *HSP70* locus. However, a TIS was not unambiguously identified in these studies and the sequences did not promote reporter gene expression in a different study. Hence, it appears that trypansomatid RNA pol II is recruited to polycistronic transcription units in an unprecedented, yet to be determined way. The low initiation specificity of trypanosomatid RNA pol II is also reflected in trypanosomatid genomic databases which lack sequences homologous to basal class II transcription factors of other eukaroytes. These factors, in contrast to their class I counterparts, are highly conserved among eukaroytes and homologs have been identified in archaebacteria. Hence, in the course of evolution, trypanosomatids may have lost these factors and the ability to specifically initiate mRNA synthesis. The RNA pol II largest subunit has been cloned and sequenced in several trypanosomatid species and is the only component of the RNA pol II transcription machinery characterized thus far. Eukaryotes typically possess a single RNA largest subunit gene. Therefore, it came as a surprise that trypanosomes which undergo antigenic variation possess two copies of this gene encoding slightly different polypeptides. The possibility that one of these enzymes transcribes VSG and procyclin ESs, however, is unlikely, because the amino acid changes are not located in the α-amanitin binding region and deletion of both alleles of one gene did not affect cell viability.

Entamoeba histolytica

Protein coding genes in *E. histolytica* are transcribed monocistronically and possess complex promoter structures indicating that gene expression in this organism is mainly controlled at the level of transcription initiation. The most detailed analyses have been carried out with the *hgl5* gene which encodes a lectin heavy subunit. The *hgl5* promoter has a tripartite core structure and five distinct upstream regulatory elements. The core promoter consists of a TATA-like element centered at position −25 and an initiator-type element at the TIS. Between these two sequences, a third element was identified containing the motif GAAC which is present in most of the known *E. histolytica* genes. Acting in concert, the three elements determine the site of transcription initiation. Moreover, the GAAC motif, which is a unique element in *E. histolytica*, is able to recruit RNA pol II to a defined TIS when placed outside of the core promoter context. Since the *E. histolytica* genome has a high AT-bias and TBP has a general affinity to AT-rich sequences, it was speculated that the additional GAAC core promoter element was evolved to localize TBP to the core promoter through the interaction with a specific GAAC-binding protein. Specific protein binding to the GAAC element was demonstrated by gel retardation assays. However, while TBP has been cloned in *E. histolytica*, a GAAC-binding transcription factor remains to be characterized. Less is known about the upstream regulatory elements. One of these elements, URE3, is present in the *hgl5* and the ferredoxin gene promoters. While in the former gene URE3 is a negative regulator, it stimulates transcription of the latter gene. A yeast one-hybrid screen revealed a 22.6 kDa polypeptide which specifically binds to URE3. Interestingly, this protein has little homology to known DNA-binding proteins. Promoters of other *E. histolytica* genes have been less well characterized, but the data are in agreement with complex promoter structures and indicate that core promoter variations exist.

Apicomplexa

In apicomplexan parasites many genes have been identified which are developmentally regulated. Monocistronic transcription units and nuclear run-on experiments suggest that regulation occurs mainly at the transcriptional level. Moreover, in *Plasmodium falciparum*, antigenic variation of the protein PFEMP1 appears to be transcriptionally controlled. PFEMP1 is located on the surface of infected erythrocytes and is responsible for the adherence of these cells to the microvascular endothelium, and is, therefore, an important virulence factor. PFEMP1 is encoded by the *var* gene family which comprises about 40 to 50 copies per haploid genome. Expression of a single *var* gene is apparently due to the activation of one *var* promoter and silencing of the others. Despite the importance of transcription in gene expression regulation, only a few promoter regions have been investigated in apicomplexan parasites thus far. This may be mainly due to the unusually large distance between important regulatory elements and the TIS, which can exceed 1 kb. Nevertheless, a few regulatory elements have been identified and analyzed. The *P. falciparum* GBP130 gene encodes a protein which is involved in erythrocyte binding and is expressed exclusively at the trophozoite stage. Examination of the promoter revealed a *cis*-activating, position-dependent element between positions −544 and −507 which enhances basal transcription efficiency fourfold and specifically interacts with nuclear factors, although in a developmentally independent manner. In another gene, *pfs25*, which is induced at the onset of gametogenesis, a regulatory promoter element was identified which binds a candidate transcription factor in a life-cycle-specific way. The 8-bp long element occurs twice, 135 and 215 bp upstream of the major TIS, is also present in the *hsp86* promoter, and binds to the mosquito-stage specific protein PAF-1. In contrast to these upstream regulatory regions, not much is known about the core promoter structure in *Plasmodium*, but TATA boxes may be important because the *P. falciparum* TBP has been discovered and putative TATA elements were recognized in a gene designated *pgs28*.

Core promoters of *Toxoplasma gondii* genes apparently lack TATA boxes. However, in NTPase genes an initiator element was identified which is essential for efficient reporter gene expression. The initiator sequence is similar to that of higher eukaryotes and is also present in the surface antigen gene 1 (SAG1). Further analysis of the NTPase 3 gene promoter showed that *cis*-activators are present upstream of position −141. NTPase gene expression is downregulated upon differentiation from the tachyzoite to the bradyzoite stage. Interestingly, life-cycle-specific regulation was still observed with a promoter 5′ truncated to position −141, indicating that it contains a determinant for the developmental control of promoter activity. The *SAG1* gene has an unusual structure. It has no TATA box and the putative initiator element is not the main TIS selector. Instead, it contains six 27-bp long conserved tandem repeats between position −70 and position −225 relative to the major TIS, which both direct efficient transcription and determine the TIS. Interestingly, this promoter element works independently of its orientation and is able to stimulate transcription from a heterologous promoter.

Finally, the identification of histone acetyltransferases and histone deacetylases in apicomplexan parasites suggests that transcription in these parasites is also regulated by histone modification. In other eukaryotes, it was shown that acetylation of lysines in the N-terminus of core histones affects the nucleosomal structure and typically is associated with transcription activation. Conversely, histone deacetylation generally leads to inactivation or transcriptional silencing.

Helminths

Our knowledge about class II transcription of protein coding genes in nematode and trematode parasites is very limited. However, helminths are phylogenetically placed between yeast and vertebrates, and most likely share the basal RNA pol II machinery characterized in these organisms. Furthermore, the relative relatedness of helminths to these organisms should facilitate the identification of transcription factors by homology. Thus far, the CCAAT-binding factor NF-Y in *Schistosoma mansoni* and the TBP-related factor 2 of *Brugia malayi* have been recognized in this way.

SL RNA AND U snRNA GENE TRANSCRIPTION IN NEMATODES AND TRYPANOSOMATIDS

Trans-splicing of a SL to nuclear pre-mRNA is an essential maturation step for all or the majority of mRNAs in trypanosomatid and nematode parasites, respectively. Since SL RNA molecules are destroyed in the *trans*-splicing process, a high rate of SL RNA synthesis is crucial for the survival of these organisms. Interference with SL RNA gene transcription seems to be a promising strategy to inhibit parasite growth. SL RNAs are small, non-polyadenylated RNAs resembling U snRNAs and, in nematodes and trypanosomatids, are synthesized by RNA pol II. In higher eukaryotes, U snRNA gene promoters which recruit RNA pol II consist in general of a proximal sequence element (PSE) located around position −55 and a short distal sequence element (DSE) 170 bp further upstream (Figure 3.2). The PSE is the core promoter and binds the multi-subunit basal transcription factor SNAPc, whereas the DSE is an enhancer element. Approximately 15 bp downstream of the U snRNA coding region, a conserved element, designated the ‘3′ box’,

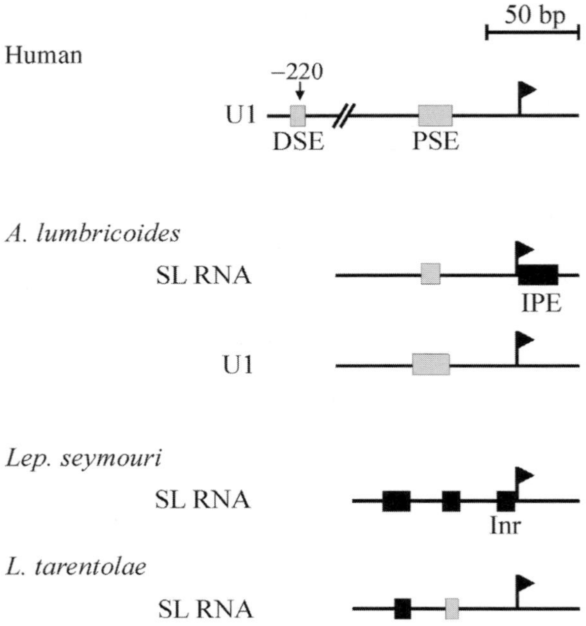

FIGURE 3.2 Comparison of SL RNA and U1 snRNA gene promoters. Schematic drawing to scale of SL RNA or U1 snRNA (U1) gene promoters of human, *Ascaris lumbricoides*, *Leptomonas seymouri* and *Leishmania tarentolae*. Promoters are aligned according to TISs, indicated by flags, and promoter elements are represented by rectangles. In the parasite promoters, elements which have been shown to specifically interact with essential transcription factors are drawn in black. DSE stands for distal sequence element, PSE for proximal sequence element, IPE for intragenic promoter element, and Inr for initiator element. The position of the DSE in the human U1 promoter is given relative to the TIS.

was found to be important for transcription termination and correct U snRNA 3′ end formation. However, recent data suggest that transcription terminates further downstream. The U6 snRNA gene promoter shares PSE and DSE with other U snRNA gene promoters but has an additional TATA-box downstream of the PSE which confers specificity for RNA pol III.

SL RNA and U1 snRNA gene transcription in *Ascaris lumbricoides*

In the nematode parasite *A. lumbricoides*, the development of an *in vitro* transcription system has enabled a detailed analysis of the SL RNA gene structure. In this organism, SL RNA and 5S rRNA genes are present on the same tandem repeat unit and are transcribed in the same orientation. The sequence between the 5S rRNA and SL RNA coding regions contains a single short SL RNA gene promoter element centered at position −50, similar to the PSE of human U snRNA genes (Figure 3.2). *In vitro* assays failed to detect a DSE in the SL RNA gene promoter but identified an essential second element just downstream of the TIS comprising the 22-bp SL sequence itself. Transcription and parallel DNase I footprinting experiments indicated that SL RNA gene transcription depends on the binding of a transcription factor to the intragenic promoter element (IPE). Accordingly, gel shift assays revealed specific band-shifts with a double-stranded DNA probe spanning the IPE and, by DNA affinity chromatography, a polypeptide of 60 kDa was isolated that binds the IPE in a zinc-dependent manner. Since the IPE is conserved among nematode species but not found in mammalian U snRNA gene promoters, it is possible that the IPE-binding protein is a nematode-specific transcription factor. Furthermore, it is likely to be a gene-specific factor as well, because *in vitro* characterization of the *A. lumbricoides* U1 snRNA gene promoter revealed a PSE at the same position as in the SL RNA gene promoter but no IPE (Figure 3.2). The PSEs of both promoters have no obvious consensus sequence and it remains to be determined whether they interact with common or specific transcription factors. The SL RNA gene contains a sequence element 12 bp downstream of its coding region which closely resembles a mammalian 3′ box. Mutation of this sequence element, however, had no influence on *in vitro* synthesis of correct-sized transcripts.

SL RNA gene transcription in trypanosomatids

SL RNA genes in trypanosomatids are tandemly repeated and transcribed monocistronically by RNA pol II. The SL RNA gene promoter has been studied in several trypanosomatid species. The most detailed analyses, which included transient reporter gene expression and *in vitro* transcription assays, were conducted in *Leptomonas seymouri, Leishmania tarentolae* and *T. brucei.* The trypanosomatid promoters share no obvious sequence homologies. However, upstream of the TIS, they contain two distinct sequence elements instead of the single, compact element present in the human and nematode U1 and SL RNA gene promoters (Figure 3.2). In *Lep. seymouri,* the upstream element, termed promoter binding protein 1 element (PBP-1E), is centered at position −65 and the proximal element PBP-2E is located around position −35. While in *L. tarentolae* the corresponding elements are located at the same positions, the proximal element of *T. brucei* is placed 10 bp further upstream. Two observations suggest that the two promoter elements have the same function in the three different trypanosomatid species. Changing the distance between these elements dramatically reduced transcription efficiency, and moving both elements closer or further away from the TIS impaired correct transcription initiation. In *L. tarentolae,* this bipartite promoter suffices to direct efficient and correct SL RNA gene transcription. Conversely, mutation of the 10-bp block just upstream of the TIS both reduced transcription efficiency and caused aberrant transcription initiation in *Lep. seymouri* and *T. brucei.* In a more detailed analysis, conducted in *Lep. seymouri,* the 6-bp sequence from position −5 to +1 was identified to be responsible for the observed effects. As specific protein binding to this sequence motif was detected by gel shift assays, it was proposed that this element corresponds to an initiator element. However, initiator elements have not been found in U snRNA genes yet. Further experiments are necessary to determine the exact role of this element in trypanosomatid SL RNA genes. In *Lep. seymouri,* specific DNA–protein complexes have also been identified at promoter elements PBP-1E and PBP-2E. Gel-shift assays and DNase I footprinting demonstrated that PBP-1E specifically and efficiently binds the putative transcription factor PBP-1. Purification of PBP-1 revealed three polypeptides of 57 kDa, 46 kDa, and 36 kDa, reminiscent of the multi-subunit SNAPc in higher eukaryotes, and photocrosslinking suggested that the 46 kDa polypeptide binds to the promoter domain. Furthermore, it was shown that binding of PBP-1 to the SL RNA gene promoter and the integrity of the second promoter element PBP-1E are prerequisites for the formation of a larger and more stable complex, suggesting that the two upstream promoter elements cooperate in the formation of a stable transcription initiation complex. Accordingly, the integrity of both elements was essential to stably bind and sequester a *trans*-activating factor in transcription competition experiments conducted in *T. brucei.* In *L. tarentolae,* however, gel-shift assays have thus far identified specific protein binding only to PBP-1E and not to the second promoter element.

A conserved feature of trypanosomatid SL RNA genes is a T-run a few basepairs downstream of the SL RNA coding region. In *L. tarentolae,* the T-run was identified as the transcription termination determinant. Hence, RNA pol II-mediated SL RNA gene transcription in trypanosomatids appears to terminate

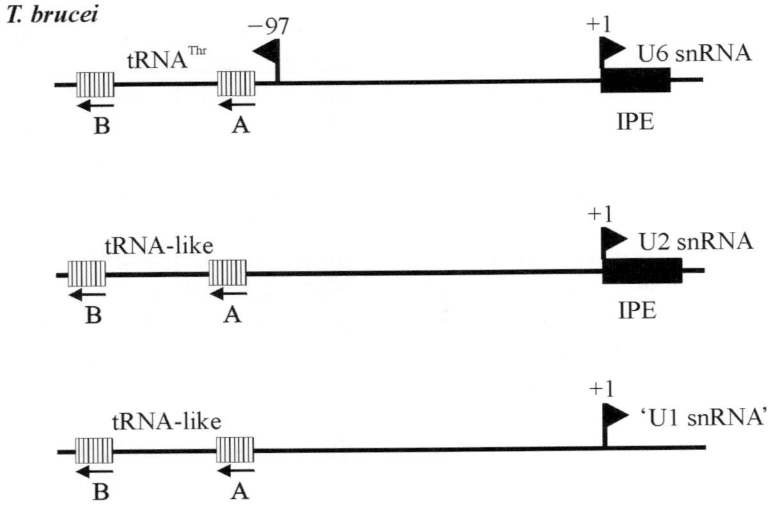

	Gene	Position of A Box	tRNA gene
T. brucei			
	'U1 snRNA'	−105	tRNA-like
	U2 snRNA	−104	tRNA-like
	U3 snRNA	−104	$tRNA^{Arg}$
	U6 snRNA	−104	$tRNA^{Thr}$
	7SL RNA	−104	$tRNA^{Lys}$
L. pifanoi			
	U6 snRNA	−105	$tRNA^{Gln}$
	7SL RNA	−107	$tRNA^{Arg}$
L. mexicana			
	U6 snRNA	−105	$tRNA^{Gln}$
Lep. seymouri			
	U6 snRNA	−106	$tRNA^{Gln}$
Lep. collosoma			
	U4 snRNA	−107	tRNA-like
	U5 snRNA	−107	$tRNA^{Cys}$
	U6 snRNA	−107	$tRNA^{Gln}$
C. fasciculata			
	U6 snRNA	−106	$tRNA^{Gln}$

in an RNA pol III-like manner. This view is supported by an experiment conducted in *T. brucei* in which transient transfection of an SL RNA gene containing an RNA pol III termination signal within the coding region prevented the production of full-size molecules. Interestingly, pre-mRNA synthesizing RNA pol II must differ to some extent from the enzyme complex transcribing SL RNA genes because it does not recognize T-runs as termination signals and reads through T-rich polypyrimidine tracts which, as essential RNA processing signals, are common motifs of polycistronic transcription units.

Functional association of tRNA genes and small nuclear or small cytoplasmic RNA genes in trypanosomatids

In trypanosomatids, low molecular weight RNAs including U snRNAs are synthesized by RNA pol III. The only exception appears to be the SL RNA gene. In the genome, class III genes are clustered and interspersed between RNA pol II transcription units. In these clusters, tRNA genes are associated head-to-head with various U snRNA genes or the 7SL RNA gene which encodes the RNA component of the signal recognition particle. The first trypanosomatid class III gene promoter analyzed was that of the *T. brucei* U2 snRNA gene. It consists of an intragenic control element comprising the first 24 bp of the transcribed region and two 11 bp-long sequence blocks located 104 and 155 bp upstream of the TIS (Figure 3.3). A comparison of the U2 promoter and corresponding sequences of other U snRNA genes and the 7SL RNA gene revealed that the upstream region of these genes was occupied by a divergently orientated tRNA gene. Promoters of tRNA genes consist of two highly conserved intragenic elements designated 'A box' and 'B box'. Interestingly, size, location, and sequence of these elements in the upstream tRNA genes corresponded to the two upstream promoter elements of the U2 snRNA gene. Furthermore, the sequence around the two U2 elements resembles a tRNA sequence and may have originated from a functional tRNA gene. Taken together, these observations suggested that the A and B boxes of the companion tRNA genes promote transcription of both the tRNA gene and the associated small RNA gene. This possibility was confirmed by a detailed mutational analysis of the linked U6 snRNA and threonine tRNA genes in *T. brucei*. Furthermore, it was found that *in vivo* expression and *in vitro* transcription of the U6 snRNA gene depended, as in the U2 snRNA gene, on an intragenic promoter element comprising the first 21 nucleotides of the U6 RNA coding region (Figure 3.3). The

FIGURE 3.3 Association of tRNA genes and small nuclear or small cytoplasmic RNA genes in trypanosomatids. At the top, a schematic drawing to scale is shown of the *Trypanosoma brucei* U6, U2, and candidate U1 snRNA gene promoters which represent the three types of U snRNA gene promoters characterized in trypanosomatids thus far. The three genes have tRNA gene A and B box elements upstream of the TIS as indicated by striped rectangles. These elements are in reverse orientation to the U snRNA transcription direction (arrows) and are either part of a functional tRNA gene as in the U6 snRNA gene ($tRNA^{Thr}$) or inside a tRNA-like sequence as in the U2 and 'U1' snRNA genes. The sequence immediately downstream of the TIS is an intragenic promoter element (IPE) in the U6 and U2 snRNA genes, but has no promoter function in the 'U1 snRNA' gene. At the bottom, a selection of U snRNA and 7SL RNA genes of *T. brucei*, *Leishmania pifanoi*, *Leishmania mexicana*, *Leptomonas seymouri*, *Leptomonas collosoma*, and *Crithidia fasciculata* is shown. For the listed genes, the upstream tRNA sequence and the location of A and B box elements have been analyzed. For each gene, the position of the A box relative to the TIS and the nature of the associated tRNA sequence is given.

intragenic promoter element, however, is not present in all U snRNA genes, because, as was recently found, A and B box elements suffice for accurate and efficient transcription of the putative U1 snRNA gene in *T. brucei* (Figure 3.3). The functional association between U snRNA genes or the 7SL RNA gene with tRNA genes is a common feature in trypanosomatids (Figure 3.3) but it does not appear to be particularly parasite-specific because B boxes have been found to be linked with U snRNA genes of other organisms. For example, U6 snRNA gene expression in yeast depends on a B box which is located downstream of the U6 coding region and which, by binding the transcription factor IIIC, is involved in remodeling the chromatin structure. However, it seems that the tRNA gene promoter elements of trypanosomatid U snRNA and 7SL RNA genes have additional functions and participate directly in the formation of a transcription initiation complex. In favor of this hypothesis are the findings that the A box but not the B box is essential for *in vitro* transcription of the *T. brucei* U6 snRNA gene, that A and B boxes suffice to direct accurate transcription of the *T. brucei* candidate U1 snRNA gene, and that the position of the A box is highly conserved (Figure 3.3) and cannot be changed without dramatically affecting U snRNA gene transcription. It is unknown how the tRNA gene promoter elements direct transcription of the associated small RNA gene. An attractive hypothesis states that binding of the transcription factors to the tRNA gene promoter elements bends the DNA so that the DNA-protein complexes of the A and B boxes come in contact with the transcription initiation region of the associated small RNA gene. In accordance with this hypothesis, it was shown in yeast that TFIIIB, the factor which binds to the A box, imposes multiple bends on DNA. Similarly, specific binding of trypanosome nuclear factors to the internal tRNA gene promoter elements resulted in DNA bending.

PERSPECTIVE

To the best of our knowledge, existing drugs against parasite infection do not interfere with gene transcription, although inhibitors such as α-amanitin, tagetitoxin, or rifampicin indicate that this is a possibility. For the development of potential parasite-specific transcription inhibitors we need to overcome our limited knowledge of transcription factors in parasites. The prospects of learning more about transcriptional processes is excellent. DNA transfection technology, *in vitro* transcription systems and growing genome databases of parasitic organisms will facilitate the identification and functional characterization of such factors in the future. Parasite-specific mechanisms of transcription and the evolutionary distance between parasites and host organisms make it likely that essential transcription factors or factor domains exist which either have no counterpart or may substantially deviate from a counterpart in the host and, therefore, may serve as suitable and specific targets against the parasite.

FURTHER READING

Blumenthal, T. (1998). Gene clusters and polycistronic transcription in eukaryotes. *Bioessays* **20**, 480–487.

Borst, P. and Ulbert, S. (2001). Control of VSG gene expression sites. *Mol. Biochem. Parasitol.* **114**, 17–27.

Campbell, D.A., Sturm, N.R. and Yu, M.C. (2000). Transcription of the kinetoplastid spliced leader RNA gene. *Parasitol. Today* **16**, 78–82.

Cross, G.A., Wirtz, L.E. and Navarro, M. (1998). Regulation of VSG expression site transcription and switching in *Trypanosoma brucei*. *Mol. Biochem. Parasitol.* **91**, 77–91.

Elmendorf, H.G., Singer, S.M., Pierce, J., Cowan, J. and Nash, T.E. (2001). Initiator and upstream elements in the α2-tubulin promoter of *Giardia lamblia*. *Mol. Biochem. Parasitol.* **113**, 157–169.

Gilchrist, C.A., Holm, C.F., Hughes, M.A., Schaenman, J.M., Mann, B.J. and Petri, W.A. Jr. (2001). Identification and characterization of an *Entamoeba histolytica* upstream regulatory element 3 sequence-specific DNA-binding protein containing EF-hand motifs. *J. Biol. Chem.* **276**, 11838–11843.

Horrocks, P., Dechering, K. and Lanzer, M. (1998). Control of gene expression in *Plasmodium falciparum*. *Mol. Biochem. Parasitol.* **95**, 171–181.

Laufer, G., Schaaf, G., Bollgönn, S. and Günzl, A. (1999). *In vitro* analysis of α-amanitin-resistant transcription from the rRNA, procyclic acidic repetitive protein, and variant surface glycoprotein gene promoters in *Trypanosoma brucei*. *Mol. Cell. Biol.* **19**, 5466–5473.

Lee, M.G. and Van der Ploeg, L.H. (1997). Transcription of protein-coding genes in trypanosomes by RNA polymerase I. *Annu. Rev. Microbiol.* **51**, 463–489.

Liston, D.R., Carrero, J.C. and Johnson, P.J. (1999). Upstream regulatory sequences required for expression of the *Trichomonas vaginalis* α-succinyl CoA synthetase gene. *Mol. Biochem. Parasitol.* **104**, 323–329.

Matkin, A., Das, A. and Bellofatto, V. (2001). The *Leptomonas seymouri* spliced leader RNA promoter requires a novel transcription factor. *Int. J. Parasitol.* **31**, 545–549.

McKnight, S.L. and Yamamoto, K.R. (eds) (1992). *Transcriptional regulation*, vols 1 and 2, Cold Spring Harbor: Cold Spring Harbor Laboratory Press.

Mechanisms of Transcription (1998). *Cold Spring Harb. Symp. Quant. Biol.* **63**.

Nakaar, V., Tschudi, C. and Ullu, E. (1995). An unusual liaison: small nuclear and cytoplasmic RNA genes team up with tRNA genes in trypanosomatid protozoa. *Parasitol. Today* **11**, 225–228.

Nakaar, V., Bermudes, D., Peck, K.R. and Joiner, K.A. (1998). Upstream elements required for expression of nucleoside triphosphate hydrolase genes of *Toxoplasma gondii*. *Mol. Biochem. Parasitol.* **92**, 229–239.

Paule, M.R. (ed.) (1998). *Transcription of Ribosomal RNA Genes by Eukaryotic RNA Polymerase I*, Berlin: Springer-Verlag.

Serra, E., Zemzoumi, K., di Silvio, A., Mantovani, R., Lardans, V. and Dissous, C. (1998). Conservation and divergence of NF-Y transcriptional activation function. *Nucl. Acids Res.* **26**, 3800–3805.

Ullu, E. and Nilsen, T. (1995). Molecular biology of protozoan and helminth parasites. In: Marr, J.J. and Müller, M. (eds). *Biochemistry and Molecular Biology of Parasites*, London: Academic Press, pp. 1–17.

Vanhamme, L., Lecordier, L. and Pays, E. (2001). Control and function of the bloodstream variant surface glycoprotein expression sites in *Trypanosoma brucei*. *Int. J. Parasitol.* **31**, 522–531.

Wade, P.A., Pruss, D. and Wolffe, A.P. (1997). Histone acetylation: chromatin in action. *Trends Biochem. Sci.* **22**, 128–132.

C H A P T E R

4

Post-transcriptional regulation

Christine Clayton
Zentrum für Molekulare Biologie, Universität Heidelberg, Heidelberg, Germany

INTRODUCTION

All medically important protist and helminth parasites have to adjust to a variety of physical and chemical conditions as they migrate to various locations in the body and are transmitted from one host to the next. To survive and multiply in disparate environments, the parasite must change its surface to combat adverse chemical and physical conditions and to minimize immune damage; it may build and destroy specialized organelles and structures needed for movement or invasion, and it needs to switch its metabolism to exploit the available substrates in aerobic or anaerobic environments.

There are many possible stages, from gene to protein, at which gene expression can be controlled. These are illustrated in Figure 4.1. For a model eukaryote, possible control points are:

1. Transcription initiation, elongation and termination
2. Splicing patterns and efficiency
3. mRNA degradation in the nucleus
4. Polyadenylation
5. Nuclear export
6. mRNA translation
7. mRNA degradation in the cytoplasm
8. Protein folding
9. Protein modification and transport
10. Protein degradation.

This chapter considers control of steps 2 to 7. Other issues were considered in Chapters 2 and 3.

WHY POST-TRANSCRIPTIONAL REGULATION?

Control of transcription initiation is essential in nearly all organisms, but there are many circumstances in which additional levels of regulation are required. In complex eukaryotes,

Molecular Medical Parasitology
ISBN 0–12–473346–8

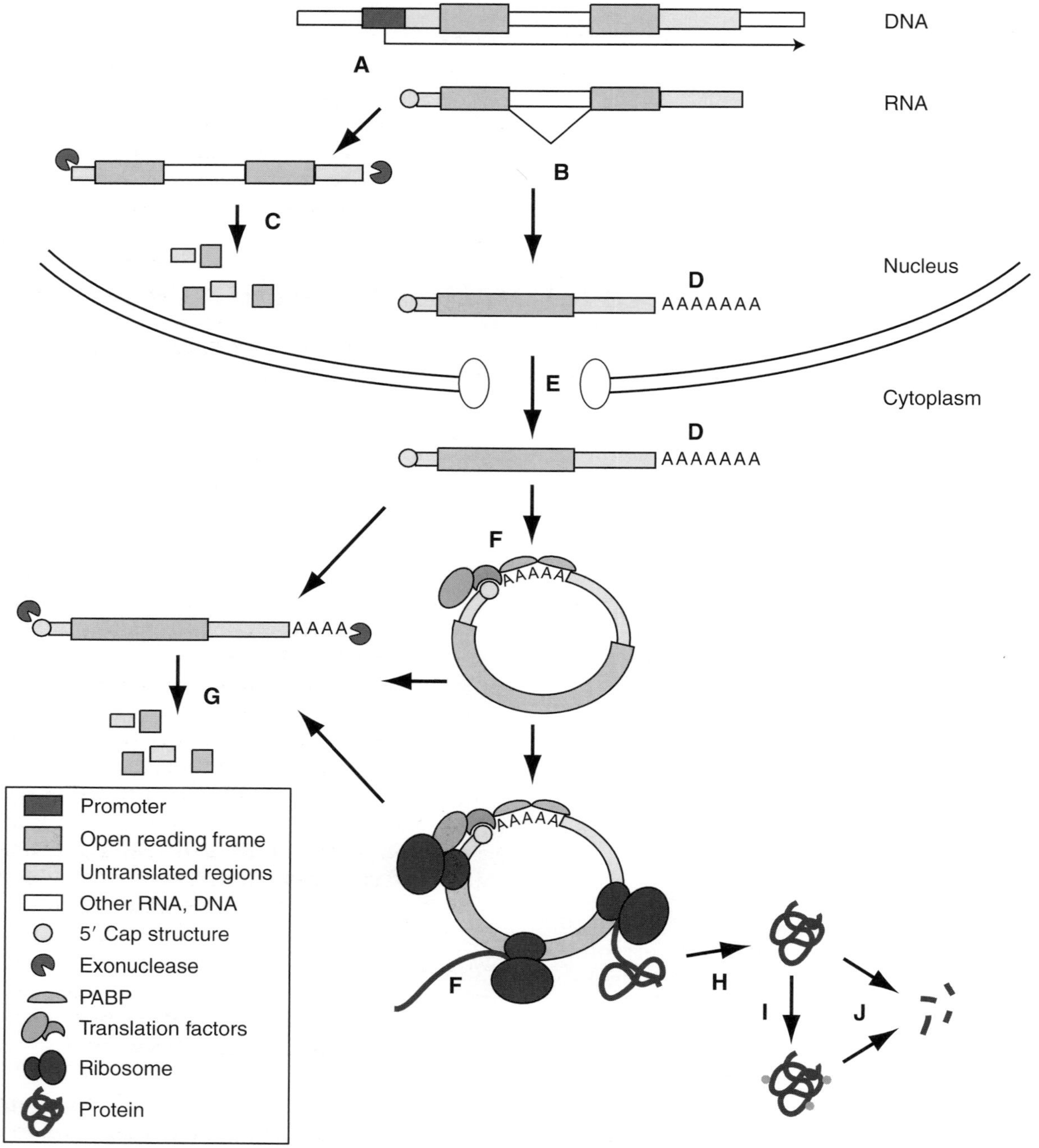

FIGURE 4.1 Options for regulation of gene expression.

a gene with several introns may be the template for a variety of mRNAs, generated by alternative splicing and encoding proteins which can have quite divergent functions. Aberrantly processed mRNAs, or mRNAs containing transcription errors, need to be eliminated before they can be translated into erroneous proteins. Some RNAs can perform their function correctly only if they are tightly localized to one part of the cell. Some proteins are needed only transiently, so that their synthesis must be abruptly halted by destruction of the corresponding mRNA. And some proteins need to be synthesized *de novo* even when no RNA synthesis is taking place. In prokaryotes also, different RNAs have different stabilities and some are translated more efficiently than others. At present, very little is known about this aspect of regulation in parasites. The only family in which it has been studied in any detail is the kinetoplastids – for the very good reason that these organisms can't control transcription of the vast majority of their genes. If all genes are being transcribed all the time, something has to be done to preserve the RNAs that are needed, and eliminate the rest. This chapter will first summarize what is known about post-transcriptional regulation in eukaryotic model systems. Next it will describe in detail the available information about the kinetoplastids, before mentioning other parasites.

POST-TRANSCRIPTIONAL REGULATION IN EUKARYOTIC MODEL SYSTEMS

RNA processing

With the recent publication of the draft human genome, it has become clear that the number of predicted genes alone does not correlate at all with the complexity of the organism. Thus the predicted number of genes in man is only around 30 000, as against an estimated 19 000 in the worm *Caenorhabditis elegans* and 6200 in the yeast *Saccharomyces cerevisiae*. The coding capacity of the genome of a multicellular organism is much higher than the predicted gene number because genes are split into multiple exons. Transcription can start and stop in different places, and the exons can be spliced in a variety of combinations to give different mRNAs and proteins. The pattern of splicing can vary during development and can determine the fate, not just of individual cells but of the whole organism. For example, sex determination in *Drosophila* is dependent on alternative splicing. In addition, the position of polyadenylation can influence the function of the encoded protein. A classic example here is the exclusion or inclusion of the membrane anchors of immunoglobulins; if the portion of the gene encoding the anchor is chopped off by use of an alternative polyadenylation site, the immunoglobulin is secreted.

Some RNAs in mammalian cells are subject to another form of post-transcriptional modification: editing by enzymatic modification of individual bases. This process, which is entirely different from the editing seen in kinetoplastid mitochondria, is for example important in determining the properties of some receptors in the brain.

Any of these processes could be important in parasite gene regulation, but until now there are no published examples.

Export from the nucleus and localization

Once an mRNA has been correctly processed, it is escorted to the cytosol by proteins. Crucial to this export is the presence of a 5′-cap structure, which is made of modified bases and is usually put in place by the transcribing polymerase, RNA polymerase II. Nearly all mRNAs also have a poly(A) tail. If the RNA has not been processed properly – for instance, if splicing is not complete – the RNA may be held back in the nucleus through recognition of intron sequences, and degraded.

The correct function of many mRNAs depends on their location. The transport of an mRNA to one part of the cell causes production of the encoded protein to be concentrated in that area. One important example is the localization of mRNAs in a developing fruit fly oocyte, which determines which end of an embryo will become the head, and which the abdomen. In polarized mammalian cells, mRNAs may be transported to the region of the cell where their product is required. And in bakers' yeast, regulation of mating type depends on the transport of specific mRNAs away from the mother cell into the developing bud.

Although at present no examples of mRNA localization in parasites are documented, its existence cannot be excluded even for the smallest unicellular organisms.

Translation

Translation initiation in eukaryotes starts by binding of translation initiation factors to the cap structure. Next, the small ribosomal subunit binds to the RNA and scans downstream to the first initiation codon (Figure 4.1F). If there is an extra initiation codon, upstream of (and out of frame with) the codon needed to make a functional protein, or a short open reading frame that terminates before the functional one, the ribosome will usually start in the wrong place. As a consequence, there will be very little translation of the open reading frame that starts further downstream. The presence of strong secondary structures, such as stem-loops, in the 5′-untranslated region before the start codon, can stop the ribosome reaching the start codon. Proteins that bind specific sequences in the 5′-untranslated region can also get in the way of the scanning ribosome. Alternatively, binding proteins can disrupt secondary structures, enabling the ribosome to scan through.

The ferritin mRNA of mammals is a classical example of translational control. Ferritin is important in sequestering iron, as too much free iron is toxic. The ferritin mRNA has an 'iron regulatory element' (IRE) sequence which forms a stem-loop in the 5′-untranslated region. The IRE is bound very specifically by the iron-regulatory protein (IRP), which stops translation initiation. IRP can only bind the RNA if there is very little iron in the cell. Then, no ferritin is made. In the presence of iron the conformation of the IRP is altered. It falls off the IRE and ferritin is produced.

Most translationally active mRNAs in eukaryotes are circularized. The poly(A)-binding protein specifically binds to the poly(A) tail, and also interacts with proteins in the translation initiation complex (Figure 4.1F). This is a very efficient arrangement as ribosomes that reach the end of an open reading frame can be rapidly recycled to restart translation. This may at least partially explain why the translation efficiency of most mRNAs is strongly enhanced by the presence of a poly(A) tail.

The cytosol of *Xenopus* oocytes contains a large store of silent 'maternal' RNAs, synthesized during fertilization but bearing extremely short poly(A) tails (20 nt). Upon fertilization, the poly(A) tails are elongated to 80–150 nt, which allows translation. When their function has been fulfilled, removal of the tail (deadenylation) initiates mRNA degradation.

Translational control is extremely important in kinetoplastid protists, and probably in other parasites. Some examples will be described below.

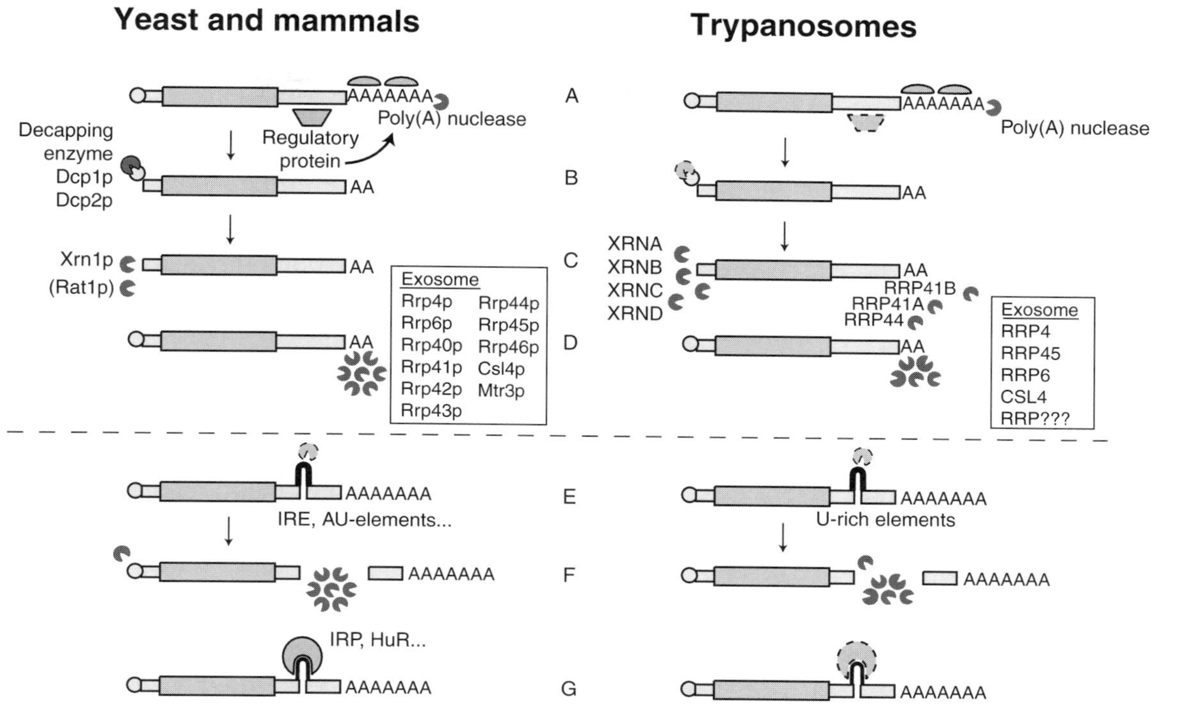

FIGURE 4.2 mRNA degradation pathways yeast, mammals and trypanosomes. Reproduced with permission from Clayton (2002) *EMBO J.* **21**, 1881–1888.

RNA decay

Many cellular processes are dependent on transient expression of proteins, or upon their rapid destruction in response to a stimulus. For example, some proteins must be present for only a short time during the cell cycle; and cytokines must be induced only transiently to prevent over-activation of the immune system and/or programmed cell death. Destruction of the protein is important (see below), but the complete absence of the protein can only be ensured if it is no longer produced – that is, the mRNA is also destroyed. Variations in mRNA stability can also provide an additional level of regulation on top of transcriptional control. Finally, mechanisms must be in place to prevent the production of aberrant proteins from faulty mRNAs.

The first checks on mRNAs occur in the nucleus. Here, mRNAs that have been incorrectly spliced can be identified and destroyed. RNAs which contain premature stop codons would be translated into abnormally short proteins, which could be toxic to the cell; these RNAs are removed by the 'nonsense-mediated decay' pathway.

One very important pathway of regulated mRNA degradation in yeast is illustrated in Figure 4.2 A–C. First, the poly(A) tail is degraded from the 3′-end by a poly(A)-specific exonuclease, till it is only about 20 nt long. Poly(A) tails may be protected from 3′→5′ degradation *in vitro* by poly(A) binding protein, but if the tail is very short the protein can no longer bind. The short tail is a signal for removal of the 5′ cap by a decapping enzyme complex. In yeast, decapping involves at least two proteins, Dcp1p and Dcp2p. The removal of the cap leaves the naked 5′-end of the RNA exposed for degradation by 5′→3′ exonucleases. Usually, the enzyme Xrn1p is the main enzyme involved. There is evidence that capping and 5′→3′ degradation enzymes are linked in a larger complex including seven other proteins, which presumably has some regulatory role. Although yeast lacking *XRN1* and *DCP1* genes are viable, they grow poorly and accumulate partially degraded RNAs. Yeast have another 5′→3′ exonuclease, Rat1p, which is normally found in the nucleus. If the nuclear targeting signal of Rat1p is removed the protein stays in the cytoplasm; then it can complement the defect in *XRN1*.

The cells that were just mentioned, which have a defective 5′→3′ degradation pathway, can only live because another system is there to eliminate unwanted mRNA. This system degrades the mRNAs from the other end. The enzymes involved in this 3′→5′ degradation are found in a complex called the exosome (Figure 4.2D). The exosome is involved in maturation of several stable RNAs, such as rRNA and U RNAs, as well as in 3′→5′ degradation of mRNAs, so mutations and deletions that eliminate exosome function are lethal. The yeast exosome has a molecular weight of 350 kDa and contains at least eleven subunits, all of which are probably 3′→5′ exonucleases in their own right. Many stable RNAs have to be very precisely trimmed to give the correct product. The association of the different enzymes may be important in regulation or coordination of these processes; mutations in the different enzymes have subtly different phenotypes. Additional proteins may be associated with the exosome in different cellular compartments, to give a further level of regulation.

In general, the basal components of the RNA degradation machinery, including the 5′→3′ exonucleases and the exosome, are conserved from yeast to mammals. However, mammals have some mechanisms that have not yet been found in yeast. Probably the best understood example is the regulation of transferrin receptor mRNA, shown in the lower part of Figure 4.2. Transferrin receptor is needed for the cell to take up iron. The transferrin mRNA has an IRE structure, just like the ferritin mRNA, but this time the IRE is in the 3′-untranslated region. Degradation of this mRNA is initiated by cleavage within this IRE. When there is little iron, the IRP binds the IRE and stops the degradation (Figure 4.1E, F, G). Thus transferrin receptor is only produced when there is little iron available – a mode of regulation complementary to that of ferritin, which is only needed when iron is in excess (see above). There are several other examples of endonuclease cleavage within 3′-untranslated regions, but despite intensive efforts the endonucleases responsible have not been definitively identified.

Many mammalian mRNAs involved in cell growth and differentiation contain destabilizing elements in the 3′-untranslated region which are rich in A and U residues. These 'AU rich elements' (AREs) often contain copies of the sequence AUUUA (or similar). Several proteins that bind to AREs have been identified (Figure 4.2G). It is suspected that some of these proteins serve to protect the ARE-containing mRNAs from degradation, whereas others may stimulate degradation, but in no case is the precise function clear. One possibility is that the ARE stimulates deadenylation of the transcript; alternatively, the ARE may be targeted by an endonuclease, like the IRE.

Protein degradation and modification

Modification and degradation of proteins are the final stages in post-transcriptional regulation. The activities of many enzymes are regulated by post-translational modifications such as phosphorylation, addition of fatty acid chains, glycosylation, and specific internal cleavage. Some proteins are needed only transiently in the cell. Some are recognized by specific proteases, but more often they are destroyed by a general protease complex, the proteasome. Proteins destined for degradation bear specific recognition sequences which are recognized by enzymes that transfer a small protein, ubiquitin, onto the protein (ubiquitin-conjugating enzymes). After this, further ubiquitin is added. The ubiquitin chain is recognized by the proteasome. To regulate degradation, the recognition sequences may, for instance, be concealed by modification or interaction with other proteins. All of this is vital to growth and survival.

POST-TRANSCRIPTIONAL REGULATION IN KINETOPLASTIDS

Structure and transcription of kinetoplastid genomes

The genome structure and transcription patterns of kinetoplastids are unique in evolution. Instead of having one strictly controlled transcriptional promoter per gene (or operon) most genes appear to have no specific transcriptional promoter at all. This is illustrated nicely by *Leishmania major* chromosome I: this chromosome has a short segment lacking open reading frames, surrounded on one side by 27 genes that are arranged head-to-tail and oriented towards one telomere, and on the other side by 50 genes that are oriented towards the other telomere. There could perhaps be transcriptional promoters in this central region, but that still would mean that all the genes are transcribed all the time.

There is abundant evidence for polycistronic transcription in kinetoplastids. Precursor RNAs containing two open reading frames, and bridging across from one 3′-untranslated region to the next 5′-untranslated region, have been seen by Northern blotting and by reverse-transcription and PCR. Transcription assays have shown that transcription of individual genes remains constant irrespective of the abundance of the mRNA product. The only known examples of transcriptional control are found in the salivarian trypanosomes, where the major surface protein genes (*VSG* and the *EP/GPEET* family) and associated genes are transcribed by RNA polymerase I and the transcriptional activity is regulated by chromatin structure (see Chapters 3 and 5).

The need for post-transcriptional regulation

Despite their lack of transcriptional control, the kinetoplastids must adapt their gene expression to a changing environment. They have various life-cycle stages – mostly spindle-shaped and flagellate, but, for amastigotes, spherical and aflagellate. Most of the organisms of human and veterinary importance are digenetic, which means that they have two hosts – the mammal and an arthropod. In the mammal, the parasites must evade the humoral immune response. For *Trypanosoma cruzi* and *Leishmania* amastigotes this means adjusting to intracellular life in the cytosol and lysosome respectively, and for *T. cruzi* trypomastigotes, extracellular survival as well. For the salivarian trypanosomes, such as the bloodstream-form trypomastigotes of *T. brucei,* it means perpetual variation of the antigenic surface coat (variant surface glycoprotein or VSG). The salivarian trypanosomes also have a non-dividing mammalian stage, the stumpy form, which is pre-adapted for initial survival in the insect

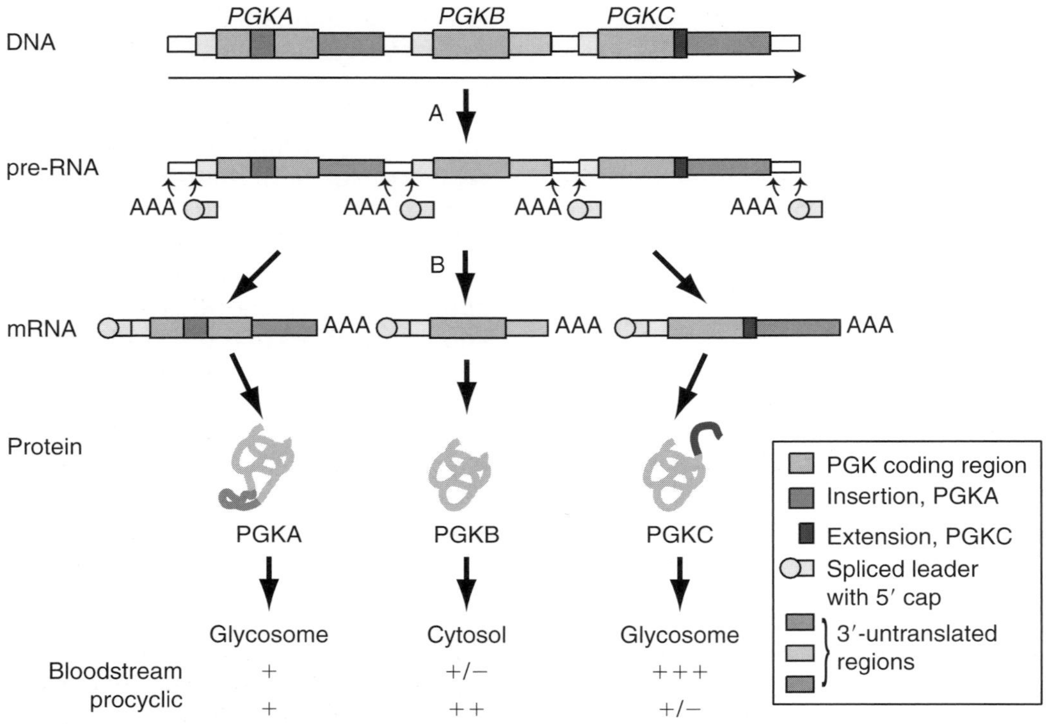

FIGURE 4.3 Phosphoglycerate kinase in *Trypanosoma brucei.* Reproduced with permission from Clayton (2002) *EMBO J.* **21**, 1881–1888.

vector and adaptive differentiation. The arthropod vector brings other challenges: an abrupt temperature change, gut proteases, different nutrients and the invertebrate defence system. Here we find procyclic trypomastigote forms of the salivarian trypanosomes, *Leishmania* promastigotes and *T. cruzi* epimastigotes. These forms all express specific surface molecules that are likely to have protective roles, such as glycoconjugates and proteases. In *T. brucei*, the VSG is replaced by the EP and GPEET procyclins, which are repetitive proteins rich in glutamate and proline. Migration to the mouth parts, adhesion, pre-adaptation for infection of the mammal, and in some cases a sexual stage involving genetic exchange, are additional challenges. The salivarian trypanosomes pass through an epimastigote stage before forming infective metacyclic trypomastigotes, *Leishmania* make metacyclic promastigotes, and *T. cruzi* makes metacyclic trypomastigotes.

The genes encoding phosphoglycerate kinase (PGK) in *T. brucei* illustrate the requirement for post-transcriptional regulation well (Figure 4.3). Bloodstream *T. brucei* are completely dependent on glycolysis for ATP production, whereas the procyclic forms have a much more highly developed mitochondrion and can use amino acids as a substrate for energy metabolism. In both forms, the first seven enzymes of glycolysis are located in a specialized glycolytic organelle, the glycosome (see Chapter 7, part 2). In bloodstream forms, PGK activity is also in the glycosome, whereas in procyclics, PGK is predominantly cytosolic. There are three *PGK* genes, in a head-to-tail arrangement on chromosome I (Figure 4.3A). Early experiments showed that all three of these genes are constitutively transcribed, and the presence of transcripts that bridged the gap from one gene to the next was demonstrated by RT-PCR (Figure 4.3A). The central gene of the three, *PGKB*, encodes a protein which is very like PGKs of other eukaryotes. PGKB is found in the trypanosome cytosol and both it and its RNA are expressed almost exclusively in procyclic trypanosomes. The first gene, *PGKA*, has an internal insertion in the protein sequence. PGKA protein is in the glycosome and is found in low abundance in both bloodstream and procyclic forms. Finally, the *PGKC* mRNA and protein are found almost exclusively in bloodstream forms; PGKC is targeted to the glycosome via a peptide extension at the C-terminus. The expression of cytosolic PGK in bloodstream forms inhibits parasite growth, presumably because the flow of intermediates in the glycosome is disrupted. Thus the mechanisms that prevent PGKB production in bloodstream trypanosomes are essential for parasite virulence.

Splicing and polyadenylation

The first task after polycistronic transcription is to separate the individual open reading frames (Figure 4.3B). In theory this could be achieved by a single cleavage in the RNA precursor. In practice, two cleavages occur: one at the 5′-end, with associated *trans*-splicing, and a further upstream cleavage with polyadenylation of the 3′-end (Figure 4.3B). The signal for these reactions is the 5′-splice site, which normally contains a polypyrimidine tract. It signals the *trans*-splicing of a short capped leader sequence (around 40 nt) to the 5′-end of the downstream RNA. The polyadenylation of the preceding mRNA is spatially and temporally (and therefore in all probability mechanistically) linked to the splicing, occuring 100–300 nt upstream depending on species.

As *trans*-splicing is essential for mRNA production, it seems an obvious candidate for developmental regulation. There are often several possible AG acceptor dinucleotides downstream of the splicing signal, which results in alternative 5′-untranslated regions of different lengths. Some trypanosome intergenic regions contain multiple polypyrimidine tracts, which may be recognized with varying efficiency by the splicing machinery. Such variable splicing can produce non-functional RNAs lacking a complete open reading frame. It can also result in mRNAs with long 5′-untranslated regions containing very short open reading frames upstream of the main reading frame. Examples of both types have been seen, but there is no firm evidence so far that differential splicing is important in developmental regulation. The paucity of examples may simply reflect the very small number of genes that has been examined, or it may be that the coupling of splicing and polyadenylation limits this type of regulation. A failure of splicing in one transcript could result in a failure to polyadenylate the preceding one, and thus preclude independent regulation of adjacent genes.

Although no examples of developmental regulation of splicing have been found, it is clear that some splice sites are more efficient than others. For example, the main reason for the low abundance of PGKA (Figure 4.3) is that the splice site of the *PGKA* mRNA is very inefficient. The regulation of the other two genes will be discussed later.

So far, the only gene that has been found to contain an intron and be subjected to *cis*-splicing is that encoding poly(A) polymerase. The mRNA is also *trans*-spliced, and the *cis*-splicing acceptor site can function as a *trans*-splicing acceptor. It is not known whether this alternative splicing has a regulatory role.

The SIRE transposable element of *T. cruzi* contains splice sites of moderate efficiency. The element is distributed throughout the genome, preferentially in intergenic regions, and SIRE sequence is present in many mature mRNAs, either 5′- or 3′- to the open reading frame. The presence of the SIRE element in an intergenic region can therefore influence the location and efficiency of both splicing and polyadenylation, influencing expression of both upstream and downstream genes.

Before leaving the issue of mRNA processing, it should be mentioned that the editing of mitochondrial transcripts exhibits considerable developmental regulation. This is described elsewhere in this volume (Chapter 2).

Cell-cycle control and the 5′-untranslated region

The insect kinetoplastid *Crithidia fasciculata* can be synchronized in culture, unlike the trypanosomes and mammalian-infective Leishmanias. This makes it possible to study the regulation of gene expression through the cell cycle. As would be expected from other systems, mRNAs linked to nuclear and mitochondrial DNA replication, such as those encoding kinetoplast topoisomerase II (*TOP2*), dihydrofolate reductase-thymidylate synthase (*DHFRTS*) and the large (*RPA1*) and middle (*RPA2*) subunits of a nuclear single-strand DNA binding protein all begin to accumulate prior to S phase then decline rapidly. At the protein level (for TOP2), this results in a stepwise accumulation of protein within the population, as maximal protein synthesis corresponds to the maximal mRNA level, but there is no subsequent protein degradation. Examination of the 5′-untranslated regions of these mRNAs revealed an octamer consensus sequence, CATAGAAA, which was present several times. Experiments with transfected reporter genes bearing 5′-untranslated regions from *TOP2* and *RPA1* showed that the sequence was necessary and sufficient for appropriate cell-cycle regulation. Intriguingly, the octamers also functioned if they were transposed to the 3′-untranslated region.

To find proteins involved in the regulation of cycling, radioactively labeled RNAs containing the octamer recognition sequence were incubated with *Crithidia* extracts. When the mixture was loaded on a polyacrylamide gel, the resulting RNA–protein complexes could be seen because they migrated more slowly than the labeled RNA alone (a 'gel-shift' experiment). The bound proteins could be covalently linked to the RNA by UV irradiation. These experiments revealed that there were indeed some binding activities present that showed the expected pattern of cell-cycle regulation. The strongest candidate so far is a 38 kDa protein. Activity was found in nuclear, not cytoplasmic extracts, suggesting that regulation operates at the level of mRNA splicing, export or degradation in the nucleus. It is also possible that the proteins also operate in the cytosol, but are not detected there because they were already bound to target RNAs in the extracts. These issues can be addressed once the genes encoding the binding proteins have been cloned.

Control by 3′-untranslated regions: RNA degradation

At the time of writing more than 30 trypanosomatid genes had been examined to find out how regulation operates. To find the sequences responsible for regulation, 5′-untranslated regions, 3′-untranslated regions, or both have been attached to reporter genes such as chloramphenicol acetyltransferase (CAT) or luciferase. For quick analyses, constructs have been transiently transfected into different life-cycle stages, and the level of reporter gene expression measured. In order to analyse the behaviour of mRNAs, however, it is usually necessary to generate permanent cell lines containing the reporter constructs, because there simply isn't enough RNA made in transient transfections. The constructs are usually integrated in the genome of *T. brucei*, or present on selectable episomes in *T. cruzi* and *Leishmania*. In each case examined, the sequences responsible for developmental regulation were found to be in the 3′-untranslated region, and low RNA abundances were caused by RNA instability. Genes with regulatory 3′-untranslated regions from *T. cruzi* include those encoding amastin, tuzin, GP72, GP85, GP82, and the genes encoding small and large mucins. Examples from *Leishmania* include protein A2 and major surface proteins MSPL, MSPS, and MSPC. The list of investigated genes is longest for the African trypanosomes, where we have strong regulation of aldolase, PGKB, PGKC, two stage-specific hexose transporters, and the major surface proteins VSG, EP and GPEET. Taking PGK as an example again, the 3′-untranslated regions of *PGKB* and *PGKC* are completely different (Figure 4.3). A *CAT* gene with the *PGKB* 3′-untranslated region is expressed in procyclics, but gives very little RNA or protein in bloodstream forms. The reverse is true for the *PGKC* 3′-untranslated region.

Although the experiments with 3′-untranslated regions and RNA degradation all came to about the same conclusion, the studies are difficult to compare quantitatively. One problem is that several of the earlier studies were done before it was known that correct polyadenylation relies on the inclusion of the next downstream splice site. The reporter constructs were therefore designed on the assumption that the polyadenylation signal was in the 3′-untranslated region, as in any other eukaryote. This could result in incorrect polyadenylation upstream of regulatory sequences, or inclusion in the mRNA of extra sequences, downstream of the normal polyadenylation site, that could influence the mRNA stability. Ideally studies should include data showing that the mRNAs produced were polyadenylated in the correct or expected position. Other factors that could influence the expression level are transcription rate and gene copy number. Thus transcriptional promoters should if possible be absolutely constitutive, independent of life-cycle stage. And if episomal vectors are used, or genes are integrated in the genome, one has to determine how many copies of the reporter are present in each cell line.

To make different experiments comparable it is extremely useful to include a control reporter transcript that is expressed constantly in all life-cycle stages examined. Examples that have been used include transcripts containing actin (*ACT*) or tubulin (*TUB*) 3′-untranslated regions. Sometimes, a reporter gene with no 3′-untranslated region at all has been used instead. This is much less satisfactory because the resulting mRNA may include plasmid sequences and there is no guarantee that these will be neutral in the assay. In transient transfection assays, the inclusion of an unregulated control is crucial, as transfection efficiencies of different life-cycle stages are rarely identical.

Once it had been shown that a 3′-untranslated region is responsible for regulation, the next step has been to attempt to identify the sequences responsible by deletion or replacement mutagenesis. This has seldom been successful. Although it is expected that the sequences involved should be quite short, usually several hundred bases have proved essential for regulation. The major reason for this may be that mRNA secondary structure is very important. Deletion of a sequence which itself has no direct role in regulation may disrupt the secondary structure of a nearby regulatory region and thereby inactivate it. Similarly, if a regulatory region is transferred from one 3′-untranslated region to another, its function is nearly always either partially or completely lost.

There are so far just four cases where short regulatory sequences have been defined. The first elements to be identified were in the 3′-untranslated region of the *VSG* mRNAs of *T. brucei*. A reporter gene bearing a *VSG* 3′-untranslated region showed about 20-fold higher expression in bloodstream forms than in procyclic forms. The sequences responsible were an octamer and a 14-mer which both stimulate bloodstream-form expression and suppress procyclic-form expression. Deletion of either of these elements both abolished the down-regulation in procyclic forms and reduced the expression threefold in bloodstream forms to the control level (a *TUB* 3′-untranslated region). *VSG* genes are not transcribed at all in procyclic forms, but the rapid degradation of *VSG* mRNA may be important as trypanosomes differentiate into procyclic forms (see also Chapters 5 and 10).

The other three identified elements are probably related, although they come from two trypanosomatid species. The *EP* and *GPEET* genes form a small family of 8–10 genes which are all expressed exclusively in the procyclic stage of *T. brucei* to form the major surface proteins. The genes are found in clusters of two or three genes, and are transcribed by RNA polymerase I. A locus containing two *EP* genes is illustrated in Figure 4.4. The 3′-untranslated regions differ between the genes, but contain two conserved sequences: a 16-mer stem-loop which enhances translation but is not involved in developmental regulation, and a uridine-rich 26-mer sequence which is responsible for rapid degradation in bloodstream forms (Figure 4.4). Poly (U) sequences with interspersed A residues also cause degradation of *PGKB* mRNAs in bloodstream-form *T. brucei* and regulation of mucin gene transcripts in *T. cruzi*. These will be discussed in more detail below.

In most prokaryotes and eukaryotes, the expression of many genes is turned on and off, or tuned up or down, according to the availability of specific nutrients. So far, few examples of this are known in trypanosomatids. Depletion of iron results in upregulation of transferrin receptor protein and mRNA in bloodstream *T. brucei*, but the difference is only two- to threefold. More dramatically, it has recently been observed that the *GPEET* genes do not follow the same expression pattern as the *EP* genes. The GPEET protein is produced mainly early after differentiation, but not in prolonged culture or in the tsetse midgut. Its production in culture can be maintained by including glycerol in the medium, and is repressed by hypoxia. The regulation is again mediated by the 3′-untranslated region but the precise sequences responsible have yet to be defined.

In every case examined so far, the 3′-untranslated-region-mediated developmental regulation of mRNA levels operated at the level of RNA degradation. Most experiments have been done by adding actinomycin D to inhibit transcription, and following the fate of mRNAs by Northern blotting. Very few mRNA half-lives have been accurately measured and different authors do not always agree. For example, two independent estimates for the half-life of *TUB* mRNAs in procyclic *T. brucei* were 35 min and 120 min! Most published results involve rather few time points with no standard deviations, so that different publications are difficult to compare. The shortest mRNA half-life measured was 7 min, for the *EP* mRNA in bloodstream trypanosomes; the

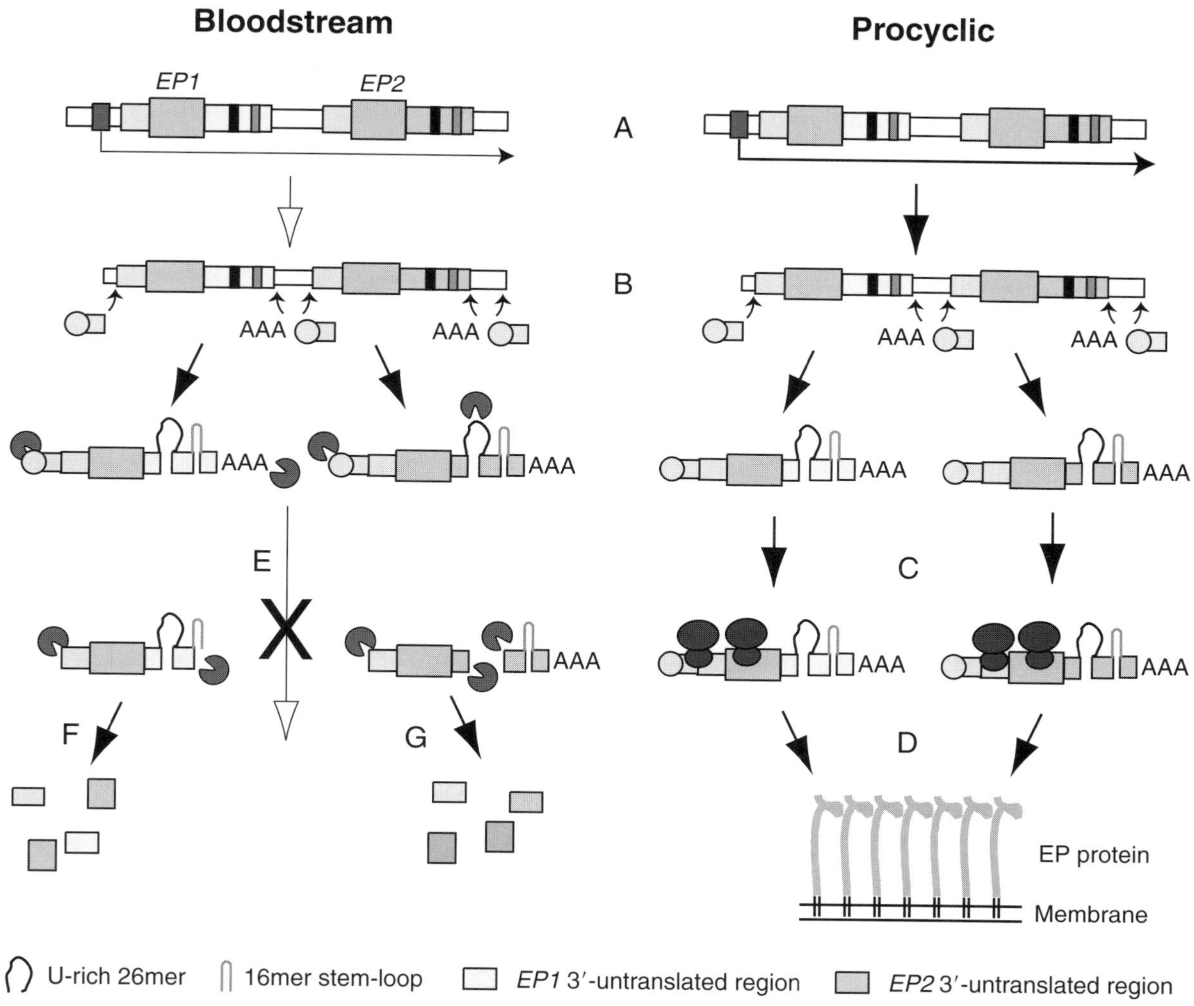

FIGURE 4.4 Regulation of EP procyclin expression in *Trypanosoma brucei*.

half-life in procyclics was about 90 min. Most half-lives measured have been in the range of 30 min to 2 h. For *T. cruzi*, half-life estimates are only available for short and long versions of the mucin RNAs: the endogenous transcripts had half-lives of at least 6 h in epimastigotes and less than 30 min in trypomastigotes. More information is available about *Leishmania* genes: here we have half-lives in log-phase promastigotes of 84 min for *MSPL*, about 1 h for *A2*, 7 h for *TUB* and less than 1 h for *HSP83* at 26°C. The *MSPL* transcripts were destabilized in stationary-phase promastigotes (half-life 17 min) as were *TUB* transcripts (180 min), and *A2* transcripts are stabilized (half-life 2 h) in amastigotes. Very short half-lives such as those seen for *EP* transcripts in bloodstream-form *T. brucei* have not been seen in the other species.

Studies of constitutively expressed genes of *T. brucei* have revealed that their mRNA half-lives are somewhat longer in procyclic forms than in bloodstream forms. How can this be reconciled with the fact that mRNA levels are the same? The discrepancy is explained by the fact that mRNA abundance is affected not only by degradation, but also by the dilution factor in the cells (the volume of procyclics is 2–3 times greater than that of bloodstream forms) and the rate at which the cells are growing and dividing.

The effects of protein synthesis inhibition

The inhibition of protein synthesis usually increases the abundance of several unstable mRNAs in trypanosomatids. Experimentally, cycloheximide is used, which arrests translating ribosomes without release of the mRNA. Under these circumstances, stabilization of RNA can be interpreted in several ways: either the stable association with polysomes inhibits degradation by preventing nuclease access, or degradation requires active ongoing translation, or degradation requires an unstable protein. In *T. brucei* the results have been confirmed using a variety of inhibitors, including puromycin, which results in chain termination and release of transcripts from the ribosome. In this case, the first explanation can be ruled out. All authors seeing such results have preferred the third explanation: that degradation is effected by an unstable protein. This could of course be a specific protein interacting with particular mRNAs. However, the disappearance of a component of the general degradation machinery would also have much more dramatic effects on unstable mRNAs than on stable ones: in yeast, cycloheximide inhibits decapping. We have found that the abundance of many mRNAs is affected to some extent by cycloheximide treatment, which leads us to prefer this less specific explanation.

Protein synthesis inhibition does not always result in increases in mRNA abundance. Indeed, opposite effects may be seen with stable and unstable mRNAs. In bloodstream *T. brucei*, *EP* mRNAs are stabilized, as are *ESAG5*, *ESAG6*, *ESAG7*, and *TUB* mRNAs, but *VSG* mRNA decreases. Conversely, in *T. cruzi* amastigotes, cycloheximide treatment resulted in a 3-fold decrease in amastin RNA and a 7-fold increase in tuzin mRNA, whereas in epimastigotes (where both mRNAs are less abundant than in amastigotes) amastin mRNA was not significantly affected and tuzin mRNA increased 4.3-fold. The *MSPL* mRNA of *L. chagasi* is very stable in virulent (recently isolated) log-phase promastigotes and less stable and abundant in stationary phase. Cycloheximide increased the abundance 4–6-fold in all growth phases but, in contrast, prevented stabilization of *L. infantum* HSP70 (genes 1–5) mRNA upon a temperature shift from 26°C to 37°C. No mechanistic conclusions can be drawn from these (and other) rather disparate results at this time.

Heat shock gene control and stress effects

Heat shock proteins (HSP) are important in protecting eukaryotic and prokaryotic cells from the effects of moderate rises in temperature. In particular, many HSPs have a chaperone function: that is, they are able to assist in the folding of newly synthesized proteins, and the refolding of proteins that have become denatured. In nearly all organisms, transcription of *HSP* genes is stimulated by a moderate (non-lethal) heat shock, and the mechanisms of the control have been extensively studied in both prokaryotes and eukaryotes. The heat shock protein (*HSP*) genes of kinetoplastids were easy to find through high sequence conservation with other *HSP* genes, and seemed to be ideal candidates for identification of inducible RNA polymerase II promoters. The response to heat shock ought to be biologically relevant, as parasites in the mammalian host can cause fever, and the insect forms are subjected to both variations in environmental temperature within the insect and a temperature rise of up to 10°C upon injection into a mammalian host. Heat shock treatments of these magnitudes have therefore been used to analyse HSP expression in kinetoplastids: for example, a switch from 37°C to 42°C to mimic fever, and elevations from 25°C or 27°C to 35°C to mimic the transfer of insect forms to the skin of a mammalian host. The first candidates to be investigated were the *HSP70* genes of *T. brucei* and *Leishmania*. These are arranged as polycistronically transcribed tandem repeats, and the levels of RNA were indeed mildly elevated upon heat shock. It was therefore hoped that there would be a promoter between each gene or, failing that, one promoter upstream of the cluster. Rather poor sequence similarities with the controlling promoter elements from higher eukaryotes were pointed out with enthusiasm. But in the end – as in all other searches for RNA polymerase II promoters upstream of protein-coding genes – it became clear that no promoters were there.

Meanwhile, additional *HSP* genes were found. Expression of mitochondrial HSP60 has been little investigated (although it is an excellent mitochondrial marker protein and shows developmental regulation in *T. brucei*) but Leishmanial *HSP85* has been analysed extensively. There is evidence from *T. brucei* that mRNA *trans*-splicing is disrupted by heat shock in bloodstream trypanosomes, but the *HSP70* mRNAs seemed little affected. Both *HSP70* and *HSP85* (sometimes also called *HSP83*) mRNAs may be more stable at high temperatures than other RNAs, and there is evidence from *L. amazonensis* that *HSP83* genes are also preferentially translated during heat shock. In *L. infantum*, there are six *HSP70* genes arranged as a tandem repeat. Nearly all the RNA comes from gene 6, but only expression from gene 1–5 mRNAs was increased by heat shock. In all cases investigated, the sequences responsible have been shown to lie in the 3′-untranslated region. But it is most unlikely that the regulation of *HSP70* or *HSP83* mRNA levels or translation has any biological relevance, as no-one has ever shown any changes in total HSP70 or HSP83 protein levels as a consequence of heat shock. HSP70 and HSP83 each make up between 2% and 3% of the total protein in unstressed *Leishmania* promastigotes, which is presumably sufficient to protect against physiological stresses.

The most interesting HSP found so far is HSP100 from *L. major* and *L. donovani*. This is upregulated during heat shock of promastigotes, and during differentiation of promastigotes into amastigotes; gene replacement experiments indicate that it plays a role in parasite virulence.

Degradation and translation of the *EP* mRNAs

The *EP* and *GPEET* genes illustrate several aspects of gene regulation in trypanosomes (Figure 4.4). Regulation of *EP* transcription was studied by inserting a CAT reporter gene, bearing a 3′-untranslated region that gave constitutive expression, into the *EP* locus (Figure 4.5A). Expression of the reporter was roughly 10-fold higher in procyclic forms than in bloodstream forms, indicating that transcription was regulated 10-fold. But *EP* RNA is extremely abundant in procyclics, and virtually undetectable in bloodstream forms, so the transcriptional

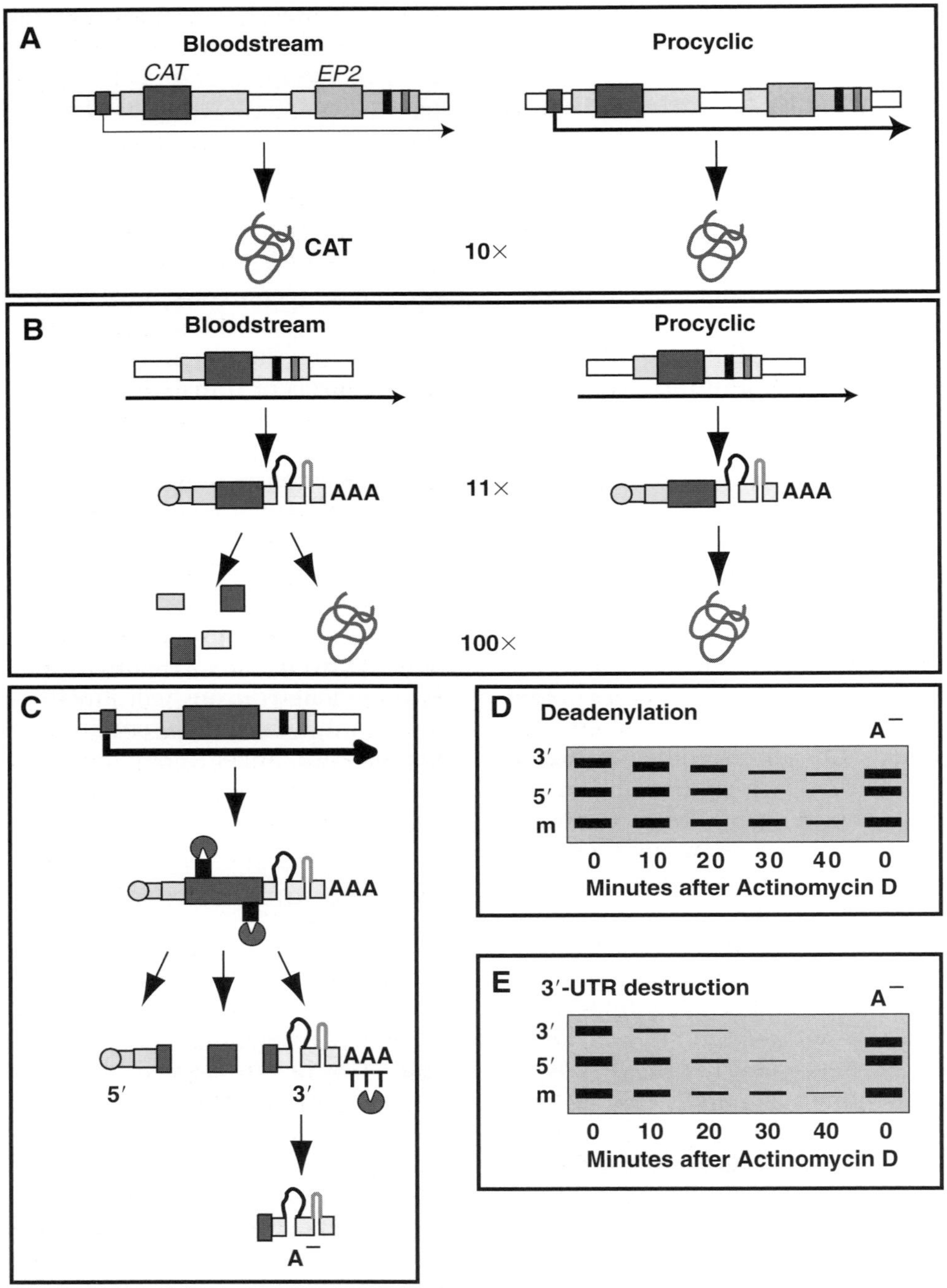

FIGURE 4.5 Experiments to determine the mechanism of EP regulation and mRNA degradation.

control alone is clearly insufficient. Next, the 3′-untranslated regions from the *EP1* or *EP2* mRNAs were linked to reporter genes encoding CAT or luciferase. Transient transfections were done, and showed that the 3′-untranslated regions gave no expression in bloodstream trypanosomes. The constructs were then transfected into bloodstream and procyclic trypanosomes in such a fashion that they would integrate into the tubulin locus, where they would be equally transcribed in both stages (Figure 4.5B). Now, there was about 100 times more expression of the reporter protein in procyclic trypanosomes than in bloodstream trypanosomes, but the reporter gene mRNA was detectable and regulated only 11-fold. That means there must also be about 9-fold regulation of translation. Thus *EP* gene expression is regulated roughly 10-fold at each of three levels: transcription, RNA degradation, and translation (Figure 4.5A, B).

The 3′-untranslated regions of the different *EP* and *GPEET* genes are not conserved, apart from a U-rich 26-mer and a 16-mer stem-loop (Figure 4.4A). The 16-mer enhances translation in procyclic forms by an unknown mechanism, but plays no role in developmental regulation. The mechanism of the translational enhancement is unknown but the stem-loop structure is important. We have attempted to find proteins that bind to this sequence using a yeast three-hybrid screen, with no success at all.

Deletions of, and point mutations in, the U-rich 26-mer abolished or reduced regulation. In particular, an *EP1* 3′-untranslated region lacking the 26-mer, *ep1Δ26*, gave mRNA and protein expression that was 30–40% of the level obtained using a control *ACT* 3′-untranslated region. The results of enzymatic and chemical probing experiments indicated that the 26-mer is normally in a single-stranded conformation. Examination of the *PGKB* 3′-untranslated region, which can also mediate procyclic-specific expression, revealed a similar U-rich sequence, with interspersed As, which is predicted to be single-stranded. Specific deletion of this U-rich domain from the *PGKB* 3′-untranslated region abolished stage-specific regulation at both the RNA and protein levels.

Both the *EP1* 26-mer and the *PGKB* regulatory sequence bear a strong resemblance to mammalian AU-rich elements (AREs). Interestingly, another control element of similar sequence, which gives preferential expression in epimastigotes and amastigotes, has been identified in the 3′-untranslated region of small mucin gene transcripts in *T. cruzi*. It will be interesting to see to what extent the degradation mechanisms are also similar.

In order to compare the degradation mechanisms of the constitutive *ACT* mRNA with the regulated *EP1* mRNA, *CAT* transgenes bearing *EP1*, *ACT* or *ep1Δ26* 3′-untranslated regions were placed downstream of a bacteriophage T7 promoter and integrated into the genome of trypanosomes that expressed the bacteriophage polymerase. This resulted in the production of sufficiently large amounts of RNA to enable detailed study of the structure of the *CAT* mRNA during degradation. To distinguish events at the 5′ end from those at the 3′ end, the resulting mRNAs were cut into specific pieces (Figure 4.5C). Oligonucleotides that hybridized to the *CAT* gene were added, together with RNAaseH, which digests RNA/DNA hybrids. Thus the RNA was cut only where the oligonucleotides were hybridized, into three pieces representing the 5′ end, 3′ end, and a middle portion. The length of the poly(A) tail was estimated by adding oligo d(T) with RNAaseH, which removed the poly(A) tail, then measuring the difference in size between poly(A)+ and poly(A)-fragments (Figure 4.5C). All three RNAs had poly(A) tails approximately 200 nucleotides long, in both bloodstream and procyclic forms. This indicated that the rate of degradation was not determined by differential polyadenylation. The pattern of degradation was followed after actinomycin D addition. Degradation of the *CAT-ACT* and *CAT-ep1Δ26* mRNAs involved deadenylation, followed by degradation by both 5′→3′ and 3′→5′ exonucleases (Figure 4.2). Thus the 3′-end fragment became progressively smaller after actinomycin D addition (Figure 4.5D). After that, the RNA was degraded from the 5′ and 3′ ends simultaneously. The same pattern was seen for *CAT-EP1* RNA in procyclic trypanosomes. In bloodstream forms, the 3′-end of the *CAT-EP1* RNA disappeared very rapidly after transcription inhibition and partially deadenylated intermediates were never seen (Figure 4.5E). Afterwards the RNA was again degraded from both ends. Although it is possible that the 26-mer instability element causes deadenylation that is so rapid that it could not be detected, it seems more likely that the 3′-untranslated region is attacked by an endonuclease, perhaps acting on the 26-mer. Products of an endonuclease cleavage could not be detected, presumably because they were very rapidly degraded.

Even though the pathway of *EP1*-mediated degradation is now fairly clear, the mechanism of developmental regulation is not. It could be that a 26-mer-specific endonuclease is expressed all the time, but in procyclic forms the *EP1* element is protected by a protein that binds the 26-mer. Alternatively, it may be that an endonuclease is expressed only in bloodstream forms. Again, there may be a protein that binds to the 26-mer in bloodstream forms and stimulates degradation by exonucleases. As the 26-mer is responsible for supressing translation as well, it seems likely that any binding proteins should interact with the translation apparatus. However, this is all highly speculative, as all attempts to identify specific binding proteins or endonucleases by biochemical means have failed so far.

In contrast, proteins that bind to the ARE in the mucin mRNA have been detected in *T. cruzi* extracts. It will be interesting to find out if homologues of these proteins exist in *T. brucei*, and whether they are involved in regulation of PGKB and EP expression. More interesting still is the parallel with degradation of mRNAs containing AREs in mammalian cells, as this too may be initiated by a specific endonuclease.

The general RNA degradation apparatus in trypanosomatids

Components of the general degradation apparatus of trypanosomatids were identified by screening the databases for sequences with similarity to known genes from other eukaryotes. The relevant proteins are shown in Figure 4.2. Poly(A) binding proteins have been identified from several species. Comparing *T. brucei* with *S. cerevisiae*, we have found four genes encoding proteins similar to the yeast 5′→3′ exoribonucleases, *XRN1* and *RAT1*, but no homologues of the decapping complex. The expression of these genes and the location and regulation of the proteins have not yet been studied. *T. brucei* has nuclear and cytoplasmic exosome complexes of about 250 kDa, which is 100 kDa smaller than the yeast complex. In accordance with its smaller size, the exosome has only five subunits that are present in stoichiometric amounts (instead of about 11 in yeast). Interestingly, some exonucleases that are in the yeast exosome are present but not exosome-associated in *T. brucei*. The functional implications of this have yet to be explored. We do not know if all these exonucleases are themselves subject to regulation, or whether their main role is to finish degradation once regulated processes (such as deadenylation or endonuclease cleavage) are complete.

Control of translation and protein processing

Translational control is probably extremely important in kinetoplastids, but has until now been almost completely neglected. Investigators looking for regulated genes have adopted two main approaches. They have cloned genes encoding proteins whose expression was already known to be regulated, such as enzymes involved in energy metabolism. Alternatively they have looked for RNAs that were preferentially expressed in one or other life-cycle stage by differential screening of libraries or, more recently, using PCR-based approaches. Overall this has resulted in a bias towards genes whose products were abundantly expressed and regulated at the mRNA level. Nevertheless indications of translational control have emerged. The control of *EP* and *HSP* translation has already been mentioned. More dramatically, cytochrome *c* and aconitase are 30–100 times more abundant in procyclic *T. brucei* than in bloodstream forms, but the differences in mRNA are very small. The *PGK* genes may also show different types of regulation in different trypanosome strains. The results so far described have been from a highly virulent, laboratory-adapted strain of parasites, but in another strain, most regulation seemed to be of protein synthesis rather than of the mRNA level.

Several trypanosomatid RNAs have long 5′-untranslated regions which include upstream open reading frames which might be expected to drastically reduce translation of the major long open reading frame. One example is the *PAG* genes, located downstream of the *EP/GPEET* genes, but it is not known if the proteins predicted for these genes are produced at all. Another is the tuzin 5′-untranslated region from *T. cruzi*, which has an upstream ORF which suppresses translation.

Protein degradation

A discrepancy between mRNA levels and protein levels can be caused not only by regulation of translation, but also by differential protein degradation. For mitochondrial proteins like cytochrome *c* and aconitase, a failure of protein import could lead to degradation; so far, though, there is no evidence that bloodstream trypanosome mitochondria are defective in protein import. Similarly, one could imagine that differences in the secretory pathway such as regulated expression of modification enzymes might result in failure to correctly process a protein and consequent degradation.

The proteasome of *T. brucei* contains, like other proteasomes, a catalytic core of about 20 S which is inhibited by lactacystin. Treatment of both procyclic and bloodstream trypanosomes with lactacystin resulted in cell cycle arrest, confirming that proteasome activity is needed for trypanosome growth. Two proteins which may be proteasome substrates are CYC3 and CYC2. These two cyclins have half-lives of less than one cell cycle, and their stability is increased by proteasome inhibitors.

POST-TRANSCRIPTIONAL REGULATION IN OTHER PARASITES

There is extraordinarily little information available about post-transcriptional regulation in parasites other than the Kinetoplastida. Work on the apicomplexans has concentrated almost exclusively on the search for transcriptional promoters and analyses of transcriptional regulation. This is doubtless partially for technical reasons, as promoter activity can be assayed in transient transfection assays, whereas work on RNA processing and turnover requires much larger amounts of parasite material, derived from permanent transformants.

Many apicomplexan genes contain introns. Potential splicing signals have the features of canonical eukaryotic signals, but have not been functionally tested. Nevertheless, at least two possible examples of alternative or regulated splicing have been documented. A study of the B7 gene of *Plasmodium berghei* showed that the transcripts expressed in the asexual and sexual stages differed in the length of the 5′-untranslated region because two different promoters and alternative splice sites were used, and analysis of the *MSP4/5* genes of *P. berghei* and *Plasmodium yoelii* revealed that both unspliced and spliced mRNA were present at steady state.

Just as for splicing, no polyadenylation signals have been functionally identified in Apicomplexa. The canonical mammalian polyadenylation signal, AAUAAA, occurs many times by chance in the AU-rich *Plasmodium falciparum* genome; in a random sequence containing 50% A and 50% T, the sequence would occur by chance once every 64 nt. Thus any assignment of functionality to this sequence has to be treated with extreme caution. For example, a 750 nt region 3′ of the pgs28 coding region of *Plasmodium galinaceum* was found to contain seven AAUAAA or AUUAAA signals, of which five were contained in the 3′-untranslated region of the mRNA and two were further downstream. Deletion of either a 155 nt poly(T) tract upstream of the fifth signal, or of the region downstream of the poly(T) tract including the fifth signal, reduced reporter gene expression in transient assays by about 90%. But it is not known whether the deletions affected the polyadenylation site. RNA degradation and translational control have not been studied in Apicomplexa and no examples of 3′-untranslated region-mediated regulation are known. Regulation of RNA processing, stability and translation in other parasites, whether unicellular or multicellular, has received almost no attention. This field is therefore wide open for future work.

POST-TRANSCRIPTIONAL REGULATION AND ANTI-PARASITIC CHEMOTHERAPY

The vital importance of post-transcriptional regulation in the kinetoplastids might suggest this process as a possible target for future anti-kinetoplastid drugs. Obviously, if the regulation were inhibited, the whole metabolism and surface structure of the organisms would become deregulated, with lethal consequences. Thus an African trypanosome might not only lose control over its energy metabolism, but also start to express the EP and GPEET surface proteins constitutively, making it a ready target for the immune response. Before any new process can be specifically inhibited, however, it is necessary to identify the proteins involved and to define their mechanism of action.

Even once this is achieved, it is not clear that post-transcriptional regulation is a good target. Many of the nucleases are conserved in evolution. To attack the direct regulatory proteins it would probably be necessary to inhibit protein–RNA interactions. It is sobering to note that although post-transcriptional regulation is of vital importance in the regulation of the immune response, and possibly in growth of cancer cells, so far no specific inhibitors of this process are (to my knowledge) available.

If the RNA degradation and stabilization machineries are not amenable to selective inhibition, it may instead be possible to affect them by attacking upstream processes. The differentiation of an African trypanosome is initiated by external environmental signals. These signals have to be carried to the degradation apparatus, in order to increase the degradation of some mRNAs and stabilize and increase translation of others. The regulation of ARE-mediated degradation in mammalian cells has been shown to involve the kinase cascades, and kinases are being extensively investigated as possible drug targets, for example for anticancer chemotherapy. If post-transcriptional regulation is to be disrupted in kinetoplastids, it is probably the signalling kinases that will be the most appropriate targets for attack.

FURTHER READING

General mechanisms

Cullen, B.R. (2000). Nuclear RNA export pathways. *Mol. Cell. Biol.* **20**, 4181–4187.

Hentze, M.W. and Kühn, L. (1996). Molecular control of vertebrate iron metabolism: mRNA-based regulatory circuits operated by iron, nitric oxide, and oxidative stress. *Proc. Natl. Acad. Sci. USA* **93**, 8175–8182.

Hentze, M.W. and Kulozik, A.E. (1999). A perfect message: mRNA surveillance and nonsense-mediated decay. *Cell* **96**, 307–310.

Lipshitz, H.D. and Smibert, C.A. (2000). Mechanisms of RNA localization and translational regulation. *Curr. Opin. Genet. Devel.* **10**, 476–488.

Lopez, A.J. (1998). Alternative splicing of pre-mRNA: developmental consequences and mechanisms of regulation. *Annu. Rev. Genet.* **32**, 279–305.

Mitchell, P. and Tollervey, D. (2000). mRNA stability in eucaryotes. *Curr. Opin. Genet. Dev.* **10**, 193–198.

Richter, J.D. (1999). Cytoplasmic polyadenylation in development and beyond. *Microbiol. Mol. Biol. Rev.* **63**, 446–456.

Sachs, A.B., Sarnow, P. and Hentze, M.W. (1997). Starting at the beginning, middle and end: translation initiation in eukaryotes. *Cell* **89**, 831–838.

van Hoof, A. and Parker, R. (1999). The exosome: a proteasome for RNA? *Cell* **99**, 347–350.

Kinetoplastids

Brittingham, A., Miller, M.A., Donelson, J.E. and Wilson, M.E. (2001). Regulation of GP63 mRNA stability in promastigotes of virulent and attenuated *Leishmania chagasi*. *Mol. Biochem. Parasitol.* **112**, 51–59.

Clayton, C.E. (2002). Developmental regulation without transcriptional control? From fly to man and back again. *EMBO J.* **21**, 1881–1888.

D'Orso, I. and Frasch, A.C.C. (2001). TcUBP-1, a developmentally regulated U-rich RNA-binding protein involved in selective mRNA destabilization in trypanosomes. *J. Biol. Chem.* **276**, 34801–34809.

Estévez, A. and Clayton, C.E. (2001). The exosome of *Trypanosoma brucei*. *EMBO J.* **20**, 3831–3839.

Irmer, H. and Clayton, C.E. (2001). Degradation of the *EP1* mRNA in *Trypanosoma brucei* is initiated by destruction of the 3′-untranslated region. *Nucl. Acids Res.* **29**, 4707–4715.

Mahmood, R., Hines, J.C. and Ray, D.S. (1999). Identification of *cis* and *trans* elements involved in the cell cycle regulation of multiple genes in *Crithidia fasciculata*. *Mol. Cell. Biol.* **19**, 6174–6182.

Mutomba, M.C., To, W.Y., Hyun, W.C. and Wang, C.C. (1997). Inhibition of proteasome activity blocks cell cycle progression at specific phase boundaries in African trypanosomes. *Mol. Biochem. Parasitol.* **90**, 491–504.

Quijada, L., Hartmann, C., Guerra-Giraldez, C., Drozdz, M., Irmer, H. and Clayton, C.E. (2001). Expression of the human RNA-binding protein HuR in *Trypanosoma brucei* induces differentiation-related changes in the abundance of developmentally-regulated mRNAs. submitted.

Vassella, E., Den Abbeele, J., Butikofer, P. *et al.* (2000). A major surface glycoprotein of *Trypanosoma brucei* is expressed transiently during development and can be regulated post-transcriptionally by glycerol or hypoxia. *Genes Dev.* **14**, 615–626.

Zilberstein, D. and Shapira, M. (1994). The role of pH and temperature in the development of *Leishmania* parasites. *Annu. Rev. Microbiol.* **48**, 449–470.

C H A P T E R

5

Antigenic variation in African trypanosomes and malaria

George A.M. Cross
Laboratory of Molecular Parasitology, The Rockefeller University, New York, NY, USA

INTRODUCTION

Two important human diseases, malaria and African trypanosomiasis (African sleeping sickness), have evolved sophisticated molecular mechanisms to evade the immune responses of their mammalian hosts, creating persistent infections that may or may not be ultimately lethal. It is not in the interests of a parasite to kill its host. Parasites that are too virulent will themselves die, if their hosts do not survive long enough for the infection to be transmitted. For both trypanosomes and malaria parasites, antigenic variation is the major but probably not the sole mechanism for persistence. Antigenic variation in trypanosomes has been long recognized, as exemplified in the following quotation (translated from the German) from a 1905 paper by E. Franke, working in Paul Ehrlich's institute, just 10 years after trypanosomes were implicated as the cause of the 'tsetse fly disease', by Surgeon-Captain David Bruce of the British Army Medical Service: 'The trypanosomes must have acquired other biological characteristics during their stay in the semi-immune body that rendered them resistant to the defensive substances.' This explanation was consistent with previous experiments in which the transfer of immune serum prolonged the life of infected animals but did not cure them, and with the relapsing nature of the parasitemia in humans. So, why do we not yet understand a phenomenon described 100 years ago?

Antigenic variation in trypanosomes is the best studied and probably the most highly developed example of an infectious microbe evolving a mechanism for countering the host's immune-diversity generator with a diversity generator of its own. Antigenic variation in the African trypanosomes appears to have arisen solely for the purpose of evading the host immune responses, although other roles for these cell-surface variations may have evaded

Molecular Medical Parasitology
ISBN 0–12–473346–8

scientific recognition so far. In contrast, variant antigens on the surface of the malaria-infected erythocyte appear to have more complex functions. These differences probably reflect the major differences in the life of these two protozoal pathogens: the African trypanosomes are extracellular throughout their occupation of the mammalian host whereas multiplication of malaria parasites occurs intracellularly, initially in the liver but predominantly in erythrocytes. Although generally referred to as 'bloodstream forms', trypanosomes generally populate the interstitial spaces, lymphatics and capillary beds at higher concentrations than they are found in the major blood vessels. As infection progresses, trypanosomes migrate to the central nervous system, leading to coma and inevitable death. Because of the state of knowledge about antigenic variation in trypanosomes, which is largely attributable to the relative ease with which they can be studied in the laboratory, this chapter will focus on trypanosomes. Investigations of the molecular genetics of antigenic variation in malaria have lagged behind those of trypanosomes, but are rapidly gaining ground. During the 1990s, trypanosomes and malaria parasites have become more amenable to cultivation and genetic manipulation. When these advances are coupled to the output of genomic sequencing projects, the prospects of understanding the ways in which these parasites evade the immune responses of their hosts look brighter than ever. Meanwhile, antigenic variation remains a major obstacle to immunization against malaria and sleeping sickness.

AFRICAN TRYPANOSOMES

Trypanosomes are members of the order Kinetoplastida, which comprises unicellular parasitic protozoa distinguished by a single flagellum and a kinetoplast, an ancient misnomer for the mitochondrial DNA (kinetoplast DNA or kDNA: see Chapter 12) that is neither a discrete organelle – although located in a specific region of the single mitochondrion – nor involved in motility of the organism, despite its juxtaposition with the basal body of the flagellum. Trypanosomes are ubiquitous parasites of invertebrate and vertebrate hosts throughout the world. They represent an evolutionary branch that diverged 500 MYA. African trypanosomiasis is a zoonosis. About a dozen *Trypanosoma* species and subspecies populate the African mammalian fauna, from small antelopes to large carnivores. Although many animals and indigenous cattle harbor the same trypanosome species that are fatal to humans and to imported livestock, most of the fauna appear to tolerate trypanosome infections without overt pathology, presumably due to natural selection during millions of years of co-evolution. Some animals, including humans, have innate immunity to the majority of trypanosome species and clones. I will return to this interesting topic in a later section. African trypanosomes are a more constant factor in animal husbandry (cattle, goats, sheep, pigs, horses, and dogs are constantly challenged) than for humans, but, in many locations, human sleeping sickness remains a serious threat, almost to the same extent as in the early part of the twentieth century. The separate evolution and primary geographic restriction of trypanosomes that cause so-called African trypanosomiasis appears to be largely dependent on the range of their specific insect vector *Glossina*, commonly known as the tsetse. Notwithstanding the several developmental stages that trypanosomes obligatorily undergo in the tsetse, transmission can occur in the absence of this vector. This is exemplified by the presence of *T. vivax* and *T. evansi* in South America and Asia, where they are transmitted by biting flies and vampire bats. *T. equiperdum* is a venereal disease of horses in Africa. The predominantly tsetse-transmitted trypanosomes can also be passed directly from mammal to mammal, by syringe, by carnivorous eating habits, and possibly transplacentally.

Trypanosoma brucei is the most widely studied of the African trypanosomes, because it is by far the most convenient laboratory model. Although it is the least prevalent of the three major species of veterinary importance (the other two being *T. congolense* and *T. vivax*), it is the species that contains the two recognized subspecies that infect humans, *T. brucei rhodesiense* and *T. brucei gambiense*. The entire life cycle of *T. brucei* can be reproduced in the laboratory, using rodents and tsetse, but not yet entirely in cell culture. Both the bloodstream forms and the tsetse midgut 'procyclic' form can be readily cultivated *in vitro*, which allows genetic manipulations to be performed in a relatively facile manner. Relatively facile means they can be genetically manipulated with about 10% the convenience of *Saccharomyces cerevisiae*, but more than ten times the ease with which similar manipulations can be performed in the human malaria parasite *Plasmodium falciparum*. Other apicomplexan parasites, especially *Toxoplasma gondii*, can be more readily manipulated than *Plasmodium*, and can be informative about the functions of some genes and organelles that are conserved between *Plasmodium* and *Toxoplasma*.

ANTIGENIC VARIATION AT THE STRUCTURAL AND FUNCTIONAL LEVEL

The variant surface glycoprotein coat

Animal-infective bloodstream and tsetse salivary-gland 'metacyclic' trypanosomes are characterized by an electron-dense 'surface coat' (Figure 5.1). The surface coat of an individual trypanosome consists of about 10 million molecules of a single molecular species of variant surface glycoprotein (VSG). Antigenic variation involves the sequential expression of coats composed of different VSGs. Mice can be totally protected against homologous challenge by

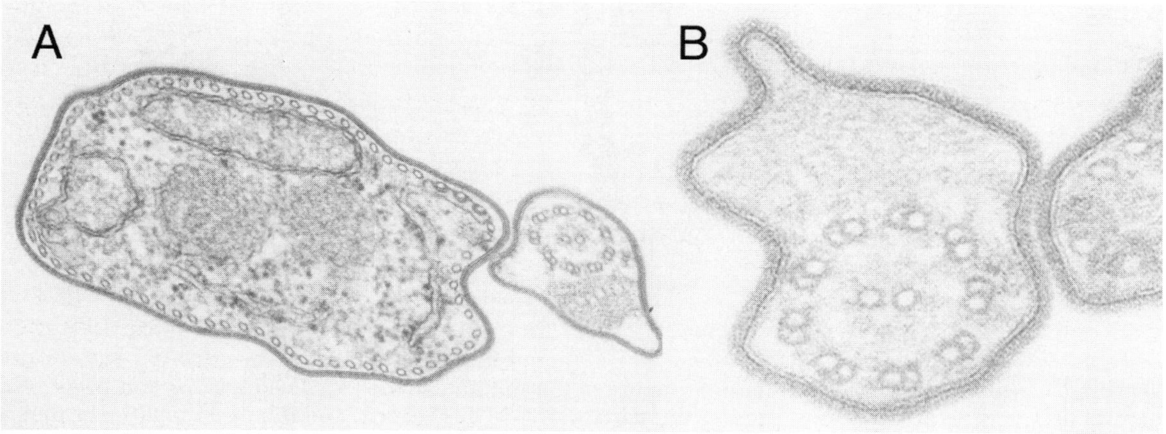

FIGURE 5.1 The surface structure of *T. brucei* bloodstream forms. A cross-section through the cell body and the flagellum shows the cell membrane with its underlying microtubular cytoskeleton and overlying surface coat at (A) low and (B) high magnifications. The approximate cross-sectional dimensions in (A) are 1 × 2 μm.

immunization with individual purified VSGs. Purification turned out to be far easier than anticipated, for two reasons. Firstly, the derivation of trypanosome clones that switch infrequently allows large quantities of individual VSGs to be purified from trypanosome populations that are >99.9% homogeneous. Secondly, although the VSG is firmly attached to the surface membrane of intact trypanosomes, it is rapidly released when the membrane is breached. For many years this convenient behavior was a paradox. Amino acid sequencing of selected regions of several VSGs led to the prediction that the carboxy terminus of purified VSG had arisen by cleavage of a larger precursor. This prediction was confirmed when the corresponding genes were cloned and sequenced. The finding that the purified protein terminated in an amide-(peptide) bonded residue of ethanolamine confirmed that proteolytic cleavage had not occurred during VSG purification and led to the identification of a covalently linked lipid, whose complete structure was ultimately determined, providing the prototype for a novel mode of protein–membrane attachment, the glycosylphosphatidylinositol (GPI) anchor (Figure 5.2). GPI anchors have now been found on proteins having a wide range of functions, in all eukaryotic cells, but they appear to be particularly abundant in parasitic protozoa, for reasons that have not been resolved. A consensus is emerging that GPI anchoring may be important for locating certain proteins to specialized subdomains of the plasma membrane. Certain aspects of GPI anchor structure are conserved in all eukaryotic cells, from

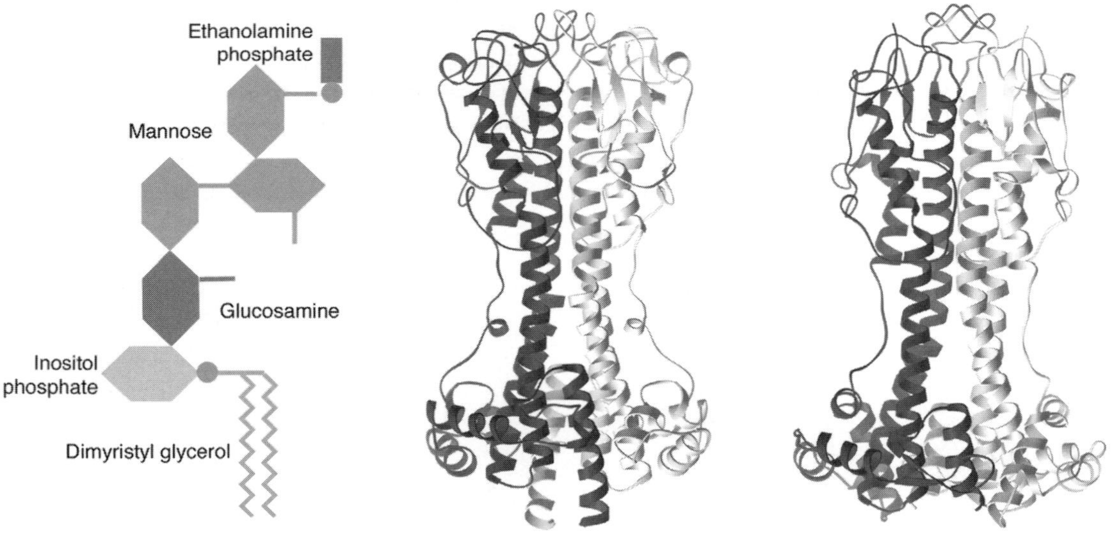

FIGURE 5.2 (See also Color Plate 2) (*Left*) Structure of the minimal ('core') VSG GPI anchor and (*right*) three-dimensional structures of the amino-terminal domains of VSG MITat 1.2 (left) and ILTat 1.24 (right). The dark and light ribbons indicate the atomic traces of the two identical units comprising the dimeric domain. The amino termini are at the top of the molecule, which forms the outer face of the coat. The bottom of this domain is linked to the carboxy-terminal domain, for which a structure is not available, which is then linked to the GPI anchor. The image was kindly provided by Ms Lore Leighton.

trypanosomes to humans. The core structure – ethanolamine–phosphate–6Manα1–2Manα1–6Manα1–4GlcNH$_2\alpha$1–6*myo*-inositol – does not vary, but the lipid constituents can vary greatly from cell to cell. The core glycan structure can be extensively decorated with additional residues of ethanolamine, sugars and sialic acids, which can vary among different proteins synthesized by the same cell. Perhaps the most unique feature of the VSG GPI anchor is its lipid moiety: dimyristyl glycerol. GPI is synthesized by sequential addition of components to cellular phosphatidyl-inositol (PI), to generate a GPI precursor that contains no myristic acid. This precursor is remodeled by the sequential replacement of each acyl chain by myristic acid. Another myristate remodeling activity works on GPI anchors already attached to VSG. Thus, for reasons that are not understood, *T. brucei* goes to quite remarkable lengths to ensure that its VSG is anchored in the membrane by this C_{14} fatty acid, which comprises only ~1% of serum lipids (see Chapter 10).

The release of GPI-anchored VSG, when the trypanosome membrane is disrupted, is due to the presence of a potent phospholipase C (PIPLC) that cleaves the diacylglycerol–phosphoinositol linkage, releasing soluble VSG that contains the hydrophilic components of the GPI anchor. One consequence of this cleavage is the formation of a terminal inositol cyclic phosphodiester, which is a major contributor to the cross-reactivity of rabbit anti-VSG antibodies. Despite much speculation and experimentation, the biological role of this PIPLC is unknown, as is the manner in which it comes into contact with VSG following membrane disruption. Its cellular location is unresolved, although it appears to be acylated and attached to the membrane of otherwise anonymous intracellular organelles. Only VSG already at the surface is susceptible; VSG in the secretory pathway appears to be inaccessible. Although it first appeared to be specific for GPI, rather than simply for PI, this may have been a misleading consequence of detergent concentrations and other components of the assay. When trypanosomes pass from the mammalian bloodstream into the tsetse midgut, the VSG coat is shed and replaced by a similarly dense coat formed from a small family of midgut-stage proteins called procyclins. These proteins, also known by the acronym PARP, for procyclic acidic repetitive proteins, consist largely of Glu-Pro or Gly-Pro-Glu-Glu-Thr repeats, and may serve to protect the procyclic stage from proteases in the insect midgut. Intriguingly, the procyclins are also GPI-anchored, but the GPI structure is distinct from that on the VSG. It is not myristylated, it contains monoacylglycerol, and the inositol is palmitylated, rendering it resistant to PIPLC. The idea that the PIPLC was involved in replacement of the VSG coat during this developmental transition was intriguing, but apparently misleading. When the PIPLC gene is deleted, trypanosomes do not appear to be impaired in their ability to complete their life cycle in the tsetse. In contrast, developmental shedding of the VSG coat appears to be mediated by proteolytic cleavage. Interestingly, disruption of an essential gene in the pathway for GPI synthesis is lethal to bloodstream-form *T. brucei* but not to the procyclic forms, which can apparently survive when deprived of all GPI-linked proteins. This observation, coupled with the many differences between GPI synthesis in trypanosomes and human cells, validates this pathway as a potential target for drug development.

Calculations and physical studies suggest that the trypanosome coat is composed of a monolayer of VSG molecules packed as tightly as possible on the surface membrane. The evidence for this view of the coat comes from various experimental observations. Where VSG glycans are located toward the carboxy terminus, they are inaccessible to lectins added to suspensions of intact trypanosomes. In some cases, living trypanosomes covered with a specific VSG are resistant to trypsinization, whereas the solubilized VSG is susceptible. This suggests that even a small protein like trypsin has problems penetrating into the deeper regions of the coat. Thirdly, there have been several studies in which panels of monoclonal antibodies were generated against purified VSGs. Only a small proportion of these antibodies react with the surface of living trypanosomes. This is consistent with the dense VSG packing, which results in the display of only a small area of the molecule on the trypanosome surface, which is also consistent with the crystal structure determined for the amino-terminal domains of two VSGs (see below).

The VSG has no known enzymatic or receptor function. The prevailing view is that its role is simply to form a physical barrier that can be exchanged, in a small fraction of the population, as the immune response develops to each VSG in turn. This barrier may have two specific roles. It may prevent antibodies to invariant or less variant membrane proteins from reaching their targets on living cells. Secondly, antibodies against VSG will not cause complement-mediated lysis because complement deposition cannot progress to membrane insertion of the C9 complex. Whether the VSG coat has a role in sequestering parasites at different sites within the infected host has been debated but little studied. Because of the extent of variation, there could be fortuitous or intentional sequestering of some trypanosomes in specific organs. Trypanosomes do not only circulate in the main vasculature, so specific retention in the tissues, or the invasion of specific VSG subtypes into the brain, are issues that could be relevant to pathogenesis.

VSG structure

The amino acid sequences of many VSGs have been deduced from the sequences of the corresponding genes. One VSG was sequenced at both the protein and DNA level, which incidentally proved that trypanosomes use the 'universal' genetic code. Partial protein sequencing of several VSGs identified key features of the amino- and carboxy-terminal signal sequences that are cleaved immediately post-translationally, the carboxy-terminal signal being replaced by the GPI anchor. Most VSGs exist as non-covalently-linked homodimers, but there are known exceptions, and maybe more that are undiscovered, that are either disulfide-linked or capable of forming higher oligomers in solution. The VSGs of *T. brucei* consist of an amino-terminal domain linked to one or two carboxy-terminal subdomains via a 'hinge' region that is exquisitely sensitive to proteolytic cleavage and whose presumed flexibility prevented crystallization of intact VSG. Consequently, the crystal structures of two amino-terminal domains were determined some time ago (Figure 5.2), but the structure of the carboxy-terminal domain is unknown.

Although only a small area of the VSG is accessible in the surface coat, amino acid sequence variation occurs throughout the molecule. If the extraordinarily conserved carboxy-terminal GPI anchor signal sequence is disregarded, the range of sequence variation among VSGs is such that some sequences would be difficult to recognize as members of the family. However, known *T. brucei* VSGs can be grouped into several classes, based primarily on their GPI signal sequences and spacing of conserved cysteine residues (Figure 5.3). Despite the wide range of primary sequence variation, VSG amino-terminal domains are predicted to fold into similar three-dimensional

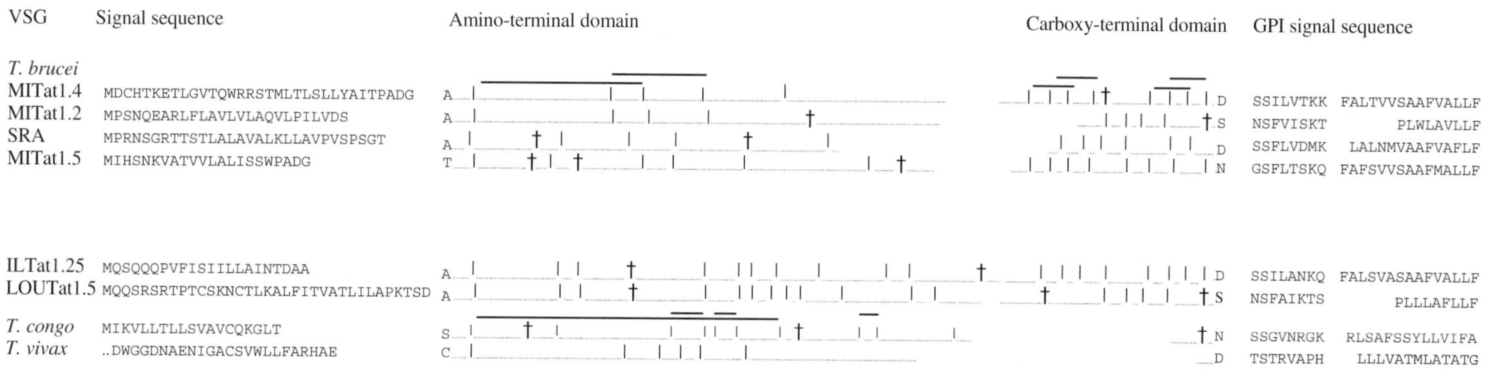

FIGURE 5.3 Conserved features of typical VSG sequences. VSG amino-terminal domains fall into two major classes, having either 4–6 or 10–12 cysteine residues (indicated by vertical bars). Carboxy-terminal subdomains contain two disulfide bonds and *T. brucei* VSGs contain either one or two subdomains. All seven sequenced *T. congolense* VSGs have a similar structure to the one example shown, but do not contain the carboxy-terminal subdomain found in *T. brucei* VSGs. The single known *T. vivax* VSG also lacks the carboxy-terminal domain and is distinguished by the absence of N-glycosylation sites. Disulfide linkages have been experimentally determined for one *T. brucei* and one *T. congolense* VSG (indicated by the overlines) but cysteine residues are so highly conserved that they are likely to adopt the same pairing in all VSGs. Breaks in the mature sequences indicate the boundary between amino- and carboxy-terminal domains, based on crystal structures or by analogy. The amino- and carboxy-terminal signal sequences highlight the extraordinary conservation of VSG GPI signal sequences, where a break has been placed before the hydrophobic domain, to emphasize the conserved bipartite motif. The carboxy-terminal GPI attachment sites have been experimentally determined for MITat 1.2, 1.4 & 1.5 and are predicted for the other sequences, which are highly conserved in *T. brucei* but less so in *T. congolense*. The mature amino-termini were experimentally determined for all sequences except VSG ILTat1.25 and SRA, which were predicted using SignalP. † indicates probable N-glycosylation sites. These have been experimentally verified for MITat 1.2, 1.4 & 1.5. Single predicted N-glycosylation sites within disulfide-linked regions of carboxy-terminal subdomains of some VSGs are not marked as they are probably not glycosylated, because of accessibility constraints. SRA is the VSG-like molecule that can confer human infectivity on *T. brucei*.

structures. The same is probably true of the smaller carboxy-terminal subdomains, which tend to have more-conserved sequences and appear to be very tightly folded. Interestingly, the few sequenced *VSG*s of *T. congolense* and *T. vivax* are quite distinct from those of *T. brucei* (Figure 5.3). They are smaller and lack any clear carboxy-terminal domain, which is probably relevant to assessing evolutionary relationships among the various African trypanosomes.

VSG synthesis, processing, secretion and turnover

Trypanosomes contain components of the conventional eukaryotic signal recognition machinery, and VSG contains an apparently conventional amino-terminal signal sequence, although there is published evidence that signal sequences are not so exchangeable as one might imagine. VSG is presumably synthesized on ribosomes that are bound to the endoplasmic reticulum, and the nascent polypeptide is translocated into the lumen of the endoplasmic reticulum, with co-translational cleavage of the amino-terminal signal sequence. Cleavage of the carboxy-terminal GPI signal sequence and covalent coupling of the pre-assembled core GPI anchor also appears to be a coupled and immediately post-translational event, concomitant with or prior to N-glycosylation. It has been proposed that the secondary glycosylation of the VSG-linked GPI anchor and N-linked glycans represents a 'space-filling' function, determined by the accessibility of the primary glycans to further glycosylation by Golgi-resident glycosyl-transferases, as VSG dimers traverse the secretory pathway. In trypanosomes engineered to express two VSGs simultaneously, the patterns of N-glycosylation and GPI glycosylation are almost identical to those found when each VSG is expressed individually in its normal cell background, which confirms that the overall structures of the VSG homodimers determine the extent and nature of the secondary glycan modifications, not the cell background in which they are expressed.

VSG half-life is more than 30 h (about 5 cell generations), yet VSG is continually and rapidly recycled, in the absence or presence of specific antibody. It is not known whether a mechanism exists for more rapid removal of the 'old' VSG, during a VSG switch. The only recognized site of endo- and exocytosis in trypanosomes is the flagellar pocket, an invagination at the exit point of the flagellum, which accounts for a small part of the total surface membrane area but turns over at a high rate. Certain 'cell-surface' proteins appear to be specifically retained in the flagellar pocket by a mechanism that may depend on a unique glycosylation pattern recognized by hypothetical trans-membrane receptor lectins in the flagellar pocket. In the context of antigenic variation, the flagellar-pocket retention of the heterodimeric transferrin receptor is intriguing, as it is also GPI anchored, but only via one of the two subunits. The locations of the many other GPI-anchored and transmembrane proteins expressed by *T. brucei* are unknown.

In recent years, much attention has been focused on the sorting of membrane proteins, with the recognition that cellular membranes are not lakes of randomly distributed lipids but consist of lipid-specific subdomains. The concept of lipid rafts has particular relevance to GPI-anchored proteins, and suggests that the very specific myristylation of the VSG GPI anchor is probably an important determinant of VSG secretion and recycling. *T. brucei* appears to exert a high degree of quality control on VSG secretion. Attempts to use bloodstream-form *T. brucei* to express high levels of alien GPI-anchored proteins have failed. Even small modifications of VSG structure are deleterious to its efficient secretion. The trypanosome appears to monitor both the overall structure of VSG, perhaps through a specific chaperone molecule, and the presence of a GPI anchor. Both components are necessary for efficient exocytosis. Further studies of trypanosomes may illuminate the general roles of GPI anchors in protein sorting, secretion and endocytosis.

GENETICS OF ANTIGENIC VARIATION

Each VSG is encoded by a unique gene. Variation is usually not generated by contemporaneous recombination events between gene segments, or by fusion of constant and variable regions. Instead, the genome of *T. brucei* contains hundreds of competent *VSG* genes, together with a large number of pseudogenes that can nevertheless contribute, through recombination events, to the expressed repertoire. The repertoire is probably evolving and expanding continuously. Although some *VSG* genes are conserved between different trypanosome isolates, others are unique to particular clones, and the total potential VSG repertoire that exists in the field may be essentially infinite.

If there are hundreds of *VSG* genes, how does the trypanosome determine that only one VSG appears on the cell surface? Part of the answer is that the expressed *VSG* is always located (or relocated) close to a telomere, in a polycistronic transcription unit that we call an expression site (ES). This appears to be a necessary but insufficient condition for the singularity of VSG expression, because there are several ESs, with very similar structures, that can be turned on and off without discernable changes in sequence or organization. The question then becomes why is only one ES transcribed at any time? This system raises many of the same issues and also represents an interesting paradigm for the general question of how many kinds of cells regulate the expression of one among many similar alleles. In no case is this kind of allelic selectivity completely understood. In mammals there are many examples of allelic exclusion, but ES switching may be a unique example of reversible allelic exclusion, perhaps because try- panosomes are continuously dividing, whereas allelic exclusion is generally an irreversible commitment that occurs at a particular stage of development or cellular differentiation. Olfactory receptors, which are encoded by the largest family – 1000 genes – in mammals, are perhaps the most extensive example of allelic exclusion, with each olfactory neuron expressing a single receptor gene.

There are two well characterized classes of VSG expression sites, which differ dramatically in their organization and which operate in the two different infectious stages of the trypanosome life cycle. Both bloodstream and tsetse metacyclic ESs contain a *VSG* in the most telomere-proximal position, about 1–3 kb upstream of the hexanucleotide (TTAGGG) telomeric repeats. The ESs used in the bloodstream stage are transcribed from a promoter that is typically 30–50 kb upstream of the VSG, and 10–50 kb tracts of a conserved '50-bp repeat' are found upstream of ES promoters (Figure 5.4). Between the promoter and the VSG, ESs typically contain eight to eleven 'expression-site-associated genes' (*ESAG*s). Some *ESAG*s are present in multiple copies in some ESs. Some *ESAG*s appear to be truly unique to ESs, but others are members of larger families that are also found outside the ES (Table 5.1). ESs can also contain fragmentary remnants of *ESAG* sequences. In contrast, metacyclic ESs appear to be monocistronic: the promoter is immediately upstream from the telomeric *VSG*. All ESs

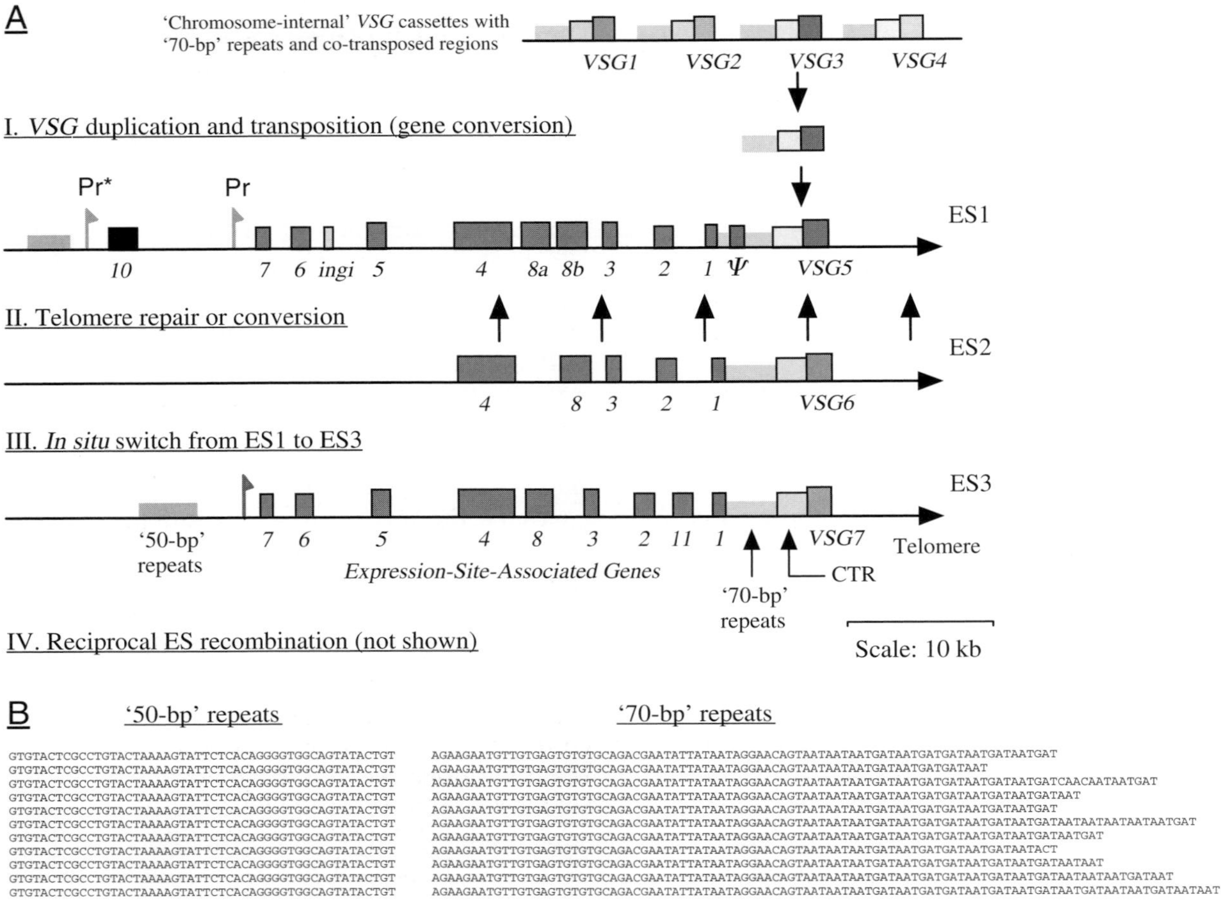

FIGURE 5.4 (See also Color Plate 3) (A) *VSG* expression site organization and alternative mechanisms for VSG switching. *ESAG*s are generally arranged in the order shown, but there can be duplications or triplications of some *ESAG*s and a duplicated promoter (Pr*) in some ESs. CTR is a region upstream of the VSG known as the co-transposed region: Ψ, a *VSG* pseudogene. (B) examples of the '50-bp' and '70-bp' repeat sequences found upstream of bloodstream-form ES promoters and *VSG*s, respectively.

are transcribed by an RNA polymerase that is resistant to α-amanitin, which is a characteristic of Pol I, the polymerase that normally transcribes rRNA. It seems likely that Pol I transcribes rRNA, VSG and procyclin genes in trypanosomes. To understand how this can be, and for further information about transcription in trypanosomes, the reader is referred to Chapter 3. Although both bloodstream and metacyclic ESs are transcribed by Pol I, the sequences of the two promoters differ, and they are probably differently regulated. What is similar is that each cell fully transcribes only one ES.

TABLE 5.1 Proteins encoded by expression-site-associated genes

ESAG	kDa[a]	Known or predicted properties of ESAG-encoded proteins[b]	#[c]
1	36	Membrane glycoprotein	92
2	50	GPI-anchored surface glycoprotein	50
3	43	Secreted glycoprotein	360
4	139	Flagellar membrane-associated receptor adenyl/guanyl cyclase	>500
5	46	Secreted glycoprotein	29
6	45	GPI-anchored transferrin receptor subunit	44
7	36	Non-GPI-anchored transferrin receptor partner of ESAG6	44
8	70	Predominantly nucleolar protein with RING zinc finger and leucine-rich-repeat motifs	8
9	34	GPI-anchored surface glycoprotein	17
10	76	Biopterin transporter	30
11	41	GPI-anchored surface glycoprotein	26

[a]The indicated sizes are those of the precursor polypeptides, before removal of signal sequences and addition of GPI and glycans.
[b]Predicted membrane glycoproteins have a candidate amino-terminal signal sequence and a potential carboxy-terminal GPI signal or membrane-spanning hydrophobic domain(s). The cellular location of proteins encoded by *ESAG*s 1, 2, 3, 5, 9 and 11 is unknown. ESAGs 3 and 5 have good candidate signal sequences but are otherwise predicted to be soluble glycoproteins, destined for secretion or for retention in the secretory pathway. Except for the heterodimeric transferrin receptor encoded by *ESAG6* and *ESAG7*, the functions of all of these predicted proteins are unknown. The adenyl cyclase activity of ESAG4 has been demonstrated. Although co-transcribed with *VSG*, most *ESAG*-encoded proteins appear to be present at less than 0.2% of the abundance of VSG.
[c]The last column (#) indicates the number of high-quality matches ($p \leq e^{-20}$) in a tBLASTn search of a *T. brucei* database (European Bioinformatics Institute) containing about 100 000 random fragments and complete genes, with a representative of each ESAG protein sequence. ESAGs 6 and 7 are themselves very similar. For ESAG8, the value given is for a search using only the first 100 amino acids, which distinguish ESAG8 from other leucine-rich-repeat proteins. Although the *p* cutoff value is somewhat arbitrary, and the search does not distinguish functional genes from pseudogenes, some trends are apparent. ESAG4 represents a large family of receptor adenyl cyclases that are not ES-restricted and the abundance of ESAG3 suggests that there are also many non-ES loci. There must also be non-ES-located copies of *ESAG10*, since this essential transporter is only found in some ESs, as are *ESAG9* and *ESAG11*. When a similar sequence database of *Leishmania* was searched, the only comparable matches were 8 with ESAG4 and 20 with ESAG10.

Mechanisms of VSG expression and switching

The expressed VSG can be changed by turning off one ES and turning on another one (*in situ* switching), by duplicative transposition (gene conversion) of silent telomeric or non-telomeric *VSG*s into the currently active ES (displacing the previously expressed *VSG*), by telomere exchange, or by telomere conversion/break-induced telomere repair (Figure 5.4A). There is very tight control of VSG expression. Even by putting different drug-selectable marker genes into different ESs, a trypanosome cannot be forced to simultaneously and stably co-express two VSGs. Regulation of singularity does not appear to be enforced at the level of VSG secretion and coat stability, since trypanosomes can be tricked into simultaneously expressing two VSGs by inserting a second *VSG* in tandem to the endogenous *VSG* in an active ES. There must be cross-talk, at some level, between ESs, but the mechanism eludes us. One possibility is that some product of an active ES is a trans-acting repressor of the other silent ESs, but this hypothesis has not yielded to extensive investigations focused on a potential regulatory protein encoded by *ESAG8*. ES *in situ* switching does not appear to involve any gene rearrangements or mobile regulatory elements. It could involve some limiting transcription factor, but experimental pressure to transcribe two ESs would probably have been able to overcome this kind of restriction. Another possibility is that there is a specific sub-nuclear location for ES transcription, which can only be occupied by one ES at a time.

VSG expression does not appear to be regulated at the level of transcription initiation, which appears to occur at both 'silent' and 'active' ESs. Several lines of evidence support this conclusion, which implies that processivity of the transcription complex must somehow be regulated. Probes for chromatin structure have failed to identify any differences that might influence processivity between silent and active ESs, but did demonstrate a major reduction in the accessibility of ES reporters after differentiation to procyclic forms, when transcription of all ESs is attenuated. In this context, it would be appropriate to mention the only identified difference between active and silent bloodstream-form ESs, and between bloodstream and procyclic trypanosomes. This is the novel DNA base J: β-D-glucosyl-hydroxymethyl uracil. J appears to be formed by a two-stage modification of thymidine residues: hydroxylation followed by glycosylation. The existence of a base modification was first indicated by differences between the DNA of active and silent ESs, in the efficiency with which certain sites could be cut by restriction enzymes. Subsequent work has shown that J replaces about 15% of the thymidine residues in telomeric repeats of bloodstream-form trypanosomes. J is found in other repetitive regions of the bloodstream-form genome, including the 50-bp repeat sequences that form the upstream boundary of the ES promoter region. However, J is absent from the transcribed region of the active ES, including the degenerate '70-bp' repeats that are generally found upstream of the *VSG*. J is not found in procyclic DNA, but J is found in other kinetoplastid protozoa, suggesting that it does not play an exclusive role in antigenic variation. Indeed, there is no evidence that it plays any role in antigenic variation. Nevertheless, it seems very likely that it would influence the binding of proteins that are expected, from precedents in yeast and mammals, to be specifically involved in telomere homeostasis.

What direct role might telomeres play in regulating antigenic variation? Apart from the presence of J, there is at least one other striking feature of *T. brucei* telomeres. Not only does the length of telomere repeats appear to be very tightly regulated in a *T. brucei* population, but telomeres grow at a steady rate: about 6 bp per generation for a silent ES telomere and 12 bp for the telomere of the active ES. This gradual increase in telomere length contrasts with the situation in *S. cerevisiae* and in mammals, which maintain their telomeres within a tightly regulated length range, and leads to the question of whether telomere growth regulates or influences ES transcription and switching in *T. brucei*. Transcription of the active ES appears to extend through the telomere and, perhaps as a consequence, these telomeres frequently break, whereas silent ones are much less likely to do so. A phenomenon known as 'telomere position effect' (TPE) in yeast, whereby a telomere-proximal reporter gene can be reversibly activated or silenced, and the corresponding state can be perpetuated through many cell divisions, is an interesting precedent for ES regulation. TPE is attributed to heritable differences in telomeric chromatin structure having an intrinsic suppressive effect on transcription, perhaps because telomeres are sequestered in particular regions of the nucleus (specifically at the nuclear envelope in *S. cerevisiae*). However, it is unclear what would allow just one telomere to escape a suppressive structure. TPE has been demonstrated in *T. brucei*, by inserting reporter cassettes into different genomic environments, but it is not clear how much of a role it has in regulating ES transcription from the telomere-distal ES promoter. TPE could play a role in suppressing transcription from telomere-proximal metacyclic ES promoters.

The use of trypanosome lines that have been highly adapted to laboratory use has received criticism from some investigators, on the basis that these lines have been selected for virulence and VSG stability, which may be incompatible with, and have led to loss or attenuation of mechanisms involved in, rapid VSG switching. On the other hand, most studies of VSG switching would not have been successful in rapidly switching lines. Because of this issue, suggestions about the relative roles of different switching mechanisms should be not be considered to be written in stone, at this stage of our knowledge. The existence of specific *VSG* recombination mechanisms is suggested by the extremely conserved sequences flanking the *VSG* conversion cassette, which form the boundaries of the gene conversion in most cases in which these have been determined. Again, there are exceptions to this statement, and one must always carefully examine the experimental conditions used in different studies before making direct comparisons of the outcome and interpretation.

When metacyclic trypanosomes enter the bloodstream, they continue to express metacyclic VSGs for several cycles of replication. Then the metacyclic ESs shut down and a bloodstream ES takes over. It has been suggested that the early bloodstream infection is a time of rapid VSG switching. Both logic and experimental evidence suggest that gene conversion is likely to be the major mechanism for VSG expression, because there are so many silent *VSG*s and only a few ESs. However, most mechanistic studies of *VSG* switching have so far focused on ES switching, probably because it is more accessible to experimental investigation. One interesting question is why multiple ESs exist? One possibility is that it is not only the VSG that needs to be exchanged on the trypanosome surface. Several *ESAG*-encoded proteins are predicted to be present on the cell surface (Table 5.1). Perhaps these proteins have specific functions that make them less tolerant of variation, or perhaps they are less exposed to the immune response, so the existence of only a few variants, encoded in different ESs, is sufficient for their survival and function. This may be fortunate because the existence of two large families of variant proteins, whose switching had to be coordinated by simultaneous recombination events, would pose a major problem for any cell. The function of only two *ESAG*s is known: the proteins encoded by *ESAG6* and *ESAG7* form a heterodimeric receptor for transferrin, whose uptake is essential for the survival of bloodstream trypanosomes. Several studies have shown that the transferrin receptors encoded by different ESs have different affinities for transferrins from different host species. It has been suggested that this variation in transferrin receptor affinity, perhaps coupled to other receptor interactions, could allow *T. brucei* to optimize its growth in different hosts, and this would determine which ES would be favored in any particular host. Although this is an interesting hypothesis, it remains a hypothesis. One interesting point about the transferrin receptor is that it is encoded by the most promoter-proximal *ESAG*s, and these are transcribed simultaneously at a low level from many 'silent' ES promoters. This became strikingly apparent when *ESAG6* was deleted from the active ES. That this was not lethal was attributed to the demonstrable transcription from 'silent' ESs. This trickle transcription of many transferrin receptors could also contribute to the parasite's flexibility in different hosts, but it could also be used as an argument against receptor–host specificity being the reason for the existence of multiple ESs. Furthermore, because many copies of *ESAG6* and *ESAG7* are being simultaneously transcribed, a large number of heterodimeric receptor combinations may be present, greatly extending the range of transferrin receptor properties.

Duplicative transposition of silent telomeric *VSG*s into the active ES seems likely to be the more quantitatively important mechanism for activating VSG genes, because telomeres appear to cluster together at several stages in the life cycle. The abundant minichromosomes of *T. brucei* appear to be a major reservoir of *VSG*s. The trypanosome may be forced into selecting less accessible (therefore activated at a lower frequency) 'chromosome-internal' silent *VSG*s at later stages of an infection. This is a time where mosaic *VSG*s have been described, arising from conversion of partial *VSG* sequences through recombination within the coding region of the active *VSG*.

Dynamics of antigenic variation: switching rates and modeling studies

Switching frequencies have been measured in different ways and, whichever methods were used, a wide range of frequencies has been reported. Switch frequency has been proposed to be a function of the trypanosome line, with more recently isolated or tsetse-transmitted lines having the highest switching rates, perhaps as high as one switch per 100 cells per generation. On the other hand, the switching rate of laboratory strains that routinely show very low switching rates ($\sim$1 switch per 10^6 cells per generation) can be dramatically increased by certain ES manipulations. There are some fairly obvious ES features that might influence switching rates, but few of these, including telomere length, have been experimentally manipulated. Early serological data suggested that the sequence of VSG expression, in different experiments, is neither totally predictable nor random. This semi-predictability may be due to the telomeric *VSG* repertoire being more accessible. Although that hypothesis has not been tested extensively and quantitatively, the trend is apparent in several recent studies.

Being largely controlled by the antibody arm of the immune system, antigenic variation is a seductive target for mathematical modelers, although the outcome of attempts to model the process have remained controversial. As one suspects is often the case, a model can be manipulated to mimic known features of a real infection, but the strength or weakness of any model ultimately depend upon knowing all the biological parameters that must be included, and on the quality of the real-world data that are selected to seed the models.

Evolution of VSG genes

T. brucei is diploid and contains 11 chromosome pairs, ranging from 1 to >5 Mb, which appear to contain the entire complement of essential genes, and about 100 'minichromosomes', ranging from 25–150 kb. The (haploid equivalent) informational content of the *T. brucei* genome, after discounting major repetitive sequences, is about 25 Mb, similar to *S. cerevisiae*. By the time this book reaches the reader, the genomic sequencing of *T. brucei* should be close to completion. Even now, sequences representing the entire gene complement are available in the public databases, in the form of 'genome survey sequences'. With only one exception reported so far, the genes of *T. brucei* do not contain introns, so genomic sequences are easy to interpret and utilize.

The number of ESs probably varies between trypanosome isolates. In one clone, the number was determined to be around 20, based on DNA hybridization with an ES promoter probe and the assumption that promoter duplications occur in about half the ESs. The number of metacyclic ESs has been estimated to be 20–30, based on a different criterion. The metacyclic form is non-dividing and represents a terminal differentiation stage in the tsetse salivary gland, where the VSG repertoire was enumerated using a library of monoclonal anti-VSG antibodies. At present, it seems likely that the total number of bloodstream and metacyclic ESs will vary from strain to strain, owing to the intrinsic instability of their telomeric location. The fitness and virulence of individual strains may depend partly upon the diversity of their ESs, either at the metacyclic stage or in the bloodstream, as discussed above in the context of the transferrin receptor.

VSG evolution almost certainly occurs by conventional processes of gene duplication and mutation, although this has not been specifically studied. The only way in which the utility of mutations can be evaluated is for the VSG to be expressed and exposed to the immune system. Because expression involves either ES switching, telomere exchange, or transposition of a 'chromosome-internal' *VSG* to the telomeric ES, which appears to be a unidirectional event, this raises the question of how the trypanosome retains a new *VSG* after 'testing' its value. One obvious way would be to use reverse transcription of the *VSG* mRNA, to recapture a copy of a useful new *VSG* into a safer chromosomal environment. It seems likely that trypanosomes did this in the past, even if they do not appear to do it today. The trypanosome genome contains hundreds of copies of apparently dead retrotransposable elements and probable reverse transcriptase pseudogenes. Many copies of these elements appear to be interspersed with *VSG* genes and pseudogenes, often in the long and essentially haploid regions that lie upstream of ESs, in the regions that appear to account for a great deal of the length variation between sister chromosomes that is a feature of *T. brucei*. Some reports have suggested that diversity could be generated concomitantly with *VSG* duplication and transposition to the ES, possibly by an error-prone repair or replication mechanism, or perhaps because mutations were introduced when replicating silent ES-located *VSG*s that were lightly modified by J. An alternative explanation of these controversial data, which has not been tested, is that these mutations were not generated concomitantly with the overt *VSG* expression event that was studied, but that this *VSG* had been expressed at subliminal levels during the infection and subjected to mutation and immune selection over a period of time. Simple calculations and computer modeling suggest that undetectable (by the usual techniques) low-level expression of many variants will occur during the course of a trypanosome infection, although each peak of parasitemia contains only a few overt variants.

IMMUNE RESPONSES TO TRYPANOSOME INFECTION

There are several pathological manifestations in trypanosomiasis, such as anemia and hypocomplementemia, whose detailed discussion is beyond the scope of this review, save to mention in passing that some of the pathology may be attributable to the escalating antibody responses to the ever-changing trypanosome antigens. There is a rapid and large increase in the size of the spleen, with the majority of the increase being caused by rise in the population of macrophages, B cells and plasma cells. The immune response to African trypanosomiasis inevitably consists of a series of primary IgM responses, leading to a huge elevation of IgM levels in the bloodstream, although IgG may be found later in infection, as VSGs with overlapping antigenic specificities are expressed. The vast rise in IgM has generally been attributed to a polyclonal B-cell response, rather than to specific responses to the plethora of trypanosome VSGs, but this issue has not been resolved. Perhaps the IgM response represents a combination of two effects: polyclonal activation and pan-VSG stimulation. It is difficult to resolve this question experimentally, because of the vast repertoire of trypanosome-specific VSG antigens involved, some of which may evoke responses that cross-react with the non-trypanosome antigens that fueled the idea that most of the IgM response was polyclonal and non-specific.

Many investigators have been able to demonstrate that animals can be protected against homologous challenge, using a range of antigen preparations, but there are no firm examples of trans-variant protection. Immunization with purified VSG is sufficient to confer immunity to challenge by trypanosomes expressing the same variant. *T. brucei* is notoriously resistant to complement-mediated lysis, for reasons that can be rationalized by our knowledge of the structure of the VSG coat, described above, which probably prevents insertion of the C9 complex into the plasma membrane. Anti-VSG antibody-dependent phagocytosis is probably the main mechanism for eliminating trypanosomes. Early studies suggested that T cells had little or no role in controlling trypanosome infections, but recent work has revisited the potential role of T-cell responses to VSG and demonstrated that VSG-specific Th1 cells are generated. Interferon-γ knockout mice have higher parasitemias, but still control them with VSG-specific antibody responses, suggesting that resistance is partly dependent upon a Th1 cell response to VSG, producing interferon-γ. It has been suggested that Th1 generation against non-surface-exposed epitopes on released VSG may provide a selective pressure for variation in 'buried' regions of the VSG sequence, although it is unclear what role such responses would have in trypanosome clearance. Recent studies suggest that *T. brucei* has an additional way of dealing with the emerging anti-VSG response. Binding of large amounts of anti-VSG antibodies *in vitro* agglutinates trypanosomes, although the clumps slowly disperse. When using sub-agglutinating concentrations, a situation which probably more accurately reflects the kinetics of antibody build-up *in vivo*, antibodies bound to VSG are rapidly internalized and degraded.

INNATE RESISTANCE, HUMAN SUSCEPTIBILITY AND SEX

This topic is included in a discussion of antigenic variation because it is important, it represents an innate mechanism that is independent of acquired immunity, and because human infection is linked to antigenic variation in two respects (see below). Trypanosomes that infect humans are classified as subspecies of *T. brucei*: *T. b. rhodesiense* and *T. b. gambiense*. These subspecies are generally distinguishable by their geographic separation, their symptoms, and their population structure, but some aspects of their separation and origin are somewhat controversial, and probably will remain so until the molecular basis of human infectivity is properly understood. The geographic separation closely follows the western edge of the African rift valley. The virulent East African *T. rhodesiense* causes acute disease that leads to death in a matter of weeks. The typical West African *T. gambiense* is less virulent, causing a chronic infection that can take a year or more to death. The issue of genetic exchange is raised here because it is relevant to the origins and population stability of human-infective lines. Genetic exchange can occur between lines of *T. brucei* under laboratory conditions and probably occurs infrequently in natural infections, where it is less easy to observe. Sex is not an obligatory stage in the trypanosome life cycle and its mechanism is unclear. Some form of 'mating' takes place in the tsetse, possibly in the epimastigote stage in the salivary gland, but no haploid stage has been identified. A still debatable but reasonable hypothesis is that *T. gambiense* and *T. rhodesiense* are human-infective clones of *T. brucei* that are substantially or completely genetically isolated. There is evidence that typical *T. gambiense* isolates are clonal and have a more limited and homogeneous *VSG* repertoire. Typing of *T. rhodesiense* from different epidemics, using various molecular markers, suggests that new and old outbreaks are caused by very similar clones.

Human resistance to infection by *T. b. brucei* has been attributed to two sub-fractions of serum high-density lipoprotein, operationally known as trypanosome (*T. b. brucei*) lytic factor (TLF). The presence of one unique component, haptoglobin-related protein (Hpr), distinguishes highly purified TLF from other HDL fractions, but the mechanism by which TLF causes lysis is unknown, beyond the fact that the difference between TLF-sensitive and resistant trypanosomes apparently depends on whether TLF is endocytosed (sensitive) or not (resistant). The genetics of Hpr itself are interesting and suggestive of ancient evolutionary selection for resistance to trypanosomiasis. Human primate ancestors may have evolved resistance to *T. b. brucei* when Hpr arose by triplication of the haptoglobin locus in Old World primates, subsequently reduced to a duplication in humans. Chimpanzees, which are susceptible to *T. b. brucei*, have a frameshift mutation in their *Hpr* gene. Intriguingly, the *Hpr* gene is amplified in a significant proportion of African Americans.

Although incompletely understood, we know more about the human infectivity of *T. rhodesiense* than of *T. gambiense*. Although it may not be the only gene that can confer human infectivity, one gene has been identified that certainly can. This so-called 'serum-resistance-associated' (*SRA*) gene encodes an intriguing and probably GPI-anchored cell-surface protein that seems to be a mosaic of a truncated VSG-like amino-terminal domain fused to a typical VSG carboxy-terminal domain (Figure 5.3). *SRA* is present in one ES, which explained previous conflicting results suggesting that human resistance was sometimes linked to VSG switching, but not necessarily so. Its presence in one ES, which is also atypical in its structure (it contains only *ESAG*s 5, 6 and 7), could also explain why serum resistance would be more readily lost than gained, and may be part of the explanation of why *T. rhodesiense* epidemics arise sporadically yet seem to be relatively clonal in nature. We do not know why and how SRA expression allows trypanosomes to evade the innate TLF-based resistance mechanism. Once we understand the structure, origin and action of the SRA protein, we may be able to better understand and even predict the chances that human-infective clones of *T. brucei* will arise from within the populations circulating in the tsetse. *T. gambiense* does not have a direct homolog of the SRA gene, and the mechanism that enables it to survive the presence of TLF is unknown.

ANTIGENIC VARIATION IN MALARIA

Having taken most of my allotted space in the discussion of trypanosomes, discussion of antigenic variation in malaria will be necessarily brief and comparative. When investigators talk of malaria in the singular, they are generally referring to studies of the most virulent human malaria pathogen, *Plasmodium falciparum*. One cannot study malaria, however, without bringing other species and other experimental models into the arena, including parasites that infect rodents, birds, and non-human primates, which all have their parts to play in unravelling the mysteries of human malaria. *P. falciparum* is the most lethal species for humans and the asexual erythrocytic stage can be cultivated *in vitro*, but is not easy to genetically manipulate. The second most important human-infective species is *P. vivax*, which cannot be cultured. Worldwide, outside of equatorial Africa, *P. vivax* is the most prevalent malaria parasite, causing relapsing but less frequently fatal infections. In endemic areas, humans are repeatedly infected with *P. falciparum*. Most fatalities occur in young children. Those who survive into adolescence acquire a degree of immunity that protects from severe disease but does not eradicate the parasites. This state of partial immunity persists only as long as the immune system is constantly boosted by exposure to infected mosquitoes.

Once established in the human host, malaria parasites multiply within the red blood cells. Different malaria species are specific for the species of host erythrocytes they will invade and multiply within. A *P. falciparum* merozoite infects a human erythrocyte, creating a so-called ring form, which grows to a trophozoite, then – close to the end of the 48-hour developmental cycle – becomes a schizont, dividing into eight or more daughter merozoites. The now grossly distorted erythrocyte bursts, releasing a new generation of merozoites, which die if they do not quickly invade naïve erythrocytes. It is the synchronicity of the erythrocytic multiplication cycle that is responsible for the periodic fever that characterizes malaria infections, at each cycle of schizont maturation.

The antigenic makeup of the merozoite surface is quite stable. Antibodies directed against the merozoite, during its transient extracellular existence, appear to be important in controlling the infection. Antigenic variation in malaria is mediated through parasite molecules that are inserted into the erythrocytic membrane. Why does the parasite modify the erythrocyte membrane, and signal its presence to the immune system? There are several reasons. By choosing to live within the terminally differentiated and relatively simple erythrocyte, the parasite is forced to insert proteins of its own into the erythrocyte membrane, to satisfy its more demanding nutritional requirements. We know that the parasite places one or more metabolite transporters into the erythrocyte membrane, and probably many more proteins that have not been recognized, making the infected erythrocyte unavoidably immunogenic. The spleen plays an important role in clearing infected erythocytes from the bloodstream. To reduce elimination by the spleen, the parasite inserts adhesins into the erythrocyte surface, resulting in sequestration of infected erythrocytes within small blood vessels. This is one mechanism for persistence, but cytoadherence to the walls of small blood vessels is a major cause of pathology, especially when the brain or placenta become involved. Although the parasite could have evolved intrinsic population limiting mechanisms, using some kind of quorum-sensing system to avoid killing its host before it can transmit its genes to a new host, a mutual tolerance treaty with the immune system could facilitate the survival of host and parasite. Differences in the severity of malaria infections can probably be attributed to genetic variation affecting both the virulence of the parasite and the susceptibility of the individual human host.

Major variant antigens

Antigenic variation in *P. falciparum* manifests itself in one well characterized family of cytoadherence proteins, and probably in other so far less characterized protein families, which are synthesized by the parasite and inserted into the surface membrane of the infected erythrocyte. There is great diversity of these major cytoadherence proteins within a single malaria clone, within an infected individual, and between different infections in the same or in different people. The best characterized protein responsible for antigenic variation in malaria is called PfEMP1, for *P. falciparum*-infected erythrocyte membrane protein 1, encoded by the *var* gene family, discovered in 1995. There are probably other variant EMPs, and corresponding gene families, that contribute to cytoadherence and antigenic variation. One additional recognized family of variant erythrocyte-surface proteins is encoded by the *rif* genes. Additional candidates will probably emerge via the *P. falciparum* genome project. Unlike VSGs, which are ~50-kDa and GPI-anchored, PfEMP1s are 200–350-kDa transmembrane proteins. They are localized on the erythrocyte surface to regions known as 'knobs', because of their physical appearance in the electron microscope. Additional proteins, on the cytoplasmic face of the membrane, are responsible for the characteristic 'knobby' appearance of trophozoite and schizont-infected erythrocytes. As with VSGs, some aspects of PfEMP1 structures are highly conserved, although there is great sequence variation in the extracellular domain, which contributes to their different host-cell receptor specificities. Three major factors delayed the identification of PfEMP1, relative to the trypanosome VSG: *P. falciparum* could not be cultivated in the laboratory until the mid-1970s, the abundance of PfEMP1 is very low, and the population of malaria-infected erythrocytes is antigenically heterogeneous, although less so in culture-propagated clones. By sequencing the DNA of subtelomeric domains, a large family of variant (*var*) genes was recently identified in *P. vivax*. Working from the sequence information, but still unable to culture the parasite, it was possible to design probes to detect the corresponding proteins on erythrocytes from infected patients, and to demonstrate the presence of corresponding circulating antibodies. There is no sequence similarity between the major variant erythrocyte surface proteins of *P. falciparum*, *P. vivax* and *P. knowlesi*.

Regulation of PfEMP1 expression

Individual malaria parasites contain about 150 *var* genes but, as with the trypanosome *VSGs*, the *var* gene repertoire of different isolates is very diverse, so the total circulating repertoire may be essentially infinite. In contrast to trypanosomes, malaria has an obligatory sexual stage in the mosquito phase of its life cycle, which increases the diversification of *var* genes. As with trypanosome telomeres, most of the telomeres of the 12 chromosomes of the haploid blood-stage *P. falciparum* have a conserved structure and harbor telomere-proximal *var* genes. The remaining *var* genes are dispersed around the genome, either singly or in small clusters. Although many, and perhaps all, *var* genes are transcribed by individual parasites in the first hours following erythrocyte invasion, all but one are subsequently silenced. Each maturing parasite transcribes only one *var* gene, whose choice is independent of chromosomal location. There is no evidence for gene conversion or any other rearrangements associated with *var* switching, and the regulatory mechanisms are unknown. On the other hand, telomere clustering has been shown to lead to higher rates of recombination among telomeric *var* genes than would be expected at chromosome-internal loci. Thus, even if not involved in regulating expression, telomeric loci can be a major contributor to the generation of *var* diversity. This is a topic that has yet to be explored in trypanosomes. As with the regulation of *VSG* expression, the study of *var* genes promises to make an interesting contribution to the general topic of epigenetic regulation.

MEDICAL APPLICATIONS

The prospects for vaccination against African trypanosomiasis appear slim, based on the role of antigenic variation and the discouraging results of early vaccine studies. There has been almost no follow-up of alternative vaccine possibilities in the last two decades, although there are concepts that have been around for a long time. One idea is that, although responses to VSG dominate the immune system, an immune response to invariant or less variant proteins could be effective if induced by strongly immunogenic antigen preparations prior to exposure to VSG. Another idea concerns the possibility of targeting vaccines to proteins that are confined to the flagellar pocket. Serum proteins must be able to enter this compartment, where the transferrin receptor resides. Finally, there have been suggestions that the potential for antigenic variation in *T. gambiense* might be more limited than in *T. brucei* or *T. rhodesiense*, so VSG-based immunoprotection might be a testable, if unlikely, option. On the negative side, however, is the ominous thought that trypanosomes have been around for a long time, and they may understand our immune system better than we do. However, some of these ideas merit further exploration, especially since *T. brucei* has many proteins with a predicted cell surface location, and the genome sequence will allow many more candidates to be identified. Some of these surface components could be investigated empirically, as potential vaccines.

In the early 1980s, there was widespread optimism that malaria vaccines would soon be developed. However, the potential for antigenic variation now seems as great for malaria parasites as for trypanosomes, which is discouraging for the development of vaccines against the bloodstream stage. There are additional issues that have to be considered, beyond the implications of the molecular mechanisms for antigenic variation, when assessing the possibilities for malaria vaccine development. Although any vaccine would be very unlikely to provide sterile immunity, a vaccine that would reduce the level of erythrocyte infection in early childhood but allow broader immunity to develop through the boosting effect of continuing natural challenge would be a major life-saver in malaria-endemic areas. Discussion of immunization possibilities against other stages of the malaria life cycle is beyond the scope of the present focus on antigenic variation. Perhaps what we need more than any vaccine is a range of cheap, safe and effective drugs for malaria and for trypanosomiasis. There is no shortage of novel targets for anti-parasite drugs and it is inconceivable that such drugs could not be readily identified. Unfortunately, there is no economic incentive and little effective public support, in the wealthy regions of the world, for the development of drugs to treat tropical parasitic diseases.

ACKNOWLEDGEMENTS

My work on antigenic variation is primarily supported by the National Institutes of Health. I would like to express my thanks to all of the anonymous investigators whose observations, along with my own, have been incorporated into this chapter. Some of them are identified by my suggestions for further reading.

FURTHER READING

General reviews on African trypanosomiasis

Pepin, J. and Meda, H.A. (2001). *Adv. Parasitol.* **49**, 71–132.
Seed, J.R. (2001). *Int. J. Parasitol.* **31**, 434–442.
Various authors (1999). In: Dumas, M., Bouteille, B. and Buguet, A. (eds). *Progress in Human African Trypanosomiasis, Sleeping Sickness,* France: Springer-Verlag.
Various authors (2001). In: *Trends Parasitol.* **17**, 1–49.

The nature of human-infective 'subspecies' of *T. brucei*

Biteau, N., Bringaud, F., Gibson, W., Truc, P. and Baltz, T. (2000). *Mol. Biochem. Parasitol.* **105**, 185–201.
Raper, J., Portela, M.P.M., Lugli, E., Frevert, U. and Tomlinson, S. (2001). *Curr. Opin. Microbiol.* **4**, 402–408.
van Xong, H., Vanhamme, L., Chamekh, M. *et al.* (1998). *Cell* **95**, 839–846.
Welburn, S.C., Fevre, E.M., Coleman, P.G., Odiit, M. and Maudlin, I. (2001). *Parasitol. Today* **17**, 19–24.

Genetic exchange and chromosome organization in trypanosomes

Gibson, W. and Stevens, J. (1999). *Adv. Parasitol.* **43**, 1–46.
Melville, S.E., Leech, V., Gerrard, C.S., Tait, A. and Blackwell, J.M. (1998). *Mol. Biochem. Parasitol.* **94**, 155–173.

VSG structure

Blum, M.L., Down, J.A., Gurnett, A.M., Carrington, M., Turner, M.J. and Wiley, D.C. (1993). *Nature* **362**, 603–609.

GPI anchor structure and cell biology

Bangs, J.D. (1998). *Curr. Opin. Microbiol.* **1**, 448–454.
Ferguson, M.A.J. (1999). *J. Cell Sci.* **112**, 2799–2809.
Ferguson, M.A.J. (2000). *Proc. Natl. Acad. Sci. USA* **97**, 10673–10675.
Muniz, M. and Riezman, H. (2000). *EMBO J.* **19**, 10–15.
Werbovetz, K.A. and Englund, P.T. (1997). *Mol. Biochem. Parasitol.* **85**, 1–7.

Antigenic variation dynamics and models

Frank, S.A. (1999). *Proc. R. Soc. Lond.* B **266**, 1397–1401.
Turner, C.M.R. (1999). *J. Cell Sci.* **112**, 3187–3192.

Antigenic variation at the molecular level: reviews

Barry, J.D. and McCulloch, R. (2001). *Adv. Parasitol.* **49**, 1–70.
Borst, P. and Ulbert, S. (2001). *Mol. Biochem. Parasitol.* **114**, 17–27.
Vanhamme, L., Lecordier, L. and Pays, E. (2001). *Int. J. Parasitol.* **31**, 522–530.

Antigenic variation at the molecular level: selected original studies

Chaves, I., Rudenko, G., Dirks-Mulder, A., Cross, M. and Borst, P. (1999). *EMBO J.* **18**, 4846–4855.
Horn, D. and Cross, G.A.M. (1995). *Cell* **83**, 555–561.
Horn, D. and Cross, G.A.M. (1997). *EMBO J.* **16**, 7422–7431.

Navarro, M., Cross, G.A.M. and Wirtz, E. (1999). *EMBO J.* **18**, 2265–2272.
Navarro, M. and Gull, K. (2001). *Nature* **414**, 759–763.
Rudenko, G., Blundell, P.A., Dirks-Mulder, A., Kieft, R. and Borst, P. (1995). *Cell* **83**, 547–553.
Vanhamme, L., Poelvoorde, P., Pays, A., Tebabi, P., Xong, H.V. and Pays, E. (2000). *Mol. Microbiol.* **36**, 328–340.

Transferrin and other receptors in *T. brucei*

Borst, P. and Fairlamb, A.H. (1998). *Annu. Rev. Microbiol.* **52**, 745–778.

Immunology and pathology of trypanosomes

Goodwin, L.G. (1970). *Trans. Roy. Soc. Trop. Med. Hyg.* **64**, 797–817.
Hertz, C.J. and Mansfield, J.M. (1999). *Cell. Immunol.* **192**, 24–32.
Losos, G.J. and Ikede, B.O. (1972). *Vet. Pathol.* **9** (suppl), 1–71.
Mansfield, J.M. (1981). In: Mansfield, J.M. (ed.). *Parasitic Diseases: The Immunology*, vol. 1, New York: Marcel Dekker, pp. 167–226.
Poltera, A.A. (1985). *Brit. Med. Bull.* **41**, 169–174.
Sternberg, J.M. (1998). *Chem. Immunol.* **70**, 186–199.

Antigenic variation in malaria

del Portillo, H.A., Fernandez-Becerra, C., Bowman, S. *et al.* (2001). *Nature* **410**, 839–842.
Freitas-Junior, L.H., Bottius, E., Pirrit, L.A. *et al.* (2000). *Nature* **407**, 1018–1022.
Scherf, A., Hernandez-Rivas, R., Buffet, P. *et al.* (1998). *EMBO J.* **17**, 5418–5426.
Scherf, A., Figueiredo, L.M. and Freitas-Junior, L.H. (2001). *Curr. Opin. Microbiol.* **4**, 409–414.

Malaria vaccine development

Duffy, P.E., Craig, A.G. and Baruch, D.I. (2001). *Trends Parasitol.* **17**, 354–356.
Good, M.F., Kaslow, D.C. and Miller, L.H. (1998). *Annu. Rev. Immunol.* **16**, 57–87.

C H A P T E R

6

Genetic and genomic approaches to the analysis of *Leishmania* virulence

Stephen M. Beverley
Department of Molecular Microbiology, Washington University Medical School, St. Louis, MO, USA

INTRODUCTION

Trypanosomatid protozoans of the genus *Leishmania* are important parasites of humans, infecting upwards of 12 million people in tropical and temperate regions of the world. These protozoans have an obligate digenetic life cycle, alternating between the flagellated promastigote form residing in the gut of the insect vector sand fly, and the intracellular amastigote stage residing within an active phagolysosome of vertebrate macrophages. How *Leishmania* carries out these developmental transitions, and the mechanisms that are employed in surviving within the host and resisting a tremendously hostile array of defenses, are key questions of interest to biologists and clinicians seeking to control these pathogens.

In the last 12 years a variety of genetic tools have been introduced that now permit manipulation of the genome of trypanosomatids, including *Leishmania*, with a high degree of specificity. These methods constitute a powerful genetic 'toolkit', allowing experimenters to take genes identified by various routes and probe their function by both gain and loss of function strategies, as well as localization using a variety of tags such as the green fluorescent protein (GFP). Complementing our ability to carry out reverse genetic manipulations has been the development of methods for 'forward' genetics and functional genetic rescue. Lastly, *Leishmania* has joined many other microbes in entering the era of genome science with a rapidly progressing genome project. When available, the complete genome sequence will

Molecular Medical Parasitology
ISBN 0-12-473346-8

provide tremendous new opportunities for gene discovery and analysis.

In this chapter I discuss genetic approaches available for the study of *Leishmania,* from the perspective of studying parasite virulence throughout the infectious cycle.

AN INTRODUCTION TO *LEISHMANIA* BIOLOGY AND VIRULENCE

Leishmania are transmitted by the bite of phlebotomine sand flies. Within the sand fly, parasites undergo a number of developmental transitions essential for survival and subsequent transmission by the time of the next bite. Initially parasites are taken up (most probably within infected macrophages) with the blood meal, which is contained within the sand fly midgut by a peritrophic matrix. There, parasites are released from the ingested macrophage and differentiate to the dividing procyclic promastigote stage. After a few days the peritrophic matrix breaks down, possibly accelerated by the secretion of chitinase from the parasite. At this point the procyclic parasites attach to the lumenal surface of the midgut, by binding of the abundant parasite surface glycolipid lipophosphoglycan (LPG). During this period of time parasites undergo extensive replication; however, as the blood meal is digested parasite growth ceases and the parasites differentiate into the highly infective metacyclic promastigote. Metacyclics undergo changes in the structure of the LPG that reduce its ability to bind to the midgut, thereby allowing release of the parasite and freeing it for migration elsewhere in the alimentary tract in preparation for transmission to the mammalian host.

Once metacyclic parasites are inoculated into the mammalian host, they must first resist the action of serum complement, and then bind and enter macrophages. There the parasite differentiates into the amastigote stage, which is adapted for life within the active phagolysosome. During the initial period while the parasite differentiates, it transiently undergoes a period in which phagolysosome maturation is inhibited, again primarily by the action of LPG. As the parasite matures to the amastigote form, LPG synthesis is shut down and phagolysosome maturation now proceeds. In some manner, the amastigotes are adapted to life in an acidified, fusogenic phagolysosome. Amastigotes also greatly affect host cell signaling pathways, in most cases downregulating or inactivating them. Which parasite molecule(s) and which host pathways are essential for survival are as yet poorly defined, although a number of candidates have been proposed.

The study of *Leishmania* biology is greatly aided by the availability of good *in vivo* and *in vitro* models relevant to host–parasite interactions. Inbred lines of laboratory mice have been intensively studied and show wide variation in susceptibility and pathology; some lines are good models for cutaneous disease, others are good models for visceral disease. This, in combination with the large collection of mouse knockouts now available, allows host genetics to be applied to the question of parasite virulence very productively, although this will not be discussed further here. Macrophages can be cultured *in vitro* and macrophage-like cell lines are available, facilitating the analysis of parasite survival in the relevant cell type. On the parasite side, excellent *in vitro* model systems are available for the study of differentiation. For the sand fly stages, *in vitro* cultured log-phase promastigotes closely resemble procyclic promastigotes, and upon entry into stationary phase they undergo differentiation into a form that closely resembles metacyclic promastigotes. For amastigotes, researchers have been able to develop conditions that enable the propagation of amastigote-like forms outside of macrophages, with relative ease and in large quantities. While perhaps not identical to authentic lesion amastigotes, most properties are faithfully reproduced, thereby making these a powerful tool for studies of amastigote biology and gene expression.

What is a *Leishmania* virulence gene? In most pathogens, investigators typically define these as ones important for survival and/or pathogenesis of the parasite within the sand fly or mammalian hosts, but not for growth in routine culture media. This definition is not ironclad and exceptions are known (one obvious example arises from the fact that there are dozens of 'routine' culture media). It also allows for the possibility that virulence genes are not 'absolute': that is, it includes genes whose loss shows a quantitative but not complete loss of virulence. In evolutionary terms, even relatively small effects in 'fitness' or virulence can have large consequences. Moreover, in the natural infectious cycle parasites often experience population bottlenecks, where even fractional decreases in virulence can have profound consequences. For example, sand flies usually transmit no more than 10–100 parasites.

The definition of virulence above is an excellent match for experimental studies of *Leishmania*, since the ability of the parasite to grow extracellularly *in vitro* makes it possible to engineer mutations in 'virulence', for subsequent assay(s) in the relevant host(s) and specifically as the intracellular amastigote stage. Obviously, this definition of virulence genes excludes 'essential' genes, defined as ones required for growth *in vitro*. However, 'essential' genes frequently can play roles in virulence beyond just enabling growth *in vitro*. Given the deadly nature of some *Leishmania* infections, and the desire of investigators to control parasite infections through chemotherapy or vaccination, both 'virulence' and 'essential' genes have great potential in efforts oriented towards the identification of new chemotherapeutic targets.

'FORWARD' VS. 'REVERSE' GENETIC APPROACHES

Genetic approaches are often divided into two types: 'forward' methods, where one begins with a mutant or variant phenotype and uses this to identify the gene(s) involved, or 'reverse' genetic approaches, where one begins with a parasite gene which is then manipulated in various ways to probe its role in the infectious cycle. In *Leishmania*, both have their advantages and are supported by the requisite genetic tools; the choice is often dictated by factors such as the availability of mutants, the inability to do sexual crosses, and insights or 'intuitions' about genes arising from various sources, etc. One advantage of 'forward' genetics is that it is based solely on the phenotype, making it likely that the genes obtained will be directly involved in the process under study, often with unanticipated roles or functions which would not have been detected by motif searches or database comparisons. While reverse genetic approaches lack this advantage, engineering null mutants is quite rapid and straightforward and thus candidate genes can rapidly be tested. The development of rapid methods for carrying out reverse genetic approaches across the genome gives them broad potential for screening for new virulence genes. Lastly, even with genes with homologs of known function in other organisms, the ability of these pathogens to utilize standard proteins/motifs in new and unique ways often allows such studies to rise above simple validation of previously known functions.

A major hurdle for forward genetic approaches in *Leishmania* is that, experimentally, these organisms are asexual diploids. Thus to generate loss of function mutations two genetic events are required, either two independent mutations or a mutation followed by a loss of heterozygosity event. The frequency of these events has been measured in *Leishmania* as $\sim 10^{-6}$ and 10^{-5} respectively, making the recovery of mutants defective in both alleles relatively rare and on the order of less than 10^{-6} even after mutagenesis. Thus, to identify mutants powerful screens or selections are required. This is especially a problem if one would like to study mutations affecting 'virulence', as the inability of 'avirulent' mutants to survive makes their recovery difficult for forward genetic selections at present (although the reverse selection towards virulence is quite powerful, as discussed below).

Sexual crossing has been used in many creatures to generate homozygotes through clever mating strategies, and genetic exchange in nature or the laboratory is extremely useful for positional cloning and classical genetic mapping. As mentioned above, experimental crosses amongst *Leishmania* strains or mutants have been unsuccessful, and there is uncertainty about the extent of genetic exchange in nature. Hybrid parasites have been observed in the field between strains or species; however, whether these represent *bona fide* sexual processes or evolutionarily 'sterile' fusions is as yet unresolved. Moreover, population genetic data suggest that in evolutionary terms, productive genetic exchange is relatively rare. Experimentally, low or non-existent levels of genetic exchange will prevent recombination from having the opportunity to dissociate changes arising through evolutionary mutations from those associated with the phenotype of interest. Clearly this will hinder and possibly eliminate the use of positional cloning strategies.

Were 'haploid' *Leishmania* available, or methods for systematically making large regions of the genome homozygous in some manner following mutagenesis, our ability to carry out many forward genetic screens or selections would increase greatly. Chromosome fragmentation approaches have been developed recently which yield parasites bearing small hemizygous regions, and this approach could be used to facilitate forward genetic analysis of these segments.

Alternatively, rather than focus on loss of function mutations one might concentrate on mutations that manifest when heterozygous, either by 'gain of function' or due to haploinsufficiency. However, the occurrence of either of these following random mutagenesis and for a given gene/phenotype is difficult to predict. A special case of dominant mutations involves the creation of gene fusions to convenient selectable markers or reporter genes. In *Leishmania* this can most conveniently be done by the creation of gene fusion libraries using transposon mutagenesis (see Forward Genetics by Transposon Mutagenesis).

EXAMPLES OF 'FORWARD' GENETIC APPROACHES TO *LEISHMANIA* VIRULENCE

Lipophosphoglycan (LPG) and surface glycoconjugates

The surface of *Leishmania* promastigotes is coated with a dense glycocalyx, a major portion of which is composed of the complex glycolipid LPG. The structure of LPG is highly conserved amongst *Leishmania* species, showing polymorphism in the composition of branching sugars attached to the disaccharide phosphate or 'phosphoglycan' repeating units which comprise the bulk of the molecule. LPG is anchored to the parasite surface by a hydrophobic glycosylphosphatidylinositol (GPI) anchor, and its outer end terminates with a capping sugar, typically a galactose residue that is capable of binding to appropriate lectins such as ricin agglutinin. Since LPG is effectively the only parasite molecule terminating in galactose, Turco and colleagues used a negative selection protocol with ricin agglutinin to isolate a variety of *L. donovani* mutants deficient in LPG biosynthesis (Chapter 10). Similar approaches have been taken with other lectins or monoclonal antibodies targeting LPG, in both *L. donovani* and *L. major*. At this point in time, more than 12 phenotypically distinguishable LPG mutants of *L. donovani* have been identified.

It should be noted that the lectin-based selection was ideally suited to the challenges posed by 'forward' *Leishmania* genetics, since it enables one to screen very large populations of mutated parasites, as mandated by the relatively low frequency of mutants. It is potentially applicable to any parasite surface molecule for which probes of suitable discrimination and specificity are available, such as GPI-anchored proteins.

The LPG-deficient mutants were the starting point for the development of methods and approaches enabling the first functional genetic rescue experiments to be performed in protozoan parasites. Initially, the R2D2 mutant was transfected with a wild-type *L. donovani* library, created in the cosmid shuttle vector cLHYG, and from this population a number of LPG+ transfectants were identified. Analysis of the cosmids present within these led to the identification of the LPG1 gene, encoding a putative Galf transferase responsible for synthesis of the LPG core. Subsequent studies have now extended this approach to the identification of more than ten genes involved in LPG biosynthesis, including genes implicated in the synthesis of the polymorphic LPG side chains.

The LPG mutants and genes have provided a resource for probing a number of questions in *Leishmania* biology, including dissection of the biosynthetic pathway of LPG and related glycoconjugates, and testing the role(s) of LPG in parasite virulence. These studies are summarized in two recent reviews (Beverley and Turco; Turco *et al.*).

The genetic rescue of avirulence induced by long-term culture *in vitro*

A powerful approach in studies of prokaryotic virulence has been the identification of avirulent mutants by mass systematic screening of randomly mutagenized populations, and identification of the relevant gene by functional rescue. Unfortunately, the low frequency of mutant recovery in diploid *Leishmania* makes the recovery of such mutants considerably more difficult. To achieve the same level of success achieved in prokaryotic pathogens by screening through thousands of mutants, *Leishmania* workers would have to screen through millions. As yet, high throughput screens for avirulent mutations in sand flies, macrophages and/or mice have not been developed, which render this infeasible. Development of such screens would clearly constitute a major advance, given the success now achievable with functional rescue methods.

While mutagenesis screens for avirulence are daunting tasks, there is one simple route for creating avirulent lines. Like many pathogens, *Leishmania* has a tendency to lose virulence during *in vitro* culture, at a rate that can be quite rapid in species such as *L. donovani* (unpublished data). Even in *L. major*, where virulence is relatively stable, many random clones that were derived from fully virulent lines by a single round of plating showed detectable losses of virulence in mouse infections. Neither the genes nor the genetic mechanisms involved in the rapid loss of virulence have been identified, although changes in abundant surface molecules such as LPG and gp63 have been variably noted by many workers.

In contrast to the difficulty of selecting *Leishmania* for loss of virulence, selections for reacquisition of virulence are readily performed, as the desired parasites are now able to survive and propagate in sand flies, macrophages or mice as appropriate. In unpublished work, our laboratory has been able to partially rescue the defect in mouse infectivity present in an avirulent clonal derivative of the Friedlin V1 strain of *L. major*, using genetic complementation. A number of loci were recovered and their mechanism(s) of action established; none affected LPG or gp63 levels. Our data suggest that the functional rescue was mediated by multi-copy suppression, rather than complementation of a defective A1 strain gene. Whether these loci are able to restore virulence in other culture-attenuated *Leishmania* strains is unknown. These data suggest that when available, attenuated *Leishmania* strains can be productive targets for functional genetic rescue approaches.

Drug resistance and its application to virulence

Selection for drug-resistant *Leishmania* has been widely used to develop parasite mutants showing a variety of genetic modifications, including gene amplification and various kinds of gene mutations and/or inactivations. Gene amplification appears to be especially common in *Leishmania*, whereas it is rare in trypanosomes. This may reflect differences in the requirement of *Leishmania* for the use of RNA polymerase II promoters or replication origins on episomal amplified DNAs.

In most instances, investigators have been able to track down the gene (or genes) responsible for drug resistance. Examples included amplified genes such as the bifunctional dihydrofolate reductase-thymidylate synthase, pteridine reductase, P-glycoproteins, ornithine decarboxylase, and IMP dehydrogenase. This has also been possible in the case of point mutations leading to loss or gain of function. Recently, selections of *Leishmania* transfectant libraries bearing multicopy episomal cosmids (each containing ~40 kb sections of the ~34 Mb genome) have been employed to identify genes able to mediate resistance when overexpressed or amplified. One advantage of this approach is that in theory each cosmid-borne gene is effectively 'pre-amplified', making their occurrence approximately 10^{-3} in the transfectant library. In contrast, the frequency of gene amplification is much less ($<10^{-6}$) and typically only occurs after several rounds of stepwise selection. The cosmid library transfectant selection approach has been applied to the recovery of genes mediating resistance to antifolates, toxic nucleosides, and drugs inhibiting sterol biosynthesis.

While drug resistance is obviously a useful tool for exploring metabolic pathways and potential parasite responses to current or prospective chemotherapy, is it relevant to the study of virulence *per se*? First, as is now clear for many pathogens, many metabolic pathways in *Leishmania* are often intimately involved in parasite virulence, beyond their known roles in viability. While the trypanothione reductase gene is essential, heterozygotes containing only a single copy grow normally *in vitro* but are attenuated in macrophage infections. As will become clear below, surprises can occur. One example is the pteridine reductase (PTR1) gene, which is essential *in vitro* in defined media, but *in vivo* appears to be dispensable, presumably because the mammalian host synthesizes reduced pteridines *de novo*. Remarkably, loss of PTR1 expression gave rise to increased pathogenesis in mouse infections, through an elevation of the rate at which parasites differentiate to the infective metacyclic form. Application of inhibitors of other pathways potentially involved in virulence may prove equally revealing.

A second reason is that aspects of the complex interplay amongst host and parasite can often be replicated (at least in part) by drug selections *in vitro*. *Leishmania* typically faces powerful chemical stresses following macrophage invasion, involving both reactive oxygen and nitrogen intermediates. These defenses can be replicated conveniently *in vitro* by agents such as hydrogen peroxide, primaquine, paraquat, xanthine/xanthine oxidase, and a variety of NO donors. We have used selections of cosmid transfectant libraries with these agents to identify several loci implicated in resistance to one or more of these agents. One could easily extend this concept to other 'chemical' defenses of the macrophage, such as low pH, toxic peptides (defensins), and proteases, all of which could be employed in selections *in vitro*. Lastly, during invasion, parasites undergo complex signaling responses that initiate and coordinate their defense mechanisms, often involving protein kinase cascades. Many of these parasite 'signaling' enzymes are known to be susceptible to inhibitors previously developed in other systems, and thus potentially amenable to identification through a drug selection approach.

Forward genetics by transposon mutagenesis

Transposon mutagenesis is an extremely powerful form of 'forward genetics', where mutations are generated by the introduction of transposable elements. Amongst its 'forward' genetic applications, transposon mutagenesis is often used for insertional inactivation, where the transposon itself can mark the relevant gene for subsequent cloning. Unfortunately, this powerful approach is poorly suited for applications in asexual diploid organisms such as *Leishmania*; since two events are required to inactivate a gene, the level of transposition needed to attain this at a given locus is such that the genome becomes quickly littered with irrelevant heterozygous transposition events.

A second common application of transposon mutagenesis is for the creation of gene fusions, for example, fusions to convenient selectable markers or reporter genes whose expression can be rapidly manipulated. While gene fusions can be generated by a variety of routes *in vitro* and *in vivo*, the extremely low frequency of non-homologous recombination in *Leishmania* and other trypanosomatids precludes the use of simple DNA transfection-based approaches *in vivo*. Transposition conveniently overcomes this limitation.

In *Leishmania*, expression of the *Drosophila mariner* transposase leads to mobilization of *mariner* elements, which can be engineered to yield relevant gene fusions that can be selected or screened for *in vivo*. The *mariner* transposase has been purified and shown to catalyze transposition *in vitro*, allowing the creation of transposon insertion libraries in cosmid shuttle vectors for subsequent analysis following transfection into *Leishmania*. Recently we have created a variety of useful modified *mariner* elements, bearing a variety of selectable markers and reporter genes that are suited for the selection of both transcriptional and translational gene fusions. *In vitro* transposition systems based upon other mobile elements such as bacterial Tn5 or yeast Ty1 have been developed, and could be similarly applied.

By one route or another, libraries of genes fused to a convenient reporter protein (such as β-galactosidase or GFP) are generated, and then scored for expression. In prokaryotes the GFP-based approach is termed differential fluorescence induction. Notably, while most gene fusion libraries emphasize regulation at the mRNA or 'transcriptional' level, one can design fusions that only work at the protein or 'translational' level. Such fusions are ideal for systematically studying expression broadly across the entire 'proteome'. Since bacteriologists began their elegant studies prior to the current 'omics' era, one could consider the application of 'translational' gene fusions to the study of protein expression as 'proteogenomics'. While characterization of gene fusions is not strictly a study of 'virulence' genes, a common expectation is that genes or proteins showing changes in expression as cells move through their infectious cycle are likely to play significant roles in that process.

REVERSE GENETICS AND *LEISHMANIA* VIRULENCE

By definition, reverse genetics starts with genes and works back towards mutants and hopefully, phenotypes. There are many criteria for picking genes to pursue: developmental regulation, abundance of encoded proteins, provocative sequence 'motifs' or phylogenetic relationships, cellular localization, or enzymatic activity have all been productively applied. Today, the fields of genomics and proteomics (and possibly other 'omics') are now providing an abundance of genes and/or proteins, many of whose functions are completely unknown and thus ripe for study.

The *Leishmania* genome project

The *Leishmania* genome comprises about 34 Mb and is estimated to encode about 10 000 genes. An international consortium of researchers is determining the genome sequence of a prototypic species, *Leishmania major*, and completion is anticipated in a few years time. As of August 2001, more than 5 Mb of finished and 25 Mb of unfinished sequence is available, allowing researchers to scan for genes of interest by a variety of approaches.

Tools for manipulating the *Leishmania* genome

Procedures required for the application of reverse genetic approaches are well developed in *Leishmania*. These include a variety of selectable markers, reporter genes and expression vectors; homologous gene replacement; inducible expression systems; transposon mutagenesis; artificial chromosomes; and chromosome fragmentation approaches. These are summarized in Table 6.1 and the reader is referred to the list of Further Reading for more information. Investigators can now readily probe gene function through effects arising from overexpression, mutation, and deletion.

One challenge remaining is the study of 'essential genes', whose loss by definition cannot be tolerated by the organism. The difficulty here is discerning when a gene is 'essential' and thus cannot be eliminated, from ordinary technical difficulties associated with unsuccessful experiments. Remarkably, attempts at knocking out 'essential' genes have yielded the planned replacements, but accompanied by expansion of the target gene number through changes in chromosomal number or ploidy. Since this is a rare event normally, the recovery of planned replacements retaining a wild-type gene through chromosome number increase has been used as a positive criterion for gene essentiality. Another possibility is the use of

TABLE 6.1 A summary of genetic tools available for use in *Leishmania*

Genetic tool	Examples
Reporter genes	*LACZ, GUS, CAT, GFP*
Transient transfection	CAT, LACZ, GUS
Stable transfection & markers	
Positive selection	*NEO, HYG, PHLEO, PAC, SAT, BSD, NAGT, DHFR-TS*
Negative selection	*TK, CD*
Expression vectors	
Circular episomes	pX series
Chromosomally integrated	pIR1SAT, pSSU-int
Regulatable (inducible)	*tet* repressor system
Homologous gene replacement	
Single copy genes	*DHFR-TS*
Gene arrays	*GP63, CYPB, αTUB*
Transposon mutagenesis	
in vivo	*mariner*
in vitro	*mariner*, Ty1
Loss of heterozygosity (LOH)	*DHFR-TS, HGPRT, APRT*
Chromosome fragmentation	*TR, PTR1*
Functional genetic rescue	*LPG* biosynthetic genes
Artificial chromosomes (LACs)	*DHFR-TS* based LAC
RNA interference (RNAi)	Not in *Leishmania* to date (many successes in trypanosomes)

regulatable systems, where genes are placed under the control of elements imported from other species such as the *E. coli* tetracycline or *lac* operon repressors. These will allow investigators the more powerful option of creating conditional mutants for the study of essential genes. Since *Leishmania* virulence assays in mice typically take months, it remains to be established whether the available regulatory systems will be up to usage over this extended time frame. Fortunately, and as noted earlier, many genes required for parasite virulence in flies or mammals are not essential in the more forgiving environment of the culture flask *in vitro*.

One approach that has proven extremely powerful in metazoans and trypanosomes has made use of the phenomenon of RNA interference (RNAi), where introduction or expression of short double-stranded RNAs leads to the rapid destruction of cognate mRNAs. This approach has a number of advantages: it is fast, requires very little sequence information, and is able to reduce expression from multiple gene copies (which is especially advantageous in asexual diploids). Thus far success with RNAi has not been reported in *Leishmania*, and there is one report showing that an RNAi mechanism was not involved in successful antisense inhibition of gene expression.

Examples of 'reverse genetic' tests of candidate *Leishmania* virulence genes

At this point in time the number of *Leishmania* genes that have been subjected to reverse genetic analysis is large and growing rapidly. A number of examples relevant to the study of virulence are included in Table 6.2, illustrating the power of this approach. Many of the studies yielded the expected phenotype: the amastigote-specific *L. donovani* A2 gene was important for macrophage survival, as were *L. major HSP100* and *L. mexicana* cysteine proteinases. There have been a number of surprises. For example, *Leishmania* gp63 is the most abundant promastigote surface protein and is encoded by a large gene family. However, targeted deletion of the gp63 gene cluster of *Leishmania major* yielded only minor phenotypic effects *in vitro* (affecting deposition of complement) but no effects in sand fly, macrophage or mouse infections. Similarly, deletion of the cluster encoding the *SHERP/HASP* genes, which encode abundant metacyclic proteins in *L. major*, had little effect, nor did loss of all GPI anchored proteins in *L. mexicana*.

TABLE 6.2 Examples of reverse genetic studies of candidate *Leishmania* virulence genes

Gene	Species	Approach	Attenuation in fly/mouse/macrophage infections
GP63	*L. major*	Knockout	Little if any
	L. amazonensis	Antisense	++
A2	*L. donovani*	Antisense	++++
Mannose biosynthesis			
PMI (phosphomannoisomerase)	*L. mexicana*	Knockout	++
GDPMP (GDP-mannose pyrophosphorylase)	*L. mexicana*	Knockout	++++
Cysteine proteinases (*CPA* + *CPB*)	*L. mexicana*	Knockout	++++
Protein GPI anchors (*GPI8*)	*L. mexicana*	Knockout	Little if any
LPG biosynthesis			
LPG1	*L. major*	Knockout	+++
	L. mexicana	Knockout	Little if any
LPG2	*L. major*	Knockout	++++
	L. mexicana	Knockout	Little if any
SHERP/HASP	*L. major*	Knockout	Little if any
		Overexpression	+++
HSP100	*L. major*	Knockout	++
TR (trypanothione reductase)	*L. donovani* *L. major*	Heterozygote (+/deletion)	++

One of the most remarkable findings is that putative virulence genes/molecules are not equally active in all *Leishmania* species. In *L. major*, deletion of *LPG1* (encoding a putative galactofuranosyltransferase required for the synthesis of the heptasaccharide core of LPG) yielded promastigotes that were specifically deficient in LPG biosynthesis. These parasites were unable to survive in sand flies or efficiently establish infections in mice or macrophages. Conversely, in *L. mexicana LPG1* deletions had little if any phenotype in mouse or macropahge infections. A similar discrepancy was reported between these two species with null mutants of *LPG2*, which are unable to synthesize all phosphoglycan repeating units due to loss of the Golgi GDP-mannose transporter. It is safe to state that most investigators expected the phenotype of LPG pathway mutations to be similar in all *Leishmania*. A similar contrast has been seen with gp63: some attenuation was observed in *L. amazonensis* but not *L. major*. While one might speculate that the differences observed reflect experimental shortcomings, these have been carefully considered and seem unlikely at present. Perhaps equally informative is the opinion of the host: just as for the parasite, mutations affecting the host immune response can have dramatically different impacts on parasite virulence in *L. major* and *L. amazonensis*.

These findings make the point that far from being monotypic, different *Leishmania* species make use of their repertoire of potential 'virulence' genes and molecules to different extents in their interactions with the host. Why this should be the case, and whether this reflects the use of convenient animal model systems, rather than the natural hosts (where the pathology can differ considerably from that of the laboratory mouse), remains to be determined. Given that these parasites differ considerably in many aspects of their biology and disease pathology, perhaps these findings should not have been a surprise.

'REVERSE GENETICS AT WARP SPEED': FUNCTIONAL GENOMICS, COMPUTATIONAL BIOLOGY AND BIOINFORMATICS

The coming availability of the entire complement of ~10 000 *Leishmania* genes poses a considerable challenge to traditional gene-by-gene reverse genetic approaches. The emerging field of 'functional genomics' is devoted to the development and application of methods which allow investigators to efficiently study many genes simultaneously. For example, changes in gene expression can be monitored through the application of DNA microarray technology across thousands of genes. Genomic resources suitable for DNA microarray approaches in *Leishmania* are now available and are being used to study developmental gene expression. Similarly, proteomic methods that allow investigators to view *Leishmania* protein expression are now beginning to be employed. Lastly, as more genes are characterized functionally in *Leishmania* and other organisms, investigators are increasingly able to bring to bear powerful computational approaches to identify candidate genes for more intensive analysis.

Functional genomics, proteomics and computational biology offer two promises for the future: the more obvious one perhaps is a way of prioritizing which genes should be the focus of more intensive studies. As these fields advance however, they promise to bring new directions that will be both distinct and complementary to the 'gene-by-gene' approach.

VALIDATION OF CANDIDATE VIRULENCE GENES: REQUIREMENTS AND CHALLENGES

Once the investigator has identified a candidate virulence gene, and modified its expression by one of the methods described above, appropriate tests to confirm its role in virulence are necessary. Stanley Falkow proposed a set of 'Molecular Koch's Postulates' that provide an excellent standard suitable for use in molecular genetic studies of *Leishmania* virulence. First, the candidate gene must be reasonably involved in processes thought to be essential to virulence; second, inactivation of the gene should lead to a loss of virulence, and third, restoration of gene function should lead to restoration of virulence. The first two are perhaps obvious to most investigators, however the importance of the last one cannot be underestimated. Like many pathogens, *Leishmania* is known to lose virulence spontaneously during culture *in vitro*, and many manipulations of the parasite are themselves mutagenic (including DNA transfection). Thus parasites may lose virulence through processes unrelated to those planned by the investigator, making fulfillment of the last criterion essential in establishing the role of a putative virulence gene. Many, but not all, of the studies summarized in Table 6.2 have employed these criteria.

Remarkably, restoration of gene function can be more challenging than one might expect, despite the availability of an extensive repertoire of *Leishmania* expression vectors (Table 6.1). Most of the available vectors lead to overexpression of the inserted gene, which in some cases is as detrimental as underexpression. This has been observed in several cases in *Leishmania*. Thus care must be taken in applying the last criterion to ensure that the expression of the restored gene is physiologically relevant. In some cases, it may be necessary to 'reintegrate' the gene back into the target locus in order to ensure proper temporal and quantitative expression.

FURTHER READING

Beverley, S.M. (1991). *Annu. Rev. Microbiol.* **45**, 417–444.

Beverley, S.M. and Turco, S.J. (1998). *Trends Microbiol.* **6**, 35–40.

Blackwell, J.M. (1996). *Parasitology* **112**, S67–S74.

Borst, P. and Ouellette, M. (1995). *Annu. Rev. Microbiol.* **49**, 427–460.

Clayton, C.E. (1999). *Parasitol. Today* **15**, 372–378.

Cruz, A.K., Titus, R. and Beverley, S.M. (1993). *Proc. Natl. Acad. Sci. USA* **90**, 1599–1603.

Cunningham, M.L., Titus, R.G., Turco, S.J. and Beverley, S.M. (2001). *Science* **292**, 285–287.

Gibson, W. and Stevens, J. (1999). *Adv. Parasitol.* **43**, 1–46.

Gueiros-Filho, F.J. and Beverley, S.M. (1997). *Science* **276**, 1716–1719.

Myler, P.J. and Stuart, K.D. (2000). *Curr. Opin. Microbiol.* **3**, 412–416.

Späth, G.F., Epstein, L., Leader, B. *et al.* (2000). *Proc. Natl. Acad. Sci. USA* **97**, 9258–9263.

Swindle, J. and Tait, A. (1996). In: Smith, D.F. and Parsons, M. (eds). *Molecular Biology of Parasitic Protozoa,* Oxford: IRL Press, pp. 6–34.

Turco, S.J. and Descoteaux, A. (1992). *Ann. Rev. Microbiol.* **46**, 65–94.

Turco, S.J., Späth, G.F. and Beverley, S.M. (2001). *Trends Parasitol.* **17**, 223–226.

SECTION

II

BIOCHEMISTRY AND CELL BIOLOGY

PROTOZOA

C H A P T E R

7

Energy metabolism

PART I: ANAEROBIC PROTOZOA

Miklós Müller
The Rockefeller University,
New York, USA

INTRODUCTION

Several groups of protists are remarkably different in their metabolic organization from most eukaryotic cells, including cells of multicellular animals. They lack an organelle with the typical energy conservation processes of mitochondria: they are 'amitochondriate'. Four groups of amitochondriate organisms contain important medical and veterinary parasites: the diplomonads, where *Giardia* species belong; the entamoebids, where *Entamoeba* species belong; the parabasalids, including *Trichomonas* and *Tritrichomonas* species; and the microsporidia with numerous species parasitic in humans and livestock. In this chapter a concise overview of amitochondriate core metabolism will be given using as examples four extensively studied species: the human parasites *Giardia intestinalis, Entamoeba histolytica* and *Trichomonas vaginalis,* and the cattle parasite *Tritrichomonas foetus*. These are extracellular parasites, which inhabit different hypoxic sites in the mammalian host. *G. intestinalis* (synonyms *G. lamblia* and *G. duodenalis*) lives in the duodenum, *E. histolytica* in the colon, and the two trichomonad species in the genitourinary tract of humans and cattle respectively. In view of their intracellular mode of life, microsporidia will not be considered. Recent work on these organisms, especially the recently completed genome of *Encephalithozoon cuniculi,* reveals far-reaching metabolic differences from the three other groups discussed here.

This chapter will focus on the biochemical aspects of the overall core and energy metabolism. Extensive biochemical and sequence information is available for a number of individual enzymes involved but these will be mentioned only if relevant to specific aspects of the overall metabolism. Phylogenetic relationships

Molecular Medical Parasitology
ISBN 0-12-473346-8

of individual enzymes will be illustrated by a few examples only.

THE TERMS 'AMITOCHONDRIATE' AND 'ANAEROBIC'

Neither of the terms 'amitochondriate' and 'anaerobic' is entirely correct to describe the organisms discussed here and many other eukaryotes without typical mitochondrial functions. Numerous objections have been raised against both terms, and therefore they are used in this chapter only as shorthand designations.

'Amitochondriate' is used in a physiological sense and applies to organisms which lack an electron-transport-associated energy metabolism and are dependent exclusively on substrate-level phosphorylations for ATP generation. Organelles (hydrogenosomes or mitosomes) that are assumed to share a common ancestor with extant mitochondria are present, however, in several or possibly all amitochondriate groups (Chapter 12). The term 'amitochondriate' does not apply to protists or multicellular organisms that dispense with oxidative phosphorylation in certain life-cycle stages (e.g. trypanosomatids and helminths) or during temporary anoxia (certain marine invertebrates).

'Anaerobic' is used in a biological sense and reflects two characteristics. It signifies that the organisms in question do not require free O_2 for survival and multiplication and are inhibited by atmospheric O_2 concentrations. They can tolerate lower O_2 concentrations however. It also signifies that their ATP-generating mechanisms do not include oxidative phosphorylation processes that depend on a mitochondrial type F_1F_0 ATPase fueled by a transmembrane proton gradient, a typical mitochondrial process. However, O_2, already at low levels (in the low micromolar range), elicits metabolic shifts toward more oxidized metabolic end-products and can result in an increased level of ATP generation by the same substrate-level phosphorylation reactions that function in the absence of O_2 (see below, 'Effects of Oxygen'). These phenomena are often cited as arguments for designating the organisms in question as 'microaerophiles'.

GENERAL FEATURES OF AMITOCHONDRIATE METABOLISM

Amitochondriate protists represent some of the most divergent types of core metabolism among eukaryotes. It is to be stressed, however, that these are still quite uniform when compared with the enormous metabolic diversity among prokaryotes. The overall map of core energy metabolism for amitochondriates is largely superimposable over that of any eukaryotic organism. Glycolysis is the main pathway of carbohydrate utilization, and most reactions involved in the formation of metabolic end-products are also known from other eukaryotes. The map shows however a major gap, indicating the absence of processes and constituents of mitochondrial energy conservation. There is no tricarboxylic acid cycle, no cytochrome-mediated electron transport, no cytochrome oxidase and no F_1F_0-ATPase, thus there is no electron transport-linked ATP generation.

Mitochondriate and amitochondriate organisms differ in a number of less conspicuous but important aspects. Amitochondriate organisms also show significant differences among each other, apparent at various levels:

- subcellular organization of core metabolism
- the nature of specific reactions and the enzymes catalyzing them

- evolutionary relationships of the structures and enzymes involved in metabolism.

Although differences at the third level, i.e. in evolutionary relationships, point to complex events in the past, they provide little insight into the physiology of the extant organisms. This chapter will therefore focus on the first two aspects.

Amitochondriate protists are essentially fermentative, i.e. they are incapable of oxidizing their energy substrates completely to carbon dioxide and water. The backbone of hexose utilization is a classical Embden–Meyerhof–Parnas glycolytic pathway with several extensions that lead to the known metabolic end-products. ATP is generated by substrate level phosphorylations without the contribution of mitochondrial-type electron transport-linked ATP production. In other words, ATP production does not depend on transmembrane proton gradients and does not require the classical cytosol/mitochondrion compartmentation separated by the inner mitochondrial membrane. Such fermentative metabolism generates only a few ATP molecules per molecule of hexose utilized; it is 'profligate'.

The main source of energy is carbohydrate, primarily glucose, its oligomers and polymers. *T. foetus* can also utilize fructose. Amino acids and lipids do not support energy metabolism, with the exception of arginine, which is metabolized by the arginine dihydrolase pathway (Chapter 8). This is a non-oxidative ATP-generating process, the contribution of which to the overall energy balance of the organisms is not clarified fully. Probably it provides only a limited part of the ATP requirements. The absence of a mitochondrial respiratory system does not mean that these organisms are unable to take up and reduce oxygen, but only that oxygen uptake is not mediated by cytochrome oxidase and that it is not linked to oxidative phosphorylation.

The main end-products of carbohydrate fermentation differ in the four species, but are primarily organic acids (acetate, succinate, and lactate), ethanol and CO_2. Alanine can also be a major end-product. The relative proportions of these depend strongly on external conditions (pCO_2, pO_2, and presence or absence of exogenous carbon sources). Usually at least two products are formed simultaneously, indicating that the fermentative pathways are branched and not linear, permitting regulation of carbon flow under various environmental conditions, as characteristic of many prokaryotes.

The amount of energy necessary for maintenance of the organism without growth, as determined in chemostat cultures of *T. vaginalis*, is about 50% of total energy production, a value much higher than observed in most microorganisms. Maintenance energy of other amitochondriates remains to be determined.

SUBCELLULAR ORGANIZATION OF AMITOCHONDRIATE ENERGY METABOLISM

While amitochondriate protists display a great biochemical and biological diversity, they can be assigned essentially to two types of metabolic organization. In Type I amitochondriate eukaryotes all processes of core energy metabolism occur in the main cytosolic compartment of the cell. Of the species included in this chapter *G. intestinalis* and *E. histolytica* belong to this metabolic type. In Type II organisms a double membrane-bounded organelle, the hydrogenosome, is present. Certain processes of extended glycolysis occur in this structure, the name of which is derived from the fact that it produces molecular hydrogen. The two trichomonads discussed are Type II amitochondriates (see below). The presence of one or the other type of metabolic organization does not imply a monophyletic origin of all Type I or all Type II organisms. The data strongly suggest that both organizational types arose independently in different evolutionary lineages several times.

STEPS OF AMITOCHONDRIATE CORE METABOLISM

As stressed above, the central core of amitochondriate energy metabolism is the classical Embden–Meyerhof–Parnas glycolytic pathway. The description of this process will be divided into four parts: (a) uptake of exogenous carbohydrate and mobilization of endogenous carbohydrate reserves (Figure 7.1); (b) conversion of hexoses into triosephosphates (Figure 7.1); (c) conversion of triosephosphates into phosphoenolpyruvate (Figure 7.1); and (d) further conversions of phosphoenolpyruvate (PEP) (Figures 7.2 and 7.3).

Uptake of exogenous carbohydrate and mobilization of endogenous carbohydrate reserves (Figure 7.1)

The organisms discussed all utilize glucose taken up by active transport [reaction b]. *Tr. foetus* stands alone in also being able to utilize fructose. They can also utilize to different extents maltose (a dimer of glucose) and glucose polymers that are hydrolyzed extracellularly. A cell-surface-bound extracellular α-glucosidase splits maltose to glucose in *T. vaginalis* [reaction a], in a process similar to the 'membrane digestion' described for intestinal cells and platyhelminths.

High levels of glycogen are the intracellular carbohydrate reserves in all four species. The mobilization of these reserves has not been studied in detail, but it probably occurs by phosphorolysis forming glucose-1-phosphate [reaction d], subsequently converted to glucose-6-phosphate by phosphoglucomutase [reaction e]. Glycogen phosphorylase has been detected in *E. histolytica* and *Tr. foetus*, and phosphoglucomutase in *E. histolytica*.

Conversion of hexoses into triosephosphates (Figure 7.1)

The entry of exogenous glucose into the glycolytic pathway is through phosphorylation by an ATP-linked kinase (glucokinase), forming glucose-6-phosphate [reaction 1]. Glucose-6-phosphate derived from the intracellular glycogen reserves enters the pathway at this level. Glucose-6-phosphate is converted to fructose-6-phosphate by glucosephosphate isomerase [reaction 2]. *Tr. foetus* also contains a fructokinase [reaction c] in agreement with the utilization of fructose by this species. These steps do not differ in their mechanism from the processes in other eukaryotes, save in the narrow substrate specificity and regulation of glucokinase.

Phosphorylation of fructose 6-phosphate to fructose-1,6-bisphosphate is catalyzed by PP_i-phosphofructokinase (PFK) in a reversible reaction using PP_i as phosphoryl donor [reaction 3]. This enzyme is present in diverse organisms, including mitochondriate ones, but is less ubiquitous than the broadly distributed ATP-PFK, which catalyses an irreversible process. The reversibility of PP_i-PFK indicates that this step is not regulated. Fructose bisphosphate is subsequently split into dihydroxyacetone phosphate and glyceraldehyde 3-phosphate by type II (metal dependent) fructose-1,6-bisphosphate aldolase in an aldol cleavage reaction [reaction 4].

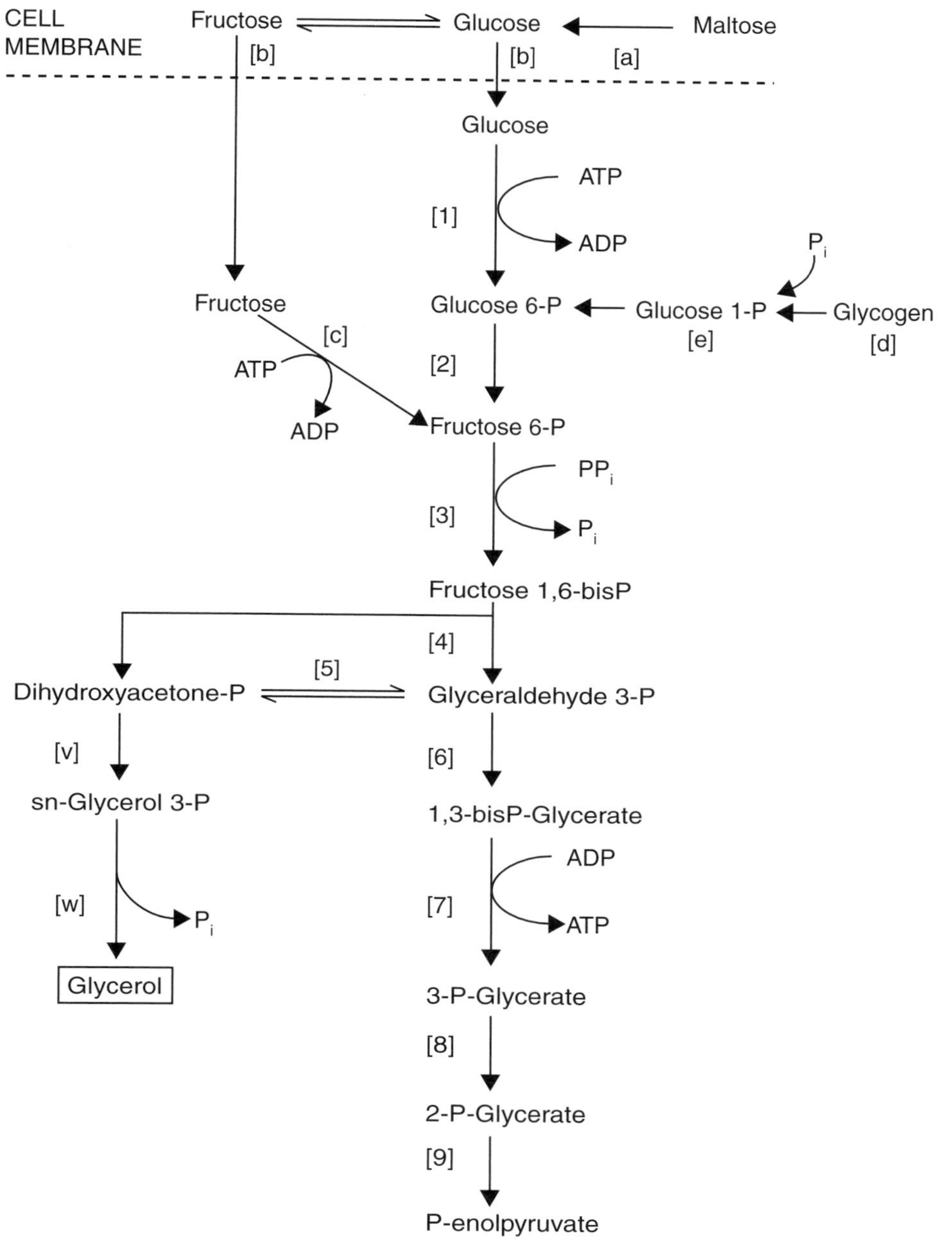

FIGURE 7.1 Map of the entry of hexoses into glycolysis and the process through the formation of phosphoenolpyruvate. Fate of reducing equivalents omitted. Metabolic end-products boxed. [a] α-Glucosidase, [b] hexose transporter, [c] fructokinase (*Tr. foetus* only), [d] glycogen phosphorylase, [e] phosphoglucomutase, [1] glucokinase, [2] glucosephosphate isomerase, [3] phosphofructokinase (PP_i-dependent), [4] fructose bisphosphate aldolase, [5] triosephosphate isomerase, [6] glyceraldehyde 3-phosphate dehydrogenase, [7] 3-glycerophosphate kinase, [8] 3-glycerophosphate mutase, [9] enolase, [v] glycerolphosphate dehydrogenase (*T. vaginalis* only), [w] glycerolphosphate phosphatase (*T. vaginalis* only). (Figures modified from illustrations in the first edition of this work.)

Conversion of triosephosphates into phosphoenolpyruvate

Dihydroxyacetone phosphate and glyceraldehyde 3-phosphate are kept in equilibrium through the action of triosephosphate isomerase [reaction 5]. Glyceraldehyde 3-phosphate dehydrogenase [reaction 6], 3-phosphoglycerate kinase [reaction 7], 3-phosphoglycerate mutase [reaction 8], and enolase [reaction 9] convert glyceraldehyde 3-phosphate to phosphoenolpyruvate (PEP). The triosephosphates represent the first branch-point of fermentation, since an alternative path leads from dihydroxyacetone phosphate to glycerol, a metabolic end-product arising through the successive actions of glycerolphosphate dehydrogenate [reaction v] and glycerolphosphate phosphatase [reaction w], a process detected in *T. vaginalis,* but not in *E. histolytica.*

Further conversions of phosphoenolpyruvate

PEP is the central, key intermediate of carbohydrate catabolism. PEP and its metabolites can be processed in different directions; thus several branch-points of metabolism are present in the pathways downstream of PEP. While Type I and Type II amitochondriates differ significantly in the processes leading from PEP to the dominant metabolic end-products and in their subcellular location, there are several common features as well (Figures 7.2 and 7.3). A main pathway leads to pyruvate by the action of pyruvate kinase [reaction 10] and/or pyruvate, orthophosphate dikinase (pyruvate dikinase) [reaction 11]. Interestingly, the presence of these enzymes is species dependent. *G. intestinalis* contains both activities, *E. histolytica* has only pyruvate dikinase, *T. vaginalis* carries pyruvate kinase, and so far neither enzyme has been detected in *Tr. foetus*. A large part of the pyruvate is subsequently processed to acetyl-CoA, as discussed below in the section on the enzymatic differences between amitochondriate and mitochondriate cells.

A putative alternative fate of PEP is carboxylation to oxalacetate [reaction 12 or 13] (Figures 7.2 and 7.3). Oxalacetate can be reduced by malate dehydrogenase [reaction 14] to malate, which in turn can be oxidatively decarboxylated to pyruvate by malate dehydrogenase (decarboxylating) or 'malic enzyme' [reaction 15]. In essence, this sequence of reactions represents a possible bypass to the pyruvate kinase/pyruvate dikinase reaction. Oxalacetate probably enters other pathways that remain to be identified. In *Tr. foetus,* cytosolic fumarate hydratase [reaction x) and fumarate reductase (reaction y) reduce oxalacetate to succinate (Figure 7.3), which is released as a metabolic end-product.

The most significant differences between Type I and Type II amitochondriates are in the metabolic steps beyond pyruvate (Figures 7.2 and 7.3). In Type I organisms all steps take place in the cytosol (Figure 7.2). Pyruvate can be transaminated to alanine [reaction 16] or oxidatively decarboxylated [reaction 17]. Acetyl-CoA formed in the latter reaction is converted either to acetate or ethanol. Acetate formation is catalyzed by a single enzyme, acetyl-CoA synthetase (ADP-forming) accompanied by substrate-level phosphorylation [reaction 18]. Ethanol is produced by the action of a fused NAD-specific aldehyde/alcohol dehydrogenase [reaction 19]. In eukaryotes, these two enzymes have been found so far only in Type I amitochondriates. Reducing equivalents produced in pyruvate oxidation are transferred to ferredoxin. The mechanism of the reoxidation of ferredoxin remains unknown, though it is likely that the reducing equivalents are used in ethanol production. Linking ferredoxin to ethanol

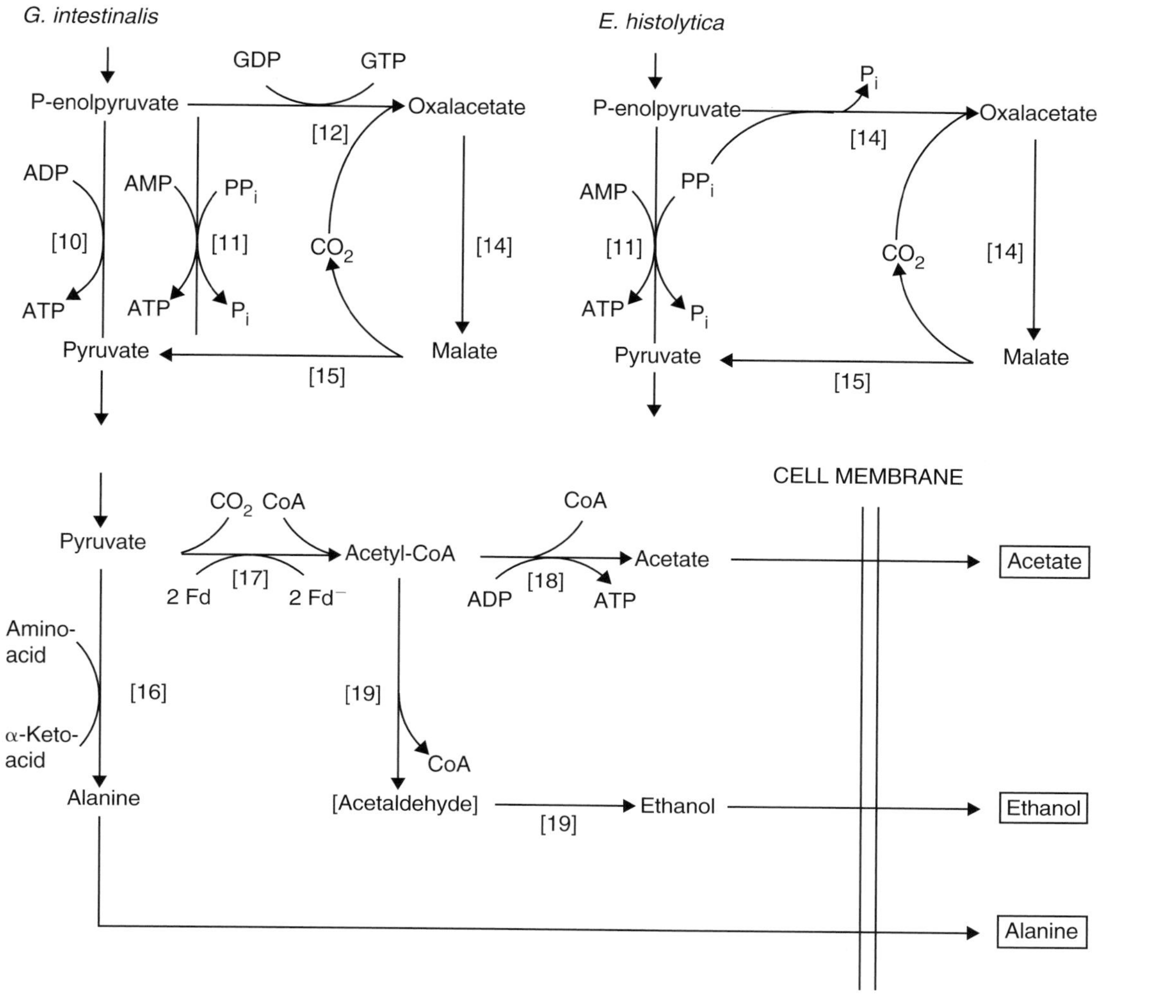

FIGURE 7.2 Map of the extensions of glycolysis beyond phosphoenolpyruvate in Type I amitochondriate parasites: *G. intestinalis* (left) and *E. histolytica* (right). Fate of reducing equivalents omitted. Metabolic end-products boxed. [10] pyruvate kinase, [11] pyruvate, orthophosphate dikinase, [12] phosphoenolpyruvate carboxykinase (GTP-dependent), [13] phosphoenolpyruvate carboxytransferase (PP$_i$-dependent), [14] malate dehydrogenase, [15] decarboxylating malate dehydrogenase (malic enzyme), [16] pyruvate aminotransferase, [17] pyruvate:ferredoxin oxidoreductase, [18] acetyl-CoA synthetase (ADP-forming), [19] aldehyde/alcohol dehydrogenase, [Fd] ferredoxin.

production presupposes the presence of a so far elusive ferredoxin:NAD oxidoreductase.

The two trichomonad species belong to Type II amitochondriates characterized by a hydrogenosomal/cytosolic compartmentalization of metabolism (Figure 7.3). Glycolysis through the formation of pyruvate (or malate) is cytosolic, as is the production of all metabolic end-products (lactate [reaction 25], ethanol [reactions 23 and 24], alanine [reaction 16] and

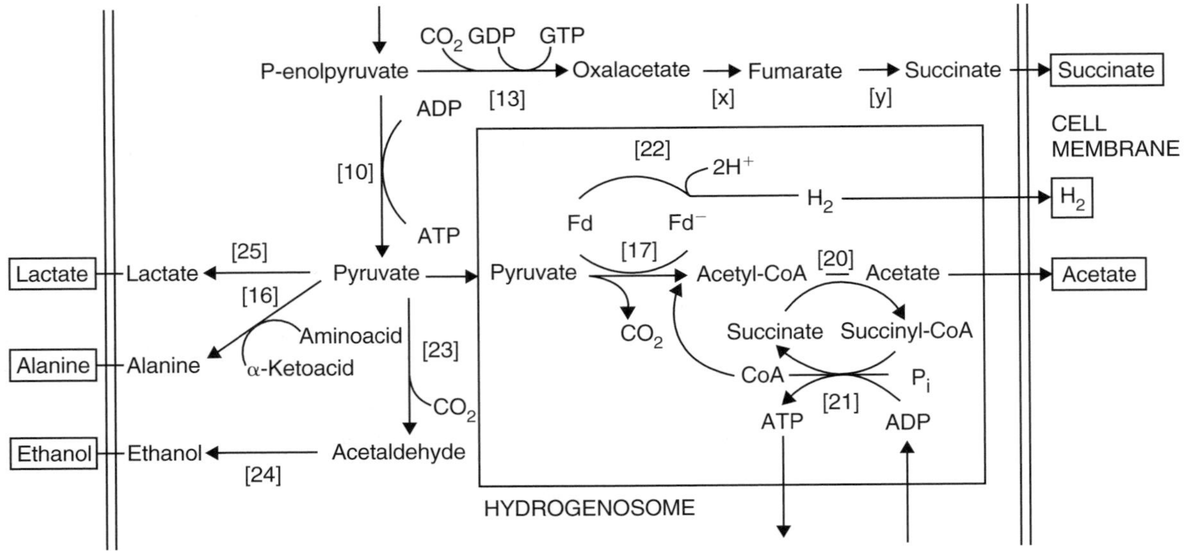

FIGURE 7.3 Map of the extensions of glycolysis beyond phosphoenolpyruvate in Type II amitochondriate parasites, *T. vaginalis* and *Tr. foetus*. Hydrogenosomal malate metabolism omitted. Solid line shows the hydrogenosomal membrane. Fate of reducing equivalents omitted. Metabolic end-products boxed. [10] pyruvate kinase, [13] phosphoenolpyruvate carboxykinase (GTP-dependent), [16] pyruvate aminotransferase, [17] pyruvate:ferredoxin oxidoreductase, [20] acetate/succinate CoA transferase, [21] succinyl-CoA synthetase, [22] hydrogenase, [23] pyruvate decarboxylase, [24] alcohol dehydrogenase, [25] lactate dehydrogenase (*T. vaginalis* only), [x] fumarate hydratase (*Tr. foetus* only), [y] fumarate reductase (*Tr. foetus* only), [Fd] ferredoxin.

succinate [reactions x and y], depending on the species) with the exception of acetate. Acetate is the only known major end-product of the hydrogenosomal metabolism. The intermediates entering the organelle are pyruvate (or malate that is converted to pyruvate by a hydrogenosomal malic enzyme, not shown). Pyruvate is oxidatively decarboxylated to acetyl-CoA by pyruvate:ferredoxin oxidoreductase [reaction 17]. The mechanism of conversion of acetyl-CoA to acetate, in contrast to the single step process seen in Type I amitochondriates, comprises two reactions. First the CoA moiety is transferred to succinate by acetate:succinate CoA transferase with the liberation of acetate [reaction 20]. In the second step, succinyl-CoA serves as a substrate for succinyl-CoA synthetase [reaction 21]. In this reaction succinate and free CoA-SH are produced, accompanied by substrate level phosphorylation of ADP to ATP.

The electrons generated by the oxidation of pyruvate are transferred to ferredoxin which, under anaerobic conditions, is reoxidized by a [Fe]hydrogenase with the production of H_2 [reaction 22]. This process gave the name 'hydrogenosome' to the organelle. If O_2 is present, H_2 evolution is inhibited and the hydrogenosome acts as a respiratory organelle. The identity of the terminal oxidase is not known.

Mechanisms of transport of substrates and products through the hydrogenosomal membrane have not been studied. ADP is imported and ATP is exported by the organelle by a transporter inhibited by atractyloside, an inhibitor of the mitochondrial ADP/ATP exchange transporter.

ENZYMATIC DIFFERENCES BETWEEN AMITOCHONDRIATE AND MITOCHONDRIATE CELLS

In addition to the absence of mitochondrial components, two characteristics are generally quoted as defining the core metabolism of amitochondriate protists. The first is the 'dominant role' of iron–sulfur center-mediated electron transfer, and the second is the privileged role of inorganic pyrophosphate (PP_i) instead of ATP in glycolysis. Since neither of these cases is as clear cut as generally held, they need to be considered in some detail.

The most significant divergence of amitochondriate from mitochondriate organisms is in the mechanism of oxidative decarboxylation of pyruvate to acetyl-CoA, a central step of core metabolism [reaction 17]. The enzyme responsible for this step in amitochondriates is the iron–sulfur protein, pyruvate:ferredoxin oxidoreductase (PFOR) and not the pyruvate dehydrogenase complex (PDH), found in mitochondria. The two enzymes are not homologous and differ in many properties (Table 7.1). The closest homologs of this enzyme are found in anaerobic and N_2-fixing eubacteria. In addition to a typical PFOR, *G. intestinalis* and *T. vaginalis* contain closely related enzymes acting on different α-keto acids (e.g. oxoglutarate) with as yet undefined roles in metabolism.

Small molecular mass iron–sulfur proteins, ferredoxins, serve as electron acceptors for PFOR. These have been isolated and characterized and belong to different ferredoxin types in the different species. In Type I organisms the main ferredoxin is of 2[4Fe4S] type, while in Type II organisms it is of the [2Fe2S] type. It is likely that additional minor types are also present, as indicated for *G. lamblia*.

TABLE 7.1 Properties of pyruvate:ferredoxin oxidoreductase (PFOR) and pyruvate dehydrogenase complex (PDH)

Property	PFOR	PDH
Molecular mass	240–280 kDa	$>10^6$ kDa
Subunit composition	homodimer	heteromer of at least three different subunits (E_1, E_2 and E_3)
Intramolecular electron transfer	[FeS] center	lipoamide
Electron acceptor	ferredoxin	NAD^+
Catalyzed reaction	reversible	irreversible

The PFOR-ferredoxin system occupies a central position in core metabolism. In Type II organisms it is linked to a further FeS protein, hydrogenase. While these iron–sulfur proteins indeed play a central role in the metabolism of amitochondriates, they are restricted to a single, but important, pathway. We deal only with a single characteristic here, and not with an all-encompassing dichotomy between amitochondriate and mitochondriate organisms.

PP_i indeed replaces ATP in some glycolytic reactions of amitochondriate protists. The assumed physiological significance of this substitution is a decreased ATP input into the process, corresponding to an increased overall ATP yield. Clearly increased energy generation could be important for amitochondriates, which depend on glycolysis as the main source of ATP. Several PP_i-linked enzymes have been detected in amitochondriates: PP_i-linked phosphofructokinase (PP_i-PFK, or more correctly, PPi-glucose-6-phosphate phosphotransferase) [reaction 3], pyruvate, orthophosphate dikinase [reaction 11], phosphoenolpyruvate carboxytransferase [reaction 13] and PP_i-acetate kinase [not in the Figures]. The complement of PP_i-linked enzymes differs in the four species studied. Only PP_i-PFK is present in all four. Pyruvate dikinase is known from the two Type I organisms. *E. histolytica* is the only eukaryote known to contain phosphoenolpyruvate carboxytransferase and PP_i-acetate kinase. Thus the statement that PP_i-linked reactions are a major characteristic of amitochondriate parasites is essentially incorrect.

Nor is the presence of PP_i-linked enzymes a character that distinguishes amitochondriate and mitochondriate eukaryotes. PP_i-PFK is present in plants and in mitochondriate protists (*Naegleria, Toxoplasma, Eimeria,* etc.), while pyruvate dikinase is present in many plants. At the same time, ATP- and PP_i-linked enzymes catalyzing the same overall reaction coexist in Type I organisms. *G. intestinalis* contains both pyruvate kinase and pyruvate dikinase, and *E. histolytica* harbors ATP-phosphofructokinase in addition to the PP_i-specific one.

EFFECTS OF OXYGEN

Although mitochondrial-type, cytochrome-mediated electron transport and a cytochrome oxidase as terminal oxidase are absent, amitochondriate protists avidly take up O_2 if it is present. In fact, much of the early biochemical exploration of these organisms was performed with respirometry. The rate of O_2 uptake is of the same order as the overall carbon flow measured under anaerobic conditions. Metabolism remains fermentative leading to the same end-products, the ratio of which, however, shifts in favor of the more oxidized compounds (acetate, succinate) and away from more reduced ones (ethanol, alanine).

Respiration exhibits a high O_2 affinity. *G. intestinalis* and *Tr. foetus* have K_m values for O_2 in the low μM range, similar to those observed for cytochrome oxidase-mediated respiration. Respiration is insensitive to most inhibitors and uncouplers of mitochondrial metabolism. In fact, the insensitivity of respiration to cyanide, indicating the absence of cytochrome oxidase, was noted in amitochondriates before mitochondria were identified as respiratory organelles in other eukaryotes.

Increased respiration in the presence of exogenous carbohydrates shows that most of the reducing equivalents are derived from carbohydrate catabolism. However, the link between extended glycolysis and O_2 uptake remains conjectural. Cytochromes are absent and only low levels of quinones have been detected. Cytosolic NADH and NADPH oxidases are present and probably contribute significantly to O_2 uptake. These enzymes are similar to H_2O-forming pyridine nucleotide oxidases found in cytochrome-deficient Gram-positive bacteria.

Isolated trichomonad hydrogenosomes act as respiratory organelles when pyruvate and ADP are added to the medium, a process probably fueled by auto-oxidation of the iron–sulfur centers in the components of the electron pathway from pyruvate to H_2.

REGULATION OF ENZYME LEVELS

Only few studies have been devoted to the regulation of enzyme levels and their experimental modification at the level of transcription and translation in *T. vaginalis*. Chemostat cultures in which growth rate and the level of an exogenous carbohydrate source can be controlled showed a complex pattern of changes in the activities of glycolytic enzymes. Iron levels in the medium correlate positively with levels in the hydrogenosomal proteins in this species. Interestingly, proteins without iron–sulfur centers were also affected, indicating that this effect is not due simply to the availability of iron for incorporation into iron–sulfur proteins.

Genetic transformation with dsRNA *Giardia* virus as vector has been used to affect the levels of PFOR and alcohol dehydrogenases in *G. intestinalis*. The results confirmed the role of PFOR in metronidazole activation, and showed that of the several alcohol dehydrogenases present in this organism only the fused aldehyde/alcohol dehydrogenase plays a significant role.

ACTION OF NITROIMIDAZOLE DRUGS

The most widely used selective drugs against amitochondriate parasitic protists are nitroimidazole derivatives (Chapter 17). Metronidazole (1-hydroxyethyl, 2-methyl, 5-nitroimidazole) was the first member of this group to be introduced and remains a much used compound. The $-NO_2$ group has to be reduced to convert these compounds into cytotoxic agents. The reduction is driven by low redox-potential electrons generated from pyruvate oxidation by pyruvate:ferredoxin oxidoreductase and passed on to ferredoxin. Other redox enzymes have been proposed as additional electron sources. Ferredoxin is generally assumed to be the proximal electron donor in the reduction of 5-nitroimidazoles. Inhibition of H_2 formation in *T. vaginalis* by nitroimidazole shows that the drug competes with hydrogenase for the electrons. The products of the reduction are short-lived cytotoxic products, and their exact identity remains uncertain despite extensive research. The one-electron product, nitro free radical, and the two-electron product, a hydroxylamine derivative, have been proposed as active compounds.

If oxygen is present, the level of metronidazole toxicity decreases. It is unclear whether this effect is due to diversion of the electrons to oxygen as an alternative acceptor or to the reoxidation of the one-electron free radical. The reoxidation of the free radical leads to the generation of reactive oxygen species (e.g. superoxide). The contribution of such secondary toxic species to the cytotoxicity of the drugs is not known. It is significant that 5-imidazoles exert their anti-parasitic activity also under strictly anaerobic conditions.

Relative 5-nitroimidazole resistance has been observed for the three human pathogens. Resistance can also be developed in cultured organisms exposed to increasing concentrations of the drugs. The levels of resistance seen in clinical cases and those obtainable under laboratory conditions are species-dependent. Moderate resistance can be easily obtained in *T. vaginalis*, less easily in *G. intestinalis* and only with great difficulty in *E. histolytica*.

The mechanism of resistance can often be attributed to decreased levels of components of the system reducing the drug. In low, clinically relevant, levels of resistance in *T. vaginalis*, the reducing system is unchanged, but the organisms are less oxygen-tolerant, indicating that modifications to the oxygen detoxifying system affect the toxicity. In laboratory strains of *G. intestinalis* with decreased drug susceptibility, decreases in the reducing system are accompanied by increased levels of enzymes involved in oxygen detoxification.

Long-term (over a year) cultivation in the presence of increasing metronidazole levels of both trichomonad species can achieve extremely high tolerance to the drug. During the process the organisms gradually lose their PFOR and hydrogenase activities and finally are unable to reduce metronidazole. Essentially these lines become hydrogenosome deficient, though the organelle itself and some of its enzymes are not lost. The lack of hydrogenosomal pyruvate metabolism is compensated by increased lactate production in *T. vaginalis* and ethanol production in *Tr. foetus*.

EVOLUTIONARY CONSIDERATIONS

This chapter deals primarily with the biochemical and functional aspects of the amitochondriate metabolic machinery. The core of this metabolism is a typical Embden–Meyerhof–Parnas glycolytic pathway, as in other eukaryotes, and none of the vast diversity of carbohydrate catabolism seen in various prokaryotes is present. The conservativism of the overall pathway indicates that it has been acquired by the common ancestor of all eukaryotes.

The evolutionary relationships of the enzymes participating in extended glycolysis in amitochondriates, however, present a complex picture. Several enzymes are of a different type from those in multicellular eukaryotes, a point alluded to above in the discussions of amitochondriate core metabolism and the differences between amitochondriate and mitochondriate cells. In some cases isofunctional enzymes of different amitochondriates belong to different types. Each of the four organisms discussed displays a different combination of enzymes with different origins. A description of these is beyond the scope of this chapter, but three examples will make the point.

The first example is glucokinase, responsible for hexose phosphorylation [reaction 1]. This activity in *G. intestinalis* and in *T. vaginalis* is catalyzed by a typical glucokinase with orthologs only in cyanobacteria and proteobacteria. The glucokinase of *E. histolytica*, in contrast, is closely related to hexokinases found only in eukaryotes. These two enzyme families derive from the same ancestral molecule but are so divergent that only a few catalytically important domains can be aligned to each other.

A second example is phosphofructokinase [reaction 3], an enzyme comprising several easily distinguished protein families and subfamilies, which probably separated early in evolution. Of the two functional types (ATP- and PP_i-linked), PP_i-linked enzymes are characteristic of the amitochondriate protists, but in different species they belong to distantly related subfamilies (clades). In addition, *E. histolytica* also contains an ATP-PFK of unknown functional significance.

The third example is glyceraldehyde 3-phosphate dehydrogenase [reaction 6]. The enzymes of *G. intestinalis* and *E. histolytica* are closely related to the glycolytic glyceraldehye 3-phosphate dehydrogenase of most eukaryotes, while the enzyme from *T. vaginalis* and other trichomonads forms a cluster separated from other eukaryotes that is nested among prokaryotic lineages.

This diversity in the nature of enzymes catalyzing formally identical reactions is difficult to interpret, but it indicates that each step of the pathway can tolerate orthologous or non-orthologous replacements of the corresponding enzyme. The ecological significance and the evolutionary processes behind such replacements are yet to be elucidated. Two major types of mechanisms can be invoked: selective retention of one or another enzyme that were present together in a common ancestor, and horizontal gene transfer from a donor to a recipient organism. Different evolutionary relationships for individual enzymes are a warning against assuming that the history of any single gene could unambiguously reflect the history of the organisms or the history of their complete metabolic machinery.

ENVOI

The description of core metabolism of a few amitochondriate parasitic protists as presented above clearly demonstrates that these organisms are highly divergent from most eukaryotes with typical mitochondria, including their mammalian hosts. This divergence is apparent not only in the metabolic consequences of the absence of mitochondria, but also in the presence of processes and enzymes found only exceptionally in other eukaryotes. In terms of applied parasitology, the most significant difference is the central role of pyruvate:ferredoxin oxidoreductase, the enzyme responsible for the high and selective activity of 5-nitroimidazole derivatives on amitochondriate parasites. It remains to be seen whether the core metabolism of these organisms will provide additional leads to new drugs (Chapter 17).

This necessarily concise presentation might give the false impression that we have a complete and coherent understanding of amitochondriate core carbohydrate metabolism. In fact, we have nothing more than a skeleton image that accounts for the bulk of carbon flow and the formation of the major metabolic endproducts. Many enzymes present have not been characterized in detail. In the ongoing genome projects new open reading frames are being recognized that code for homologs of metabolic enzymes of other organisms, but their putative functions cannot be accommodated within our current understanding. Another significant shortcoming is that published metabolic carbon and redox balance data for all amitochondriate organisms are incomplete. While this circumstance is easily explained by technical difficulties presented by organisms that cannot be grown on simple defined media, it hampers the development of a comprehensive metabolic map and makes forays into explorations of metabolic regulation difficult.

Almost all the evidence presented here has been collected from *in vitro* studies of organisms maintained in bacteria-free (axenic) cultures. Such cultures, however, do not reflect closely the conditions in their natural habitats. In addition it cannot be excluded that prolonged cultivation under artificial conditions can provoke changes in metabolism. While we have rather clear ideas of certain metabolic characteristics of the organisms studied, there is always the caveat that we might have studied only a restricted and possibly modified subset of their metabolic capabilities.

The absence of mitochondria is often equated with an 'ancestral' status of these organisms. While this issue is much debated, increasing evidence indicates that a mitochondrion-related organelle was present in the ancestors of amitochondriate eukaryotes and that the lack of mitochondria is a derived character, due to regressive evolution by functional losses (Chapter 12). The 'simplification' of the metabolic machinery is probably not a result of the establishment of a parasitic lifestyle by the ancestors of amitochondriate organisms, since similar metabolic characters are present in their free-living relatives that live in anaerobic or hypoxic sediments of diverse water bodies. This simplification probably was advantageous in the transition to parasitism in the metazoan gut, where hypoxic conditions prevail. These events are, however, shrouded in a veil of a distant past.

The data also disclose the existence of characters that are difficult to relate to functional losses, and indicate that these have been acquired by horizontal gene transfers from other, probably prokaryotic, organisms. The picture is rather confusing at present. Some clarification can be expected from genome projects on these parasites and their free-living relatives, as well as on an ever-increasing number of diverse bacteria. But progress made so far already has had significant impact on our understanding of the parasites in question and also contributed to deciphering the evolutionary history of the eukaryotic cell itself.

ACKNOWLEDGMENTS

I thank all members, past and present, of the Laboratory of Biochemical Parasitology at the Rockefeller University, who provided many original results mentioned in this chapter and also gave invaluable comments and discussions. Space does not permit me to list all of them, but I have to mention my current collaborators: Dorothy V. Moore, Oksana Ocheretina, Lidya B. Sánchez, Gang Wu and Jan Tachezy from the Charles University of Prague. I also have to express my sincere thanks to Prof. Bill Martin of the University of Düsseldorf for his insights and decisive contributions to the development of our views on amitochondriate eukaryotes, and his unfailing support and generous hospitality during my three-month visiting professorship in his department in 2001, where parts of this chapter were written. Original research in my laboratory has been supported since the seventies by grant AI11942 from the National Institutes of Health, USPHS and also by grants from the National Science Foundation.

FURTHER READING

Adam, R.D. (2001). Biology of *Giardia lamblia*. *Clin. Microbiol. Rev.* **14**, 447–475.

Benchimol, M. (1999). The hydrogenosome. *Acta Microsc.* **8**, 1–22.

Brown, D.M., Upcroft, J.A., Edwards, M.R. and Upcroft, P. (1998). Anaerobic bacterial metabolism in the ancient eukaryote *Giardia duodenalis*. *Int. J. Parasitol.* **28**, 149–164.

Danforth, W.F. (1967). In: Chen, T.-T. (ed.). *Research in Protozoology*, vol. 1, Oxford: Pergamon Press, pp. 201–306.

Fenchel, T. and Finlay, B.J. (1995). *Ecology and Evolution in Anoxic Worlds,* Oxford: Oxford University Press.

Hackstein, J.H.P., Akhmanova, A., Boxma, B., Harhangi, H.R. and Voncken, F.G.J. (1999). Hydrogenosomes: eukaryotic adaptations to anaerobic environments. *Trends Microbiol.* **7**, 441–447.

Jarroll, E.L., Manning, P., Berrada, A., Hare, D. and Lindmark, D.G. (1989). Biochemistry and metabolism of *Giardia lamblia*. *J. Protozool.* **36**, 190–197.

Jarroll, E.L. and Paget, T.A. (1995). Carbohydrate and amino acid metabolism in *Giardia*: a review. *Fol. Parasitol. (Praha)* **42**, 81–89.

Lindmark, D.G., Eckenrode, B.L., Halberg, L.A. and Dinbergs, I.D. (1989). Carbohydrate, energy and hydrogenosomal metabolism of *Tritrichomonas foetus* and *Trichomonas vaginalis*. *J. Protozool.* **36**, 214–216.

McLaughlin, J. and Aley, S. (1985). The biochemistry and functional morphology of *Entamoeba*. *J. Protozool.* **32**, 221–240.

Müller, M. (1988). Energy metabolism of protozoa without mitochondria. *Annu. Rev. Microbiol.* **42**, 465–488.

Müller, M. (1989). Biochemistry of *Trichomonas vaginalis* In: Honigberg, B.M. (ed.). *Trichomonads parasitic in humans*, New York: Springer, pp. 53–83.

Müller, M. (1993). The hydrogenosome. *J. Gen. Microbiol.* **139**, 2879–2889.

Müller, M. (1998). Enzymes and compartmentation of core energy metabolism of anaerobic protists – a special case in eukaryotic evolution? In: Coombs, G.H., Vickerman, K., Sleigh, M.A. and Warren, A. (eds). *Evolutionary relationships among protozoa*, Dordrecht: Kluwer, pp. 109–131.

Petrin, D., Delgaty, K., Bhatt, R. and Garber, G.E. (1998). Clinical and microbiological aspects of *Trichomonas vaginalis*. *Clin. Microbiol. Rev.* **11**, 300–317.

Reeves, R.E. (1984). Metabolism of *Entamoeba histolytica* Schaudinn, 1903. *Adv. Parasitol.* **23**, 105–142.

Saavedra-Lira, E. and Perez-Montfort, R. (1996). Energy production in *Entamoeba histolytica*: new perspectives in rational drug design. *Arch. Med. Res. (Mexico)* **27**, 257–264.

Samuelson, J. (1999). Why metronidazole is active against bacteria and parasites. *Antimicrob. Agents Chemother.* **43**, 1533–1541.

Schofield, P.J., Edwards, M.R. and Krantz, P. (1991). Glucose metabolism in *Giardia lamblia. Mol. Biochem. Parasitol.* **45**, 39–47.

Upcroft, P. and Upcroft, J.A. (2001). Drug targets and mechanisms of resistance in the anaerobic protozoa. *Clin. Microbiol. Rev.* **114**, 150–164.

PART II: AEROBIC PROTISTS – TRYPANOSOMATIDAE

Fred R. Opperdoes and Paul A.M. Michels
Christian de Duve Institute of Cellular Pathology,
Catholic University of Louvain, Brussels, Belgium

INTRODUCTION

This chapter focuses on the carbohydrate metabolism of the Trypanosomatidae, protozoan parasites responsible for a number of important diseases of man. Many enzymes of glycolysis and related pathways in trypanosomatids are sequestered inside microbodies, which renders them interesting not only as a drug target but also for basic research. The potential for chemotherapeutic exploitation of carbohydrate metabolism is significant. Questions such as: which enzymes are the best to target; what further information is required to allow their use for rational drug development; what compounds would constitute the best inhibitors and which of the enzymes of the hexose-monophosphate pathway are present inside the glycosomes, are addressed (see also Chapter 17).

The biochemistry of trypanosomes has been studied in great detail, mainly owing to the fact that these organisms harbor many peculiarities which have attracted biochemists interested in fundamental aspects of these organisms. The fact that trypanosomes are amongst the few parasitic organisms that can easily be grown in large numbers in the blood of infected rodents has facilitated their study. The advent of molecular biology, where the number of cells available for research is no longer a limiting factor, has led to an enormous growth of the literature over recent years. The expected availability of the complete genome sequence of *Trypanosoma brucei* is expected to give a tremendous boost to our understanding of trypanosome biology.

T. brucei is one of the best studied representatives of the so-called African trypanosomes. *T.b. brucei, T.b. rhodesiense* and *T.b. gambiense* are responsible for nagana in cattle, and human sleeping sickness in East and West Africa, respectively. One of the most exciting areas of research has been the remarkable organization of the trypanosome's carbohydrate metabolism, where many enzymes of this pathway are sequestered within a peroxisome-like organelle called the glycosome. For a more detailed description of the earlier work the reader is referred to a number of older reviews (Further Reading). The more recent information about the bloodstream-form trypanosomes will be discussed here, together with what is known about the procyclic insect stage.

THE EMBDEN–MEYERHOF–PARNAS (EMP) PATHWAY OF GLYCOLYSIS

In the bloodstream-form trypanosome, glucose serves essentially as the sole source of carbon and energy, but fructose, mannose and glycerol can be metabolized as well. This stage dwells in the blood and tissue fluids of its mammalian host, where it has access to an unlimited source of glucose which is maintained relatively constant at a concentration of 5 mM. Glucose is metabolized by a form of aerobic fermentation at a rate exceeding that found in most other eukaryotic cells. Whereas in most cells the glycolytic pathway takes place in the cytosol, in the trypanosome its first seven enzymes are localized within a membrane-bound organelle. In addition, two enzymes of glycerol metabolism and several enzymes of the hexose-monophosphate pathway are present in these organelles as well. Although this organelle resembles the peroxisomes found in other cell types, it lacks typical peroxisomal enzymes such as peroxidase, catalase and acylCoA oxidase. By contrast 90% of its protein represents glycolytic enzyme. Therefore, these highly specialized peroxisomes have been called 'glycosomes'. Because neither a functional tricarboxylic acid cycle, nor a classical respiratory chain, are present in the organism's single mitochondrion, the end-product of this metabolism is pyruvate, which is excreted directly into the host's bloodstream rather than being metabolized to lactate or carbon dioxide plus water. Hence, little ATP is produced and that only by substrate-level phosphorylation. This low ATP yield explains the high rate of glucose consumption (Figure 7.4). Bloodstream forms contain neither carbohydrate stores, such as glycogen or other polysaccharides, nor any energy reserves of any significance, such as creatine phosphate or polyphosphates. Thus inhibition of the glycolytic pathway or depletion of an energy substrate such as glucose immediately results in a rapid drop of cellular ATP levels, a total loss of motility and disappearance of trypanosomes from the bloodstream of the infected host.

The NADH produced during glycolysis in the glyceraldehyde-phosphate dehydrogenase reaction is reoxidized indirectly by molecular oxygen via a glycerol-3-phosphate (G3P):dihydroxyacetone-phosphate (DHAP) shuttle comprising the glycosomal NAD-linked G3P dehydrogenase and a FAD-linked G3P dehydrogenase present in the mitochondrial inner membrane, which catalyses the reversed reaction. Reducing equivalents produced by this cycle are supposed to react with ubiquinone present in the mitochondrial membrane. The ubiquinol thus formed is oxidized by a ubiquinol:oxygen oxidoreductase, also called trypanosome alternative oxidase (TAO), present in the mitochondrial inner membrane. This enzyme resembles the alternative oxidases of plants and certain fungi in that it is insensitive to cyanide, but is sensitive to salicyl hydroxamic acid (SHAM). Moreover, in a phylogenetic analysis its amino-acid sequence branched close to the alga *C. reinhardtii* and formed a single clade with the plant alternative oxidases with a high bootstrap support. This enzyme has also been detected in the plant trypanosomatid *Phytomonas*, but seems to be absent from *Leishmania* spp., where neither the activity nor the gene encoding the enzyme were detected.

Inhibition of the TAO by SHAM mimics the effect of a lack of oxygen on carbohydrate metabolism of the bloodstream form. NAD is now regenerated by the reduction of DHAP to G3P, followed by the formation of glycerol which is excreted by the organism. The ATP synthesized in the latter reaction compensates for the loss of one mole of ATP in the glycosomal phosphoglycerate kinase reaction, because now one mole of phosphoglycerate, rather than two, is produced per mole of glucose consumed. As a consequence, glucose is dismutated into equimolar amounts of pyruvate and glycerol, with net synthesis of one molecule of ATP. Under these conditions bloodstream forms survive and remain motile, while cellular ATP levels drop to about 50%. Thus, inhibition of respiration alone is not sufficient to kill the organism and the trypanosomes survive for a prolonged time as long as glycerol does not accumulate in the medium. This probably explains why of the many new and effective inhibitors of the trypanosome alternative oxidase which have

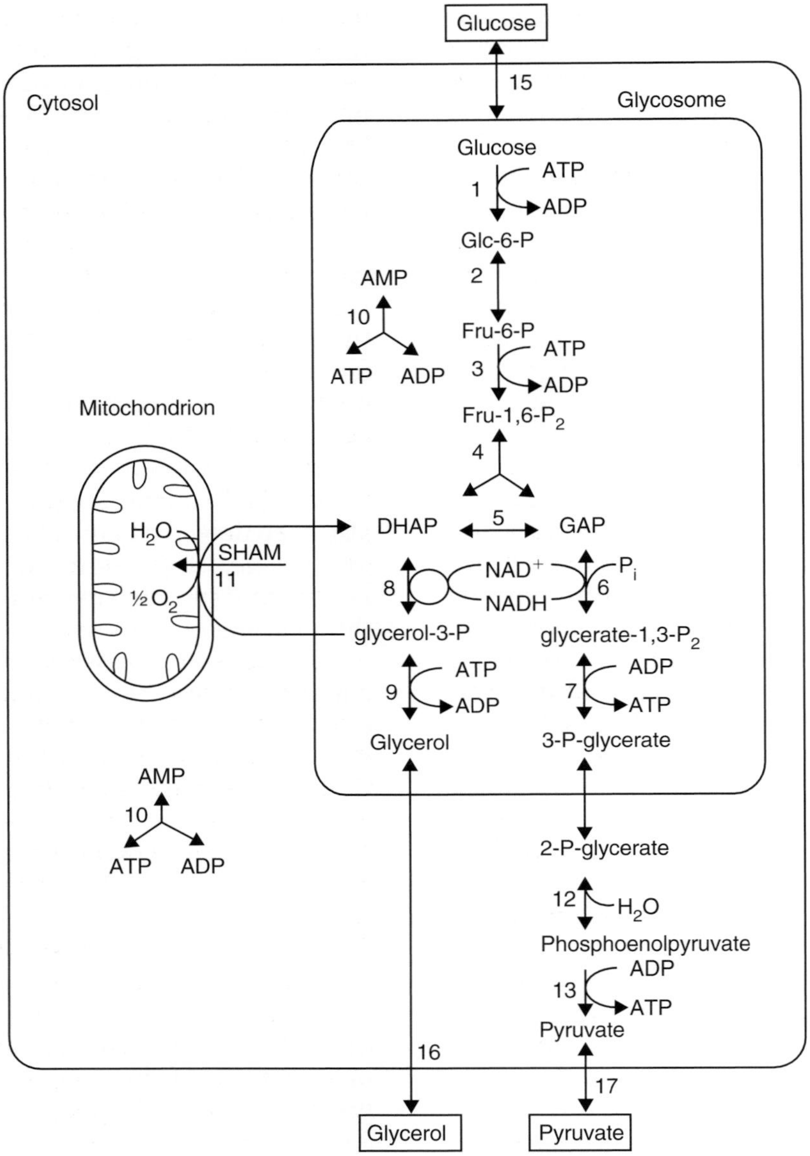

FIGURE 7.4 Glycolysis in the bloodstream-form trypanosome. Boxed metabolites are nutrients or end-products of metabolism. Enzymes: (1) hexokinase; (2) phosphoglucose isomerase; (3) phosphofructokinase; (4) aldolase; (5) triosephosphate isomerase; (6) glyceraldehyde-3-phosphate dehydrogenase; (7) phosphoglycerate kinase; (8) glycerol-3-phosphate dehydrogenase; (9) glycerol kinase; (10) adenylate kinase; (11) glycerol-3-phosphate oxidase; (12) phosphoglycerate mutase; (13) enolase; (14) pyruvate kinase; (15) glucose transporter; (16) glycerol transporter; (17) pyruvate transporter.

been developed, and of which ascofuranone was the most potent, none has been developed into an effective anti-trypanosome drug. Once glycerol accumulates above several millimolar in the presence of a TOA inhibitor, this metabolite becomes highly toxic to the trypanosome. Due to mass action the reversal of the glycerol-kinase reaction comes to a halt and glycolysis stops. Since trypanosomes lack energy stores, cellular ATP levels rapidly drop to zero, leading to their complete immobilization. Under these conditions trypanosomes are rapidly cleared from the blood of infected animals.

Based on the above observations, there is little doubt that the enzymes of glycolysis in the bloodstream form of *T. brucei* constitute excellent chemotherapeutic targets, but few of these have been validated by gene knockout experiments. Moreover, there is much discussion as to which of the enzymes constitute the better targets within the pathway. Mathematical modeling experiments, using all the available information on the individual enzymes of this pathway, were able to predict successfully the experimentally determined fluxes and metabolite concentrations in trypanosomes. This has allowed us to get information about the degree of control each of the enzymes exerts on the overall carbon flux through the pathway. The glucose transporter protein that is present in the parasite's plasma membrane may confer 50% or more of the control over the pathway. However, the enzymes aldolase, glyceraldehyde-3-phosphate dehydrogenase (GAPDH), glycerol-3-phosphate dehydrogenase (G3PDH) and phosphoglycerate kinase (PGK) exert a considerable amount of control, especially under conditions where the blood levels of glucose are not limiting. Tight-binding inhibitors of enzymes that confer control over the pathway, with affinities in the nanomolar range, as well as irreversible inhibitors of enzymes that do not confer any control, may be effective in killing the organism, provided they succeed in inhibiting the flux to a large extent. That this approach is feasible has been illustrated by the use of conditional knockouts of *T. brucei* cell lines. When the activity of triosephosphate isomerase (TIM) was decreased to 15%, the rate of cell division was halved. Lower TIM levels were lethal. This result was surprising since TIM was considered a non-essential enzyme for trypanosome glycolysis. Its inhibition (or removal) would force the trypanosome into the formation of equimolar amounts of pyruvate and glycerol with a net production of only one ATP per molecule of glucose consumed, rather than the two molecules of ATP produced under normal conditions. Yet the removal of TIM leads to arrest of growth. Apparently the reduced amount of ATP produced under these conditions may not be sufficient to sustain cell division, but other explanations, such as the accumulation of certain toxic metabolites, cannot be excluded.

Similarly, *in vitro* growth in the presence of low concentrations of SHAM, which imposes an anaerobic type of glycolysis upon the organism, turned out to be impossible over extended periods as well. Upon lowering the level of TAO expression by a conditional gene knockout, the oxygen consumption was reduced fourfold and the rate of trypanosome growth was halved. This indicates that any glycolytic inhibitor that could exert a considerable reduction of the overall glycolytic flux, and/or a reduction of the ATP yield of glycolysis, would arrest growth completely and thus could be a promising drug candidate. Bearing this in mind, each of the individual steps of the glycolytic pathway will be discussed.

Metabolite transporters

Glycolysis proceeds at very high rates (up to 100 nM of glucose consumed per min and per mg of protein) in the bloodstream form of

T. brucei. Transport across the plasma membrane of the substrate of the pathway, glucose, and its end-products, pyruvate, and under anaerobic conditions, glycerol, are all mediated by facilitated diffusion transporters. No information is available about metabolite transporters in the glycosomal membrane.

Glucose transport

T. brucei utilizes glucose, fructose and mannose as substrates for the glycolytic pathway, and the glucose transporter present in the plasma membrane of bloodstream forms is capable of transporting all three substrates. The glucose transporter is considered not only an ideal target for drugs that would directly interfere with glucose transport, but the transporter could also be instrumental in the import of drugs into the trypanosome's cytosol by complexing them to sugar analogs. Detailed studies on the transporters of *T. equiperdum*, *T. rhodesiense* and *T. brucei* have been carried out. They have revealed important differences in specificity between parasite and host transporters and these have been exploited for the synthesis of specific inhibitors of the trypanosome glucose transport, as well as for targeting drugs to the trypanosome's cytosol. Unfortunately, sugar analogs containing various bulky substituents at carbon positions that were not required for recognition of the transporter were still able to bind to the transporter, but not to enter the cell by facilitated diffusion. This inability of the glucose transporter to transport such bulky analogs will render it less suitable for this interesting type of approach.

Pyruvate transporter

The plasma membrane of *T. brucei* contains a facilitated diffusion carrier which can be saturated by pyruvate ($K_m = 2$ mM). The activity of the carrier is sufficient to account for the rate of pyruvate production by the cell and is specific for pyruvate in that it does not transport lactate. It is inhibited by a number of other monocarboxylic acids including pyruvate analogs, but not L-lactate, which makes it unique among the monocarboxylate transporters. Transport inhibitors like di-isothiocyanostilbene disulfonate, quercetin and phloretin all inhibit pyruvate transport. The pyruvate carrier is rather insensitive to the monocarboxylate carrier inhibitor α-cyano-4-hydroxycinnamate ($K_i = 10$ mM) but could be completely blocked by UK5099 [α-cyano(1-phenylindol-3-yl)acrylate] ($K_i = 55$ μM). Inhibition of pyruvate transport results in the accumulation of pyruvate within the trypanosomes causing acidification of the cytosol and lysis of the cells, indicating that carrier-mediated pyruvate export in the bloodstream-form trypanosome is of vital importance to their survival.

Glycerol transporter

In *T. brucei* bloodstream forms, glycerol is transported across the plasma membrane by both simple diffusion and by a saturable facilitated-diffusion carrier. At low concetrations of glycerol, the carrier, with a K_m of 0.17 mM for glycerol, is responsible for most of the transport of glycerol, but at higher concentrations simple diffusion predominates. Transport is neither sodium-dependent nor proton-motive-force driven. The transporter is sensitive to phloretin and cytochalasin B and is also inhibited by the substrate analog glyceraldehyde. In procyclic insect forms glycerol is taken up by simple diffusion only.

Hexokinase (HK)

Hexokinase, the first enzyme of the glycolytic pathway, was one of the last enzymes for which sequence information became available. HK has a low percentage of identity (36%)

with the human enzyme, which increases the chances of developing specific inhibitors. The latter enzyme is also twice as large owing to a gene duplication that must have taken place after the separation of the trypanosomatid lineage from the main line of eukaryotic evolution. Inspection of the sequence also suggests that HK is routed to the glycosome by an N-terminally located peptide that is reminiscent of a peroxisomal import signal, also called PTS2. The enzyme, which has a low specificity towards sugars, is neither regulated by glucose 6-phosphate nor by glucose 1,6-bisphosphate, as are the HKs from most other sources. Interestingly, the enzyme has a low specificity for ATP, and is also able to utilize other nucleotide triphosphates such as UTP and CTP. This absence of specificity probably reflects the absence of such nucleotides from the matrix of the glycosome and thus may render the nucleotide-binding pocket of this enzyme an interesting target for drug development. The glucose-binding pocket offers an interesting target as well. A number of glucosamine analog inhibitors have been synthesized. The most potent inhibitor, *m*-bromophenyl glucosamide, did not affect the activity of HK from yeast but inhibited the growth of live trypanosomes *in vitro*. The mode of action of the *m*-bromophenyl glucosamide inhibitor on the trypanosome HK has been elucidated by modeling the structure of the *T. brucei* enzyme.

Glucose-6-phosphate isomerase (PGI)

In the bloodstream form PGI is mainly associated with the glycosome. It catalyses an equilibrium reaction and is often considered not to confer any significant control over the glycolytic flux. However, any efficient PGI inhibitor would be able to cause arrest of growth. In this respect the potent transition-state analog 5-phospho-D-arabinonate is an interesting candidate. This compound binds to the enzyme with a K_i of 50 nM, while a K_i/K_m value of 2000 was obtained. Such a tight binding inhibitor may compete effectively with the high concentration of intermediates inside the glycosome. Its mode of binding to the active site of the homologous rabbit enzyme has been solved at 1.9 Å resolution, and the compound mimics the *cis*-enediol intermediate of the catalytic reaction. Unfortunately, due to the nature of such an inhibitor, there was little specificity for the trypanosome enzyme when compared to the yeast PGI.

Phosphofructokinase (PFK)

T. brucei PFK, as well as the *Leishmania donovani* enzyme, has been cloned, sequenced and overexpressed. The deduced polypeptide sequences both contain a peroxisome targeting signal present at the C-terminus (so-called PTS1). Although their respective sequences predict a typical inorganic pyrophosphate (PP_i)-dependent PFK, both have an absolute requirement for ATP and are completely inactive with PP_i as substrate. Apparently, during evolution this trypanosomatid enzyme must have changed its specificity. Another difference from the PFKs from other eukaryotes is that it is not regulated by fructose 2,6-bisphosphate. The two trypanosomatid enzymes are inhibited by AMP and furthermore have similar properties. Contrary to the situation in mammalian cells, the *T. brucei* PFK exerts hardly any control over the glycolytic flux. While from a rate-controlling point of view PFK would be less interesting for drug design, its remarkable structural difference from the host enzyme renders it a very interesting candidate indeed.

Aldolase

This enzyme belongs to the class I aldolases that catalyse the reversible cleavage of the fructose

1,6-bisphosphate, via the formation of a Schiff base, into the trioses dihydroxyacetone phosphate (DHAP) and glyceraldehyde 3-phosphate (GA3P) without the need of a divalent metal ion. The enzyme is abundantly present in *T. brucei* and exerts a significant amount of control over the glycolytic flux, which does not seem to be the case for the red blood cell, where deficiencies seem to have little or no effect. The aldolase gene of both *T. brucei* and *L. mexicana* has been cloned and sequenced and the predicted polypeptide sequence contains an N-terminal PTS2. In a recent phylogenetic analysis the two trypanosomatid sequences formed a monophyletic group with the aldolases found in the chloroplasts from both algae and plants, where the trypanosomatid occupied a sister position relative to the green and red algae.

Triose-phosphate isomerase (TIM)

The trypanosome TIM was the first parasite enzyme to be crystallized and its three-dimensional structure solved. This enzyme catalyses the reversible conversion of the triosephosphates DHAP and GA3P. In the bloodstream form the enzyme is mainly found in the glycosome, and it represents less than 0.04% of the total trypanosome protein. In general few inhibitors of TIM have been described. The trypanosome enzyme is inhibited by sulfate, phosphate, arsenate, and by 2-phosphoglycolate. Suramin also seems to bind to *T. brucei* TIM, and there is some evidence that suramin may interfere with the formation of the active homodimeric form of the enzyme. Although there is no theoretical argument to suggest that inhibition of the enzyme would lead to the death of the parasite, knockout experiments have shown that TIM is a vital enzyme and thus may be a good drug target.

TIM is the only glycolytic enzyme found in glycosomes for which no (putative) peroxisome-import signal has been reported so far. Import studies into glycosomes have shown that a 21 amino-acid-long internal peptide has the ability to route a reporter protein to glycosomes. Replacement of this peptide by a corresponding peptide of the yeast TIM completely abolished import. It is believed that this peptide, which is exposed at the surface of the native protein, mediates the interaction of TIM with another protein that does carry a peroxisome targeting signal (PTS). Import of TIM into glycosomes would then be a piggyback process.

Glyceraldehyde-3-phosphate dehydrogenase (GAPDH)

All Trypanosomatidae studied, including *T. brucei*, have two separate GAPDH isoenzymes, one in the cytosol and one in the glycosome. While the glycosomal enzyme is directly involved in glycolysis, the function of the cytosolic isoenzyme is not yet clear. The glycosomal isoenzyme is inhibited by the epoxide-containing GAPDH inhibitor pentalenolactone and by suramin, gossypol and agaricic acid. Although its substrate-binding site is well conserved, it has been possible to synthesize epoxide or alpha-enone containing analogs that selectively and irreversibly inhibit not only the enzyme but also the growth of the trypanosome in an *in vitro* culture system. The three-dimensional structures of the *T. brucei*, *T. cruzi* and *L. mexicana* enzymes have been solved, and this has revealed a binding site for the NAD cofactor which has some interesting differences from the corresponding binding site in the human enzyme, responsible for a reduced affinity for this cofactor. A number of highly selective adenosine analogs have been synthesized that tightly fit the NAD-binding pocket of the trypanosome enzyme, but do not bind to the corresponding pocket of the mammalian homolog. Some of these compounds have a high selectivity for the glycosomal enzymes of the three trypanosomatids and inhibit their activity at micromolar to submicromolar concentrations. In *T. brucei* these compounds interfere directly with glycolysis as revealed by the rapid inhibition of pyruvate production and motility, and they also inhibit *in vitro* growth in the low micromolar range. Interestingly, these compounds are equally effective on *T. brucei* bloodstream forms, *L. mexicana* promastigotes and amastigote-like cells, and on intracellular amastigotes of *T. cruzi*. No toxicity for the host cells could be demonstrated below their solubility limit in aqueous medium. These results demonstrate for the first time that, despite the fact that *T. cruzi* and *L. mexicana* are thought to be less dependent on glycolysis owing to the presence of an active mitochondrion, inhibition of glycolysis is lethal to each of these parasites.

Glycerol-3-phosphate dehydrogenase (G3PDH)

This NAD-dependent enzyme catalyzes the reversible oxidation of DHAP to glycerol 3-phosphate. In *T. brucei* it plays an important role in the DHAP:G3P shuttle between glycosome and mitochondrion. A comparison of the *T. brucei* sequence with that of other G3PDH sequences revealed an identity of the trypanosome enzyme with that of eukaryotes ranging from 23–25%, while the identity with a number of prokaryotic G3PDHs ranged from 32–36%. Moreover, the trypanosome G3PDH has typical prokaryotic signatures. This and the fact that the enzyme exerts a considerable amount of control over the glycolytic flux (see above) renders G3PDH a very interesting drug target. Both the *T. brucei* and *L. mexicana* enzyme have been overexpressed as active proteins and the three-dimensional structure of the *Leishmania* enzyme was recently solved. Since the *T. brucei* and the *L. mexicana* enzyme share 70% positional identity this opens interesting perspectives for the development of very specific inhibitors for both enzymes. Both trypanosomatid enzymes are inhibited by agaricic acid and by the trypanocidal drugs suramin, melarsen oxide and cymelarsen, while the corresponding enzyme from rabbit muscle is much less sensitive to these inhibitors.

Glycerol kinase (GK)

The *T. brucei* glycerol kinase has been cloned, sequenced and overexpressed. The enzyme has been purified and its kinetic properties examined. This enzyme has a tenfold reduced affinity for both glycerol and G3P as compared with GKs from other sources. Sequence comparisons and site-directed mutagenesis studies suggest that a specific substitution, only found in the GK of African trypanosomes capable of the anaerobic dismutation of glucose into pyruvate and glycerol, contributed to this reduced affinity. A serine that forms a hydrogen bond with the backbone N of glutamic acid in the glycerol-binding pocket, and which is present in most GKs, was found to be changed into an alanine in each of these trypanosomes. The mutation of this alanine to a serine indeed resulted in an enzyme with an increased affinity for glycerol.

Phosphoglycerate kinase (PGK)

This enzyme has been extensively studied in the past. Three isoenzymes have been described. A cytosolic enzyme is active in insect stages of *T. brucei* and is also found in the cytosol of *Leishmania*. A major glycosomal isoenzyme is active in the bloodstream-form trypanosome, but is also found in the glycosomes of *Leishmania* promastigotes. A minor glycosomal isoenzyme has also been found in *T. brucei*, but this enzyme seems to be absent from *Leishmania*. *T. brucei* PGKs have an almost absolute requirement for ATP, contrary to the enzymes from yeast and rabbit muscle, which accept GTP and ITP as well. The two major trypanosome isoenzymes are inhibited by suramin, but the glycosomal enzyme, with its higher positive charge, is much more sensitive to this inhibitor. The glycosomal isoenzyme has been crystallized after removal, by genetic engineering, of the C-terminal extension with PTS1, responsible for its import into the glycosome, and its structure solved. Comparison of the protein sequences of the trypanosomatid enzymes with those of other organisms has revealed a typical prokaryotic origin of the former. The enzyme must have entered the trypanosomatid ancestor by horizontal transfer, but the donor organism cannot, at present, be identified.

Phosphoglycerate mutase (PGAM) and enolase

These two enzymes of *T. brucei* are present in the cytosol and therefore they have received less attention. The genes coding for both these proteins now have been cloned and sequenced and the corresponding proteins have been overexpressed. Two classes of non-homologous PGAM exist. All mammalian PGAMs use as cofactor 2,3-bisphosphoglycerate, while many bacterial and plant PGAMs are cofactor-independent. The *T. brucei* PGAM gene has been cloned and sequenced and the enzyme expressed in *E. coli*. The trypanosome enzyme turns out to be a typical cofactor-independent protein and thus is totally different from the isofunctional enzyme of the host. It is closely related to the cofactor-independent PGAMs found in the cytosol of plants. Modeling of its three-dimensional structure suggests that this class of proteins is related to the family of alkaline phosphatases.

Pyruvate kinase (PYK)

The last enzyme of the glycolytic pathway is pyruvate kinase. Like PGAM and enolase, PYK is a cytosolic enzyme. The PYKs from *T. brucei* and *L. mexicana* have been cloned, sequenced, overexpressed in *E. coli* and purified. The enzyme is activated by various sugar phosphates, most notably fructose 2,6-bisphosphate, but also by fructose 1,6-bisphosphate, glucose 1,6-bisphosphate and ribulose 1,5-bisphosphate. Its activity is modulated by adenine nucleotides and inorganic phosphate, while it is insensitive to regulation by most of the metabolites that modulate PYK activity from other sources. The structure of the *L. mexicana* enzyme has been solved and reveals a pocket that contains a sulfate molecule. It is this pocket that is thought to be responsible for the binding of its allosteric regulator fructose 2,6-bisphosphate. Based on the comparison of the three-dimensional structure of *Saccharomyces cerevisiae* PYK crystallized with fructose 1,6-bisphosphate present at the effector site (R-state) and the *L. mexicana* enzyme crystallized in the T-state, two residues (Lys453 and His480) have been proposed to bind the 2-phospho group of the effector. These predictions have subsequently been confirmed by site-directed mutagenesis.

Pyruvate phosphate dikinase (PPDK)

In addition to an ATP-dependent PYK, *T. brucei* also has a PP_i-dependent pyruvate, phosphate dikinase, or PPDK. The enzyme is encoded by a single-copy gene. The gene product, a 100-kDa protein, is expressed in procyclic forms but not in the bloodstream forms of *T. brucei*. The AKL motif at the C-terminal extremity of PPDK has been shown to be essential for glycosomal targeting. This contrasts with the cytosolic PYK which is mainly active in the bloodstream form. PPDK has been detected in every trypanosomatid tested – *T. congolense, T. vivax, T. cruzi, Phytomonas, Crithidia* and *Leishmania* – with a good correlation between the amount of protein and enzymatic activity.

The role of PPDK in glycosomes is not yet clear. In addition to PPDK, there are at least four other enzymes in the glycosome that produce PP_i: fatty acyl-CoA synthetase, hypoxanthine-guanine phosphoribosyl transferase, xanthine phosphoribosyl transferase and orotate phosphoribosyl transferase (Chapter 9). Each of them catalyzes reactions with a $\Delta G'_0$ close to zero and thus PP_i needs to be hydrolysed in order to assure a continuation of the phosphoribosyl transferase reactions of the purine salvage and pyrimidine biosynthesis pathways. While a pyrophosphatase activity has been found in the cytosol of *T. brucei*, such activity could not be detected in its glycosomes, and this suggests that trypanosomes may have developed an alternative route to hydrolyse glycosomal PP_i. PPDK could do so by converting PEP into pyruvate while producing ATP. Whether PPDK does in fact contribute in this way to the glycosomal PP_i and ATP balance remains to be determined. The fact that PPDK is not present in bloodstream-form glycosomes would also suggest that the other four PP_i-producing enzymes may be absent from bloodstream-form glycosomes as well. So far only the absence of fatty-acid oxidation activity from bloodstream-form trypanosomes has been confirmed experimentally.

The presence of both a glycosomal ATP-dependent PFK with PP_i-dependent characteristics (see above), together with a PP_i-dependent PPDK, suggests that an ancestral trypanosomatid may have possessed a PP_i-dependent type of glycolysis, similar to what is still encountered in some anaerobic protists, such as *Giardia* and *Entamoeba* (Chapter 7, part 1). A subsequent adaptation to an aerobic type of metabolism, where PFK changed its phospho-substrate specificity from PP_i to ATP and where PPDK was replaced by PYK for the conversion of PEP into pyruvate, may have resulted in the replacement of a PP_i-dependent glycolysis by a completely ATP-dependent one. PPDK would then have been retained after having acquired its new function in maintaining the PP_i and ATP balance inside glycosomes.

Since PPDK does not exist in higher eukaryotes, the enzyme could be considered as a good target for the design of drugs against *Leishmania* and *T. cruzi*, provided that this enzyme were essential for cell viability in these trypanosomatids. The enzyme from *T. brucei*, which has been crystallized and the structure of which is being solved, would be a less appropriate target, because it is not expressed in the bloodstream form.

The hexose-monophosphate pathway

Glucose 6-phosphate (G6P) is also metabolized by the hexose-monophosphate pathway (HMP), also known as the pentose-phosphate shunt (Figure 7.5). While the role of glycolysis is the generation of ATP and pyruvate, the HMP is mainly involved in maintaining a pool of cellular NADPH, which may serve as a hydrogen donor in reductive biosynthesis, and in defense against oxidative stress. The HMP also serves to convert G6P to ribose 5-phosphate (R5P), which is required for nucleotide biosynthesis. While the glycolytic pathway has been extensively investigated in *T. brucei*, there have been only a limited number of studies on HMP and its contribution to carbohydrate metabolism.

The oxidative branch of the pathway converts G6P to ribulose 5-phosphate (Ru5P), a process which leads to the production of two moles of NADPH per mole of G6P consumed. Recent biochemical evidence, as well as sequence information, suggests that several enzymes of this

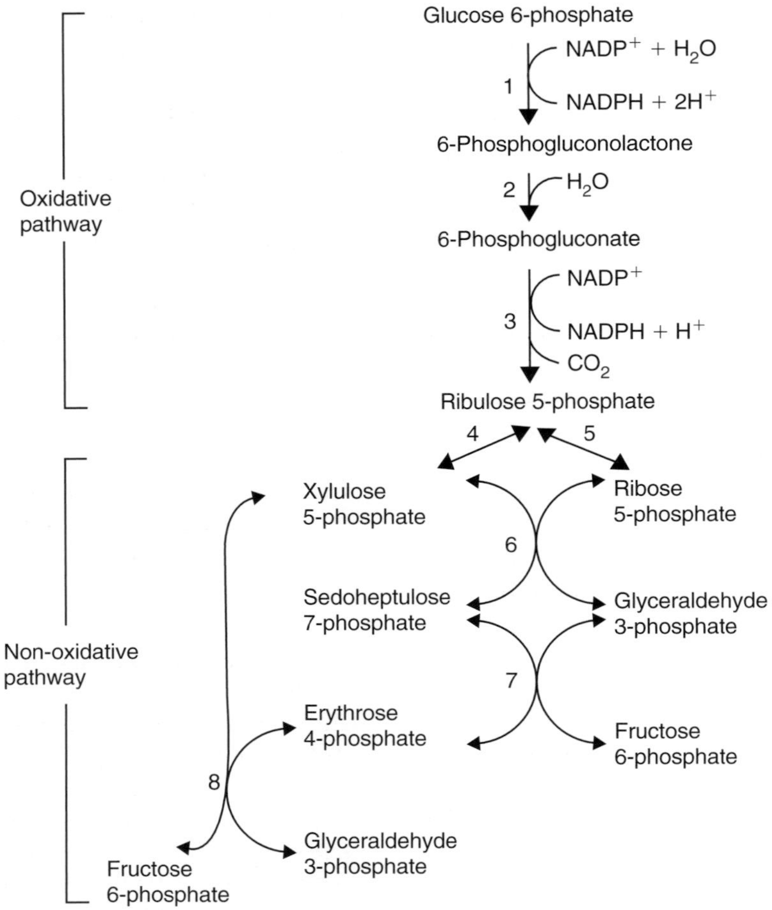

FIGURE 7.5 The hexose-monophosphate pathway. Enzymes: (1) Glucose-6-phosphate dehydrogenase; (2) 6-phosphogluconate lactonase; (3) 6-phosphogluconate dehydrogenase; (4) ribulose 5-phosphate 3-epimerase; (5) ribulose 5-phosphate isomerase; (6) transketolase; (7) transaldolase; (8) transketolase.

pathway may be associated with glycosomes as well. While 50% of the activity of glucose-6-phosphate dehydrogenase (G6PDH) in *T. brucei* is found in glycosomes, its inferred protein sequence did not reveal a peroxisome-targeting signal. However, the *T. brucei* 6-phosphogluconolactonase (6PGL), of which some 15% of the activity is glycosomal, contains a C-terminal motif showing similarity to the established type of targeting signal, in agreement with its subcellular localization. The corresponding genes from *L. mexicana* have also been cloned and sequenced and here the G6PDH contains a PTS 1 import signal, while the 6PGL did not. However, the sequence of the third enzyme in the pathway, 6-phosphogluconate dehydrogenase (6PGDH), does not contain a detectable PTS. Interestingly, the sequences of 6PGL and 6PGDH in both organisms are of prokaryotic origin and therefore most likely entered the trypanosomatid ancestor by a mechanism of horizontal transfer. The G6PDH, by contrast, is of true eukaryotic origin, but this enzyme is closely affiliated with its homologs found in the plant cytosol.

Until recently, the role of 6PGL in metabolism was still questionable, since 6-phosphogluconolactones were believed to undergo rapid spontaneous hydrolysis. Studies using the *T. brucei* 6PGL together with ^{13}C- and ^{31}P-nuclear magnetic resonance spectroscopy have now characterized the chemical scheme and kinetic features of the oxidative branch of this pathway. The δ-form of the lactone is the only product of G6P oxidation. It leads to the spontaneous formation of the γ-lactone by intramolecular rearrangement. However, only the δ-lactone undergoes spontaneous hydrolysis, the γ-form being a 'dead-end' of this branch. The δ-lactone is the only substrate for 6PGL. Therefore, this enzyme significantly accelerates hydrolysis of the δ-form, thus preventing its conversion into the γ-form. Furthermore, 6PGL guards against the accumulation of δ-6-phosphogluconolactone, which may be toxic through its reaction with endogenous cellular nucleophiles.

Suramin is a known inhibitor of *T. brucei* 6PGDH. A number of its analogs and several aromatic dyes act as inhibitors of this enzyme. The PGI transition-state inhibitor 5-phospho-D-arabinonate also inhibits 6PGD with a K_i of 50 μM.

The non-oxidative branch generates from various pentose-phosphates, the substrates for the enzymes transketolase and transaldolase, which transfer, respectively, two- and three-carbon units between a variety of phosphorylated carbohydrates. End-products of the non-oxidative branch are the glycolytic intermediates fructose 6-phosphate (F6P) and glyceraldehyde 3-phosphate (GA3P). The latter branch may also function in the other direction, where GA3P and F6P are used to generate other phosphorylated sugars. Several of the respective genes (e.g. transketolase) have been cloned and sequenced, or partial gene sequences have been identified in the *Trypanosoma* genome database. None of these genes shows evidence of horizontal gene transfer events. While in *T. brucei* the enzymes of the oxidative branch are present in both the life-cycle stages, in the bloodstream form two enzymes of the non-oxidative pathway, Ru5P epimerase and transketolase, were not detected. Transaldolase was present.

WHY DO TRYPANOSOMATIDAE HAVE GLYCOSOMES?

Most eukaryotes and certainly all aerobes have peroxisomes. In general these organelles are involved in various activities such as protection against oxidant stress, β-oxidation of fatty acids and ether-lipid biosynthesis. Moreover, when required, peroxisomes may adapt to specific metabolic conditions as is the case for the specialized peroxisomes of alkane- or methanol-oxidizing fungi. However, only in glycosomes of the Trypanosomatidae have seven enzymes of the glycolytic pathway been described, together with other enzymes of carbohydrate metabolism. Until now, no clear answer has been provided to the question why these organisms, contrary to other eukaryotes, have sequestered a major part of their glycolytic pathway inside glycosomes. With the accumulation of more and more protein sequences in the database and the possibility of comparing them with their homologs from other organisms, it has become clear that in all representatives of the Trypanosomatidae analyzed thus far, a number of glycolytic enzymes, but also some enzymes involved in other pathways, such as the HMP, differ much more from their respective homologs found in other eukaryotes than can be explained by a normal line of descent. Moreover, some of these enzymes are clearly more related to their prokaryotic than to their eukaryotic counterparts (e.g. G3PDH, PGK, 6PGL, 6PGDH). Other enzymes, such as aldolase, PGAM, TAO and G6PDH, more closely resemble their homologs from plants and algae. Thus, most likely these enzymes entered the trypanosome ancestor by mechanisms of horizontal transfer.

Trypanosomatids and the euglenids together belong to the group of the Euglenozoa. Within this group many euglenids are photosynthetic and contain chloroplasts, while others have lost these photosynthetic organelles. Trypanosomatids are non-photosynthetic, and there is no evidence for the presence of chloroplasts in these organisms. The present day chloroplasts of the euglenids are supposed to have been derived from an algal endosymbiont, which once was phagocytosed by the euglenid ancestor and then lost again, except for its chloroplasts which were retained by the host. The presence in the trypanosomatid genome of genes of either prokaryotic or of typical plant or algal origin, which code for key enzymes of carbohydrate metabolism, may suggest that it was a common ancestor of both Euglenids and Trypanosomatids that once harbored such an endosymbiont. While in *Euglena* the chloroplasts were retained, the trypanosomatids lost them completely, but only after some of the algal and chloroplast genes had been transferred to the nucleus of the trypanosomatid.

Apart from the cytosolic PGAM and the mitochondrial TAO, most of these enzymes have acquired a peroxisome-targeting signal and are now found inside glycosomes. These glycosomal enzymes, which in the trypanosomatid are now involved in carbohydrate catabolism, may have fulfilled an essential role in the photosynthetic Calvin cycle or in carbohydrate synthesis in the algal endosymbiont. Apparently enzymes of chloroplast/cyanobacterial origin originally involved in the synthesis of carbohydrates in a common ancestor of kinetoplastids and euglenids were all relocated to the kinetoplastid peroxisome when the chloroplast was lost. This relocation of a large number of algal enzymes could then have triggered the formation of the glycosome, which then adopted carbohydrate catabolism as its major new function.

SUMMARY AND CONCLUSIONS

Through the combined efforts of many researchers in various disciplines over many years, we have now obtained a good understanding of the glycolytic pathway of the African trypanosome. These efforts have led to the cloning, sequencing and overexpression of all the enzymes of the pathway and the resolution of the crystal structure of many. Theoretical calculations and gene knockout experiments have demonstrated that glycolysis constitutes a valid drug target. Most of the enzymes involved have been fully characterized and in some cases, such as GAPDH, this already has led to the design of very potent inhibitors that exert their effect not only *in vitro*, but also *in vivo*, and not only on *T. brucei*, but also on *T. cruzi* and *Leishmania*. Recently a start has been made with a similar approach to the enzymes of the HMP. The possibility that chloroplast- or algal-derived enzymes and, perhaps, complete plant-derived pathways are still present in trypanosomatids opens new possibilities for the identification of drug targets. Since such enzymes should not be present in mammals, drugs directed against those targets could provide high selectivity. Moreover, there is a wide range of herbicides that has been developed for use in agriculture and which are now potentially useful against trypanosomatids.

FURTHER READING

Azema, L., Bringaud, F., Blonski, C. and Périé, J. (2000). *Bioorg. Med. Chem.* **8**, 717–722.

Bakker, B.M., Michels, P.A.M., Opperdoes, F.R. and Westerhoff, H.V. (1997). *J. Biol. Chem.* **272**, 3207–3215.

Bakker, B.M., Walsh, M.C., Ter Kuile, B.H. *et al.* (1999). *Proc. Natl. Acad. Sci. USA* **96**, 10098–10103.

Chevalier, N., Rigden, D.J., Van Roy, J., Opperdoes, F.R. and Michels, P.A.M. (2000). *Eur. J. Biochem.* **267**, 1464–1472.

Conception, J.L., Adjé, C.A., Quinones, W., Chevalier, N., Dubourdieu, M. and Michels, P.A.M. (2001). *Mol. Biochem. Parasitol.* **118**, 103–113.

Grady, R.W., Bienen, E.J., Dieck, H.A., Saric, M. and Clarkson, A.B. Jr. (1993). *Antimicrob. Agents Chemother.* **37**, 1082–1085.

Hannaert, V. and Michels, P.A.M. (1994). *J. Bioenerg. Biomembr.* **26**, 205–212.

Heise, N. and Opperdoes, F.R. (1999). *Mol. Biochem. Parasitol.* **99**, 21–32.

Helfert, S., Estevez, A.M., Bakker, B., Michels, P. and Clayton, C. (2001). *Biochem. J.* **357**, 117–125.

Michels, P.A., Hannaert, V. and Bringaud, F. (2000). *Parasitol. Today* **16**, 482–489.

Opperdoes, F.R. (1987). *Annu. Rev. Microbiol.* **41**, 127–151.

Opperdoes, F.R. and Michels, P.A.M. (1993). *Biochimie* **75**, 231–234.

Willson, M., Sanejouand, Y.H., Périé, J., Hannaert, V. and Opperdoes, F.R. (2002). *Chem. Biol.* **9**, 839–847.

PART III: ENERGY METABOLISM IN THE APICOMPLEXA

Michael J. Crawford, Martin J. Fraunholz and David S. Roos
Department of Biology, University of Pennsylvania,
Philadelphia, PA, USA

OVERVIEW

The phylum Apicomplexa consists entirely of obligate intracellular parasites, including the causative agents of malaria (*Plasmodium* spp.), toxoplasmosis (*Toxoplasma gondii*), coccidiosis (*Eimeria* spp.), and several other diseases of clinical and veterinary importance. Recent years have seen great advances in our knowledge of the pathways responsible for apicomplexan energy production and metabolism. These include enhanced understanding of the mechanisms of nutrient uptake, basic aerobic and anaerobic processes, structure and biochemistry of the enzymes responsible for these activities, insights into metabolic regulation, and improved understanding of the host–parasite relationship.

The earliest biochemical experiments on apicomplexans were carried out during the 1930s and '40s, and focused on carbohydrate and energy metabolism. These studies demonstrated a primary dependence on glycolytic substrate-level phosphorylation, with limited or absent capacity for oxidative phosphorylation. Since the life-cycle stages of most apicomplexans require low levels of oxygen for survival and growth, these parasites are perhaps best described as 'microaerophilic'. This places them biochemically between the protozoan groups discussed in the previous two sections of this chapter. Enzymatically, apicomplexan energy production is an amalgam of aerobic and anaerobic pathways and can vary considerably from species to species. The presence of anaerobic enzymes, such as pyrophosphate-dependent phosphofructokinase and pyruvate: ferredoxin oxidoreductase, is reminiscent of *Entamoeba* and *Trichomonas*, but canonical aerobic pathways, including the TCA cycle and mitochondrial ATP production, are also present (in the Kinetoplastida, for example).

A reference sequence for the *Plasmodium falciparum* genome is now essentially complete. Extensive sequence information also is available for *Cryptosporidium parvum* and two *Theileria* species. Genome projects are underway for *Toxoplasma gondii* and several species of *Plasmodium*. A total of nearly 100 000 EST sequences are available from various strains and various life-cycle stages of several apicomplexan parasites. These data identify numerous (putative) enzymes and, where complete sequence is available, also permit the identification of pathways that are lacking. With the ever-accelerating pace of 'omics' projects (genomics, transcriptomics, proteomics, metabolomics, etc.), it soon should be possible to produce a complete blueprint of metabolic pathways for most of the major apicomplexan groups. While the results from *in silico* analysis require validation, in an effort to keep this review current we have noted several probable enzymes and activities strongly suggested by genomic sequencing results.

The opening sections of this chapter focus on energy metabolism in *Plasmodium falciparum*, due to the importance of this parasite as a human pathogen, the extensive biochemical information available for this species, and the comprehensive catalog of genome sequence data now entering the public domain. *P. falciparum* provides an adequate 'type specimen' for considering many aspects of apicomplexan biochemistry, including energy metabolism, but this phylum includes several thousand species, exhibiting considerable diversity. In particular, apicomplexan parasites infect a bewildering array of host species and tissues, which places diverse constraints on intracellular survival. The red blood cell, for example, is a highly inhospitable environment, with little endogenous metabolic activity and high potential for oxidative damage to the parasite. Most Apicomplexa reside within the far more hospitable environment provided by the cytoplasm of nucleated cells. The gut pathogen *Cryptosporidium parvum* most closely resembles anaerobic protozoa during its energy production, and appears to harbor only a relict mitochondrion.

Most of the Apicomplexa also undergo distinct morphological and physiological changes in the course of their complex life cycle, and differences in these developmental stages are also noted. For example, coccidian parasites such as *Toxoplasma* and *Eimeria* differ from *Plasmodium* in their ability to metabolize carbohydrates during certain stages of growth. The two developmental phases of *T. gondii* within mammalian hosts (tachyzoites and bradyzoites) vary markedly not only in their glycolytic pathways but also in their ability to store energy in the form of polymerized sugars (amylopectin). Energy metabolism during the sporulation stage of *Eimeria* species is highlighted by an accumulation of the carbohydrate mannitol, revealing a branch of the glycolytic pathway previously unknown in the protozoa.

CARBOHYDRATE METABOLISM

The earliest biochemical studies on *P. falciparum* characterized a robust glycolytic pathway during intra-erythrocytic replication. This parasite harbors no energy reserves in the form of polysaccharides or fatty acids, and therefore relies upon an extracellular supply of glucose as its primary energy source. Uninfected human erythrocytes utilize ~5 micromoles of glucose during a 24-hour period. In contrast, *P. falciparum*-infected erythrocytes consume ~150 micromoles of glucose per day, with peak levels approaching two orders of magnitude above the uninfected red cell background. Although it is conceivable that the host glycolytic machinery could operate at an accelerated rate during parasite infection, all evidence indicates that parasite pathways are responsible for this increased activity. *P. falciparum* enzymes are less sensitive than the host cell glycolytic machinery to the lower pH found in parasitized erythrocytes, and surveys of gene expression patterns using DNA microarrays show strong transcriptional stimulation of key parasite glycolytic enzymes during the early stages of infection. Although seemingly at odds with its parasitic lifestyle, recent studies indicate that the malaria parasite even provides the host cell with major products from its own carbohydrate metabolism, including ATP and reduced glutathione. A complete system for the uptake and consumption of glucose, and for the removal of the lactic acid end-product of anaerobic metabolism, has been described biochemically and confirmed by genomic analyses.

Nutrient uptake

As with other intracellular pathogens, *P. falciparum* faces the challenge of importing

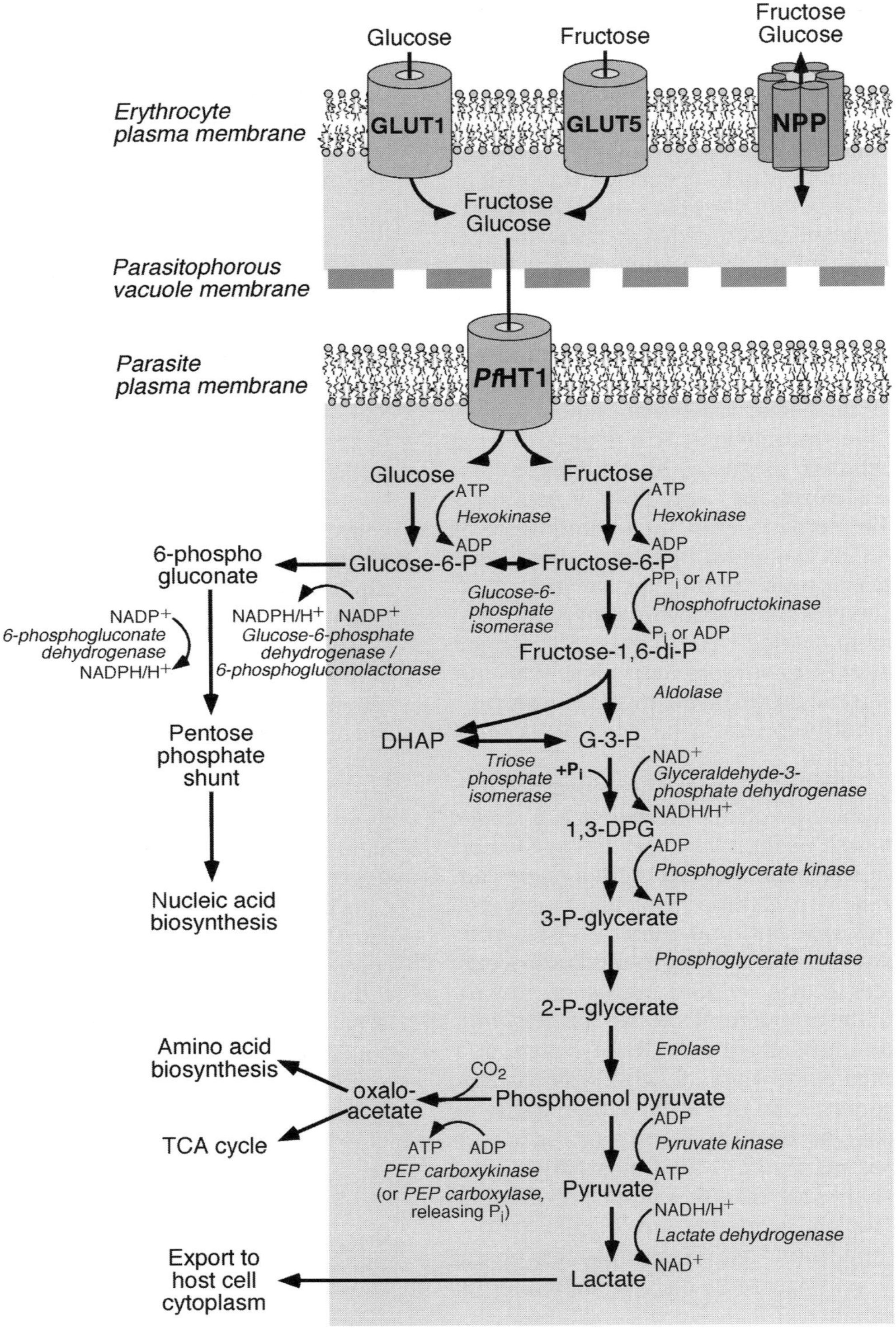

Glucose
Fructose
Fructose
Glucose
Erythrocyte plasma membrane
GLUT1
GLUT5
NPP
Fructose
Glucose
Parasitophorous vacuole membrane
Parasite plasma membrane
PfHT1
Glucose
Fructose
ATP
Hexokinase
ADP
ATP
Hexokinase
ADP
6-phospho gluconate
Glucose-6-P
Fructose-6-P
NADPH/H+ NADP+
Glucose-6-phosphate dehydrogenase / 6-phosphogluconolactonase
NADP+
6-phosphogluconate dehydrogenase
NADPH/H+
Glucose-6-phosphate isomerase
PPi or ATP
Phosphofructokinase
Pi or ADP
Fructose-1,6-di-P
Aldolase
Pentose phosphate shunt
DHAP
G-3-P
Triose phosphate isomerase
+Pi
NAD+
Glyceraldehyde-3-phosphate dehydrogenase
NADH/H+
1,3-DPG
ADP
Phosphoglycerate kinase
ATP
Nucleic acid biosynthesis
3-P-glycerate
Phosphoglycerate mutase
2-P-glycerate
Enolase
Amino acid biosynthesis
CO2
oxalo-acetate
Phosphoenol pyruvate
ADP
Pyruvate kinase
ATP
TCA cycle
ATP ADP
PEP carboxykinase
(or PEP carboxylase, releasing Pi)
Pyruvate
NADH/H+
Lactate dehydrogenase
NAD+
Export to host cell cytoplasm
Lactate

essential nutrients across multiple membrane barriers, including the host cell plasma membrane, the parasitophorous vacuole, and the plasma membrane of the parasite. The predominant pathway for hexose uptake into normal human erythrocytes is through an archetypal member of the major facilitator superfamily of integral membrane transporters, the *Homo sapiens* glucose transporter 1 (GLUT1) (Figure 7.6). This protein mediates the simple and rapid equilibration of glucose between the extracellular and intracellular media. GLUT1 is present on the erythrocyte membrane at high density, insuring that the intraerythrocytic glucose concentration is not rate limiting. Indeed, the capacity of GLUT1 exceeds the rate of glucose consumption by normal (uninfected) erythrocytes by orders of magnitude, providing sufficient glucose for even a *P. falciparum*-infected cell. Removing glucose from the medium or inhibiting GLUT1 by treatment with the fungal metabolite cytochalasin B causes an immediate decline in ATP levels within both the host cell and the parasite.

Another potential avenue for glucose uptake from the extracellular medium is provided by a parasite-induced increase in the permeability of the host-cell membrane. This phenomenon, known as the new permeation pathways (NPP), mediates the uptake of a variety of low molecular-weight solutes into infected red blood cells. Given the potency of GLUT1, the contribution of the NPP to glucose transport is probably small in human erythrocytes, but this pathway could be essential for malaria parasites infecting species where glucose uptake into the red cell is substantially lower (such as mice or birds).

After transport across the host-cell plasma membrane, the next potential barrier for glucose entry is provided by the parasitophorous vacuole membrane (PVM), which surrounds the dividing parasites. The PVM is endowed with a large number of high capacity, non-selective pores, of sufficient size to accommodate molecules up to 1400 Da. Although the passage of glucose through these channels has not been measured directly, the similarly-sized gluconate is freely diffusible across the PVM, suggesting that glucose in the erythrocyte cytoplasm should gain easy access to the parasite plasma membrane.

Goodyer and colleagues provided evidence for a facilitative, cytochalasin B-sensitive activity that is required for glucose transport directly into *P. falciparum* parasites, indicating that a transporter similar to human GLUT1 is expressed on the parasite plasma membrane. The recently published sequence of *P. falciparum* chromosome 2 includes a homolog of GLUT1, designated *P. falciparum* hexose transporter 1 (PfHT1) that localizes to the parasite plasma membrane. Recombinant expression of PfHT1 in a heterologous system provides evidence for a stereospecific hexose transporter that can be inhibited by cytochalasin B. PfHT1 also has the capacity to transport fructose, prompting experiments which have now

FIGURE 7.6 Hexose uptake and metabolism in *Plasmodium falciparum*-infected erythrocytes. Both glucose and fructose are transported into the host cell via high-capacity hexose transporters (GLUT1 and GLUT5, respectively). After diffusion through the parasitophorous vacuole, a single transporter (PfHT1) is responsible for hexose import into the parasite cytoplasm. Most apicomplexans express a pyrophosphate-dependent phosphofructose kinase, which leads to a net production of three ATP/hexose rather than the usual two. A potent lactate dehydrogenase activity produces lactic acid and regenerates NAD^+. Branching pathways include the pentose phosphate shunt, which generates NADPH critical for oxidative stress protection, and carbon dioxide fixation into oxaloacetate, a TCA-cycle intermediate.

shown that fructose is able to substitute for glucose in supporting *P. falciparum* growth. Other hexose transporters found on the erythrocyte cell membranes, such as the fructose transporter GLUT5, could therefore be relevant for parasite survival, in addition to GLUT1.

Glycolysis

After entry into the *P. falciparum* cytoplasm, most glucose is degraded to lactate via the anaerobic Embden–Meyerhoff–Parnas (EMP) pathway (Figure 7.6). Many of the genes encoding parasite EMP enzymes have been cloned and characterized, and those not previously identified have now been revealed through analysis of the *P. falciparum* genomic sequence. Several of the glycolytic pathway enzymes associate with the parasite cell membrane/cytoskeleton, possibly serving to increase pathway efficiency by facilitating substrate channeling. This provides an interesting parallel to the Kinetoplastida, where glycolytic enzymes are concentrated within a membrane-bounded organelle, the glycosome (Chapter 7, Part 2). The intense study of glycolysis in malaria parasites is reflected in the fact that glycolytic enzymes account for three of the four protein crystal structures currently available for *P. falciparum.*

Hexokinase, the first enzyme in the *P. falciparum* glycolytic pathway, has been cloned and shares 26% identity with human hexokinase. A long, hydrophobic C-terminal extension in the *P. falciparum* protein (relative to the human enzyme) may mediate association of hexokinase with the parasite plasma membrane/cytoskeleton, as noted above. Although the isolated parasite enzyme exhibits lower binding affinity for glucose than that of the host, a 25-fold increase in the conversion of glucose to glucose 6-phosphate is nevertheless observed upon infection, attributable to the high levels of parasite hexokinase expression. The high production of glucose 6-phosphate also increases substrate entry into the parasite pentose phosphate pathway, as discussed below.

Infected erythrocytes exhibit a 4–9-fold increase in glucose phosphate isomerase (GPI) activity when compared with normal erythrocytes, with enzyme levels reaching a maximum in schizont stage parasites. *P. falciparum* GPI shares 34% identity with the human isoform, and was the first malarial enzyme to be cloned by complementation of an *E. coli* mutant. Hexokinase also provides a likely alternate source of fructose 6-phosphate, through the phosphorylation of any fructose transported by PfHT1.

Phosphofructokinase (PFK) provides the first committed step in the *Plasmodium* glycolytic pathway, and is thus a potential control point. Most of our knowledge regarding *Plasmodium* PFK has been obtained from work on *P. berghei,* a rodent parasite. The malarial PFK displays atypical kinetic properties with respect to allosteric interactions and feedback inhibition. Unlike the red blood cell PFK, which becomes rate-limiting at low pH, the parasite enzyme is still active under the low pH conditions that accompany parasite infection (due to production of lactic acid). Like the human enzyme, *Plasmodium* PFK activity is enhanced by the substrate fructose 6-phosphate, but the positive effects of AMP and ADP and the negative influence of ATP are marginal in comparison to mammalian PFKs. The insensitivity of the parasite PFK to low pH and ATP, both of which normally depress glycolysis, may be the single major factor explaining the highly increased glycolytic flux found in *Plasmodium*-infected erythrocytes.

PFK enzymes of *Toxoplasma, Eimeria,* and *Cryptosporidium* have been shown to rely on pyrophosphate rather than ATP for full activity. Pyrophosphate dependence has not been assessed for the *Plasmodium* enzyme, but this appears likely, given that *Plasmodium* PFK exhibits greatest sequence similarity to that of other apicomplexans. A primary dependence on pyrophosphate could explain the observed lack of nucleotide regulation, and would also increase the efficiency of ATP production for the entire glycolytic pathway by 50%.

A three-dimensional structure is available for *P. falciparum* fructose bisphosphate aldolase, providing the potential for structure-based design of selective inhibitors. Although both proteins are well conserved throughout evolution, some structural differences are evident in the *Plasmodium* enzymes. The C-terminal tail of aldolase is critical for enzyme activity, and displays significant divergence from the human enzyme. Expression of antisense oligonucleotides specific for parasite aldolase has been reported to reduce enzyme activity, slowing ATP production and inhibiting *P. falciparum* growth in culture.

The crystal structure of triose phosphate isomerase (TPI) has also been solved, revealing significant differences from the human enzyme, including the presence of different amino acids adjacent to the catalytic residues and at the homodimer interface. These structural variations hold promise for development of small molecule inhibitors at the active site and/or peptide (or other inhibitors) that disrupt the dimerization required for function of this enzyme.

In keeping with the common themes noted above, transcription of *P. falciparum* glyceraldehyde 3-phosphate dehydrogenase (GAPDH) is induced ~35-fold within the parasitized red cell. Recombinant *P. falciparum* GAPDH specifically reduces NAD^+ rather than $NADP^+$ (as in the human enzyme), providing the NADH necessary to drive lactate dehydrogenase activity at the end of the glycolytic pathway (see below).

The specific activity of *Plasmodium* phosphoglycerate kinase (PGK) is greatest (~7-fold higher than host-cell PGK activity) during the later phases of intraerythrocytic growth. Activity is enhanced by the addition of a mild detergent, supporting the concept that malarial glycolytic enzymes may be associated with membranous or cytoskeletal components of the cell, as noted above.

All of the subsequent enzymes in the glycolytic pathway – which constitute the ATP-generating steps – can be recognized in the *P. falciparum* genome database, although only enolase has been characterized in any detail. Interestingly, both the GAPDH and enolase proteins (of *Plasmodium* and other apicomplexans as well) contain amino acid insertions and deletions (indels) that are otherwise restricted to organisms that harbor endosymbiotic plastid organelles (Chapter 12). These amino acid insertions provide valuable information concerning the evolutionary relationship of apicomplexans with plants, dinoflagellates, algae, and other plastid-bearing species, and also define variations in enzyme structure that may be exploited for pharmacological intervention.

The pyruvate kinase (PK) of other apicomplexan species, including *Toxoplasma gondii*, *Eimeria tenella* and *Cryptosporidium parvum*, has been characterized in some detail. Unlike mammalian and other PK enzymes, the apicomplexan enzyme is not significantly influenced by fructose 1,6-bisphosphate, but is strongly activated by the upstream monophosphate derivatives of glucose and fructose. This unusual regulation likely reflects the lack of regulation at the PFK step, as noted above. As a result, fructose 1,6-bisphosphate levels are a poor indicator of glycolytic load in apicomplexan parasites.

Rather than contributing to mitochondrial processes, most of the NADH produced by GAPDH is consumed by a potent lactate dehydrogenase activity in malaria parasites. Unlike other organisms, the activity of *Plasmodium* lactate dehydrogenase is insensitive to pyruvate or lactate, ensuring NAD^+ regeneration and the continued flux of glucose regardless of substrate or product concentration. *Plasmodium* lactate dehydrogenase is also distinguished by the ability of 3-acetylpyridine adenine dinucleotide (APAD, an NAD analog) to enhance activity *in vitro.* These unique kinetics and cofactor specificities have been exploited to develop diagnostic tests for parasitemia, and to assess drug efficacy. The crystal structure of *P. falciparum* LDH reveals significant differences from the host enzyme, presumably explaining these differing biochemical properties. A five-amino acid insertion (KSDKE) near the pyruvate contact region creates a unique active-site architecture and has generated considerable interest as a possible chemotherapeutic target.

The large quantities of lactic acid produced in the parasite cytosol might be expected to greatly reduce intracellular pH. While malarial glycolytic enzymes are resistant to acidic conditions, excessively acidic conditions could threaten the osmotic stability of the cell. Studies on *P. falciparum* have demonstrated the presence of a monocarboxylate transporter on the parasite surface that couples the efflux of both lactate and protons into the extracellular space. These moieties are thought to freely diffuse out of the parasitophorous vacuole into the red cell, where they may be excreted into the serum via erythrocyte transporters and/or the new permeation pathways (NPP) established by the parasite on the host cell surface.

In summary, the complete pathway of apicomplexan glycolytic catabolism produces either two or three molecules of ATP per mole of hexose (depending on whether PFK utilizes ATP or pyrophosphate). This energy production pathway is easily demonstrated in the cultured stages of *Plasmodium* parasites through the rapid and efficient conversion of radiolabeled glucose to lactate (up to 85% in *P. falciparum*), and by the rapid depletion of intracellular ATP upon glucose withdrawal. Inhibitors of mitochondrial electron transport or ATP synthetase induce only a minor, transient drop in parasite ATP pools, but the mitochondrion may well be essential for energy production in other phases of parasite growth, as discussed below.

Branching pathways

Although flux through the anaerobic Embden–Meyerhoff–Parnas pathway predominates during malarial intraerythrocytic growth of *Plasmodium* parasites, it is clear that not all glycolytic intermediates are funneled to lactate. The hexose monophosphate (or pentose phosphate) shunt utilizes glucose 6-phosphate to generate both reductive energy in the form of NADPH, and to supply ribose 5-phosphate, a precursor in the synthesis of nucleic acids (Chapter 9). In *P. falciparum*, the first steps in this shunt are performed by a single protein, glucose-6-phosphate dehydrogenase-6-phosphogluconolactonase, a novel bifunctional enzyme presently known only in *Plasmodium* species. Both this enzyme and the subsequent 6-phosphogluconate dehydrogenase step provide NADPH for glutathione reductase, an enzyme crucial for oxidative damage protection. As noted above, reduced glutathione generated within the parasite is exported into the erythrocyte, where it may serve to reduce oxidative stress within the host cell.

Plasmodium and other apicomplexan parasites require CO_2 for survival, and are able to fix carbon dioxide, presumably for the production of amino acids and citric acid-cycle intermediates. Two enzyme activities capable of reversibly fixing CO_2 to phosphoenolpyruvate (PEP) to yield oxaloacetate have been detected in

Plasmodium extracts. PEP carboxykinase transfers a phosphate to ADP, while PEP carboxylase releases free phosphate. Expression studies indicate that maximal PEP carboxykinase expression takes place during the sexual stages of parasite development in the mosquito, probably correlating with mitochondrial activation (see below). Oxaloacetate can then be converted into malate by the cytosolic malate dehydrogenase, followed by transport into the mitochondria and entry into the citric acid cycle.

MITOCHONDRIAL METABOLISM

While the cytoplasm of a generic eukaryotic cell (e.g. a mammalian fibroblast) may contain several dozen ATP-generating mitochondria, most apicomplexan parasites appear to harbor only a single one of these endosymbiotic organelles. The ~6 kb linear organellar genome of *Plasmodium* spp. is the smallest known mitochondrial DNA, encoding only three proteins: cytochrome *b*, and cytochrome oxidase subunits I and III (Chapter 12). Apicomplexan parasites possess at least a partial electron transport system and are capable of oxygen consumption. Mitochondrial proteins missing from the organellar genome are encoded in the nucleus, translated in the cytoplasm, and imported post-translationally. The mitochondrion occupies approximately the same fraction of subcellular volume as typically observed in mammalian cells; the presence of a only single organelle per cell probably reflects architectural constraints on organellar distribution during schizogony rather than reduced functional capacity.

Microscopic analyses of *C. parvum* in various intracellular life-cycle stages initially failed to recognize a mitochondrion, and attempts to detect mitochondrial enzyme activities have proven fruitless. Genomic sequence data reveal dozens of genes encoding probable mitochondrial proteins, however. These open reading frames, such as the mitochondrial chaperone CPN60 and several iron–sulfur cluster proteins, suggest that *C. parvum* may harbor a mitochondrion after all. Riordan *et al.* have identified a ribosome-studded organelle enclosed by a double membrane, posterior to the nucleus. Although acristate, this organelle accumulates electron potential-sensitive dyes, and is sensitive to cyanide and other electron-transport inhibitors.

Biochemical evaluation from various malarial species has produced conflicting evidence concerning the presence of oxidative phosphorylation or a complete citric acid (TCA) cycle during the blood stages of growth. Avian parasites (e.g. *Plasmodium gallinaceum*) are able to stimulate oxygen uptake in the presence of lactate, pyruvate and TCA intermediates, and can generate radioactive carbon dioxide from radiolabeled glucose. In contrast, the blood stages of mammalian malarial species appear devoid of such activities. Consistent with these data, morphological studies on avian malaria species show numerous cristae, the membranous folds that organize cytochromes to form a functional electron transport chain. Mammalian malaria species exhibit relatively few cristae during the blood-cell stages, suggesting limited activity.

Electron transport

The classical electron transport pathway consists of NADH dehydrogenase (complex I), succinate dehydrogenase (complex II), ubiquinone (coenzyme Q), cytochrome bc_1 oxidoreductase (complex III), cytochrome c, and cytochrome oxidase (complex IV) (Figure 7.7). Proton translocation occurs at complexes I, III, and IV, providing a gradient to drive ATP synthesis.

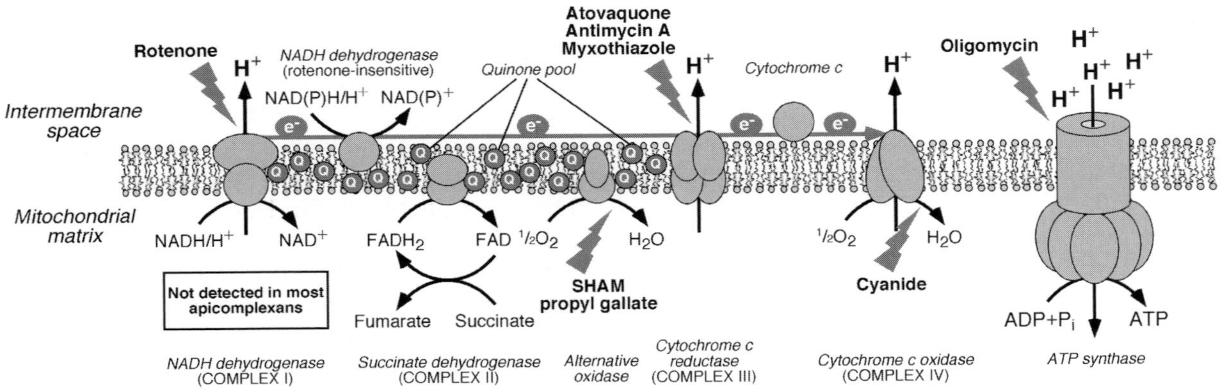

FIGURE 7.7 Electron transport and ATP production in *Plasmodium* mitochondria. The quinone pool (Q) is reduced from ubiquinone to ubiquinol by NADH dehydrogenase(s) or various flavoenzymes, including succinate dehydrogenase (Complex II), glycerol 3-phosphate dehydrogenase, dihydroorotate dehydrogenase, or malate:quinone oxidoreductase (not shown). The presence of a proton-pumping NADH dehydrogenase (Complex I) has been reported, but remains controversial. The presence or absence of this activity may differ between species. No evidence for this gene is found in the *P. falciparum* genome database, but a single-subunit NAD(P)H dehydrogenase has been identified in several apicomplexan genomes. Electrons are transported to molecular oxygen either through the cytochrome chain, which contributes to a proton gradient and subsequent ATP production, or to an alternative oxidase, terminating electron flow. This process is the target of various compounds demonstrated to inhibit *P. falciparum* growth (lightning bolts). The ubiquinone analog atovaquone is highly specific for the apicomplexan Complex III, and is currently used for anti-parasitic chemotherapy (in combination with other drugs, to avoid the emergence of drug resistance mutations).

In *P. falciparum*, a core pathway of electron transfer from ubiquinone to oxygen is now well established. Electrons can enter the pathway by transfer to ubiquinone from various flavin (FAD)-containing proteins, including succinate dehydrogenase (complex II), glycerol 3-phosphate dehydrogenase, or dihydroorotate dehydrogenase. *Plasmodium* mitochondria are capable of producing and maintaining a proton gradient through the actions of complex III and IV, as indicated by the mitochondrial localization of cationic fluorescent dyes that concentrate in regions of negative membrane potential.

Parasite respiration is 10- to 100-fold less sensitive than mammalian mitochondria to typical complex III inhibitors, such as antimycin A and myxothiazol, but ~1000 times more sensitive to hydroxynaphthoquinones (ubiquinone analogs). Atovaquone, for example, rapidly collapses the mitochondrial membrane potential and reduces oxygen consumption. This drug is effective against a variety of apicomplexans in the clinic, probably due to two atypical amino acids in the ubiquinone binding site of the parasite cytochrome *b*. As might be expected, however, drug-resistance mutations in cytochrome *b* emerge rapidly, and atovaquone is therefore recommended for antimalarial chemotherapy only in combination with other drugs.

Although oxygen consumption is greatly reduced after the addition of atovaquone or the complex IV inhibitor cyanide, ~30% of total parasite respiratory activity is resistant to these drugs. Residual oxygen consumption is sensitive to salicylhydroxamic acid (SHAM) and propyl gallate, well-characterized inhibitors of an alternative oxidase activity in plant (and certain fungal) mitochondria. This enzyme transfers electrons directly from ubiquinol to oxygen, circumventing the cytochrome complexes. SHAM and propyl gallate inhibit parasite growth and exacerbate the activity of cyanide and atovaquone, suggesting that both electron-transport pathways are important for mitochondrial function. Because the alternative oxidase bypasses the proton pumping machinery, this pathway does not contribute to energy production. Instead, the alternative oxidase pathway is thought to play a role in stress protection by channeling electrons safely to oxygen during conditions of excess or improper electron flow within the cytochrome chain.

Inhibitor studies suggest that electron flow from ubiquinone to oxygen is essential for apicomplexan survival, but mitochondrial contributions to the total parasite ATP pool remain uncertain. The entire *P. falciparum* electron transport system may simply serve to dispose of electrons generated by the mitochondrial flavoenzyme reactions noted above, particularly for dihydroorotate dehydrogenase, an enzyme critical for pyrimidine biosynthesis (Chapter 9). The import of substrates and proteins for other essential mitochondrial processes (such as heme biosynthesis) requires a proton motive force, and could explain the continued maintenance of a membrane potential in *Plasmodium*.

Oxidative phosphorylation

The presence or absence of the canonical 'bookends' of the electron transport system – a proton-pumping NADH dehydrogenase (complex I) and an ATP synthase complex – has been controversial, but both biochemical evidence and genomic sequence data challenge the traditional notion that *Plasmodium* mitochondria are less sophisticated than their mammalian counterparts. Using low concentrations of digitonin to selectively increase the permeability of *P. berghei* plasma membranes without affecting the functional integrity of parasite mitochondria, Docampo *et al.* were able to show that flavoprotein-linked substrates entering the system at ubiquinone (succinate, glycerol 3-phosphate, dihydroorotate) enhance ADP phosphorylation. Further, a collapse of membrane potential was observed in the presence of rotenone (a complex I inhibitor), and ATP production was sensitive to oligomycin, a well known inhibitor of mitochondrial ATP synthase. Assuming that these data are not attributable to contaminating host-cell activity, these results suggest that at least some *Plasmodium* mitochondria contain both NADH dehydrogenase and ATP synthase activity, despite the failure of many groups to detect rotenone sensitivity or ATP synthesis in *P. falciparum* mitochondria.

Bolstering these biochemical data, components of an F_1F_0 ATP synthase have been detected in the genomes of *P. yoelii* (a rodent malaria species closely related to *P. berghei*) as well as *P. falciparum*. The *P. falciparum* genome does not encode a rotenone-sensitive, proton-pumping complex I, but shows evidence for a single-subunit 'alternative' NAD(P)H: ubiquinone oxidoreductase (Figure 7.7). This enzyme could potentially couple NAD(P)H formed during the TCA cycle or other reactions to the electron transport chain, but is predicted to be rotenone-insensitive and would not translocate protons. The presence of a mitochondrial ATP synthase indicates that *Plasmodium* parasites are likely to engage in oxidative ATP production during at least part of the parasite life cycle. Although there is no convincing evidence for oxidative phosphorylation in erythrocytic stage *P. falciparum*, the marked

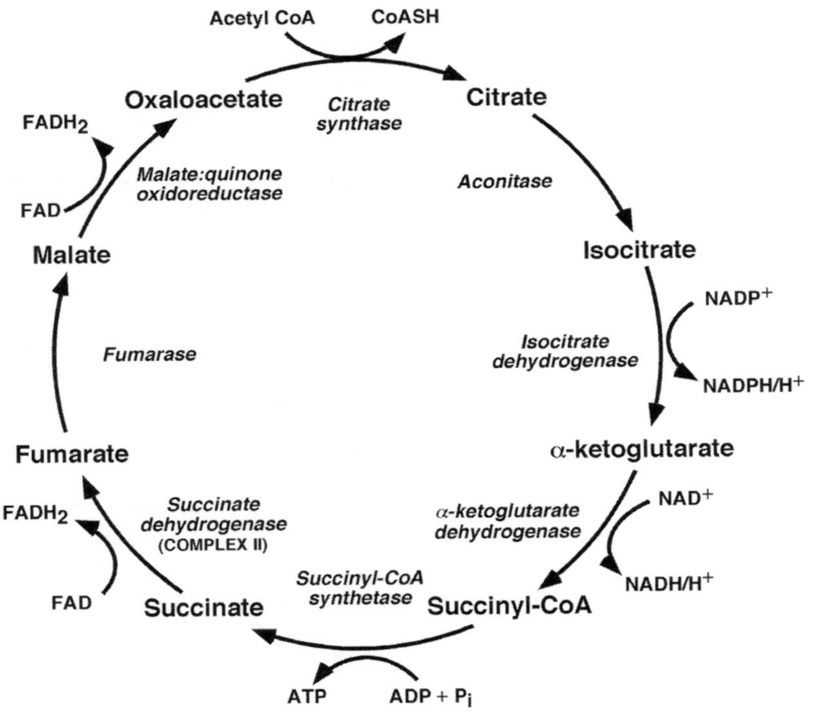

FIGURE 7.8 Proposed citric acid (TCA) cycle in *Plasmodium falciparum*. Although only a few of the enzyme activities required for this process have been measured in *P. falciparum* extracts (e.g. succinate dehydrogenase, isocitrate dehydrogenase), the parasite genome harbors genes for all eight proteins. The cycle may fully engage in a clockwise fashion (aerobic) in stages of the parasite life cycle that are difficult to isolate and characterize (e.g. within the mosquito midgut). At other times, portions of the cycle may flow counterclockwise to provide substrates for biosynthetic pathways (such as succinyl-CoA for heme production). The malate:quinone oxidoreductase is a membrane-bound flavoenzyme that, like succinate dehydrogenase, is capable of reducing ubiquinone directly.

increase in cristae formation observed during gametogenesis and in sexual stage parasites suggests a requirement for more robust mitochondria in the mosquito midgut.

Citric acid (TCA) cycle

In mammalian systems, the eight-enzyme TCA cycle produces NAD(P)H and $FADH_2$, which feed into the electron transport chain, leading to a highly productive output of up to 15 molecules of ATP per cycle. TCA enzyme activity has been detected in *P. falciparum* (Figure 7.8), and studies with digitonin-permeabilized *P. berghei* (see above) have shown that many TCA cycle intermediates are able to stimulate ATP production. Succinate dehydrogenase is clearly mitochondrial, but other activities, including malate dehydrogenase, appear to be located in the cytoplasm. Genomic data indicate that both *P. yoelii* and *P. falciparum* contain genes encoding all eight enzymes required to complete the TCA cycle. The presence of a malate: ubiquinone oxidoreductase (MQO) is of particular interest, as this enzyme has not previously been detected in any eukaryotic system.

Unlike the soluble malate dehydrogenase, MQO is a membrane-bound flavoenzyme capable of transferring electrons directly to ubiquinone rather than to NAD^+.

The mechanism of acetyl coenzyme A production in apicomplexan parasites is not entirely clear. The *P. falciparum* genome encodes a pyruvate dehydrogenase, but this enzyme is likely associated with the plastid (apicoplast) rather than cytoplasmic or mitochondrial metabolism. Acetyl CoA is probably synthesized from acetate using acetyl CoA synthetase enzymes, two of which are encoded in the parasite genome.

Interestingly, the *C. parvum* genome reveals a pyruvate:ferredoxin oxidoreductase (PFO) or pyruvate:$NADP^+$ oxidoreductase (PNO) homolog – a protein unique to this organism within the apicomplexan lineage, and once thought to be restricted to amitochondrial eukaryotes such as *Giardia lamblia* and *Trichomonas vaginalis*. This enzyme catalyzes the anaerobic oxidation of pyruvate, and subsequent transfer of electrons to ferredoxin (PFO) or $NADP^+$ (PNO). Pyruvate is decarboxylated in the process, yielding acetyl coenzyme A.

DEVELOPMENTAL VARIATION AND CARBOHYDRATE STORAGE

The phylum Apicomplexa includes >5000 species, and this diverse group encompasses many metabolic pathway variations. As discussed above for *Plasmodium* spp., most stages of all apicomplexans probably garner their energy supply from glycolysis under normal circumstances, secreting lactate as the major end-product. Mitochondrial electron transport has been detected in many species, and is undoubtedly important for survival, but as noted above, it is probable that electron transport usually plays only a marginal role in parasite ATP production. Glycolysis may be differentially regulated at different life-cycle stages however. Enzymes for carbohydrate storage and polysaccharide utilization are also required at specific stages of the life cycle in many apicomplexan parasites that exhibit a period of prolonged dormancy or encystation.

Stage-specific pathway variation

Studies on the regulation of energy metabolism in *Toxoplasma gondii* have been aided by the ability to cultivate both tachyzoite and bradyzoite stage parasites *in vitro*. Rapidly dividing tachyzoites are the primary cause of morbidity and mortality in humans and animals. Tachyzoites spontaneously differentiate into bradyzoites at low frequency, and bradyzoites also are capable of re-emerging as tachyzoites. The host immune response destroys *T. gondii* tachyzoites, but fails to eliminate the slowly dividing bradyzoite 'tissue cyst' form, which may remain viable in the brain, muscle, and other tissues for the life of the host. Bradyzoites (in undercooked meat) are probably the major form of transmission to adult humans, and recrudescence of latent tissue cysts is a serious complication associated with immunodeficiency (in AIDS, or in patients given immunosuppressive treatments for organ transplantation, or cancer chemotherapy).

Inhibitors of mitochondrial electron transport, such as antimycin and myxothiazol, cause the expression of bradyzoite-specific antigens on the cell surface, suggesting that this stage may rely on anaerobic glycolysis. Supporting this view, succinate dehydrogenase activity is not detected in bradyzoites. These observations probably also explain the relatively low sensitivity of bradyzoites to atovaquone.

Several glycolytic enzymes are differentially expressed in tachyzoites vs. bradyzoites.

Stage-specific isoforms of lactate dehydrogenase (71% identical at the amino acid level) are regulated at the level of transcription or mRNA stability. Studies on the recombinant enzymes indicate that the tachyzoite-specific isoform (TgLDH1) is inhibited by high concentrations of pyruvate, while TgLDH2 is not, possibly reflecting the importance of regenerating NAD^+ to fuel glycolysis during encystation. Stage-specific isoforms of enolase have also been identified; these proteins share ~74% identity, including a highly conserved amino acid insertion found previously only in *P. falciparum* and plant enolases. The bradyzoite-specific isoform exhibits low specific activity relative to the tachyzoite-specific isoform, but is also more resistant to elevated temperature, and possibly other stress factors as well. Glucose 6-phosphate isomerase (GPI) transcripts are observed in both tachyzoites and bradyzoites, but Western blotting reveals that the protein is present only in bradyzoite parasite extracts, suggesting post-translational regulation. The *T. gondii* EST database suggests the presence of a tachyzoite-specific GPI, but this gene has yet to be fully characterized.

Carbohydrate storage

The encysted stages of apicomplexan parasite development are commonly associated with the presence of large polysaccharide granules. In *T. gondii,* the bradyzoite tissue cyst wall can be stained with specific lectins (Figure 7.9), providing an obvious morphological feature distinguishing tachyzoites from bradyzoites. In *Eimeria* spp., several studies have shown a strong correlation between the size of polysaccharide granules and parasite viability, and an inverse relationship has been noted between granules and lactic acid production, consistent with the use of polysaccharides as an energy reserve in these parasites.

Although originally thought to be composed of glycogen (as found in bacteria, fungi, and animals), later analysis has shown that the polysaccharide granules in apicomplexans are composed of amylopectin, as found in plants and green algae. Chromatographic and NMR studies identified the substance as a variant of amylopectin known as 'floridean starch', found previously only in red algae. This similarity suggests an origin related to the origin of the apicoplast, a secondary endosymbiotic organelle acquired by engulfment of a eukaryotic alga, and retention of the algal plastid (Chapter 12).

A cytoplasmic starch synthase has been detected in *E. tenella* extracts, and this enzyme is able to transfer glucosyl donors to either mammalian glycogen or amylopectin. The enzyme prefers UDP-glucose (substrate for floridean starch) rather than ADP-glucose (substrate for amylopectin). The phosphorylase responsible for the first step in the mobilization of starch reserves has also been identified in *E. tenella* extracts, producing glucose 1-phosphate. As might be expected, the activity of this enzyme in parasite extracts increases during life-cycle stages that utilize polysaccharide. The phosphoglucomutase step, converting glucose 1-phosphate into the glycolytic substrate glucose 6-phosphate, has not yet been characterized.

The mannitol cycle (Figure 7.10) provides another noteworthy feature of carbohydrate metabolism in *E. tenella,* and probably other coccidian parasites as well. This pathway was once thought to be restricted to fungal species, but several stages of parasite development are capable of converting exogenous glucose to mannitol. High concentrations of mannitol (up to 300 mM) are present in fully developed *E. tenella* oocysts, accounting for up to 25% of the dry mass of the parasite.

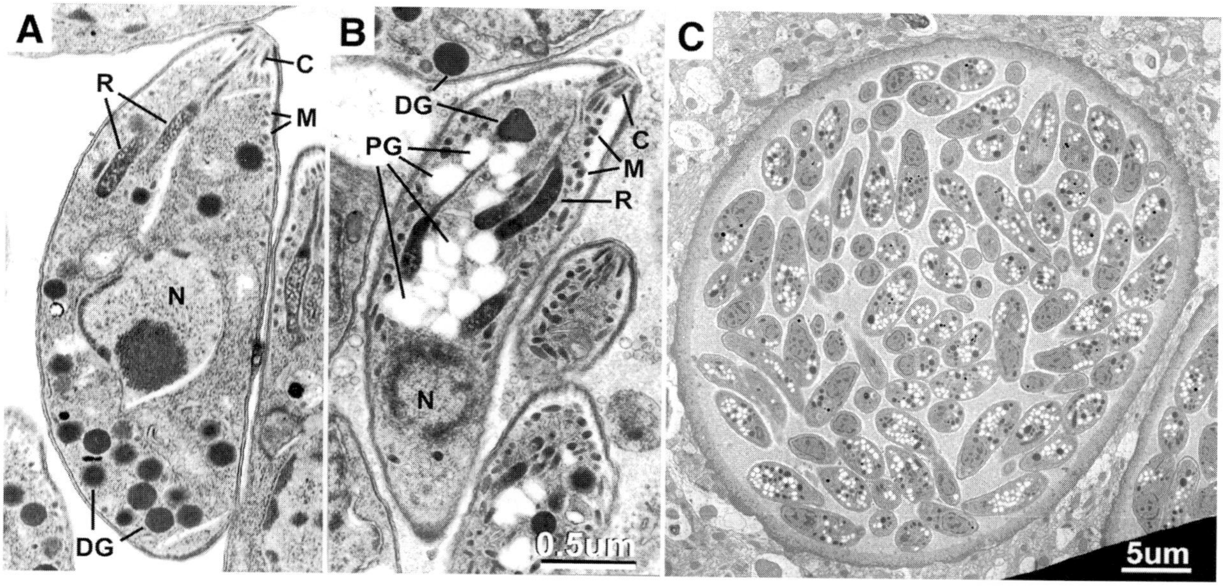

FIGURE 7.9 Electron micrographs of *Toxoplasma gondii*. (A) *T. gondii* tachyzoite, showing a nucleus (N) and various organelles of the 'apical complex': *C, conoid; R, rhoptries; M, micronemes*. Dense granules (DG) are involved in constitutive secretion, releasing parasite proteins into the parasitophorous vacuole. (B) The bradyzoite stage displays a similar complement of organelles (albeit organized in a slightly different manner), but is characterized by the appearance of amylopectin/floridean starch granules (PG). (C) Dozens of *T. gondii* enclosed within a single tissue cyst. Note the prevalence of starch granules within individual bradyzoites, and the proteoglycan wall surrounding the bradyzoite cyst. Electron micrographs provided courtesy of David J. Ferguson, Oxford University.

The first committed step in the mannitol pathway is carried out by mannitol 1-phosphate dehydrogenase (M1PDH), which converts fructose 6-phosphate into mannitol 1-phosphate. M1PDH is developmentally regulated by a homodimeric protein, which inhibits activity. Removal of this homodimer by denaturation, pH shift, or addition of a polyclonal antibody raised against the inhibitory protein restores M1PDH activity. A mannitol 1-phosphatase irreversibly converts mannitol 1-phosphate to mannitol and completes half of the cycle. This phosphatase is highly specific for its substrate, and has also been detected in *Toxoplasma gondii* and *Cryptosporidium* oocysts.

Temperature acts as an interesting regulator of mannitol production. Both M1PDH and the phosphatase are highly active at 41°C in *E. tenella* (the typical body temperature of the avian host), but essentially inactive at 25°C (outside of the host). In contrast, enzymes involved in mannitol catabolism – including mannitol dehydrogenase (MDH) and hexokinase (the same enzyme initiating the glycolytic pathway) – are substantially more active at ambient temperatures, correlating with the degradation of mannitol upon oocyst sporulation.

Specific inhibition of M1PDH by a 3-nitrophenyl disulphide (nitrophenide) leads to defective sporulation, validating the pathway as essential for completion of the parasite life cycle. Despite the formidable energy storage capabilites of starch granules, the

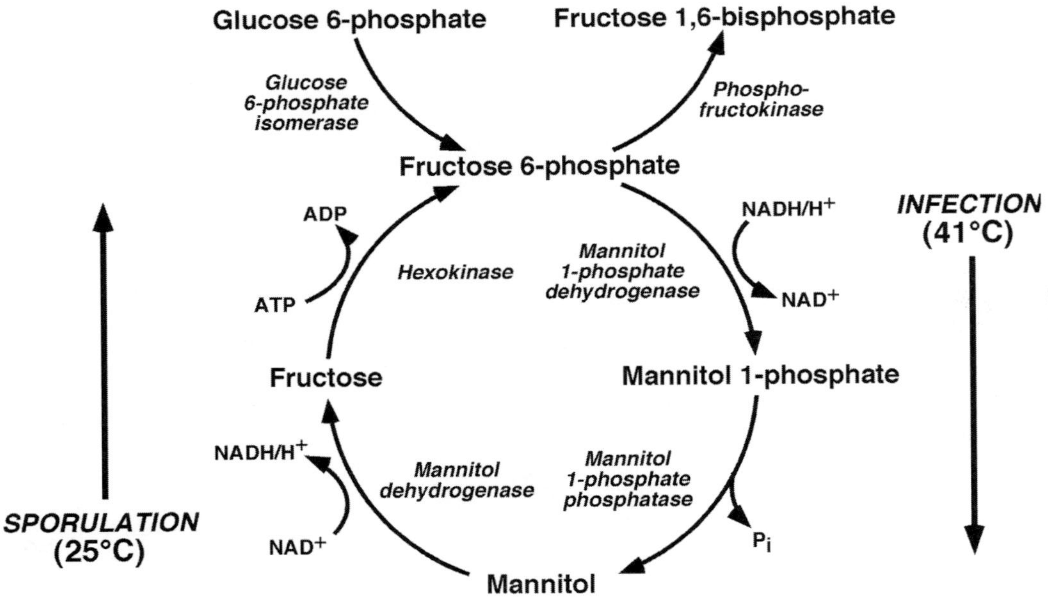

FIGURE 7.10 The mannitol cycle in *Eimeria tenella*. This cycle constitutes a major branching point from glycolysis in *E. tenella*. Copious amounts of mannitol are created within the avian host (body temperature 41°C) where it may function as an energy source during sporulation in the environment (25°C). Given its absence from the host, this cycle has been the subject of interest as a drug target, leading to the production of nitrophenide, a mannitol 1-phosphate dehydrogenase inhibitor that prevents mannitol production and leads to defective *E. tenella* sporulation.

function(s) of the mannitol cycle in *E. tenella* and other apicomplexans are not entirely clear (as is also the case for fungi). It is interesting to speculate that possible roles for mannitol production may include the production of NADH for electron transport and oxidative phosphorylation. Alternatively, mannitol might act as an osmoregulator or as a protectant under conditions of oxidative or salt stress.

MEDICAL SIGNIFICANCE

Although virtually all of the activities involved in apicomplexan carbohydrate metabolism and electron transport are well known from other systems, the pathways as a whole diverge significantly from host species, and have already proven useful as a target for chemotherapeutic intervention. Features that distinguish the Apicomplexa range from simple amino acid changes in enzyme cofactor- or substrate-binding sites, to the presence of pathways known from other organisms that are missing in the human host. Malate:quinone oxidoreductase was previously known only in a subset of prokaryotes, floridean starch in red algae, and the mannitol cycle in bacteria and fungi.

The selective interaction of atovaquone with the apicomplexan Complex III is already being exploited in the clinic, and has proven particularly useful in combination with the antifolate proguanil (Chapter 17). Synthetic peptides that mimic the homodimer interface of *P. falciparum* triosephosphate isomerase, and cottonseed oil derivatives (gossypols) that selectively bind to *P. falciparum* lactate dehydrogenase, have shown efficacy against malaria parasites in culture. Apicomplexan parasites harbor an unusually large intracellular supply of pyrophosphate, which is used (at least in part) in place of ATP for several critical reactions (phosphofructokinase, PP_i-dependent proton pump); analogs of pyrophosphate (bisphosphonates) have demonstrated significant activity against a variety of apicomplexans. The discovery of a PFO/PNO homolog in *C. parvum* is surprising, highlighting the need for further studies on energy metabolism in this organism, particularly in light of the poor selection of potential targets for chemotherapy currently available. The rapidly expanding genome and EST sequence databases for apicomplexan parasites, coupled with biochemical characterization and pharmacological studies, will surely yield new drugs that are active against these important pathogens.

FURTHER READING

Ajioka, J., Boothroyd, J.C., Brunk, B.P. *et al.* (1998). Gene discovery by EST sequencing in *Toxoplasma gondii* reveals sequences restricted to the apicomplexa. *Genome Res.* **8**, 18–28. See also:<http://www.cbil.upenn.edu/ParaDBs/Toxoplasma/ index.html>.

Bahl, A., Brunk, B.P., Coppel, R.L. *et al.* (2002). PlasmoDB: the *Plasmodium* genome resource. An integrated database providing tools for accessing and analyzing mapping, expression and sequence data (both finished and unfinished). *Nucl. Acids Res.* **30**, 87–90. See also: <http://PlasmoDB.org>.

Coombs, G.H., Denton, H., Brown, S.M.A. and Thong, K.-W. (1997). Biochemistry of the Coccidia. *Adv. Parasitol.* **39**, 141–225.

Krishna, S., Woodrow, C.J., Burchmore, R.J.S., Saliba, K.J. and Kirk, K. (2000). Hexose transport in asexual stages of *Plasmodium falciparum* and Kinetoplastidae. *Parasitol. Today* **16**, 516–521.

Lang-Unnasch, N. and Murphy, A.D. (1998). Metabolic changes of the malaria parasite during the transition from the human to the mosquito host. *Annu. Rev. Microbiol.* **52**, 561–590.

Riordan, C.E., Langreth, S.G., Sanchez, L.B., Kayser, O. and Keithly, J.S. (1999). Preliminary evidence for a mitochondrion in *Cryptosporidium parvum*: phylogenetic and therapeutic implications. *J. Eukaryot. Microbiol.* **46**, 52S–55S.

Roth, E. Jr. (1990). *Plasmodium falciparum* carbohydrate metabolism: a connection between host cell and parasite. *Blood Cells* **16**, 453–460.

Schmatz, D.M. (1997). The mannitol cycle in *Eimeria. Parasitol.* **114**, S81–S89.

Sherman, I.W. (1998). Carbohydrate metabolism of asexual stages. In: Sherman, I.W. (ed.). *Malaria: Parasite Biology, Pathogenesis and Protection*, Washington, DC: ASM Press, pp. 135–143.

Srivastava, I.K., Rottenberg, H. and Vaidya, A.B. (1997). Atovaquone, a broad spectrum antiparasitic drug, collapses mitochondrial membrane potential in a malaria parasite. *J. Biol. Chem.* **272**, 3961–3966.

Tomavo, S. (2001). The differential expression of multiple isoenzyme forms during stage conversion of *Toxoplasma gondii*: an adaptive developmental strategy. *Int. J. Parasitol.* **31**, 1023–1031.

Uyemura, S.A., Luo, S., Moreno, S.N. and Docampo, R. (2000). Oxidative phosphorylation, Ca(2+) transport, and fatty acid-induced uncoupling in malaria parasite's mitochondria. *J. Biol. Chem.* **275**, 9709–9715.

CHAPTER

8

Amino acid and protein metabolism

Juan José Cazzulo
Instituto de Investigaciones Biotecnológicas, Universidad Nacional de General San Martín, Buenos Aires, Argentina

INTRODUCTION

The metabolism of proteins and amino acids in parasites has been the subject of a number of studies, from different research groups, over the last two decades. As usual when studying metabolic pathways in parasitic organisms, most of the research has been focused on those single enzymes, or pathways, which differ sufficiently between the parasite and its host. This chapter will deal mostly with such enzymes and pathways, and therefore, since many aspects of amino acid metabolism in helminths are similar to those in mammals, we will be concerned essentially with parasitic protozoa.

Figure 8.1 summarizes the general features of protein and amino acid metabolism in most organisms, including parasites. Amino acids can be obtained from one or more of three possible sources, namely uptake from the medium where the parasite lives, proteolysis of host proteins, or biosynthesis. The last is, in general restricted, and many parasites, particularly the Protozoa, require most of the common amino acids necessary for protein synthesis. The free amino acids can then be used for the synthesis of parasite proteins, or in the biosynthesis of other compounds of physiological importance, such as polyamines, or as fuel, for energy generation. In the latter case, normally catabolism starts with the transfer of the amino group to a 2-keto acid, most frequently 2-ketoglutarate, to yield L-glutamate and the 2-keto acid corresponding to the original amino acid, in a reaction catalysed by an aminotransferase (also known as transaminase). The amino group is then liberated from L-glutamate as ammonia, in the reaction catalyzed by L-glutamate dehydrogenase; it can be excreted as such (ammonotelic organisms), or converted into urea (ureotelics) or uric acid (uricotelics). The first type of excretion is the most common in parasites, although a number produce urea,

Molecular Medical Parasitology
ISBN 0-12-473346-8

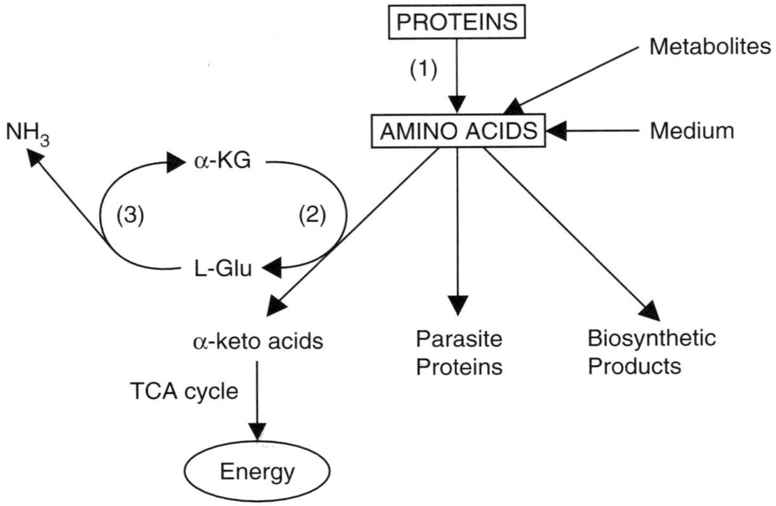

FIGURE 8.1 General metabolism of proteins and aminoacids. (1) Proteolytic enzymes; (2) aminotransferases; (3) glutamate dehydrogenases.

probably as a by-product of L-arginine catabolism and not through the urea cycle. The carbon skeleton of the 2-keto acid is further degraded, in general by pathways leading to intermediates of the tricarboxylic acid or Krebs cycle (TCA cycle), finally yielding CO_2 and water.

AMINO ACID SOURCES

Amino acid transport

Since all parasites are able to utilize exogenous amino acids, they must possess uptake systems to incorporate them from the medium. The uptake of proline has been studied in particular detail in *Leishmania* spp., mostly by the groups of A.J. Mukkada and D. Zilberstein, and is also important in other trypanosomatids. The insect stages of several hemoflagellates utilize proline preferentially, in good agreement with the presence of large quantities of this amino acid in the hemolymph. Proline transport has been proposed to be an active process, dependent on the proton motive force across the plasma membrane, and seems to be present in both promastigotes and amastigotes. However, the pH optimum of the process differs in the two stages and is related to the environment in which these *Leishmania* stages are found. Recent studies with axenic *Leishmania donovani* amastigotes and promastigotes indicate that the latter have two transport systems, one fairly specific for proline, the other, like the one present in amastigotes, with broad specificity. In contrast, proline transport in procyclic trypomastigotes of *Trypanosoma brucei* is not driven by a proton motive force, but instead depends on ATP and is independent of Na^+ or K^+ co-transport.

Active transport of L-arginine has been reported in promastigotes of *L. donovani*; its most distinctive feature is the lack of competition by lysine or ornithine. Diamidines, such as pentamidine and some derivatives, on the other hand, competitively inhibit arginine transport, and diamidines may be incorporated through this arginine transporter. The transport systems for arginine have also been characterized in *Trypanosoma cruzi* and in *Trichomonas vaginalis*.

The intraerythrocytic stages of *Plasmodium falciparum* require glutamine, the most abundant amino acid in plasma, and this requirement is satisfied through a parasite-induced increase in the permeability of the red blood cell to glutamine. Normal erythocyte membranes are relatively impermeable to this amino acid.

Cysteine is essential for growth of *Giardia lamblia*, which cannot synthesize cysteine *de novo* or from methionine. This parasite has two transport systems for cysteine, one non-saturable and probably representing passive diffusion, and a second specific for SH-containing amino acids. Cysteine also is essential for growth of *T. brucei* bloodstream trypomastigotes, but the transport system responsible for its uptake has not been characterized. *G. lamblia* has a functional L-alanine transporter, probably an antiport system catalyzing the exchange of alanine, serine, glycine and threonine. This antiport seems to function for both the influx and efflux of alanine. Parasitic protozoa frequently contain high concentrations of free amino acids. L-alanine is usually the most abundant, as reported in some trypanosomatids, *G. lamblia* and *T. vaginalis*. In contrast, glutamate, leucine, proline and valine are the most abundant amino acids in *Entamoeba histolytica*.

Helminths have amino acid pools that are usually larger than those in vertebrates, and these pools frequently consist of only a few amino acids, normally related to TCA intermediates, such as alanine, glutamate, proline, serine and glycine. In addition, a number of non-protein amino acids, such as 3-alanine, 2-, 3- and 4-aminoisobutyrate, 3- and 4-aminobutyrate, citrulline, ornithine, taurine, creatinine, homoserine, norleucine and norvaline, have been detected in helminths.

Proteolysis

The proteolytic enzymes in parasites have been widely studied during the last decade. In addition to their expected role in the nutrition of the parasite at the expense of host proteins, proteinases are relevant to important aspects of the host–parasite relationship, to the point of being considered virulence factors for some parasites. In some cases, proteinases are major parasite antigens, and thus participate in the development of the immune response of the host. In other cases, proteinases act to protect the parasite against the immune response, by degrading immunoglobulins bound to surface antigens. Proteinases are fundamental for the invasion of tissues by some parasites, either from outside the host's body, as in *Schistosoma* spp., or from the digestive tract into organs like the liver, in the case of *E. histolytica*. Proteolytic enzymes may participate in the penetration of intracellular parasites into the mammalian cell, and/or in the release of the parasites from the infected cell, to invade new cells and perpetuate the infection. Proteolytic degradation for nutritional purposes can be either intracellular (as in the malaria parasites) or extracellular, in the digestive tract (in some helminths).

Proteolytic enzymes

The general name of peptidases is applied to all the enzymes able to hydrolyze the peptide bond. Exopeptidases hydrolyze peptide bonds located at one end of the polypeptide chain. Aminopeptidases start hydrolysis from the N-terminus, and carboxypeptidases cut from the C-terminus. Endopeptidases hydrolyze internal peptide bonds and are frequently referred to as proteinases or proteases. The modern classification of peptidases, as developed by A.J. Barrett, divides them first by catalytic type (according to the chemical group mainly responsible for catalysis), then by molecular structure, and finally the individual peptidases. Two major catalytic types have been identified: those enzymes in which the nucleophilic attack of the peptide bond is performed by a group in the protein (the sulfhydryl of cysteine in the cysteine peptidases, and the hydroxyl of serine or threonine in the serine or threonine peptidases), and those enzymes in which the nucleophile is water, through a mechanism involving two aspartyl residues (aspartic peptidases) or a metal ion (metallopeptidases). The first approach for the classification of a new peptidase within a catalytic class is usually based on its pattern of response to inhibitors, but frequently a definitive classification requires knowledge of its full amino acid sequence. Within each of the five catalytic classes, enzymes are classified by molecular structure into clans (proteins with a similar folding pattern) and, within clans, into families (enzymes with significant sequence homology). To date, more than 130 peptidase families have been described, including some unclassified peptidase families whose reaction mechanisms are still undetermined. Peptidases belonging to all the catalytic classes have been described in parasites. Most of the enzymes studied so far are proteinases, but in some cases exopeptidases have been well characterized.

Proteolytic enzymes of protozoa

Cysteine proteinases

These enzymes are abundant in protozoa and often seem to be the major proteolytic activities present. In most cases where the genes encoding these enzymes have been cloned, they predict the synthesis of a pre-pro-enzyme, similar to the precursors of the papain family (C1 according to the Barrett nomenclature); in the case of trypanosomatids, a number of enzymes with an additional C-terminal extension have also been described.

Cysteine proteinases are particularly abundant in amastigotes of *Leishmania mexicana*, in epimastigotes of *T. cruzi*, and in bloodstream trypomastigotes of *T. brucei*. Coombs and co-workers have classified these enzymes into three types: Types I and II, both cathepsin L-like, and Type III, cathepsin B-like. The genes encoding Type I cysteine proteinases predict a long C-terminal extension, are present in multiple copies, are arrayed in tandem, and frequently several genes are simultaneously expressed, generating a mixture of isoforms. The expression of Type I genes is developmentally regulated at different levels in different stages of the parasite life cycle. Type I enzymes have been found in all Trypanosomatids studied to date, including the Leishmania, *T. cruzi* (cruzipain, also known as GP57/51 or cruzain), *T. brucei* (trypanopain), *Trypanosoma congolense* (congopain), *Trypanosoma rangeli*, and *Crithidia fasciculata*. In the case of the Type I cysteine proteinases of *L. mexicana* (CPb) and of cruzipain, a complete tandem array has been sequenced; the gene at the 3′ end, which has a shorter C-terminal extension, is not expressed. Type I cysteine proteinases (CPb in *L. mexicana*) are cathepsin L-like endoproteinases with good gelatinolytic activity. The C-terminal extension, which has substantial homology among different trypanosomatids, and contains eight conserved cysteine residues, is lost during processing in *T. brucei* and *L. mexicana*; in these cases, the mature proteinases are similar in size to papain. In contrast, cruzipain, congopain and the Type I cysteine proteinases from *T. rangeli* and *C. fasciculata* retain the C-terminal extension in the mature enzyme, and thus have a molecular weight appreciably higher than those of the papain-like cysteine proteinases. The longest C-terminal extension described so far, that of cruzipain (130 amino acids) undergoes a number of post-translational modifications, including N-glycosylation. Cruzipain and congopain are strongly antigenic. In the case of cruzipain, the C-terminal extension is responsible for antigenicity.

A truncated recombinant form of cruzipain (cruzainΔc), lacking the C-terminal domain, has been crystallized and its three-dimensional structure, complexed with several synthetic peptidic inhibitors, has been determined. The single polypeptide chain of 215 amino acids folds into two distinct domains which define the active site cleft, as well as seven substrate-binding subsites, a structure similar to that found in other members of the C1 family. The active site cleft contains the catalytic triad, formed by Cys25 and His159, which form a thiolate–imidazolium ion pair essential for the enzymatic activity, and Asn175. The major structural differences between cruzipain and papain are found in the regions of the loops and turns. Processing of the pro-enzymes to their mature forms may be due to self-activation, but at least in the case of the CPb enzymes from *L. mexicana*, it appears that CPa can process the CPbs. The CPb enzymes are specific to the amastigote stage, and their expression is induced in promastigotes when the pH drops to about 4, accompanying differentiation to the amastigote stage.

Type II and Type III cysteine proteinases detected in *Leishmania* (CPa and CPc) are encoded by single copy genes (*lmcpa* and *lmcpc*), and lack gelatinolytic activity. Mature CPa has a characteristic three-amino acid insertion close to the N-terminus, and a very short C-terminal extension. CPc completely lacks a C-terminal extension, is amphiphilic and possibly associated with membranes.

Trypanosomatid Type I and II cysteine proteinases are lysosomal enzymes, and recently a role for the pro-domain in their targeting to the lysosome has been proposed. At least in the case of *L. mexicana*, trafficking to the lysosomes seems to occur via the flagellar pocket. In addition to the lysosomal forms, which certainly account for most of the activity, membrane-associated forms of Type I and Type III enzymes also appear to be present.

P. falciparum and other malarial parasites contain a complex array of proteolytic enzymes, representing all catalytic classes. Plasmodial proteinases are thought to be involved in red blood cell invasion and rupture, and are responsible for the digestion of hemoglobin inside the food vacuole. Several cysteine proteinases have been described in *P. falciparum* and other *Plasmodium* species, specifically falcipains 1 and 2. Falcipain 1 is a cathepsin L-like enzyme. The gene sequence predicts a 67 kDa protein which contains a signal peptide and a large pro-enzyme domain, about 280 amino acids long, and thus larger than the mature cysteine proteinase, which has an apparent M_r of 26.8 kDa. Falcipain 2 is the major cysteine proteinase in the trophozoite.

E. histolytica secretes a number of cysteine proteinases into the medium. Those best characterized have apparent molecular masses of 22–27 kDa (amoebapain), 26–29 kDa (histolysain), 16 kDa (cathepsin B) and 56 kDa (neutral cysteine proteinase). At least seven genes encoding enzymes from this class have been identifed from *E. histolytica*; *Eh-CPp1* encodes amoebapain whereas *Eh-CPp2* encodes histolysain. The mature proteins are 86.5% identical and are cathepsin B-like. As in most other parasitic protozoa, these enzymes are found in organelles similar to lysosomes, and also, to a lesser extent, at the outer cell surface.

T. vaginalis secretes a complex mixture of cysteine proteinases, with apparent molecular masses ranging from 20 to 96 kDa. However, cloning and sequencing of several of these cysteine proteinase genes has not predicted mature enzymes larger than 24 kDa. These sequences are more similar to mammalian cathepsin L than to cruzipain or the Type I and II cysteine proteinases from *L. mexicana*. *T. vaginalis* enzymes have both cathepsin L-like and cathepsin B-like specificity. The cysteine proteinases in the cattle parasite *Tritrichomonas foetus* appear to be similar to those in *T. vaginalis*. *T. foetus* produces and secretes several cysteine proteinases into the medium, and seven different genes have been cloned and sequenced.

G. lamblia trophozoites contain at least 18 proteolytic activities with apparent molecular mass values from 30 to 211 kDa, with responses to inhibitors suggesting that many are cysteine proteinases. These activities appear to be lysosomal. Recently, McKerrow and co-workers have cloned and sequenced a cysteine proteinase, with homology to cathepsin B, that is required for excystation of the parasite.

Serine proteinases

These enzymes have been reported much less frequently than cysteine proteinases in parasitic protozoa. *T. cruzi* oligopeptidase B has homology to members of the prolyl oligopeptidase family (S9) of serine peptidases, and appears to be indirectly involved in the penetration of trypomastigotes, by activating a factor which induces a transient increase of cytosolic Ca^{2+} in the host cell. A *T. brucei* enzyme with 70% identity to the *T. cruzi* enzyme has no activity with typical prolyl oligopeptidase substrates, but instead prefers substrates characteristic of trypsin-like enzymes. Serine oligopeptidases with similar specificity also have been purified and partially characterized from *T. congolense* and *L. amazonensis*. A prolyl endopeptidase, reported to belong to the prolyl oligopeptidase family, has been purified and partially characterized from *T. cruzi* (Tc80), is able to hydrolyze human collagens I and IV at neutral pH, and its substrate specificity clearly indicates that it is different from the oligopeptidases B mentioned above. *T. cruzi* also contains a lysosomal serine carboxypeptidase, belonging to the S10 family, that is specific for C-terminal hydrophobic amino acids.

P. falciparum serine and cysteine proteinases participate together in the invasion and/or rupture of red blood cells. The 75 kDa serine protease is bound as an inactive precursor to the membranes of schizonts and merozoites by a glycosylphosphatidyl inositol (GPI) anchor and is activated after release by a specific phospholipase C. Since the purified enzyme from merozoites is able to cleave two major red blood cell proteins, glycophorin A and Band III protein, it may be involved in the formation of the parasitophorous vacuole from the erythrocyte plasma membrane. PfSUB1, a Ca^{2+}-dependent, membrane-bound, subtilisin-like serine proteinase, processes the major merozoite surface protein, MSP-1. The polypeptides produced seem to be involved in the receptor-ligand interactions which initiate the invasion of the red blood cell. The precursor of PfSUB1 seems to be a Type I integral membrane protein. Both PfSUB1 and a second member of the group, PfSUB2, are located in the dense granules of the merozoites, and are processed and secreted after completion of the red cell invasion.

Metalloproteinases

These enzymes have been described in a number of parasitic protozoa, but only those present in promastigotes of *Leishmania* spp.

have been thoroughly characterized. Leishmanolysin or gp63 (63 kDa glycoprotein) is bound to the promastigote membrane by a GPI anchor, and is the most abundant protein exposed at the promastigote membrane (about 5×10^5 molecules per cell). A homologous enzyme with an acidic optimal pH, found as a soluble protein in lysosomes, is present at lower levels in amastigotes. The membrane-bound enzyme is a Zn^{2+}-dependent HEXXH metalloproteinase, able to degrade components of the extracellular matrix, like fibrinogen. Leishmanolysin is *N*-glycosylated at its three potential sites, and contains three domains; two have novel folds, while the N-terminal domain is similar to the catalytic domain of the metzincin class of zinc proteinases (family M8). Leishmanolysin is produced as a proenzyme and appears to be activated by a Cys-switch mechanism, as proposed for matrix metalloproteinases. Recently, genes presumably encoding leishmanolysin homologs have been cloned and sequenced from *C. fasciculata, Herpetomonas samuelpessoai, T. brucei* and *T. cruzi*. Several membrane-bound metalloproteinases have been detected in *T. cruzi* and several other Trypanosomatids, but have not been characterized. Two soluble metalloproteinases, a carboxypeptidase and an aminopeptidase, also have been identified in *L. major*.

P. falciparum contains a recently characterized metalloproteinase, falcilysin, which has homology with pitrilysin and other members of the M16 family. Falcilysin is located in the food vacuole, and seems to be involved in the degradation of the peptides derived from the action of the falcipains and plasmepsins. A gene encoding an aminopeptidase belonging to the M1 family of metallopeptidases also has been cloned from *P. falciparum*. Antibodies raised against a peptide predicted from the sequence recognizes two schizont proteins with M_r values of 96 and 68 kDa.

Aspartic proteinases

These enzymes have been described and studied in most detail in malarial parasites. In contrast, no members of this class have yet been identified in Trypanosomatids. *P. falciparum* contains aspartic proteinases with molecular masses ranging from 10 to 148 kDa. Plasmepsins I and II have 73% amino acid identity, but cleave hemoglobin with different specificities. However, both enzymes cleave the Phe 33/Leu 34 bond in the α-chain of native hemoglobin, which destabilizes hemoglobin, leading to its unraveling and further proteolysis. The crystal structure of a complex of plasmepsin II with the inhibitor pepstatin is similar to that of mammalian and fungal aspartic proteinases. The pro-plasmepsins are Type II integral membrane proteins that are cleaved to yield the soluble mature enzymes. The interaction of the pro-domain of pro-plasmepsin II with the catalytic moiety is unusual, since, instead of blocking the active site cleft, the pro-domain interacts with the C-terminal domain keeping both domains apart and preventing the formation of the active site. Genes encoding aspartic proteinases have also been identified in other Apicomplexa, including *Toxoplasma, Eimeria* and *Cryptosporidium*.

Threonine proteinases

The 20S proteasome, belonging to the threonine peptidase class, is a high molecular weight complex which presents, in eukaryotes, three proteolytic activities with different specificity, and is the major proteolytic activity in the cytosol and within the nucleus. 20S proteasomes have been recently identified in *E. histolytica, T. brucei, T. cruzi, L. mexicana, P. falciparum* and *P. berghei, G. lamblia* and *Entamoeba invadens*. The 20S proteasome has an apparent M_r of about 670 kDa, and is made up of a number of subunits with apparent M_r values ranging, in different parasites, from 22 to 35 kDa, and with pI values ranging from 4.5 to 8.5. Genes encoding proteasome α subunits have been cloned from most of these protozoa and in general exhibit considerable homology to proteasome subunits from higher animals. A proteasome-activating protein, PA26 from *T. brucei,* polymerizes spontaneously into a heptamer ring and is able to activate the 20S proteasomes from the rat as well as *T. brucei.* The active form of the proteasome involves, in addition to the 20S catalytic complex, two 19S activating complexes, made up of a number of subunits, some of which possess ATPase activity. A gene encoding a proteasome S4 ATPase, a member of the 19S regulatory complex, has been cloned from *P. falciparum.* Recently a 26S proteasome with ATP-dependent chymotrypsin-like activity has been purified from *T. cruzi.* The purified complex consists of about 30 proteins, with molecular masses ranging from 25 to 110 kDa. The 26S proteasome participates in protein turnover during differentiation from trypomastigotes to amastigotes, and its inhibition leads to the accumulation of ubiquitinated proteins. In addition to the 20S proteasome, *E. histolytica* trophozoites contain an 11S proteinase, which seems to be a hexamer of identical 60–65 kDa subunits, consisting of two trimers which are still active under weakly denaturing conditions.

Proteolytic enzymes of helminths

Schistosomes have a complex life cycle, involving a snail as intermediate host. Proteinases are involved in the active skin penetration of cercariae liberated by the snail during their transformation into schistosomula. A 28 kDa soluble serine proteinase, with similarities to pancreatic elastases, is released by the acetabular glands during penetration. A similar antigenically cross-reactive proteinase is GPI-anchored to the tegument of mechanically transformed schistosomulae. A metalloproteinase activity associated with the surface of schistosomula and adults also has been described. Adult schistosomes feed on erythrocytes, which are lysed in the gut; the hemoglobin is then digested. Two immunodominant schistosome antigens, Sm31 and Sm32, are cysteine proteinases. Sm31 is a cathepsin B-like cysteine proteinase, whereas Sm32 is an asparaginyl endoproteinase, homologous to the *Canavalia ensiformis* legumain, a cysteine proteinase belonging to the C13 family, quite different from the papain family. In addition, two cathepsin L-like enzymes have been recently described. L_1 is more closely related to cruzipain, and L_2 is more closely related to human cathepsin L. *Fasciola hepatica* also has a snail as intermediate host, but its cercariae infect man by ingestion and penetration through the intestinal mucosa. A number of cysteine proteinase genes have been amplified from cDNAs of adult worms, including those encoding a cathepsin L-like enzyme which contains unusual 3-hydroxyproline residues and a cathepsin B-like enzyme found in newly excysted juveniles and in adults. A secreted serine peptidase, with dipeptidylpeptidase activity, has been identified in juveniles and adults. Aspartic proteinases are also present. Recently, a 28 kDa cruzipain-like cysteine proteinase has been characterized in *Paragonimus westermani*; the enzyme is located in the intestinal epithelium, suggesting that it may be secreted and could be involved in nutrient uptake.

Proteinases belonging to all catalytic classes have been described in parasitic nematodes. *Ancylostoma caninum* secretes 90 and 50 kDa metalloproteinases important for skin penetration, and contains a number of cathepsin B- and L-like cysteine proteinases. Third and fourth stage larvae of the ovine parasite *Haemonchus contortus* contain a 46 kDa metalloproteinase able to digest fibrinogen and fibronectin, and a cysteine proteinase is involved in the degradation of extracellular matrix proteins by adults. Adult *Dirofilaria immitis* contains a 42 kDa aspartic proteinase, which is present in a number of different tissues, and a subtilisin-like serine proteinase, furin. Metalloproteinases and cysteine proteinases are present in secretions and extracts of L3 and L4 larvae, and cysteine proteinase inhibitors inhibit the third to fourth larval stage molt.

Proteinase inhibitors

Mammals possess proteinase inhibitors capable of inhibiting all catalytic classes; most are class-specific, but a few, like α_2-macroglobulin, have broad specificity. These proteinase inhibitors have important regulatory functions, and the disruption of the balance between a proteinase and its inhibitor(s) is frequently the cause of disease. During co-evolution with their hosts at least some parasites seem to have developed proteinase inhibitors that can protect them from destruction by host proteolytic activities, such as proteinases released from leukocytes in the case of bloodstream parasites, or digestive juices in the case of intestinal parasites.

Little is known about proteinase inhibitors in parasitic protozoa. Recently a serpin (serine proteinase inhibitor) has been cloned and sequenced from *T. gondii*. This inhibitor belongs to the Kazal family, inhibits trypsin, chymotrypsin, elastase and tryptase, and is located in the dense granules of the tachyzoites. Cystatin-like proteinase inhibitors act on C1 family cysteine proteinases, and have been detected, but not characterized, in several protozoan parasites.

Far more is known about proteinase inhibitors from helminths. Nematodes contain a number of serine proteinase inhibitors, some of which belong to the serpin family, the others to a different family of small proteinase inhibitors, for which the name smapins has been recently proposed. Related inhibitors also are found in trematodes, such as *Schistosoma* spp. Much less is known about cysteine proteinase inhibitors.

Little is known of the actual functions of these inhibitors, although they are presumably important in the host–parasite relationship. A microfilarial serpin specifically inhibits two human neutrophil enzymes, cathepsin G and elastase, and may protect the parasite in the bloodstream from neutrophil attack. With reference to smapins, their function seems clear in at least two cases: in *Ascaris* they protect against the proteolytic environment of the host's gut, and in the hookworm *Ancylostoma caninum* they act as anticoagulants, inhibiting several of the serine proteinases involved in the blood clotting cascade.

Functions of parasite proteolytic enzymes

Digestive functions

Proteinases from different parasites participate in the digestion of host proteins for nutrition. Protein digestion is usually extracellular in the helminths, with the proteinases secreted from the gut epithelium in those worms with a digestive tract. In protozoa most protein digestion occurs in lysosomal compartments. Hemoglobin degradation has been extensively studied in two bloodstream parasites, *Plasmodium* spp. and *Schistosoma* spp. *P. falciparum* has three major proteolytic systems in the food vacuole, a lysosome-like organelle responsible for the massive hemoglobin digestion characteristic of malarial trophozoites: the falcipains, the plasmepsins, and falcilysin. Cysteine proteinase inhibitors kill malarial parasites, both *in vivo* and *in vitro*, with the concomitant accumulation of hemoglobin in the digestive vacuole, suggesting a major role for cysteine proteinases in hemoglobin degradation. Specific inhibition of the plasmepsins also results in the killing of malarial parasites, again with the concomitant accumulation of hemoglobin in the digestive vacuole. However, despite a number of thorough studies, the actual role of the falcipains as hemoglobinases, as well as the relative roles played by the falcipains and the aspartic proteinases plasmepsins I and II in plasmodial nutrition, are still controversial points. Although plasmepsins are able to cleave native hemoglobin, falcipain may be required for the activation of the pro-plasmepsins. Hemoglobin digestion is completed by the action of two metalloproteinases, falcilysin and an aminopeptidase. Falcilysin has a specificity complementary to that of falcipains and plasmepsins and acts within the vacuole, while the aminopeptidase degrades in the cytosol the peptides arising from the concerted action of the endopeptidases.

As in the case of *Plasmodium* spp., *Schistosoma* spp. do not appear to have a single 'hemoglobinase'; hemoglobin degradation again is the result of the concerted action of both cysteine and aspartic proteinases. Legumain (Sm32) may be involved in the processing and activation of the cathepsins B (Sm31) and L, the aspartic proteinase cathepsin D, and the dipeptidylpeptidase cathepsin C. Although there is still controversy, it seems likely that cathepsins B, L, and D all participate in the process of hemoglobin digestion.

In trypanosomes, proteinases appear to be localized in specialized digestive organelles. For example, in *T. cruzi* epimastigotes cruzipain is located in reservosomes, organelles of a lysosomal nature that are present only in this parasite stage and fill with proteins from the external medium during epimastigote growth. These proteins are consumed during the differentiation to metacyclic trypomastigotes, leading to the complete disappearance of the reservosomes. In *L. mexicana* the major cysteine proteinases are present in the megasomes, which are also lysosomal organelles.

Developmental regulation

Most protozoan parasites have complex life cycles, involving several differentiation steps, and proteolytic activities appear to be important in these processes. In *T. cruzi* cruzipain appears to be important in the differentiation of epimastigotes to metacyclic trypomastigotes. Cysteine proteinase inhibitors that can inhibit cruzipain block differentiation, and cruzipain overexpression is linked to an increase in the differentiation to metacyclics. Differentiation during the *T. cruzi* life cycle is much more sensitive to cruzipain inhibitors than the multiplication of the replicative stages, epimastigotes and amastigotes. The proteasome also is involved in differentiation. Specific inhibition of proteasome activity by lactacystin prevents the differentiation of trypomastigotes to amastigotes, and the intracellular differentiation of amastigotes to trypomastigotes. Moreover, a role for the proteasome in protozoan cell remodeling is not restricted to *T. cruzi*, since recent evidence suggests a similar role in *T. brucei*, *P. berghei* and *P. falciparum*, as well as in *E. invadens*. Lactacystin also inhibits the progression of the cell cycle in *T. brucei*, as well as the intracellular replication of *T. gondii*.

Protozoan proteinases as virulence factors

Although the cysteine proteinases from Trypanosomatids seem to be important for growth and differentiation, as suggested by the effect of irreversible inhibitors *in vivo*, their actual functions are far from being demonstrated. However, several lines of evidence indicate that they act as virulence factors in some parasites. The most promising approach to understand the role of these enzymes is the generation of null mutants. In *L. mexicana*, knockout mutants lacking each of the three types of cysteine proteinases have been generated. Null mutants for the gene encoding CPa cysteine proteinase exhibit a wild-type phenotype, suggesting that the enzyme is not essential for viability. Null mutants for CPb cysteine tandem arrays also are able to grow and differentiate, although their macrophage infectivity *in vitro* is considerably reduced, and lesions caused in mice are much smaller than those caused by the wild type. Re-incorporation of one of the Type I genes to the null mutant restores full infectivity. Double-null mutants for CPa and CPb have a phenotype similar to that of the CPb null mutants. Mutants null for CPc also are considerably less infective to macrophages *in vitro,* but only a little less infective than wild type parasites to BALB/c mice. Since the low infectivity ΔCPa/CPb mutants did not produce an established disease, but were able to elicit a protective immune response, these mutants have been proposed as candidates for live attenuated vaccines against leishmaniasis.

E. histolytica pathogenesis requires penetration of the trophozoite through the intestinal mucosal cells and the basal membrane. The amoebal cysteine proteinases are highly active with collagen types I, IV and V, fibronectin and laminin, which is consistent with the participation of these enzymes in penetration. Although several of the cysteine proteinase genes present in pathogenic *E. histolytica* are also present in the non-pathogenic *Entamoeba dispar,* there seems to be a clear correlation between the secretion of cysteine proteinases and virulence. The gene encoding CP5, an *E. histolytica* cysteine proteinase able to interact with membranes but still retain its proteolytic activity, is missing in *E. dispar.* Neutral cysteine proteinases are able to degrade human IgGs, in addition to their participation in tissue invasion, thus protecting the parasite against the immune response of the host.

The role of the cysteine proteinase activities present in *T. vaginalis* is still not well understood, but some membrane-bound forms have been linked to adhesion to epithelial cells and cytotoxicity. Some of the secreted proteinases seem to participate in the degradation of immunoglobulins (serum and secreted IgA and IgG), and may be involved in defensive mechanisms of the parasite against the immune system of the host. A *G. lamblia* cathepsin B-like cysteine proteinase seems to be involved in the excystation of the trophozoite.

Parasite proteinases as targets for chemotherapy

Over the last ten years, a number of studies have suggested that some parasite proteinases may be excellent targets for the development of new drugs for malaria, leishmaniasis, and Chagas disease. Cruzipain inhibitors are effective in killing the parasite *in vitro,* and also in effectively treating the disease in experimental animals. McKerrow and his co-workers have tested two vinyl sulfone inhibitors, designed to increase selectivity towards cruzipain, increase their half-life *in vivo,* provide oral bioavailability and reduce toxicity, as compared with the previously used dipeptide-based inhibitors, on *T. cruzi*-infected mice. These compounds rescued all animals from lethal acute infection, and those exhibiting chronic infections were cured by a second treatment. It is noteworthy that these inhibitors also selectively arrest replication of other protozoan pathogens, like *L. major* and *P. falciparum,* without untoward toxicity to the host.

Similarly, inhibitors of both falcipains and plasmepsins kill the intraerythrocytic stages of *P. falciparum,* by inhibiting the intravacuolar proteolysis of globin. Recent studies have suggested that the use of both cysteine and aspartic proteinase inhibitors has a synergistic effect, both *in vitro,* on cultured parasites, and *in vivo*, in infected mice. The activity of cysteine proteinase inhibitors against *T. cruzi*, and also against other parasitic protozoa, observed in cultured systems or in animal models of infection, correlates with the direct inhibition of the proposed targets. Most of the inhibitors are also able to inhibit the mammalian cathepsins, yet mammalian cells are not adversely affected by concentrations which effectively kill the parasite. This selective effect may be due to the redundancy of proteolytic activities in higher eukaryotic cells compared with parasitic protozoa. In *T. cruzi*, cruzipain inhibitor-resistant parasites were not also resistant to the established drugs, nifurtimox and benznidazole, used for treatment of Chagas disease, indicating that their mechanisms of action are different. These studies point to parasite proteinases as valid targets for the development of much needed alternative drugs against tropical diseases.

AMINO ACID METABOLISM

The disposal of amino nitrogen: aminotransferases and glutamate dehydrogenase

Aminotransferases and glutamate dehydrogenases have a central role in amino acid metabolism, both in catabolic and biosynthetic processes. Aminotransferases catalyze the exchange of the amino and keto groups between amino acids and α-ketoacids, and act frequently as a first step in amino acid degradation, by transferring the amino group to α-ketoglutarate to yield L-glutamate. L-glutamate is the substrate of the glutamate dehydrogenases, which deaminate glutamate to α-ketoglutarate and ammonia, with the concomitant reduction of NAD or NADP (Figure 8.1). Mammals have a glutamate dehydrogenase that can use both coenzymes, whereas protozoa, like plants, fungi and bacteria, have usually one or two coenzyme-specific enzymes. Aminotransferases have been reported in all protozoa, and frequently several enzymes, with different specificities, are present. For example, *T. vaginalis* contains at least four aminotransferases: one specific for ornithine and lysine; a second for the branched amino acids; a third for aspartate and aromatic amino acids; and a fourth for alanine. The most frequently identified enzymes are alanine aminotransferase and aspartate aminotransferase; the presence and usually high levels of the former enzyme are associated with the production of L-alanine, which is a frequent final product of glucose catabolism in protozoa (see below).

Glutamate dehydrogenases also are found in most parasitic protozoa. *T. cruzi* has two enzymes, one NADP-linked and the other NAD-linked. The NADP-linked enzyme is strictly cytosolic, and has been proposed to have a biosynthetic role. However, enzyme levels are developmentally regulated, and are higher in the gut stages present in the insect, where amino acids can be expected to be the main source of energy. NADP-linked glutamate dehydrogenases have also been characterized from *T. vaginalis*, *Plasmodium chabaudi*, *G. lamblia*, and *P. falciparum*. The *T. cruzi* NAD-linked glutamate dehydrogenase is mitochondrial, although there seems to be also a cytosolic isoform, and it probably has a catabolic function. The functions of the glutamate dehydrogenases are probably different in different protozoa. The NADP-linked enzyme appears to function in NADPH generation in *P. falciparum*, whereas in *T. cruzi* and *G. lamblia* it appears to be involved in NADPH reoxidation and is essential, together with alanine aminotransferase, for the production and excretion of alanine. The NADP-linked glutamate dehydrogenase from *P. falciparum* appears to be a circulating antigen in infected humans, and may become useful for the immunodiagnosis of malaria.

Aminotransferases and glutamate dehydrogenases also have been described in many helminths. Although many different substrate specificities have been detected for aminotransferases, most of the enzymes have not yet been characterized. Alanine and aspartate aminotransferase activities seem to be predominant, as in protozoa. Ammonia, a product of the glutamate dehydrogenase reaction, is usually excreted directly into the medium. Urea is excreted by a number of parasites; however, it apparently arises from arginine catabolism and not from the functioning of the urea cycle, which has never been properly demonstrated in any parasite.

Amino acid catabolism

Proline

This amino acid is an important source of energy for some protozoa, particularly those stages of hemoflagellates present in the insect vector, where these parasites must live on the products of protein digestion in the gut. In addition, proline is a major energy source for insects, and is present in large quantities in the hemolymph. Proline catabolism leads to the formation of glutamate, which, by transamination, is able to enter the TCA cycle as α-ketoglutarate (Figure 8.2). Since the cycle is frequently incomplete, or the respiratory chain is not able to cope with the reducing equivalents generated by catabolism, succinate, alanine and aspartate often accumulate as the final products of proline catabolism. In *T. brucei*, bloodstream forms are incapable of metabolizing proline, in contrast to insect forms of the parasite. Although proline oxidase has not been directly demonstrated in *Plasmodium* spp., plasmodial mitochondria are able to oxidize proline.

Proline metabolism has been studied in less detail in helminths. All the enzymes of proline catabolism have been detected in the nematodes *Heligmosomoides polygyrus* and *Panagrellus redivivus*, with activities similar to those present in rat liver. Proline decarboxylation has been detected in *Ancylostoma ceylanicum* and *Nippostrongylus brasiliensis*. Proline oxidation appears to be stage-dependent. Cell-free extracts of *Brugia pahangi* microfilariae are much more active in proline oxidation, measured as $^{14}CO_2$ formation, than are homogenates of the adults. Some helminths contain significant amounts of hydroxyproline, formed by the post-translational modification of proline. In *F. hepatica*, hydroxyproline is found in a number of soluble proteins, including a cysteine proteinase. In *Onchocerca volvulus* hydroxyproline is found in the adult cuticle, but is absent in eggs and nodular microfilariae, suggesting that extensive proline hydroxylation accompanies cuticular maturation of the parasite in the human host.

Arginine

The three major pathways for arginine catabolism are illustrated in Figure 8.2. In most parasites, as well as in higher animals, arginine is catabolized through the concerted action of arginase and ornithine aminotransferase yielding glutamate semialdehyde, which is then converted to glutamate. Ornithine can be used for polyamine biosynthesis (see below). The γ-guanidinobutyramide pathway in *L. donovani* leads to the formation of succinate which enters the TCA cycle. Arginine decarboxy-oxidase uses O_2 in the presence of Mg^{2+}, Mn^{2+}, FMN and pyridoxal phosphate, to produce CO_2 and γ-guanidinobutyramide. The arginine dihydrolase pathway, present in *G. lamblia*,

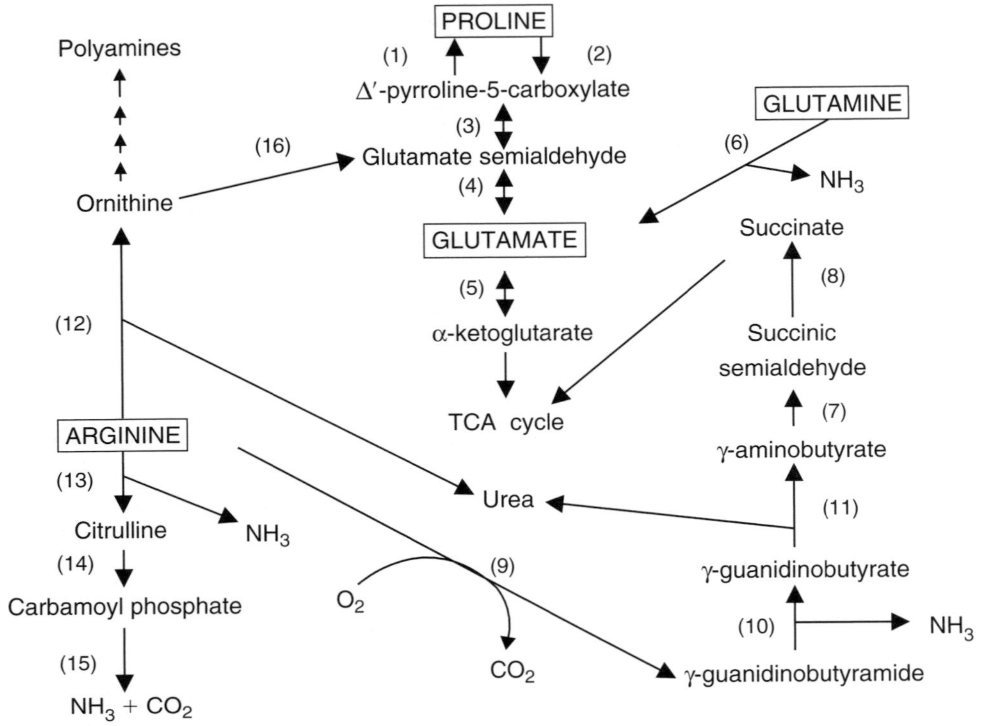

FIGURE 8.2 Catabolic pathways of proline, glutamine, glutamate and arginine. (1) proline oxidase; (2) Δ′-pyrroline-5-carboxylate reductase; (3) non-enzymatic reaction; (4) glutamate semialdehyde dehydrogenase; (5) glutamate dehydrogenase; (6) glutaminase; (7) γ-aminobutyrate aminotransferase; (8) succinic semialdehyde dehydrogenase; (9) arginine decarboxyoxidase; (10) γ-guanidinobutyramide aminohydrolase; (11) γ-guanidinobutyrate amidinohydrolase; (12) arginase; (13) arginine deiminase; (14) catabolic ornithine transcarbamoylase; (15) carbamate kinase; (16) ornithine aminotransferase.

T. vaginalis and *T. foetus*, ultimately leads to ammonia and carbon dioxide formation. This pathway is unusual in eukaryotes; it is usually restricted to prokaryotes. In trichomonads, ornithine transcarbamoylase and carbamate kinase are predominantly cytosolic, whereas the first enzyme of the pathway, arginine deiminase, is particulate. The arginase and the γ-guanidinobutyramide pathways lead to the generation of ATP under aerobic conditions via the TCA cycle. On the other hand, the dihydrolase pathway is operative under anaerobic conditions, and results in the direct formation of ATP in the carbamate kinase reaction. Indeed, the dihydrolase pathway appears to be responsible for a significant proportion of the ATP generated by these protozoa, although glucose remains as the major fuel, at least in the case of *T. vaginalis*. The enzymes of the dihydrolase pathway are present in *G. lamblia* at considerably higher levels than in *T. vaginalis*.

The arginase pathway is operative in many helminths; in some trematodes, glutamate semialdehyde enters the biosynthetic pathway of proline, which is then the final product of arginine catabolism.

Glutamine

Apart from the requirement of glutamine for optimal intraerythrocytic growth of *P. falciparum*, mentioned in the section on Amino acid Transport, little is known about the relevance of glutamine and its catabolism in parasitic protozoa. Helminths, such as filarids and schistosomes, oxidize glutamine to CO_2 through the TCA cycle and the respiratory chain. This oxidation implies the formation of α-ketoglutarate, which can be obtained by transamination of the glutamate produced in the glutaminase reaction (Figure 8.2), or succinate, if glutamine is first metabolized to γ-aminobutyrate and succinic semialdehyde. Although the latter pathway has been proposed, evidence for a functional γ-aminobutyrate bypass is incomplete, since succinic semialdehyde dehydrogenase has not been identified to date.

Methionine, cysteine and homocysteine. Methionine recycling

The catabolism of methionine is intimately connected with biosynthetic processes, participating as *S*-adenosyl methionine; these processes include methylation and polyamine biosynthesis (see section on Arginine, Methionine and Synthesis of Polyamines). Part of the carbon chain of methionine can be used to generate energy, but most organisms contain systems for the recycling of the metabolites arising from biosynthetic reactions back to methionine (Figure 8.3).

Methionine can be degraded in mammals to α-ketobutyrate, methanethiol and NH_4^+ by the concerted action of two enzymes, catalyzing first a transamination reaction leading to 2-keto-4-methylthiobutyrate, and then a dethiomethylation reaction. In *T. vaginalis*, however, this process is accomplished by a single enzyme, methionine γ-lyase; methanethiol is produced and excreted by the parasite. This pyridoxal 5′ phosphate-requiring enzyme also has homocysteine desulphurase activity, degrading homocysteine to α-ketobutyrate, SH_2 and NH_4^+. α-ketobutyrate and α-hydroxybutyrate are not excreted by *T. vaginalis*, and α-ketobutyrate may be used in the production of propionyl-CoA in the pyruvate:ferredoxin oxidoreductase reaction, leading to the excretion of propionate and to energy conservation in the form of ATP.

Recycling of methionine can occur starting from *S*-adenosylhomocysteine, after transfer of the methyl group in a transmethylation reaction, or from 5′-methylthioadenosine (MTA), after transfer of the aminopropyl group for polyamine biosynthesis (Figure 8.3). In the former case, most of the methionine carbon is recovered, whereas in the latter only the methyl group in the recovered methionine originates in the original molecule, the rest of the carbon atoms coming from the ribose in the nucleoside. In general, these pathways are similar to those operative in the mammalian host. For example, *S*-adenosylmethionine synthetase and *S*-adenosylmethionine decarboxylase, as well as *S*-adenosylhomocysteine hydrolase and the cobalamin-dependent, tetrahydrofolate-utilizing methionine synthase, have been characterized in *P. falciparum*. However, some parasitic protozoa exhibit one important difference at the level of conversion of MTA to 5-methylthioribose-5-phosphate. In mammals, as well as in *T. brucei* and in *P. falciparum*, this reaction is catalyzed by a single enzyme, MTA phosphorylase. In *G. lamblia* and *E. histolytica*, this reaction is performed by the concerted action of two enzymes, MTA nucleosidase and 5-methylthioribose kinase. Since the latter enzyme is found only in bacteria and these primitive protozoa, it may be a suitable target for chemotherapy.

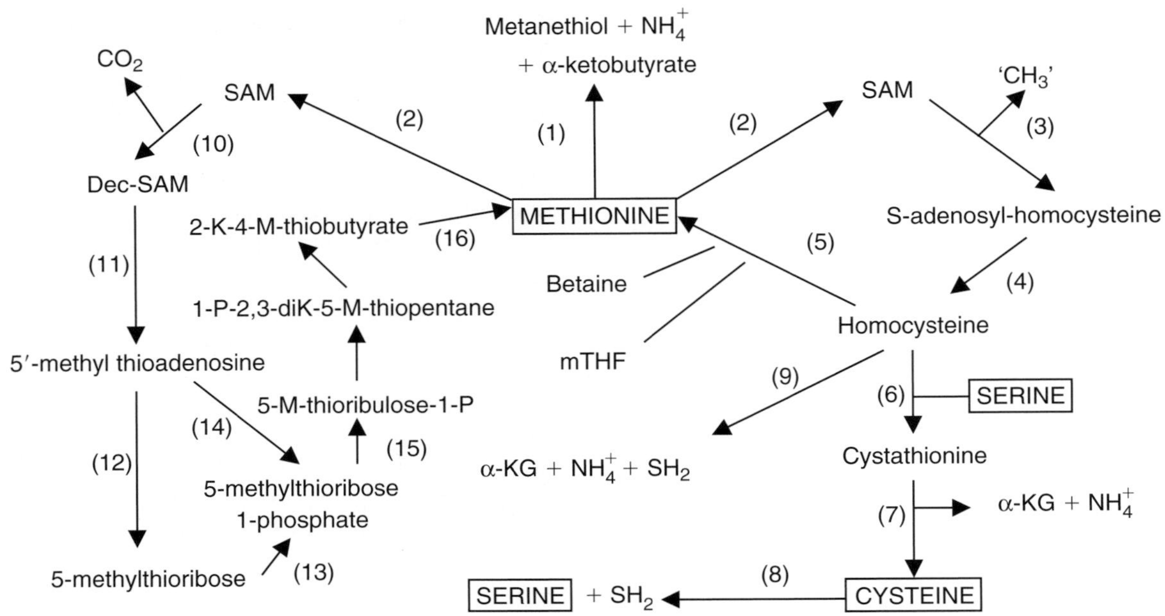

FIGURE 8.3 Methionine salvage pathways and catabolism. SAM, *S*-adenosylmethionine; dec-SAM, decarboxylated *S*-adenosylmethionine; α-KG, α-ketoglurate; 'CH_3', methyl group transferred to several substrates; 5-M-thioribulose-1-P, 5-methylthioribulose-1-phosphate; 1-P-2,3-diK-5-M-thiopentane, 1-phospho-2,3-diketo-5-methylthiopentane; 2-K-4-M-thiobutyrate, 2-keto-4-methylthiobutyrate; mTHF, methyltetrahydrofolate. (1) methionine γ-lyase; (2) *S*-adenosylmethionine synthetase; (3) *S*-adenosylmethionine-linked methyl transferases; (4) *S*-adenosylhomocysteine hydrolase; (5) methionine synthase; (6) cystathionine β-synthetase; (7) γ-cystathionase; (8) L-serine desulfhydrase; (9) homocysteine desulfurase; (10) *S*-adenosylmethionine decarboxylase; (11) spermidine synthase; (12) 5-methylthioadenosine nucleosidase; (13) 5-methylthioribose kinase; (14) 5-methylthioadenosine phosphorylase; (15) 5-methylthioribose 1-phosphate isomerase; (16) methionine aminotransferase.

Cysteine can be obtained from methionine, through the reactions of cystathionine-β-synthetase and γ-cystathionase (Figure 8.3). Cystathionine-β-synthetase has been studied in *T. vaginalis* and in nematodes; like the mammalian enzyme, both parasite enzymes have L-serine sulfhydrase activity which catalyzes the reversible interconversion of cysteine and serine, with the exchange of SH_2. In contrast to the mammalian enzyme, the parasite enzymes have 'activated L-serine sulfhydrase' activity catalyzing the reaction of cysteine with a number of R-SH compounds, resulting in the formation of a cysteine thioether and the liberation of SH_2. Both the nematode and the *T. vaginalis* enzymes are inhibited by a number of compounds, such as dichlorophene and hexachlorophene, which have both antitrichomonal and anthelmintic effects.

Threonine and serine

Two pathways of threonine catabolism have been described in some protozoa (Figure 8.4).

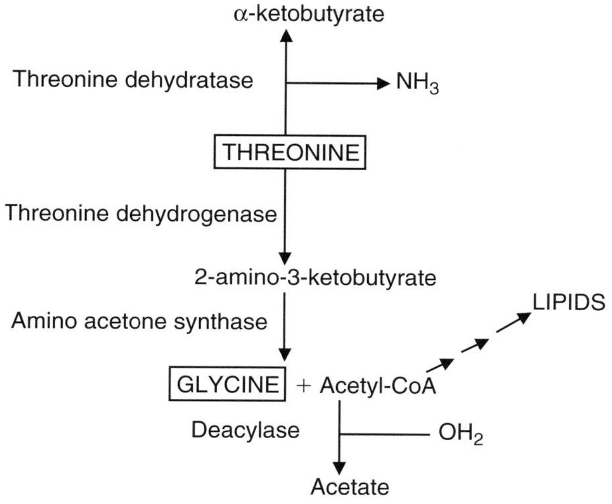

FIGURE 8.4 Catabolic pathways for threonine.

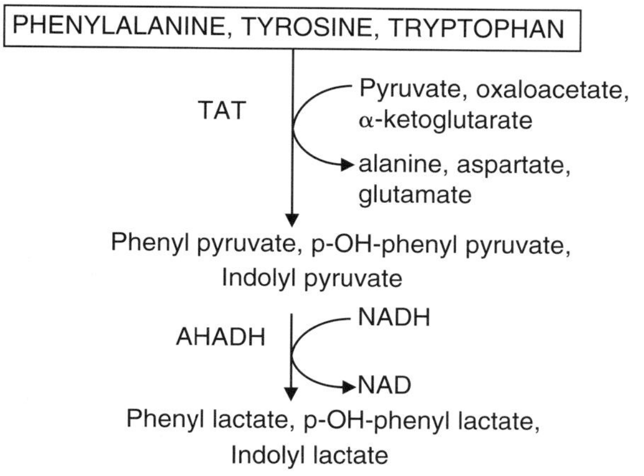

FIGURE 8.5 Catabolism of aromatic amino acids in *Trypanosoma cruzi*. TAT, tyrosine aminotransferase; AHADH, aromatic L-α-hydroxy acid dehydrogenase.

The pyridoxal 5′-phosphate-dependent enzyme threonine dehydratase, present in trichomonads, *Entamoeba* and *Giardia*, dehydrates and deaminates threonine yielding α-ketobutyrate and ammonia. In *T. brucei*, this enzyme is absent and a different pathway is operative. Threonine dehydrogenase produces 2-amino-3-ketobutyrate, which is split into glycine and acetyl-CoA in the reaction catalyzed by aminoacetone synthase (acetyl-CoA:glycine C-acetyl transferase). This acetyl-CoA is not oxidized in the TCA cycle, but instead the acetate moiety is incorporated into fatty acids. The dehydrogenase is strongly inhibited by tetraethylthiouram disulfide (disulfiram), which also inhibits parasite growth.

As noted above, serine can be interconverted with cysteine in some protozoan parasites. Serine dehydratase has been detected in trophozoites of *E. histolytica*. Serine and glycine can also be interconverted in the serine hydroxymethyltransferase reaction (see below).

Aromatic amino acids

The catabolism of aromatic amino acids in *T. brucei* results in the excretion of the aromatic lactate derivatives into the medium; indeed, in animals with heavy parasitemias, some of these catabolites can be detected in blood and urine. Epimastigotes of *T. cruzi* also excrete into the medium small amounts of phenyl lactate and *p*-hydroxyphenyl lactate, as well as a derivative of indolyl lactate. Tyrosine aminotransferase (TAT) and aromatic L-α-hydroxy acid dehydrogenase (AHADH) (Figure 8.5), have been characterized from epimastigotes of *T. cruzi*.

The *T. cruzi* TAT has high homology with the mammalian TAT, although, in contrast to the mammalian enzyme, also exhibits substantial alanine aminotransferase activity. AHADH, the first enzyme described with this stereospecificity, has high homology with cytosolic malate dehydrogenases (cMDHs). *T. cruzi* has two MDH activities, one mitochondrial and the other glycosomal, and no evidence for a cMDH has been found. This has led to the proposal that AHADH is derived from a cMDH no longer present in the parasite, through a limited number of point mutations; indeed, directed mutagenesis experiments have shown that replacement of only two amino acid residues in AHADH results in the expression of an enzyme with MDH activity as high as that of AHADH, which is actually increased in the double mutants. The enzymes responsible for aromatic amino acid catabolism have not been characterized in *T. brucei,* although this parasite has AHADH activity. An enzyme with similar specificity has been described in a *Phytomonas,* again with high homology to the *Phytomonas* glycosomal MDH and low homology with the cMDHs. It seems, therefore, that *T. cruzi* and *Phytomonas,* although belonging to different subfamilies, have both used an MDH as a scaffold to build a new specificity for aromatic keto acids. *L. donovani* excretes indolyl lactate and contains aminotransferase and dehydrogenase. In *L. mexicana,* the aminotransferase responsible for aromatic amino acid catabolism is not a TAT, but a broad-specificity aspartate aminotransferase; evidence for an AHADH-like enzyme is lacking. It appears that Trypanosomatids have a simple catabolic pathway for aromatic amino acids, but seem to have adapted different enzymes on the evolutionary road leading to its development.

Biosynthesis of amino acids

Parasites synthesize few amino acids, taking most of them from the environment, or, in cases like *Plasmodium* spp. and *Schistosoma* spp., from the intracellular or extracellular digestion of host proteins. In general, the amino acids synthesized are only those arising from short pathways, starting from metabolic intermediates, such as pyruvate in the case of L-alanine, or some TCA cycle intermediates, in the cases of aspartate and glutamate. Some amino acids can be obtained by interconversion of other amino acids, as in the case of serine and cysteine, or from the catabolic pathway of another amino acid, as cysteine from methionine. Two special cases are the synthesis of alanine and cysteine.

Alanine is an end-product of glucose catabolism in some Trypanosomatids, *G. lamblia* and *T. vaginalis.* Production and excretion of alanine has been associated with the re-oxidation of glycolytic NADH, through alanine aminotransferase and the glutamate dehydrogenases. In the case of *T. cruzi,* both the NADP-linked and the NAD-linked glutamate dehydrogenases have been proposed to participate in a mechanism which additionally involves the cytosolic malic enzyme isoform (Figure 8.6). The amino group of alanine may come from a number of different amino acids, transaminated to α-ketoglutarate to yield glutamate, whereas the carbon skeleton is the end-product of glycolysis. Alanine has also been reported as a final product of anaerobic glycolysis in some helminths.

Unlike its mammalian host, and also with other parasitic protozoa, *E. histolytica* and *E. dispar* have a cysteine biosynthetic pathway similar to that present in bacteria and plants, and are thus able to perform *de novo* synthesis of this amino acid, which is very important as an antioxidant for these organisms, which lack glutathione. The two enzymes in the pathway, serine acetyltransferase, and cysteine synthase, which replaces the *O*-acetyl group by a sulfhydryl arising from sulfide, have been characterized, and the genes encoding them cloned and sequenced. Sulfide is obtained by reduction of the inorganic sulfate incorporated in the reaction of ATP sulfurylase, an enzyme also cloned and sequenced from *E. histolytica,* having sulfite as an intermediate.

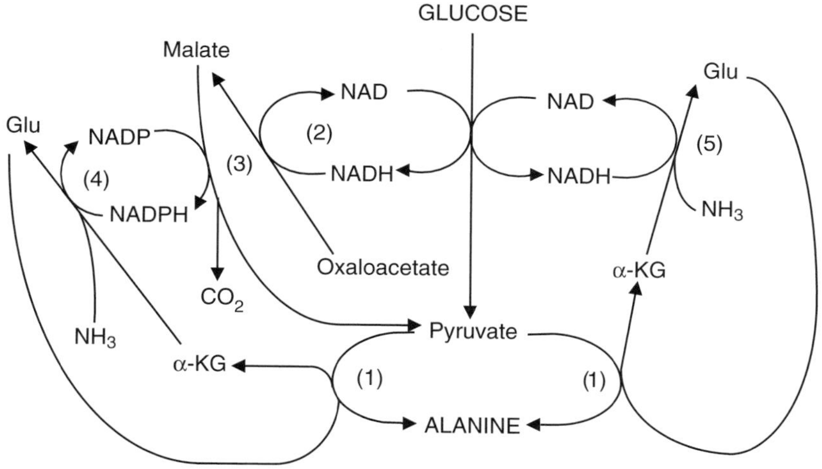

FIGURE 8.6 Synthesis of alanine and reoxidation of glycolytic NADH in *Trypanosoma cruzi*. (1) alanine aminotransferase; (2) malate dehydrogenase; (3) NADP-linked malic enzyme; (4) NADP-linked glutamate dehydrogenase; (5) NAD-linked glutamate dehydrogenase.

Role of some amino acids in the biosynthesis of other compounds

Arginine, methionine and synthesis of polyamines

Polyamines are low molecular weight diamines and triamines, among which the most important are putrescine, spermidine and spermine. They have a number of important functions, participating in cell multiplication and differentiation, macromolecular synthesis and membrane stabilization in most living cells. Parasites are no exception to this rule, and their cells contain putrescine and spermidine, and in some cases also spermine. The synthesis of polyamines starts from two common amino acids, arginine, through the provision of ornithine, and methionine, through the active intermediate *S*-adenosyl methionine (Figure 8.7). This biosynthetic pathway in parasites has been the subject of many studies over the last twenty years, since polyamines are a potential target for chemotherapy. The first enzyme in the pathway, the pyridoxal 5′-phosphate-dependent ornithine decarboxylase (ODC), has unusual properties in some parasites, which distinguish it from the mammalian enzyme. The latter has a very short half-life and is degraded by the proteasome after binding to a specific protein, antizyme. Trypanosomatids like *T. brucei* and the Leishmanias, on the other hand, have ODCs with a considerably longer half-life, which may explain why irreversible inhibitors of ODC, such as α-difluormethylornithine (DFMO), inhibit parasite growth, and indeed cure animals infected with *Trypanosoma gambiense*. This is not the case for all Trypanosomatids; the monogenetic parasite *C. fasciculata*, as well as a *Phytomonas* spp., have ODCs with short half-lives; *T. cruzi* lacks ODC and depends instead on the uptake of putrescine or spermidine for survival. The genes encoding several ODCs from Trypanosomatids have been cloned and sequenced; they predict proteins that lack the PEST sequence, supposedly responsible

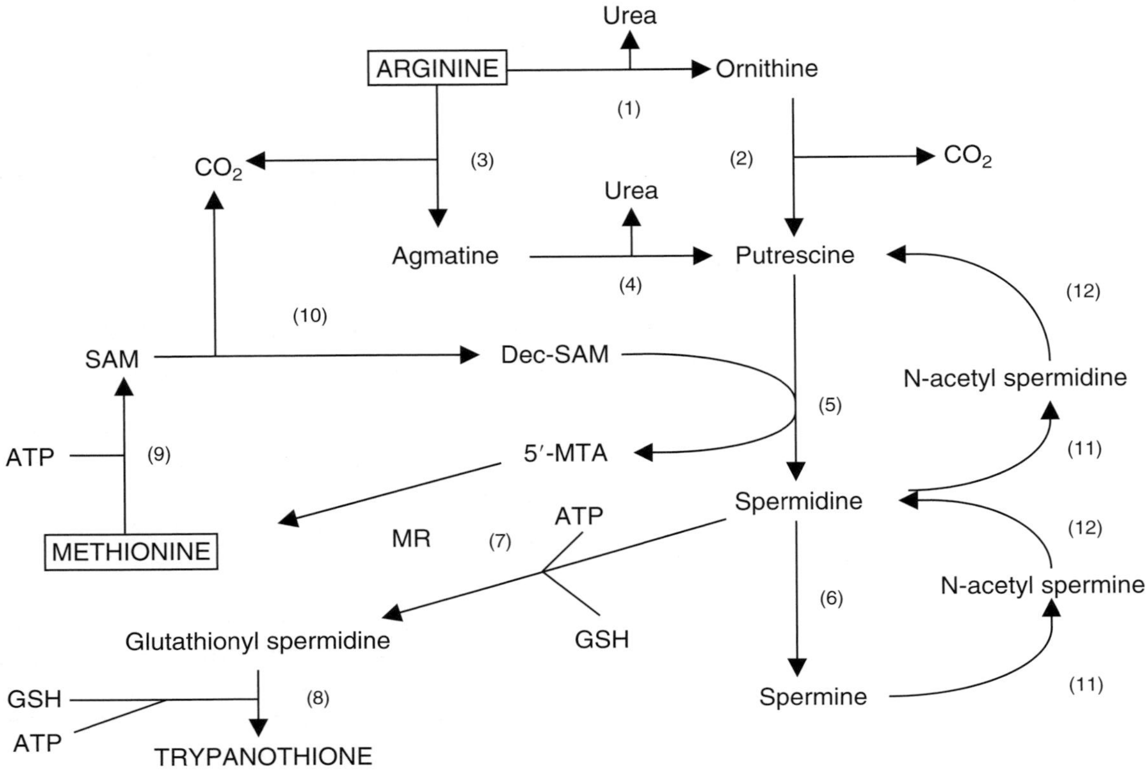

FIGURE 8.7 Role of methionine and arginine in polyamine biosynthesis. SAM, *S*-adenosylmethionine; Dec-SAM, decarboxylated *S*-adenosylmethionine; MR, methionine recycling (Figure 8.3); GSH, reduced glutathione. (1) arginase; (2) ornithine decarboxylase; (3) arginine decarboxylase; (4) agmatine deiminase; (5) spermidine synthase; (6) spermine synthase; (7) glutathionylspermidine synthetase; (8) trypanothione synthetase; (9) *S*-adenosylmethionine synthetase; (10) *S*-adenosylmethionine decarboxylase; (11) polyamine acetylase; (12) diamine oxidase.

for the metabolic instability of mammalian ODC. However, the *C. fasciculata* ODC, which has a short half-life, also lacks this sequence, so that there is clearly more to this story. The crystal structure of a recombinant ODC from *T. brucei* indicates that the active site is identical to that of the mammalian enzyme. *L. donovani* ODC deletion mutants behave as polyamine auxotrophs. An alternative to ODC is arginine decarboxylase, which decarboxylates arginine to agmatine, which is then converted to putrescine by agmatine deiminase. This enzyme has been recently described in *Cryptosporidium parvum*. Although arginine decarboxylase activity has been reported in *T. cruzi*, this is still controversial, and the parasite usually behaves as a polyamine auxotroph.

The second enzyme in the biosynthetic pathway, spermidine synthase (Figure 8.7) transfers the aminopropyl group from decarboxylated *S*-adenosyl methionine (decSAM) to putrescine to yield spermidine. Although *P. falciparum* contains spermine, it is probably taken up from the medium.

Methionine adenosyltransferase, the enzyme which synthesizes *S*-adenosylmethionine (SAM) from methionine and ATP, is present in some parasitic protozoa, although a number of parasites can take up SAM from the medium. Methionine adenosyltransferase has been detected in *T. brucei*, *T. vaginalis* and partially characterized in *Leishmania braziliensis panamensis* and *Leishmania infantum*. SAM decarboxylase is also present in parasitic protozoa, including *T. b. brucei* and *T. b. rhodesiense*, and in *T. vaginalis*. Further, the gene encoding the *T. cruzi* enzyme has been recently cloned, sequenced and expressed in *E. coli*. It is only 25% homologous to the human SAM decarboxylase. The *P. falciparum* SAM decarboxylase has the unique characteristic of being present as part of a bifunctional protein, the other half possessing ODC activity. The N-terminal SAM decarboxylase is connected to the C-terminal ODC by a hinge region. SAM decarboxylases are normally heterotetrameric proteins, with two different subunits arising from self-cleavage of the pro-enzyme. The tetrameric structure of the SAM decarboxylase moiety, as well as the dimeric structure of the ODC moiety, are conserved in the bifunctional enzyme.

Spermidine synthase has been recently cloned and sequenced from *L. donovani*, and has 56% amino acid identity with the human enzyme. Null mutants are polyamine auxotrophs, a requirement that can be satisfied by spermidine, but not by putrescine or other diamines, nor by spermine. This latter finding, consistent with that of *L. donovani* ODC deletion mutants, implies that the pathway from spermine to spermidine, present in mammalian cells, is not present in these organisms.

Helminths seem to take up polyamines from the medium; the biosynthetic enzymes described above are apparently absent. Helminths have, on the other hand, enzymes for the interconversion of spermine back to spermidine, and from spermidine back to putrescine, that seem to be lacking in protozoa. Another interesting aspect of helminth polyamine metabolism is that, unlike most other eukaryotes, including protozoa, they usually contain much less putrescine than spermidine and spermine.

Arginine and synthesis of nitric oxide

The production of nitric oxide (NO) from arginine by nitric oxide synthase (NOS) is involved in the protective responses of the host against a number of parasites. On the other hand, mounting evidence indicates that, at least in some cases, the parasites themselves produce NO, probably involved in their regulatory cascades. NOS has been partially purified from *T. cruzi*, is activated by Ca^{2+}, and is stimulated by glutamate and *N*-methyl-D-aspartate. Cyclic GMP appears to be controlled by L-glutamate through a pathway mediated by NO. NOS seems to be localized in the inner surface of cell membranes and in free cytosolic clusters in the body, flagellum and apical extreme. A soluble NOS purified 2800-fold from *L. donovani* is a dimer made up of 110 kDa subunits, requires NADPH and is inhibited by EGTA. Several lines of evidence suggest that it is similar to the mammalian NOS isoform I. NOS have also tentatively been identified in the intraerythrocytic forms of *P. falciparum*, and in neuromuscular tissue of the nematode *Ascaris suum*. The malarial NOS activity, present in lysates of infected red blood cells, is independent of Ca^{2+}, inhibited by three well known NOS inhibitors, and has a molecular weight of about 100 kDa, suggesting that it is different from the isoforms present in mammals.

Threonine and leucine in lipid synthesis

The role played by threonine in producing acetate for incorporation into fatty acids has been described (section on Threonine and Serine). Recent studies with *Leishmania mexicana* indicate that, whereas acetate is readily incorporated into fatty acids both in promastigotes and amastigotes, leucine was the only substrate efficiently incorporated into sterols by both parasite stages. Promastigotes of *L. mexicana, L. major* and *L. amazonensis* produced about 75% of their sterols from leucine, whereas the same stage from other *Leishmania* spp. (*L. braziliensis, L. donovani* and *L. tropica*) produced much less, around 30%, of their sterols from leucine. *Endotrypanum monterogeii* also produced a large proportion of its sterols (77%) from leucine, whereas *T. cruzi* produced only 8%, the rest arising from acetate. All these Trypanosomatids produced fatty acids from acetate. ^{13}C-NMR studies indicated that promastigotes of *L. mexicana* incorporate the intact carbon skeleton of leucine into the isoprenoid pathway leading to sterol, without previous conversion to acetyl-CoA, as in plants and animals. However, the conversion to acetyl-CoA occurs, since carbon from leucine is also incorporated into fatty acids, which are optimally synthesized from acetate. The presence of this hitherto unknown pathway for sterol biosynthesis in Trypanosomatids opens up a new possibility for chemotherapy.

Serine and folate metabolism

Serine has an important role in folate metabolism, through its participation in the reaction catalyzed by serine hydroxymethyltransferase (SHMT), which transfers the β-carbon of serine to tetrahydrofolate (THF) to give 5,10-methylene tetrahydrofolate (CH_2THF) and glycine. CH_2THF then transfers its one-carbon unit to dUMP to give dTMP, or, with the previous formation of methyl-THF, to homocysteine to give methionine (Figure 8.8). SHMT has been partially purified and characterized from *Plasmodium lophurae* and *P. chabaudi*. Recently, the gene encoding the enzyme has been cloned and sequenced from *P. falciparum*, and exhibits 38–47% identity to SHMTs from other organisms. In addition, most of the other genes involved in folate synthesis have been cloned and sequenced from *P. falciparum*. The shikimate pathway is present in Apicomplexa, and has previously been reported in bacteria, algae, fungi and higher plants, but is absent in mammals. This pathway seems to be essential for the parasites because of its involvement in folate synthesis, through the formation of *p*-aminobenzoate, and not for the synthesis of aromatic amino acids. Growth of *P. falciparum*, *T. gondii* and *Cryptosporidum parvum* is inhibited by the herbicide glyphosate, and this inhibition is reversed by addition of *p*-aminobenzoate. A homolog of *p*-aminobenzoate synthase has been recently detected in *P. falciparum*. SHMT has also been detected in *C. fasciculata* and *T. cruzi*. *C. fasciculata* has three isoforms, localized in the mitochondrion, glycosome and cytosol. All three isoforms are tetrameric, like SHMTs from other organisms. The *T. cruzi* enzyme differs from the others in being monomeric, with an apparent molecular weight of 69 kDa.

Cysteine and synthesis of glutathione

Glutathione (GSH) is a tripeptide (γ-glutamyl-cysteinyl-glycine) present at fairly high concentrations in most cells, where it acts as a protecting agent against oxidative damage. In Trypanosomatids glutathione is present primarily as trypanothione (*bis*-glutathionyl spermidine). Tripanothione synthesis is mediated by two enzymes, γ-glutamyl-cysteine synthetase and glutathione synthetase, with the utilization of two moles of ATP per mole of GSH synthesized. Most parasites contain GSH,

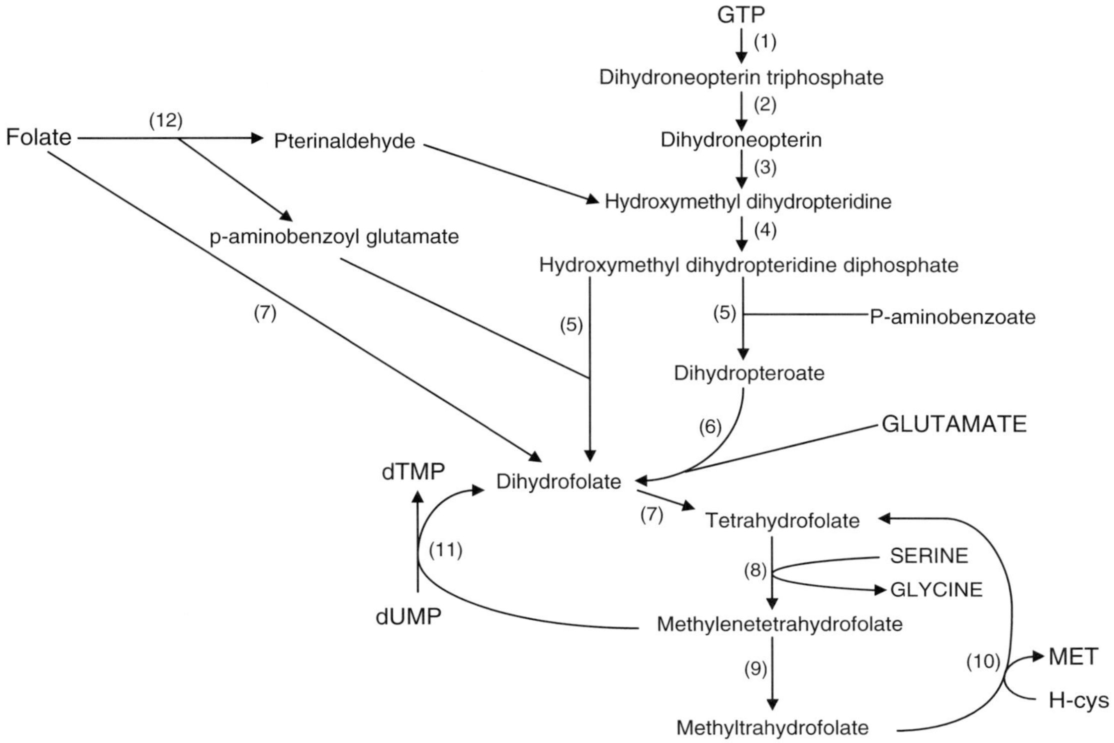

FIGURE 8.8 Participation of serine in the folate cycle. MET, methionine; H-cys, Homocysteine. (1) GTP cyclohydrolase; (2) dihydropterin triphosphate pyrophosphohydrolase; (3) dihydroneopterin aldolase; (4) hydroxymethyl dehydropteridine pyrophosphokinase; (5) dihydropteroate synthase; (6) dihydrofolate synthase; (7) dihydrofolate reductase; (8) serine hydroxymethyltransferase; (9) 5,10-methylenetetrahydrofolate reductase; (10) methionine synthase; (11) deoxythymidilate synthase; (12) pteridine 6-methylhydrolase. Most of the reactions shown have been demonstrated in *Plasmodium* spp.; some reactions are proposed, because of the ability of folate or some of its metabolites to antagonize sulfonamide inhibition of parasite growth.

with the notorious exception of *E. histolytica*, which is unable to synthesize the tripeptide; however, very low amounts are present in the trophozoites, taken up from the medium. In *E. histolytica*, cysteine seems to play the antioxidant role of GSH (section on Biosynthesis of amino acids). Buthionine sulfoximine, an inhibitor of γ-glutamyl-cysteine synthetase, also inhibits the growth of *T. brucei* and *T. cruzi*, demonstrating the relevance of the GSH biosynthetic pathway.

Amino acids as osmoregulators

The studies of J.J. Blum and his co-workers with *L. major* and *L. donovani* indicate that amino acids, particularly alanine, are involved in responses to osmotic stress. Hypo-osmotic swelling of the promastigotes leads to release of alanine and, to a lesser extent, other amino acids; on the other hand, under hyperosmotic conditions alanine concentration increases, most probably due to synthesis even in the absence of glucose, although production by proteolysis cannot be completely excluded. Under iso-osmotic conditions, part of the large alanine pool is catabolized, and this catabolism stops altogether under hyperosmotic conditions. *Crithidia luciliae* shows a somewhat similar behaviour: under hypo-osmotic conditions, proline, valine, alanine and glycine are released, with preference for the acidic amino acids; the cationic amino acids are not released. In both organisms amino acid release seems to be mediated by a swelling-activated amino acid channel, showing some specificity.

CONCLUDING REMARKS

Although many aspects of protein and amino acid metabolism in parasitic protozoa and helminths are similar to those in the mammalian host, there are a number of differences, which allow us to consider some individual enzymes or pathways as suitable targets for the development of new chemotherapeutic agents. So far, the best targets which may lead to the development of drugs are the cysteine proteinase inhibitors, active against some trypanosomatids; the cysteine and aspartic proteinase inhibitors active against *Plasmodium* spp.; and the polyamine synthesis inhibitors active against *T. brucei*. New very promising targets arise from recent research, such as the leucine-dependent sterol biosynthesis in *Leishmania* spp., and the shikimate pathway (only marginally related to amino acid metabolism) in Apicomplexa. The Apicomplexa contain a symbiotic alga-derived, plastid-like organelle, the apicoplast, and contain a number of proteins, encoded in the nuclear DNA, that exhibit high sequence homology to plant proteins. A similar situation, except for the absence of a plastid-like organelle, has recently been suggested for trypanosomatids, linked to the presence of the glycosome. Plant-like enzymes, absent or very different from those present in the host, may be excellent targets for chemotherapy, in addition to the already known bacterial-like enzymes or pathways present in some parasitic protozoa.

FURTHER READING

Amino acid transport

Luján, H.D. and Nash, T.E. (1994). *J. Eukaryot. Microbiol.* **41**, 169–175.

Mazareb, S., Fu, Z.Y. and Zilberstein, D. (1999). *Exp. Parasitol.* **91**, 341–348.

Proteolytic enzymes

Barrett, A.J., Rawlings, N.D. and Woessner, J.F. (eds) (1998). *Handbook of Proteolytic Enzymes*, London: Academic Press. Introduction and Chapters 127, 201, 202, 203, 204, 205, 208, 254, 286, 287, 288 and 383.

Barrett, A.J. (1999). In: Turk, V. (ed.). *Proteases, New Perspectives*, Basel: Birkhäuser Verlag, pp. 1–12.

Cazzulo, J.J., Stoka, V. and Turk, V. (2001). *Curr. Pharmaceut. Design* **7**, 1143–1156.

de Diego, J.L., Katz, J.M., Marshall, P. *et al.* (2001). *Biochemistry* **40**, 1053–1062.

Eggleson, K.K., Duffin, K.L. and Goldberg, D.E. (1999). *J. Biol. Chem.* **274**, 32411–32417.

Francis, S.E., Sullivan, D.J. Jr. and Goldberg, D.E. (1997). *Annu. Rev. Microbiol.* **51**, 97–123.

Mottram, J.C., Brooks, D.R. and Coombs, G.H. (1998). *Curr. Opin. Microbiol.* **1**, 455–460.

Pszenny, V., Angel, S.O., Duschak, V.G. *et al.* (2000). *Mol. Biochem. Parasitol.* **107**, 241–249.

Que, X. and Reed, S.L. (2000). *Clin. Microbiol. Rev.* **13**, 196–206.

Semenov, A., Olson, J.E. and Rosenthal, P.J. (1998). *Antimicrob. Agents Chemother.* **42**, 2254–2258.

Tort, J., Brindley, P.J., Knox, D., Wolfe, K.H. and Dalton, J.P. (1999). *Adv. Parasitol.* **43**, 161–266.

Zang, X. and Maizels, R.M. (2001). *Trends Biochem. Sci.* **26**, 191–197.

Amino acid catabolism

Barrett, J. (1991). *Adv. Parasitol.* **30**, 39–105.
Cazzulo Franke, M.C., Vernal, J., Cazzulo, J.J. and Nowicki, C. (1999). *Eur. J. Biochem.* **266**, 903–910.
Coombs, G. and North, M. (eds) (1991). *Biochemical Protozoology*, London: Taylor and Francis. Chapters 9, 40, 41, 42, 43 and 46.

Biosynthesis of amino acids

Nozaki, T., Tokoro, M., Imada, M. *et al.* (2000). *Mol. Biochem. Parasitol.* **107**, 129–133.

Arginine, methionine and synthesis of polyamines

Müller, S., Coombs, G.H. and Walter, R.D. (2001). *Trends Parasitol.* **17**, 242–249.
Sharma, V., Tekwani, B.L., Saxena, J.K. *et al.* (1991). *Exp. Parasitol.* **72**, 15–23.

Threonine and leucine in lipid synthesis

Ginger, M.L., Chance, M.L., Sadler, I.H. and Goad, L.J. (2001). *J. Biol. Chem.* **276**, 11674–11682.

Serine and folate metabolism

Lee, S.C., Salcedo, E., Wang, Q., Wang, P., Sims, P.F. and Hyde, J.E. (2001). *Parasitology* **122**, 1–13.
Roberts, F., Roberts, C.W., Johnson, J.J. *et al.* (1998). *Nature* **393**, 801–805.

Amino acids as osmoregulators

Blum, J.J. (1994). *J. Bioenerg. Biomembr.* **26**, 147–155.

CHAPTER

9

Purine and pyrimidine transport and metabolism

Nicola S. Carter, Nicolle Rager and Buddy Ullman
Department of Biochemistry and Molecular Biology,
Oregon Health & Science University,
Portland, OR, USA

INTRODUCTION

Purine and pyrimidine nucleotides are phosphorylated, intracellular metabolites vital to virtually every biochemical and metabolic process within the living cell. Functions for which purines and pyrimidines are absolutely requisite include: (1) ATP, as the universal currency of cellular energy; (2) cyclic AMP and cyclic GMP, as second messenger molecules in cell signaling pathways; (3) adenine nucleotides, as component parts of $NAD(P)^+$, FAD, and CoA, each a vitamin-derived coenzyme required for many of the key enzymes of intermediary metabolism; (4) nucleotides, as constituents of activated intermediates involved in many biosynthetic pathways, e.g. UDP-glucose in glycogen synthesis, CDP-choline and CDP-diacylglycerol in phospholipid metabolism, and *S*-adenosylmethionine in essential transmethylation reactions; and (5) purine and pyrimidine nucleoside triphosphates, as the monomeric precursors of DNA and RNA metabolism. Somewhat ironically, it is the synthetic processes of replication/repair and transcription that actually deplete cellular nucleotide pools, thereby necessitating their regeneration by *de novo* or salvage pathways.

Investigations from many laboratories have definitively demonstrated that the pathways by which parasites generate purine nucleotides can be highly discrepant from those of their infected hosts. Whereas virtually all nucleated mammalian cells express a long and complex pathway for synthesizing purine nucleotides from small molecules (i.e. amino acids, CO_2, and 1-carbon folate derivatives), all protozoan and helminthic parasites studied to date are incapable of synthesizing the purine heterocycle *de novo*. Consequently, each genus of parasite has evolved a unique complement of purine transporters and salvage enzymes that

Molecular Medical Parasitology
ISBN 0–12–473346–8

enable them to scavenge preformed purines from the host. Thus, purine salvage serves an indispensable nutritional function for the parasite, and offers a plethora of potential targets for selective therapeutic manipulation using inhibitor approaches. These purine salvage pathways, which will be discussed in detail, can be quite simple, as for example in *Giardia lamblia*, or myriad, intertwined, and complex, as in *Leishmania donovani*.

The pathways by which pyrimidine nucleotides are synthesized in parasites have been less well studied, primarily because most, although not all, parasites synthesize pyrimidine nucleotides *de novo*, using essentially the same enzymatic machinery that is found in mammalian cells. The exceptions are the amitochondrial protozoan parasites, *Trichomonas vaginalis*, *Tritrichomonas foetus*, and *G. lamblia*, which are obligatory scavengers of exogenous pyrimidines. Consequently, the mechanisms by which parasites salvage pyrimidine bases and nucleosides tend to be less elaborate than those for purine salvage, since pyrimidine salvage is, with few exceptions, not essential for parasite survival and proliferation. Additionally, *T. vaginalis* and *G. lamblia* also appear to lack a functional ribonucleotide reductase (RR) activity and are thus dependent upon exogenous sources of deoxynucleosides in order to produce the dNTP precursors necessary for DNA synthesis and repair (see Table 9.1 for abbreviations).

Between 1975 and 1995, classical biochemical approaches, particularly enzyme activity and metabolic flux measurements, were employed to attain an understanding of the basic enzymatic machinery available to protozoan and helminthic parasites for their purine and pyrimidine metabolism. The more recent advances in our knowledge of purine and pyrimidine transport and metabolism in parasites, however, originate from an interdisciplinary amalgamation of approaches and techniques from molecular biology, genetics, structural biology, and immunocytochemistry. Many genes encoding parasite purine and pyrimidine salvage enzymes have been cloned, overexpressed in *Escherichia coli*, and the recombinant proteins purified to homogeneity and characterized. A number of these proteins have been crystallized and their three-dimensional structures determined in atomic detail by X-ray crystallography. Genes encoding purine and pyrimidine transporters also have been isolated and functionally characterized after expression in *Saccharomyces cerevisiae*, nucleoside transport-deficient *L. donovani*, and *Xenopus laevis* oocytes. Finally, antibodies have been raised to a number of parasite nucleoside transporters and salvage enzymes and exploited to localize these proteins to subcellular compartments. Despite these advances in our knowledge of the mechanisms by which parasites synthesize and acquire purine and pyrimidine nucleotides, a complete characterization of these pathways in any parasite has yet to be accomplished.

Because the sequencing of several parasite genomes is well under way, many purine and pyrimidine salvage and interconversion enzymes and transporters from parasites have been identified. As the genome sequencing projects of some parasites are completed and others expanded or initiated, most of the genes encoding proteins affecting purine and pyrimidine transport and metabolism within a given organism will be identified. The challenge will be the functional characterization of these genes and proteins in intact parasites. Targeted gene replacement strategies have proven to be informative tests of gene and protein function in some genera.

A comprehensive review of what is currently known on this subject is not possible within the confines of this chapter. Rather, we will

TABLE 9.1 Purine and pyrimidine enzymes

Number[1]	Enzyme	Abbreviation[2]
1	Adenine deaminase	AD
2	Adenosine deaminase	ADA
3	Adenosine kinase	AK
4	AMP deaminase	AMPD
5	Adenosine phosphorylase	AP
6	Adenine phosphoribosyltransferase	APRT
7	Adenylosuccinate lyase	ASL
8	Adenylosuccinate synthetase	ASS
9	Aspartate transcarbamoylase	ATC
10	Cytidine phosphorylase	CP
11	Carbamoyl phosphate synthetase II	CPSII
12	CTP synthetase	CTPS
13	Deoxycytidine deaminase	dCD
14	Deoxycytidine kinase	dCK
15	Deoxycytidylate deaminase	dCMPD
16	Deoxyguanosine kinase	dGK
17	Dihydrofolate reductase	DHFR
18	Dihydrofolate reductase–thymidylate synthase	DHFR–TS
19	Dihydroorotase	DHO
20	Dihydroorotate dehydrogenase	DHODH
21	Guanine deaminase	GD
22	GMP reductase	GMPR
23	GMP synthetase	GMPS
24	Guanine phosphoribosyltransferase	GPRT
25	Hypoxanthine-guanine phosphoribosyltransferase	HGPRT
26	Hypoxanthine-guanine-xanthine phosphoribosyltransferase	HGXPRT
27	IMP dehydrogenase	IMPDH
28	Nucleoside hydrolase	NH
29	Nucleoside phosphotransferase	NPT
30	Nucleotidase	NT
31	Nucleoside triphosphatase	NTPase
32	OMP decarboxylase	ODC
33	Orotate phosphoribosyltransferase	OPRT
34	Purine nucleoside phosphorylase	PNP
35	Phosphoribosyltransferase	PRT
36	PRPP synthetase	PS
37	Ribonucleotide reductase	RR
38	Thymidine kinase	TK
39	Thymidine phosphorylase	TP
40	Thymidylate synthase	TS
41	Uridine-cytidine kinase	UK
42	Uridine phosphorylase	UP
43	Xanthine oxidase	XO
44	Xanthine phosphoribosyltransferase	XPRT

[1] Numbers used to denote enzymes in Figures 9.3–9.10.
[2] Abbreviations used in the text.

provide overviews of the key pathways in the more prominent model parasites and emphasize the new knowledge that has accumulated over the last decade. The reader is referred to the selected readings given at the end of this chapter for more background information. Because of their lengthy names and the frequency by which they will be referred in this review, all enzymes will be designated in the text by their acronyms, listed in Table 9.1. When referring to an enzyme from a specific species, the abbreviated enzyme will be preceded by initials describing the genus and species, e.g. *Plasmodium falciparum* hypoxanthine-guanine-xanthine phosphoribosyltransferase will be PfHGXPRT.

TRANSPORT SYSTEMS

The salvage of exogenous purines and pyrimidines is initiated by the translocation of either the nucleoside or nucleobase across a single or, in the case of intracellular parasites, multiple membranes. Since nucleosides and nucleobases are hydrophilic and consequently cannot diffuse across the lipid bilayer, parasites have devised specialized translocation systems, each uniquely adapted to their hosts' environments. Until recently, most of our knowledge derived about transporters has arisen from observations on transport into intact parasites or parasitized host cells. However, molecular genetic tools and the burgeoning parasite genome databases have led to the identification and functional characterization of a number of parasite translocation proteins. Thus far, all of the parasite genes identified appear to belong to a family of transporters, termed the Equilibrative Nucleoside Transporters (ENTs), after their mammalian homologs. These transporters are distinguished by an overall similarity in predicted topology (11 transmembrane domains) and possess a number of conserved or signature residues, located primarily within predicted transmembrane domains (Figure 9.1). Many specialized mammalian cell types, such as kidney and intestine, also contain another class of nucleoside transporter that couples nucleoside uptake to the cellular Na^+ or H^+ gradient. These Concentrative Nucleoside Transporters, CNTs) have yet

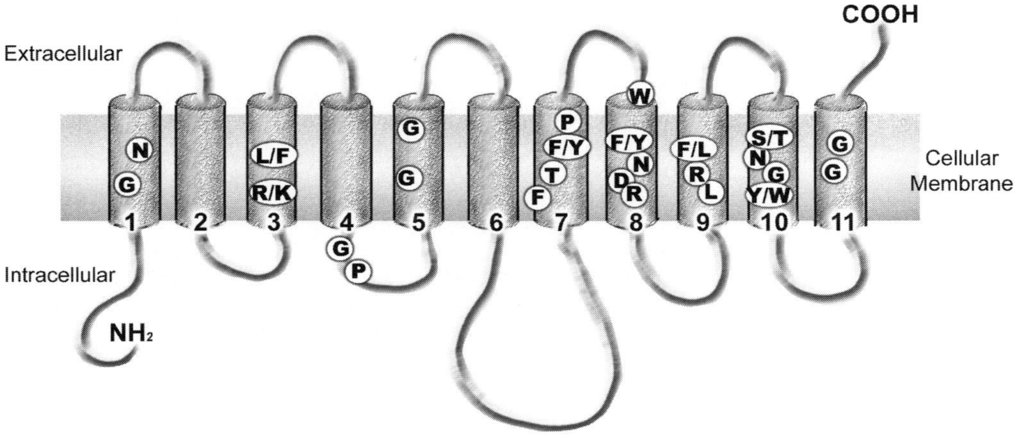

FIGURE 9.1 Predicted topology and conserved residues of the equilibrative nucleoside transporter (ENT) family.

to be discovered in parasitic protozoa, although it is probable that helminths possess CNT-type proteins.

Apicomplexa

Plasmodium

Most of our knowledge about purine transport in plasmodia is based upon observations on the intraerythrocytic stage of *P. falciparum*. Manipulation of the exogenous purine source in the cultivation medium suggests that, while hypoxanthine is the preferred purine source, the parasite also can salvage the purine nucleobases, adenine and guanine. It also can salvage the purine nucleosides, adenosine, guanosine, and inosine. This implies that the parasite possesses the necessary machinery to transport purines from the extracellular milieu. Intraerythrocytic plasmodia reside in a parasitophorous vacuole (PV) within the host erythrocyte. Mature erythrocytes are unable to synthesize the purine ring *de novo* and contain limited amounts of endogenous purines, so preformed purines must be salvaged from the host plasma. Therefore, the parasite has evolved an intricate purine translocation pathway (Figure 9.2).

The uninfected erythrocyte possesses a broad capacity equilibrative nucleoside transporter, hENT1, in its plasma membrane. Shortly after invasion the parasite appears to induce a new nucleoside transport activity in the infected erythrocyte, which can be distinguished from hENT1 since it is non-saturable, transports both D- and L-adenosine, and is impervious to the hENT1 inhibitor, 4-nitrobenzyl-6-thioinosine. This induced activity shares many of the properties of the 'new permeation pathway' proposed by Kirk and coworkers, since it is non-saturable, lacks stereo-specificity, and is exquisitely sensitive to anion transport inhibitors such as furosemide.

Parasites are separated from the erythrocyte cytosol by the parasitophorous vacuolar membrane (PVM). Thus, purines must traverse the erythrocyte membrane, cytoplasm and PVM. A nutrient channel within the PVM, which appears to accommodate small solutes of <1400 Da or <23 Å, may function as a molecular sieve, enabling nucleosides and nucleobases to freely permeate the parasitophorous vacuolar space. In addition, the mature stages of the parasite possess a tubovesicular membraneous network, which is contiguous with the PVM and has been implicated in nucleoside translocation. This may serve to increase the surface area of the PVM, allowing for greater permeation of nutrients.

P. falciparum mRNA extracted from intraerythrocytic stages and expressed in *Xenopus* oocytes identified an adenosine transporter

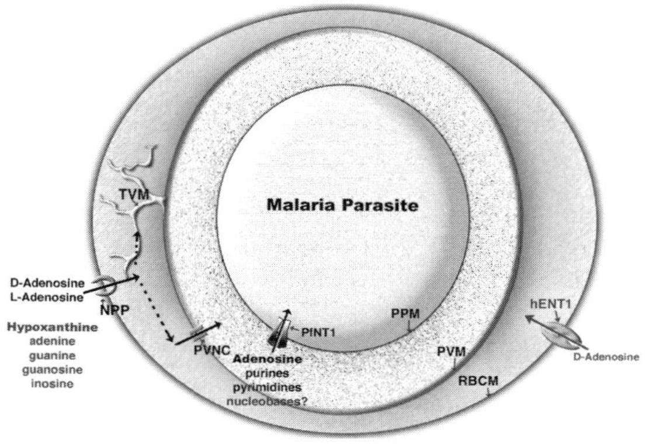

FIGURE 9.2 Transport of nucleosides and/or nucleobases by the intraerythrocytic malaria parasite for purine salvage. NPP, new permeation pathway; TVM, tubovesicular membrane; RBCM, red blood cell membrane; PVM, parasitophorous vacuole membrane; PPM, parasite plasma membrane; PfNT1, *P. falciparum* nucleoside transporter 1; PVNC, parasitophorous vacuole nutrient channel; hENT1, human erythrocyte nucleoside transporter 1.

activity that was distinct from the induced activity at the erythrocyte membrane because it was both stereo-selective and saturable. Recently, the gene that likely encodes this nucleoside transport activity was identified in *P. falciparum*. PfNT1 (PfENT1) is a broad specificity purine and pyrimidine nucleoside, and possibly purine nucleobase, transporter, although it appears to have a significantly higher affinity for adenosine. The preference of PfNT1 for adenosine is reasonable since ADA activity is augmented in intraerythrocytic *P. falciparum* and is likely the preferred salvageable nucleoside. PfNT1 also recognizes L-nucleosides at millimolar affinity. However, PfNT1 is an unlikely candidate for the activity induced at the erythrocyte membrane since it is saturable and localizes primarily to the parasite plasma membrane. During the intraerythrocytic lifecycle, *PfNT1* is constitutively expressed, but expression is significantly upregulated in the trophozoite and schizont stages when the demand for nucleic acid precursors is acute. *Plasmodium* has a V-type H^+-pump on its plasma membrane, which is predicted to result in a significant negative inward proton gradient, but whether PfNT1 acts as a proton symporter is not yet known. Thus far *PfNT1* is the only nucleoside transporter gene to be cloned and functionally characterized from *Plasmodium*. Another related nucleoside transporter sequence (*PfNT2*) also has been identified within The Institute for Genomic Research database, although the protein has not been functionally validated.

Toxoplasma

The actively dividing mammalian form of *T. gondii* also is intracellular. Unlike *Plasmodium*, it can invade and propagate in virtually any human or avian nucleated cell type. *T. gondii* resides within a parasitophorous vacuole (PV) and is dependent upon exogenous purines that are supplied by the host cell (see Figure 9.3). Thus, preformed purines must permeate into the parasitophorous vacuolar space. The parasitophorous vacuolar membrane (PVM) in *T. gondii* is strikingly similar to that described for *Plasmodium*. It also is thought to be derived from the host-cell membrane, and appears to act as a molecular sieve with a size exclusion limit similar to the PVM of *P. falciparum*. Until recently, the hydrolysis of adenine nucleotides to adenosine within the PV by parasite-secreted NTP hydrolases was thought to provide the primary purine source for *T. gondii*. However, it now appears that the activity for the terminal hydrolysis (the release of adenosine from AMP) is absent within the PV. From direct transport measurements into tachyzoites and drug sensitivity profiles with toxic purine analogs, it appears that *T. gondii* transports the nucleobases hypoxanthine, xanthine, and uracil, and the nucleosides adenosine, inosine, and guanosine. A low-affinity adenosine transporter, which also recognizes inosine and guanosine, and possibly the nucleobases hypoxanthine and adenine, has been characterized in *T. gondii* tachyzoites. There appears to be a second route of uptake for inosine distinct from the adenosine transporter.

Recently the cDNA for an adenosine transporter (*TgAT*) has been cloned by disruption of the *TgAT* locus by insertional mutagenesis to generate adenine arabinoside-refractory parasites and subsequent identification of the *TgAT* cDNA by hybridization with part of the disrupted locus. *TgAT* encodes an ENT-type protein of 462 amino acids, which shares limited sequence identity with the human ENTs 1 and 2. Functional expression of *TgAT* in *Xenopus* oocytes indicates that it shares most of the characteristics of the adenosine transporter, although it does not appear to recognize adenine. Like *Plasmodium*, *T. gondii* has on its plasma membrane a V-type H^+-pump that extrudes H^+ from the cytosol into the PV, but whether this gradient is used by TgAT to drive transport is not yet known. At present the genes or cDNAs for other nucleoside and nucleobase transporters have not been identified within the *T. gondii* genome, but with the advent of a genome sequencing initiative it is likely that other nucleoside transporter sequences will emerge.

Eimeria, Cryptosporidia, *and* Babesia

The purine translocation pathways of these parasites have received little attention. Like *Plasmodium*, *Babesia bovis* induces a new permeation pathway within the infected erythrocyte, but very little is known about uptake into the parasitophorous vacuole and translocation across the parasite plasma membrane. Virtually nothing is known about purine and pyrimidine translocation into *Eimeria* and *Cryptosporidium*. A recent screen of the relevant parasite genome databases suggests that both genera accommodate ENT homologs.

Kinetoplastida

Leishmania *and* Crithidia

Leishmania parasites have a digenetic lifecycle in which the flagellated extracellular promastigote persists in the midgut of the sandfly and the aflagellate amastigote resides within the acidic phagolysosome of the mammalian host macrophage. Most of our knowledge about the available transport of purines and pyrimidines in *Leishmania* has been derived from studies with the promastigote form. Two distinct nucleoside transport loci have been genetically defined in *L. donovani*. These two transporters display non-overlapping substrate specificities and recognize either adenosine and pyrimidine nucleosides (LdNT1) or inosine and guanosine (LdNT2) with high affinity. These observations have also been extrapolated to other *Leishmania* spp. and *Crithidia* spp. Transfection of a *L. donovani* cosmid library into nucleoside transport-deficient *L. donovani* provided an avenue for cloning the *LdNT1* and *LdNT2* genes by functional rescue of the nucleoside transport-deficient phenotype. The *LdNT1* locus, which consists of two nearly identical copies (*LdNT1.1* and *LdNT1.2*), and *LdNT2* encode ENT-type proteins of 491 and 499 amino acids, respectively, and display limited sequence identity to each other and to the human ENTs. The functional expression of these genes within nucleoside transport-deficient *L. donovani* confirmed that these proteins are mutually exclusive, high-affinity transporters for adenosine–pyrimidine nucleosides and inosine–guanosine, respectively. Although LdNT1and LdNT2 share similarity with hENT1 in their overall predicted topological structures and signature sequences, electrophysiological studies in *Xenopus* oocytes indicate that they are not equilibrative but rather function as proton symporters. The homologous genes have now been cloned from *C. fasciculata*.

It is clear from manipulation of the cultivation medium with various exogenous purine nucleobases and nucleosides that *Leishmania* is capable of transporting and utilizing any preformed purine or purine nucleoside, with the exception of xanthosine. Adenine and hypoxanthine transport has been demonstrated in *L. braziliensis panamensis*, but these activities were much reduced compared to the transport of nucleosides. Since purine base transport in *Crithidia* spp. and other Trypanosomatidae is significantly enhanced under conditions of purine depletion, it is entirely possible that at physiological purine concentrations nucleobase transport is repressed. Purine nucleoside transport also is affected by conditions of purine stress, although to a much lesser degree. Recently, an ENT homolog, *LmNT3*, was identified within the *L. major* genome database. *LmNT3* encodes a 501 amino acid ENT-type transporter that has 33% identity to LdNT1 and 35% to LdNT2. When functionally characterized in *Xenopus* oocytes, LmNT3 was found to be a high-affinity transporter for adenine, hypoxanthine, guanine, and possibly guanosine. Surprisingly, LmNT3 does not transport xanthine, suggesting that at least one other nucleobase transporter gene is encoded within the leishmanial genome.

It is entirely feasible that other purine and pyrimidine transporters are present within the leishmanial genome. Some of the major purine salvage enzymes in Kinetoplastida are compartmentalized in the glycosome, probably requiring the translocation of purine bases into this unique organelle. Localization studies with Green Fluorescent Protein-coupled transporters have shown that these transporters localize primarily to the parasite plasma membrane (and possibly to the flagellum in the case of LdNT2). In addition to these other intracellular transporters, it is possible that different nucleoside and nucleobase transporters are expressed within the amastigote life stage of the parasite. Recent studies on intact *L. donovani* amastigotes revealed two distinct adenosine transport activities, an *LdNT1*-like activity, and a second activity that was inhibited by the purine nucleosides inosine and guanosine, but not by pyrimidines. Whether *LdNT1* and *LdNT2* are expressed in amastigotes has not been determined, although a partial cDNA sequence from *L. major* amastigotes with high homology to *LdNT2* has been deposited into the GenBank database.

As mentioned above, purine transport is upregulated in several species of Trypanosomatidae under conditions of purine depletion. Transport is also modulated in *Leishmania* by the <u>TO</u>xic nucleoside <u>R</u>esistance protein (TOR), named because it was first identified in *L. donovani* selected for resistance to the toxic adenosine analog, tubercidin. TOR comprises some 478 amino acids and is similar to the mammalian transcriptional activator OCT-6. When overexpressed in *Leishmania* TOR represses purine nucleoside, and most significantly nucleobase, transport. The mechanism for this phenomenon is unknown. TOR can also function as a transcriptional activator in yeast, although it is notable that transcriptional enhancers have not been described for *Leishmania*.

Trypanosoma cruzi

Very little is known about nucleoside and nucleobase transport into *T. cruzi*. Most of our knowledge is derived from studies on tubercidin-resistant epimastigote-stage parasites. These have shown that, like *Leishmania* and *Crithidia,* these parasites possess an adenosine–thymidine transporter, and possibly other adenosine, inosine and uridine transporters. Expression Site Tagged (EST) sequences homologous to LdNT1 and LdNT2 have been found for *T. cruzi* epimastigotes.

Trypanosoma brucei *and other African trypanosomes*

Unlike other kinetoplastids, African trypanosomes contain a plethora of purine and pyrimidine transporters with overlapping substrate specificities that are stage-specifically expressed in the procyclic (insect stage) and bloodstream (mammalian infectious stage) forms of the parasite. Unlike *Leishmania* and *T. cruzi*, *T. brucei* persist in an extracellular form throughout their entire life cycle, and the purine and pyrimidine environment that they encounter is influenced by plasma concentrations.

Two types of high-affinity adenosine transport activities have been identified in *T. brucei* and other species of *Trypanosoma*: P1, which transports adenosine, inosine and guanosine, and P2, which exhibits a restricted specificity for adenine and adenosine. The P2 transporter is distinguished by its ability to transport two classes of anti-trypanosomal drugs with high affinity: the diamidines and the melaminophenyl arsenicals. Trypanosomes resistant to these classes of drugs either lack or have significantly impaired P2 transport activity. Recently the P2 gene, or *TbAT1*, was cloned by functional complementation using yeast purine auxotrophs that are naturally deficient in adenosine transport. *TbAT1* is a single-copy gene and encodes an ENT-type protein of 436 amino acids that shares limited sequence identity with LdNT1 and LdNT2 and the human ENTs 1 and 2. When expressed in yeast, *TbAT1* confers sensitivity to the melaminophenyl arsenical drug, melarsen oxide, although TbAT1-mediated adenosine transport was not susceptible to the diamidine drugs pentamidine and berenil. Isolation of the *TbAT1* allele from a melaminophenyl arsenical-resistant isogenic strain revealed nine point mutations within the mutant *TbAT1* allele, six of which result in amino acid substitutions. Subsequent analysis of 65 melaminophenyl arsenical-resistant field isolates identified eight of the same nine mutations with an additional deletion of an entire codon.

A gene encoding P1 activity has also been cloned from *T. brucei*. This gene, *TbNT2*, shares considerable homology to *TbAT1* (58% identity at the amino acid level), but upon expression in *Xenopus* oocytes displays a P1-type activity that transports adenosine, inosine, and guanosine with high affinity. Southern analyses and searches of the *T. brucei* genomic database have revealed a surprising molecular complexity at the *TbNT2* locus. *TbNT2* is part of a multigene cluster on chromosome 11, comprising some six members that all share 81–96% sequence identity at the amino acid level.

Nucleobase transport appears to be mediated by several stage-specifically expressed transport activities with overlapping ligand specificity. In bloodstream *T. brucei*, two activities have been delineated, H2 and H3. H2 is a broad-specificity purine and pyrimidine nucleobase transporter, which also recognizes guanosine and possibly inosine. H3 is a high-affinity transporter with a substrate specificity restricted to the purine nucleobases. In procyclic *T. brucei*, two nucleobase transport activities have been demarcated, H1 and U1. H1 is strikingly similar to the H3 activity described in the bloodstream stage of the parasite. Indeed, H1 and H3 can only be differentiated by their respective affinities for xanthine and the pyrazolopyrimidine, allopurinol. U1 is a high-affinity, uracil-specific transporter, which does not appear to efficiently recognize any other pyrimidine nucleobase or nucleoside, although it may transport uridine with low affinity.

Although electrophysiological measurements have not been performed, it appears that all of the *T. brucei* nucleobase and nucleoside transporters couple transport to the translocation of a proton. Destruction of the proton gradient across the parasite plasma membrane with protonophores or metabolic inhibitors results in a loss of transport capability for all of the transporters described above.

Amitochondriates

Giardia lamblia and *Trichomonas vaginalis*

G. lamblia cannot synthesize pyrimidines and deoxynucleosides, and must be able to access a broad range of host nucleosides. Saturable transport activities have been described for both nucleobases and nucleosides in *G. lamblia*. These have been classified as type 1, specific for thymidine, uridine, and possibly their nucleobases; type 2, a broad-specificity transport activity for purine and pyrimidine nucleosides and deoxynucleosides but which discriminates against nucleobases; and type 3, a low-affinity nucleobase transporter. A type 2 activity has also been described in *T. vaginalis*. Currently, none of these transport activities has been analyzed at the molecular level, although the sequence for a porin-like transporter that was identified by functional complementation of nucleoside transport-deficient *Escherichia coli* has been deposited in GenBank.

Entamoeba histolytica

A saturable adenosine transport activity inhibited by other nucleosides has been demonstrated in *E. histolytica*. Sequences homologous to ENT-type transporters have been deposited within GenBank, but these genes have not been functionally validated.

Helminths

There are no molecular clones available to study purine and pyrimidine transporters in worms. Most of our knowledge about nucleoside and nucleobase transport systems in helminths is based upon incorporation studies undertaken several decades ago with intact parasites. Since worms are multicellular and also incapable of synthesizing purines *de novo*, it is likely that these parasites possess multiple nucleobase and nucleoside transport mechanisms. Thus, conclusions based on uptake in whole parasites are complicated by the presence of overlapping activities.

Cestoda

In the cestode *Hymenolepis diminuta*, three transport foci have been delineated for nucleobases. These comprise a thymine–uracil carrier, a transport activity that appears specific for purine nucleobases, and a third type of activity that is apparently specific for hypoxanthine. Little is known about nucleoside transporters in these parasites. Uridine uptake appears to be carrier-mediated and is inhibited by both purine and pyrimidine nucleosides. Whether there are several permeases or one broad-specificity carrier has not been determined. Likewise, the effect of nucleobases on uridine transport has not been determined.

Trematoda

Most information on nucleoside and nucleobase transport in trematodes is based upon early work with *Schistosoma mansoni*. Uptake of radiolabeled pyrimidine bases is nonsaturable and unaffected by high exogenous concentrations of pyrimidines. Uptake of purine nucleobases is apparently carrier-mediated. It is possible that this transporter also may recognize the purine nucleosides adenosine and guanosine.

Nematoda

Relatively little is known about the transport mechanisms in nematodes, although they can salvage both purine and pyrimidine nucleobases and nucleosides. Recently, however, with the completion of the *Caenorhabditis elegans* genome and the emergence of partial mRNA sequences for several parasitic nematodes, multiple homologs have been identified from both the ENT and CNT families.

PURINE METABOLISM

Purine ribonucleotides are synthesized by animal cells from glutamine, glycine, aspartate, ribose-5-phosphate, CO_2, and tetrahydrofolate cofactors. They can also be generated by salvage of preformed purine bases, hypoxanthine, guanine, and adenine (but not xanthine), and adenosine (but not inosine or guanosine). The biosynthetic pathway is highly conserved among phylogenetically diverse organisms, including most prokaryotes, yeast, plants, and birds.

However, all parasitic protozoa and worms studied to date are incapable of synthesizing purines *de novo*, and thus are completely dependent on salvage of host purines. This conclusion was based upon the nutritional dependence of parasites on an exogenous purine supply and the inability of these organisms to incorporate radiolabeled precursors (i.e. [^{14}C]formate or [^{14}C]glycine) into purine nucleotides. It should be noted that mammalian cells might also not incorporate [^{14}C]formate or [^{14}C]glycine into nucleotides under conditions in which purines are present in the environment, primarily because of feedback mechanisms on the early steps of purine biosynthesis. The lack of an intact purine pathway in parasites is strongly supported by the failure to detect a complete set of homologs of purine biosynthesis enzymes within parasite genomes.

In addition to the biosynthetic pathway, mammalian cells also express a variety of purine salvage and interconversion enzymes (Figure 9.3). The salvage pathway functions to take up purines obtained from nucleic acid and nucleotide turnover, as well as from the diet. The process of purine salvage is advantageous to the cell since it is less energetically expensive than the *de novo* pathway, which requires hydrolysis of seven high-energy phosphate bonds just to synthesize AMP or GMP. Purine salvage also traps diffusible purine bases and nucleosides as anionic, phosphorylated metabolites, thereby ensuring

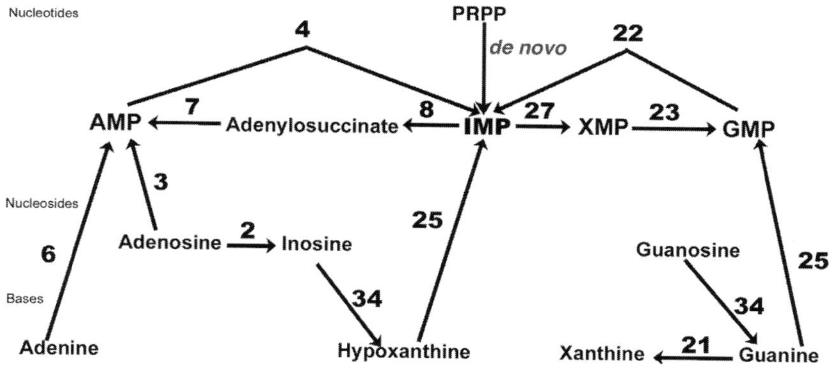

FIGURE 9.3 Mammalian purine salvage and interconversion pathways. Enzymes are designated numerically and can be identified from Table 9.1.

their accumulation in environments in which purines are scarce. Although many of the enzymatic activities involved in mammalian purine salvage are shared by parasites, the overall salvage pathways between an individual parasite genus and its host and among parasite genera can be different.

In mammalian cells purine nucleotides obtained by nuclease digestion of nucleic acids are converted to nucleosides by nucleotidases or non-specific phosphatases. Presumably a similar mechanism is operative in most parasites, but these activities have not been well characterized except in *Leishmania* and *Crithidia*. Purine nucleosides can be directly salvaged to the nucleotide level via either relatively specific kinase or non-specific phosphotransferase enzymes. There are no known nucleoside phosphotransferases in mammalian cells, and the only kinase for ribonucleosides that is operative in mammalian cells is AK (Figure 9.3). Inosine and guanosine are not phosphorylated by AK or any other enzyme in mammals, and purine deoxyribonucleosides are inefficient substrates for both AK and dCK, but are probably not salvaged in great quantity. Nucleosides can be cleaved, however. In mammalian cells, PNP catalyzes the reversible phosphorylytic cleavage of inosine, guanosine, deoxyinosine, and deoxyguanosine to hypoxanthine/guanine, while parasites express a battery of unique phosphorylases and NHs that convert nucleosides to nucleobases. Adenosine can also be deaminated to inosine, a reaction catalyzed by ADA, an enzyme that is found in mammalian cells and some parasites.

Purine bases are salvaged by PRT enzymes that catalyze the 5-phosphoribosyl-1-pyrophosphate (PRPP) dependent conversion of purine bases to the nucleotide level. Both mammalian cells and parasites possess PRTs that are specific for either adenine or 6-oxypurines. The latter class varies considerably in its substrate specificities. The human HGPRT enzyme is specific for hypoxanthine and guanine but does not recognize xanthine, unlike some of its parasite counterparts, which are HGXPRTs, and *G. lamblia* which has a GPRT activity. Adenine and guanine can also be deaminated via AD and GD, respectively. The latter is expressed in many parasites and in humans, whereas AD is found exclusively in microorganisms. The PRPP substrate for PRT reactions is produced by PS and is required for both the synthesis of purine and pyrimidine nucleotides *de novo* and the salvage of purine and pyrimidine nucleobases. Any parasite that relies on purine base salvage must have a PS activity.

Mammalian cells and many parasites can also interconvert adenylate and guanylate nucleotides, a process in which inosine monophosphate (IMP), the first nucleotide formed by the *de novo* pathway, is the focal point. IMP can be converted to AMP by ASS and ASL, while GMP is formed from IMP by the sequential actions of IMPDH and GMPS (Figure 9.3). These 'branchpoint' enzymes in animal cells are reciprocally regulated by feedback inhibition by the nucleoside monophosphate end-product of each branch and stimulation by the nucleoside triphosphate end-product of the opposite branch, i.e. GTP is essential for AMP synthesis, and ATP is required for GMP production. This ensures a balanced supply of adenylate and guanylate nucleotides for the cell. IMP can also be regenerated from AMP and GMP by AMPD and GMPR, respectively (Figure 9.3). AMPD and GMPR are also tightly controlled by regulatory mechanisms.

The end-product of purine catabolism in humans is urate, although other organisms further metabolize urate to allantoin, allantoic acid, urea, and ammonia. Purine waste products are excreted when the amount of purine available to the cell is in excess of its requirements. Urate is synthesized from hypoxanthine and xanthine, an oxidation reaction that is catalyzed by XO. XO is absent in both protozoan and helminthic parasites. Because of the nutritional necessity of purines and the relative scarcity of purines in the environment, it is unlikely that parasites would have a great need for eliminating dispensable purine.

The purine deoxynucleotide substrates for DNA synthesis are derived from ribonucleotides. Most RRs are two-subunit enzymes that catalyze the reduction of the 2′-hydroxy moiety of all four naturally occurring purine and pyrimidine ribonucleoside disphosphates to the corresponding deoxynucleotides. RR is the only known mechanism for *de novo* production of deoxynucleotides, and the enzyme is subject to complex allosteric regulatory mechanisms that ensure a balanced supply of all four dNTPs for DNA synthesis. RR activity in mammalian cells is also cell-cycle dependent; the enzyme is active during the S phase of the cell cycle when DNA is replicated and relatively inactive during the quiescent (G1 and G2) and mitotic (M) phases of the cell cycle. Deoxynucleotides can also be generated, albeit inefficiently, via kinase enzymes. AK, dCK, and several mitochondrial deoxynucleoside kinases can metabolize deoxyadenosine and/or deoxyguanosine to the nucleotide level. Most parasites are thought to generate deoxynucleotides directly via ribonucleotide reduction. The existence of RR is supported by the ability of parasites to grow in medium lacking deoxynucleosides, to incorporate radiolabeled bases and nucleosides into DNA, the detection of RR activities in crude parasite lysates, the existence of parasite clones encoding RR subunit homologs, and RR subunit sequences in the parasite genome sequencing databases.

Apicomplexa

Plasmodium

Because infection of erythrocytes with *Plasmodium* spp. dramatically augments their capacity to salvage purines, most of the early knowledge on purine salvage capabilities of malarial parasites was gleaned from comparisons of uninfected and infected erythrocytes. Like the mammalian red blood cell, *Plasmodium* species cannot synthesize purines *de novo,* and their growth is contingent on salvage. Any of a number of purines, including hypoxanthine, guanine, adenine, inosine, and adenosine, can satisfy the purine requirements of *P. falciparum.* Hypoxanthine is likely the preferred source of malarial nucleotides, and can be acquired directly from blood or from erythrocytes through nucleotide degradation. Furthermore, addition of XO, which converts hypoxanthine to urate, to the growth medium inhibits *P. falciparum* proliferation, strongly implying that extracellular hypoxanthine is the critical source for parasite nucleotides. ADA and PNP activities necessary for adenosine deamination and inosine cleavage are also present in both the red blood cell and parasite and are likely major contributors to purine salvage as well (Figure 9.4). Similar findings were obtained with rodent and avian malarias. Because a single purine can fulfill all of the parasite's nucleotide requirements, the enzymatic machinery to convert IMP to adenylate and guanylate nucleotides must be operational in the parasite.

A number of purine salvage and interconversion enzymes from *P. falciparum* have been characterized at the biochemical level (Figure 9.4). Most prominent among these are HGXPRT, PNP, and ADA. The *PfHGXPRT* gene has been cloned, sequenced, and overexpressed in *E. coli,* and the recombinant protein purified to homogeneity. Preferred substrates for

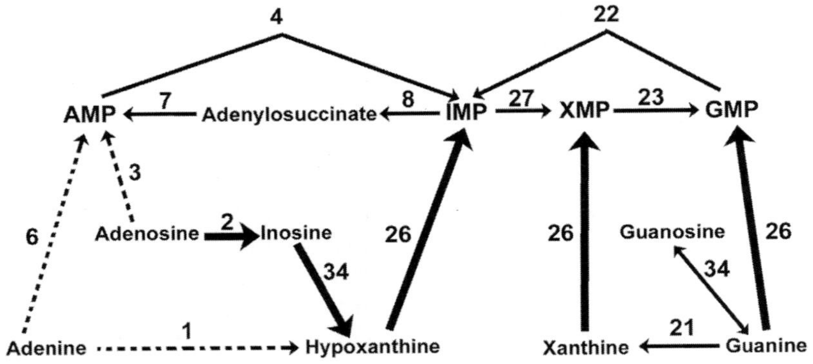

FIGURE 9.4 Malaria purine salvage and interconversion pathways. Key pathways are designated by thick lines. Dashed lines are for published pathways for which current evidence is unsupportive. Enzymes are designated numerically and can be identified from Table 9.1.

PfHGXPRT are hypoxanthine and guanine with micromolar K_m values, while the K_m value for xanthine is two orders of magnitude higher. The malarial enzyme, like the human HGPRT, is inhibited by nanomolar concentrations of two transition state analogs of the enzyme, immucillin-HP and immucillin-GP. The immucillins are analogs of naturally occurring purines in which the ribose is replaced with a 1-(9-deazapurin-9-yl)-1,4-dideoxy-4-iminoribitol moiety (immucillin-HP is the IMP analog, immucillin-GP is the GMP analog). A high resolution crystal structure of recombinant PfHGXPRT has been solved in the presence of immucillin-HP. Whether *PfHGXPRT* is an essential gene remains to be proven. Gene replacement strategies are now available for *P. falciparum,* but whether creation of δ*pfhgxprt* knockouts in malaria is feasible has not been tested. Because the malaria genome sequence databases do not appear to encompass *AK* or *APRT* genes, it is quite possible that PfHGXPRT represents the lone route for purine salvage in the parasite. If so, the enzyme should be an attractive target for inhibitor design.

PfPNP has also been thoroughly studied and is another potential therapeutic target. The *PfPNP* gene has been cloned and overexpressed in *E. coli* and the recombinant enzyme purified to homogeneity and characterized kinetically. Immucillin-H, the desphosphorylated derivative of immucillin-HP, is a potent transition state inhibitor of the malarial enzyme with a K_i of 0.6 nM. A battery of other immucillin nucleoside analogs also inhibit PfPNP at nanomolar concentrations. Additionally, immucillin-H induces a purine-less death in *P. falciparum* with an IC_{50} value of 35 nM, and this toxicity is reversed by hypoxanthine but not by inosine. These data imply that the major and possibly only route of purine salvage by *P. falciparum* is either through inosine by the sequential actions of PfPNP and PfHGXPRT, or by phosphoribosylation of hypoxanthine directly by PfHGXPRT. They further suggest that PfPNP is a particularly attractive target for inhibitor design, perhaps more than PfHGXPRT, because of the ability of the 'lead compound' immucillins to traverse the multiplicity of membranes in which the parasite is enclosed.

Other purine enzyme activities have also been detected in *P. falciparum*. These include ADA, the gene sequence for which has been deposited in PlasmoDB, and APRT, a gene that is not currently present in the database. PfADA appears to differ from its human counterpart by its refractoriness to erythro-9-(2-hydroxy-3-nonyl)adenine. It is unclear whether the parasite has a functional APRT activity. It is possible that the gene has yet to be discovered or that the early biochemical evidence for the existence of the enzyme was influenced by the presence of AD activity. The ability of *P. falciparum* to salvage adenine, however, intimates that either APRT or an AD/HGXPRT pathway must be operative. PlasmoDB also contains DNA sequences encoding ASS, ASL, IMPDH, GMPS, AMPD, and GMPR. Metabolic flux measurements prove the functionality of the first four of these six enzymes, but none of the six has been studied in detail. The current model for purine salvage in *P. falciparum* is depicted in Figure 9.4.

Toxoplasma gondii

T. gondii also cannot synthesize purine nucleotides *de novo* but can proliferate without added purine. Thus, it can derive all its purine needs from the host cell. Early studies suggested that host adenylate nucleotides could access the parasitophorous vacuole and be dephosphorylated by the tachyzoite, the actively proliferating stage of the parasite. AMP appeared to be the most efficiently utilized nucleotide, although both ATP and ADP were also incorporated. The vacuole itself contains high concentrations of NTPase (apyrase) activity that is capable of hydrolyzing ATP to ADP and AMP. A number of different isoforms of the NTPase were discovered and shown to exhibit slightly different substrate selectivities. Despite this finding, *T. gondii* lacks a 5′-NT activity that would dephosphorylate AMP to adenosine and allow purine entry into the cell. Furthermore, a nucleoside transport-deficient *T. gondii* strain was greatly compromised in AMP-uptake capability. This implies that AMP was dephosphorylated to adenosine by contaminant host 5′-NT activity. These data argue against a role of NTPase in purine acquisition by the parasite.

Extracellular tachyzoites are capable of taking up a number of purines including adenosine, inosine, guanosine, adenine, hypoxanthine, guanine, and xanthine. Adenosine was incorporated at a rate 10–25-fold greater than any of the other purines, suggesting that adenosine was the primary nutritional purine for the parasite. Biochemical studies further revealed that AK is the most active purine salvage enzyme present in tachyzoite extracts, more than ten-fold more active than any other enzyme (Figure 9.5). No other nucleoside kinase or phosphotransferase activities were detected. *T. gondii* extracts were also capable of cleaving guanosine and inosine, but not adenosine, implying that PNP activity was present. Both AD and GD activities were found, as were phosphoribosylating activities for all four bases. However, the observed adenine incorporation is likely imputed to AD-mediated deamination and subsequent hypoxanthine salvage through TgHGXPRT. Interestingly, radiolabeled inosine, hypoxanthine, and adenine labeled both adenylate and guanylate nucleotides, while guanine, guanosine, and xanthine only labeled guanylate nucleotide pools. These data imply that *T. gondii* lacks GMPR activity, although all the branchpoint pathways from IMP to AMP and GMP must be intact (Figure 9.5).

The genes encoding TgAK and TgHGXPRT have been cloned using insertional mutagenesis strategies, and both recombinant proteins have been purified and characterized. TgAK is specific for adenosine, while TgHGXPRT recognizes hypoxanthine, guanine, and xanthine, although xanthine binds less well than the other bases. High resolution crystal structures for both TgAK and TgHGXPRT have been obtained,

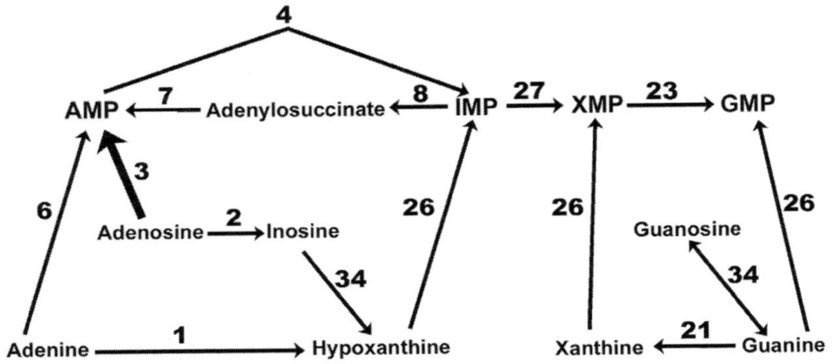

FIGURE 9.5 *Toxoplasma gondii* purine salvage and interconversion pathways. The key enzyme, adenosine kinase, is indicated by a thick line. Enzymes are designated numerically and can be identified from Table 9.1.

each in a variety of mechanistic states, i.e. apoenzyme, substrate- and product-bound forms. These structures have provided considerable insights into the mechanisms by which PRT enzymes stimulate catalysis.

T. gondii deficient in either TgAK or TgHGXPRT have been obtained by both chemical and insertional mutagenesis. Neither enzyme alone is essential for parasite viability or proliferation. It has not been possible to generate a double mutant devoid of both TgAK and TgHGXPRT activity, suggesting that *T. gondii* accommodates only these two routes of purine salvage. Given the discrepancies in the specific activities of the two enzymes in extracts, it can be reasonably conjectured that AK and HGXPRT represent the primary and secondary purine salvage mechanisms of this parasite. Thus, the purine salvage pathway of *T. gondii* can be primarily distinguished from that in *P. falciparum* by its high AK activity.

Eimeria tenella

Unlike *P. falciparum* and *T. gondii*, which can be cultured, *E. tenella* is not amenable to simple biochemical or genetic manipulations. There is little recent information on purine salvage by this parasite. Radiolabel incorporation experiments suggest that purine salvage by *E. tenella* is similar to that in *T. gondii*. Both sporozoites and merozoites incorporate adenine, adenosine, hypoxanthine, and inosine into adenylate and guanylate nucleotides, but guanine and xanthine are only converted to guanylates. This implies that guanine and xanthine cannot satisfy the purine needs of the parasite. The metabolic labeling results indicate that GR activity is likely absent, whereas the branchpoint pathways from IMP to AMP and GMP are intact. An HGXPRT activity has also been isolated and affinities for the three oxypurine substrates determined.

Kinetoplastida

Leishmania *spp.*

The purine salvage pathway of *Leishmania* is interwoven and complex. The key enzymes of purine salvage are PRT enzymes. *L. donovani* express three different PRT activities: HGPRT, APRT, and XPRT (Figure 9.6). XPRT is an unusual enzyme. Mammalian cells, which contain both HGPRT and APRT enzymes, lack XPRT. All three

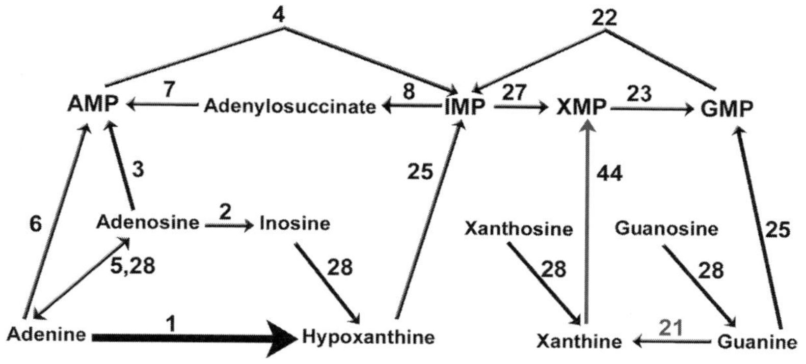

FIGURE 9.6 Kinetoplastid purine salvage and interconversion pathways. The white line indicates that AD enzyme (1) is present in *Leishmania* promastigotes but not in *T. cruzi* or *T. brucei*. Enzyme identities are listed in Table 9.1.

PRT genes have been cloned, and the recombinant proteins have been purified and characterized. LdHGPRT and LdAPRT exhibit absolute specificity for the purines for which they are named, whereas LdXPRT, for which xanthine is the preferred substrate, can also recognize hypoxanthine and guanine. Pyrazolo (3,4-d)pyrimidine bases, e.g. allopurinol and novel antileishmanial compounds that are nontoxic to mammalian cells, require HGPRT for their conversion to toxic phosphorylated metabolites, but allopurinol is still a relatively inefficient LdHGPRT substrate. LdHGPRT and LdXPRT also contain a COOH-terminal tripeptide that is a topogenic signal for targeting proteins to the glycosome, a unique kinetoplastid organelle. Both LdHGPRT and LdXPRT have been localized to the glycosome by immunofluorescence, immunoelectron microscopy, and subcellular fractionation. Conversely, LdAPRT is a cytosolic enzyme. Other *L. donovani* enzymes that can convert purines to the nucleotide level are LdAK and possibly LdNPT. LdNPT has been implicated in the metabolism to the nucleotide level of the antileishmanial pyrazolopyrimidine nucleoside analogs, allopurinol riboside, formycin B, and 4-thiopurinol riboside, but whether the enzyme recognizes naturally occurring purine nucleosides is unknown.

Gene replacement studies have shown that Δ*ldhgprt*, Δ*ldaprt*, and Δ*ldxprt* knockouts are all viable as promastigotes, proving that none of the three LdPRTs is by itself essential for parasite growth. LdAK is also non-essential, since AK-deficient *L. donovani* has been created after mutagenesis. Combinations of mutations at the *LdHGPRT*, *LdAPRT*, and *LdAK* loci have revealed that even a triple mutant deficient in all three activities can survive on any purine nucleobase or nucleoside tested. Further gene replacement experiments have failed to generate a double Δ*ldhgprt*/Δ*ldxprt* knockout providing powerful, but only suggestive, negative evidence that *L. donovani* promastigotes require either a functional HGPRT or XPRT activity for growth.

The PRPP and nucleobase substrates for PRT enzymes are derived from PS and NHs, respectively. *L. donovani* has at least three NH activities that cleave nucleosides to their corresponding bases and ribose: one specific for inosine, uridine, cytidine, and xanthosine (IUNH); one for deoxynucleosides; and one for inosine and guanosine. The IUNH from *L. major* has been cloned, and the recombinant protein evaluated for substrate selectivities and affinities. Moreover, a 2.5 Å crystal structure has been determined. The IUNH appears to be non-essential, because *Δldiunh* knockouts are viable as promastigotes. *L. donovani* also possesses an AP activity that cleaves adenosine to adenine. To produce NH substrates from nucleotides and nucleic acids, *L. donovani* has surface membrane 3′ NT/nuclease and 5′ NT activities, as well as an active surface phosphatase activity. The *L. donovani* 3′ NT/nuclease and acid phosphatase genes have been cloned, and their proteins localized and studied in detail.

Metabolic flux experiments with *L. donovani* promastigotes have demonstrated that adenine and inosine are funneled to hypoxanthine through AD and NH, respectively, whereas guanine is converted to xanthine by GD. Moreover, mutationally derived APRT-deficient *L. donovani* take up adenine as efficiently as wild-type cells. Thus, APRT does not appear to play an important role in purine uptake by promastigotes. A more important role for APRT can be inferred from studies with amastigotes, which lack AD, but possess ADA, activity. This implies that adenine is salvaged via APRT in amastigotes, whereas adenosine can either be deaminated to inosine which is then cleaved by NH or phosphorylated directly by AK. AK also plays a prominent role in adenosine uptake by the promastigote, since AK-deficient parasites do not take up adenosine as efficiently as parental cells.

The ability of the parasite to interconvert adenylate and guanylate nucleotide pools demonstrates that GR and AMPD are both expressed in *L. donovani*. The LdGR can be distinguished from its human counterpart by several kinetic features and by its increased sensitivity to inhibition by pyrazolopyrimidine nucleotides. The LdASS is also unique in that it can accept allopurinol ribonucleotide monophosphate and formycin B monophosphate as substrates. The unique properties of LdASS and LdGR likely contribute to the selective actions of pyrazolopyrimidines on the parasite.

Trypanosoma cruzi

Purine metabolism in *T. cruzi* is quite similar to that in *Leishmania* with few exceptions (Figure 9.6). *T. cruzi* epimastigotes and amastigotes incorporate bases and nucleosides into their nucleotide pools, and adenylate and guanylate nucleotides are interconvertible. Bases are preferred over nucleosides. The epimastigotes appear to phosphorylate adenosine through AK, although the enzyme has not been isolated, and the PRT activities appear to be similar to those of *L. donovani*, although only TcHGPRT has been well studied at the molecular level. Elegant structural studies promoting TcHGPRT as a target for inhibitor design have been accomplished, although whether TcHGPRT is a good target could be complicated by XPRT- and APRT-mediated purine salvage. Whether TcHGPRT is essential awaits the construction of a *Δtchgprt* null mutant. There have been conflicting reports over whether *T. cruzi* epimastigotes express AD activity. Thus, adenine may be phosphoribosylated directly to AMP or deaminated to hypoxanthine, depending on whether APRT or AD is most active. Epimastigotes appear to have AK activity, and a *TcAK* cDNA is present in the *T. cruzi* database. *T. cruzi* also expresses purine nucleoside-cleaving enzymes, including a phosphorylase, an inosine/guanosine NH, and a 2′-deoxyinosine NH. Finally, the *T. cruzi* ASS and ASL proteins are akin to their leishmanial equivalents. Overall, the purine salvage pathway of *T. cruzi* is similar to that of *L. donovani*, except that the substrate specificities of the nucleoside-cleaving activities are different, and AD expression is unknown.

Trypanosoma brucei

Metabolism of purines by *T. brucei* is qualitatively similar to that by *T. cruzi*. Bloodstream-form *T.b. gambiense* and *T.b. rhodesiense* and procyclic *T.b. gambiense* all salvage bases and nucleosides efficiently. The observation that all four bases are incorporated into nucleotides implies the existence of three distinct PRT activities. Only TbHGPRT, both in its native and recombinant forms, has been purified and studied in some detail. The enzyme, like its leishmanial counterpart, is localized to the glycosome, is specific for hypoxanthine and guanine, and recognizes allopurinol. *T.b. gambiense, T.b. rhodesiense* and *T.b. brucei* all can incorporate allopurinol into 4-aminopyrazolopyrimidine nucleotides and RNA, implying intact ASS and ASL activities. Purine nucleoside salvage proceeds either through direct phosphorylation via kinase or phosphotransferase activities and by cleavage to the base. AK activity has been detected. Nucleoside cleavage activities in procyclic forms include two NHs, one specific for purine ribonucleosides and one for pyrimidine nucleosides, and an adenosine/methylthioadenosine phosphorylase activity. Bloodstream-form *T.b. brucei* appears to contain two nucleoside phosphorylase activities, one for inosine, adenosine, and guanosine (IAGNH), and one specific for adenosine and 5′-methylthioadenosine. The gene encoding TbIAGNH has been cloned, and the properties of both the native and recombinant IAGNH enzyme have been investigated in detail. *T.b. gambiense* procyclic and *T.b. brucei* bloodstream parasites lack both ADA and AD activities but can deaminate guanine. *T. congolense* and *T. brucei* appear to metabolize purines similarly. The lack of AD appears to be a defining difference between most trypanosomes and *Leishmania*. The exception is *T. vivax*, which has been reported to have AD.

Amitochondriates

Tritrichomonas foetus

The ability of *T. foetus* to salvage all four purine bases, as well as adenosine, inosine, and guanosine, into both adenylate and guanylate nucleotides, indicates that the parasite contains intact branchpoint and AMPD and GMPR activities. TfIMPDH is a well-characterized enzyme that differs from its mammalian counterpart in a manner that could be therapeutically exploited, and whose gene has been cloned and crystal structure determined. Competition studies for nucleoside uptake by the corresponding nucleobases implies that inosine is primarily cleaved to hypoxanthine, whereas adenosine and guanosine can be directly phosphorylated, but this model is unconfirmed by enzymological studies. A single phosphoribosyltransferase, TfHGXPRT, that can phosphoribosylate all three 6-oxypurines, has been studied in detail, and the crystal structure of the recombinant enzyme has been solved to 1.9 Å resolution. Wang and coworkers have dissected the amino acid determinants that govern substrate specificity of TfHGXPRT. Specific site-directed mutations have converted TfHGXPRT into an HGPRT which cannot recognize xanthine and into an HGXAPRT which can also recognize adenine. Structure-based drug design approaches have also identified novel classes of TfHGXPRT inhibitors that can inhibit the enzyme at nanomolar concentrations without affecting the human HGPRT. These enzyme inhibitors also impede parasite growth, and this growth inhibition could be circumvented by supplementing the medium

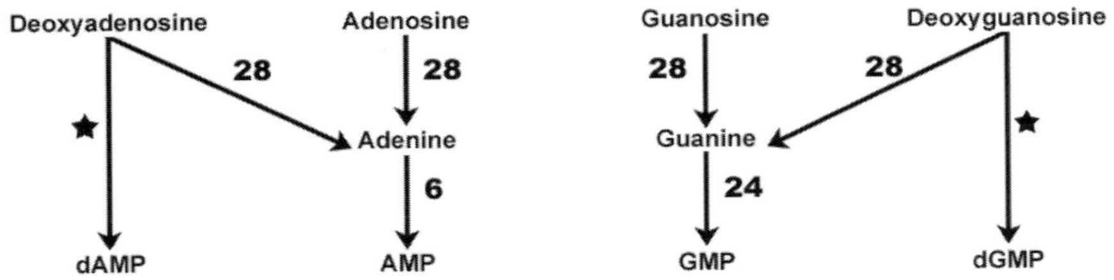

FIGURE 9.7 *Giardia lamblia* purine salvage and interconversion pathways. ★ indicates a deoxynucleoside kinase activity. Enzyme identities are listed in Table 9.1.

with either hypoxanthine or guanine. These data provide powerful suggestive evidence that the sole route of purine salvage in *T. foetus* is through TfHGXPRT and that the parasite lacks other routes of purine salvage (e.g. APRT or AK), although such activities have been proposed. Other purine enzymes that have been detected include AD, GD, and PNP.

Giardia lamblia, Trichomonas vaginalis, *and* Entamoeba histolytica

The purine salvage pathways of *G. lamblia, T. foetus,* and *E. histolytica* all differ dramatically from those of the Apicomplexa, Kinetoplastida, and *T. foetus* in that they cannot interconvert AMP, IMP, and GMP. Thus, none of these parasites would be expected to contain ASS, ASL, IMPDH, GMPS, GMPR, or AMPD activities, and they should not be able to survive or proliferate in an environment without a source of both adenine and guanine rings. Furthermore, *G. lamblia* and *T. vaginalis*, but not *E. histolytica*, do not appear to have a mechanism to generate deoxynucleotides *de novo*. Thus, they also require sources of environmental purine and pyrimidine deoxynucleosides.

G. lamblia cannot salvage hypoxanthine, xanthine, or inosine but can incorporate adenine, guanine, and their corresponding ribonucleosides (Figure 9.7). However, adenine and adenosine are only incorporated into the adenylate pool, while guanine and guanosine are metabolized exclusively to guanylate nucleotides. Thus, *G. lamblia* has no HPRT or XPRT activities, and IMP is not the common salvage intermediate for AMP and GMP synthesis. Radioisotope incorporation experiments have revealed that adenosine and guanosine are first cleaved to their respective bases, which are, in turn, phosphoribosylated by APRT and GPRT, respectively (Figure 9.7). Both *GlAPRT* and *GlGPRT* have been cloned and the recombinant enzymes shown to exhibit strict substrate specificities for their nucleobase substrates. The three-dimensional structure of GlGPRT has also been solved and provides some important insights into the restricted base specificity of the enzyme. Despite the lack of evidence for direct phosphorylation of naturally occurring nucleosides, a number of analogs appear to be phosphorylated by intact *G. lamblia*. Perhaps there exists a NPT activity. There is no detectable RR activity in *G. lamblia*, and exogeneous ribonucleosides are not converted to DNA. Thus, *G. lamblia* seems to obligatorily scavenge purine, as well as pyrimidine, deoxynucleosides from the environment, a process that requires direct phosphorylation by kinase enzymes. A deoxynucleoside kinase activity has been detected in *G. lamblia* extracts.

Metabolic labeling studies have demonstrated that *T. vaginalis* and *E. histolytica* also do not interconvert AMP, IMP, and GMP, but the pathways differ from those in *G. lamblia* in

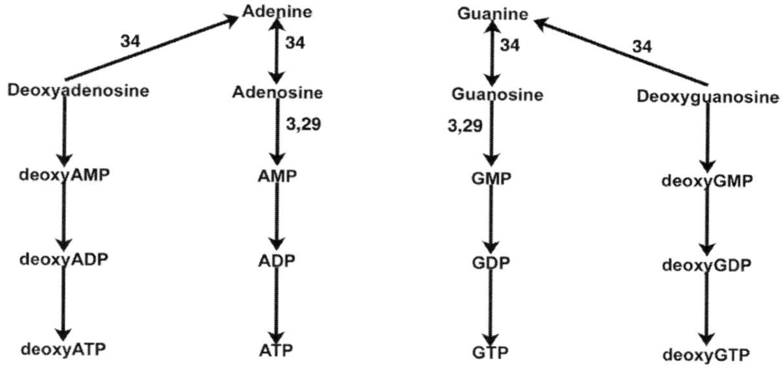

FIGURE 9.8 *Trichomonas vaginalis* purine salvage and interconversion pathways. * Deoxynucleoside phosphotransferase. All other enzyme identities are listed in Table 9.1.

that adenine and guanine are first converted to their respective nucleosides, which are then phosphorylated to nucleotides (Figure 9.8). Neither hypoxanthine nor guanine is metabolized by either parasite. Thus, both *T. vaginalis* and *E. histolytica* lack PRT enzymes. Enzyme measurements in lysates confirm the results of the radiolabeling experiments. *T. vaginalis*, like *G. lamblia*, also lacks RR activity, and the parasite, therefore, is a deoxynucleoside auxotroph (Figure 9.8). A deoxynucleoside-specific NP activity that does not recognize ribonucleosides has been detected in *T. vaginalis*. Although it is conjectured that *E. histolytica* has an RR activity, the evidence is indirect. The parasite is sensitive to growth inhibition by hydroxyurea, a free radical scavenger that inhibits RR in other systems. Whether RR is the cellular target for hydroxyurea is unclear.

Helminths

Purine metabolism in worms has been investigated in only a few parasites and in a relatively cursory fashion. Like protozoan parasites, *Schistosoma mansoni* adults and schistosomules lack the capacity to synthesize purine nucleotides *de novo*. The schistosomules can salvage hypoxanthine, guanine, adenine and their corresponding nucleosides, but not xanthine. Interconversion among the adenine and guanine nucleotide pools is minimal. The schistosomules express AP and APRT activities, which are likely routes for adenosine/adenine metabolism. They also express low, but sufficient, amounts of AK to initiate tubercidin metabolism and toxicity. In contrast to the schistosomules, adult worms have high levels of AK and ADA activities, in addition to the AP/APRT routes of salvage found in the schistosomules. The *S. mansoni HGPRT* gene has been cloned, and the protein purified, characterized biochemically, and its three-dimensional structure determined. The enzyme is similar to the human HGPRT in structure and substrate specificity. There is no separate XPRT, although GD activity, which produces xanthine, is high.

Purine metabolism in nematodes and cestodes is largely uncharted. The dog heart worm, *Dirofilaria immitis*, and the cat and primate filarial worm, *Brugia pahangi*, both lack the *de novo* pathway, but not much else is known. *Ascaris lumbricoides* possesses a number of purine salvage enzymes and the capacity to break down AMP, adenine, xanthine, urate, and allantoin to urea and ammonia, a purine

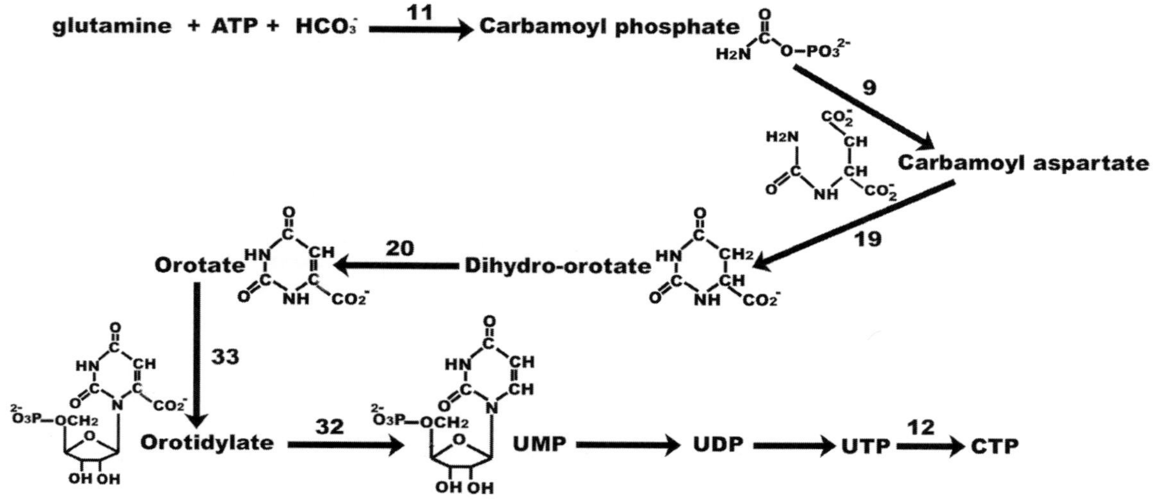

FIGURE 9.9 *De novo* pathway for pyrimidine biosynthesis. Enzyme identities are shown in Table 9.1.

catabolic pathway far more intricate and complex than that in humans. Much work on helminthic purine pathways remains to be done.

PYRIMIDINES

Most parasites, with the noted exceptions of *T. foetus*, *T. vaginalis*, and *G. lamblia*, are capable of synthesizing pyrimidine nucleotides *de novo*. This biosynthetic pathway requires six reactions to generate UMP from CO_2, glutamine, aspartate, and PRPP (Figure 9.9). UMP then serves as the precursor for all other pyrimidine nucleotides.

In mammalian cells, UMP synthesis is catalyzed by two cytosolic multifunctional proteins and a single mitochondrial enzyme. The first three enzymes, CPSII, ATC, and DHO, are part of a trifunctional protein encoded by a single transcript. The CPSII is the major site of regulation in mammalian cells and is inhibited by UTP and CTP. The fourth enzyme, DHODH, is a mitochondrial flavoprotein that generates orotate from dihydroorotate in an NAD^+-dependent reaction. The synthesis of UMP from orotate is then accomplished by a bifunctional protein that accommodates both OPRT and ODC activities. The former adds the ribose phosphate moiety to the base in a PRPP-dependent reaction, while the latter decarboxylates OMP to form UMP.

UMP synthesis is much less energetically expensive than purine nucleotide synthesis. Two ATP molecules are consumed in the first reaction and an additional two ATP molecules are required for the synthesis of PRPP. However, the NADH product of the mitochondrial reaction can be used by the electron transport chain to produce three additional ATP molecules. Thus, there is a net requirement for only a single high-energy phosphate in the pyrimidine biosynthetic pathway. It has been conjectured that the large energy expenditure for the purine biosynthetic pathway precipitated its loss in parasites residing in energetically

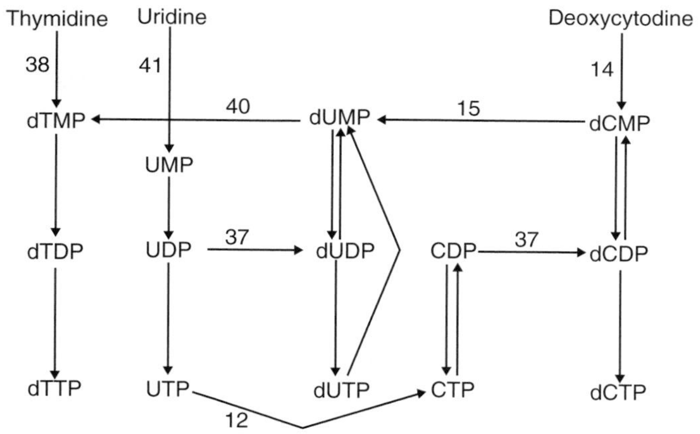

FIGURE 9.10 Mammalian pyrimidine salvage and interconversion pathways. Enzyme identities are listed in Table 9.1.

unfavorable environments. Consistent with this supposition is the example of the amitochondriates that cannot regenerate the three ATP molecules via oxidative phosphorylation and also are unable to synthesize pyrimidine nucleotides *de novo*.

Cytidylate, deoxycytidylate, and thymidylate nucleotides are generated from uridylate nucleotides via three specific reactions (Figure 9.10). The first is CTPS, which converts UTP to CTP. The second is RR, which synthesizes dCDP and dUDP (as well as dGDP and dADP) from CDP and UDP, respectively. The third enzyme, TS, catalyzes the reductive methylation of dUMP using 5,10-methylenetetrahydrofolate to produce TMP. TS is a well characterized enzyme in many organisms and has been targeted in anti-neoplastic chemotherapies. The enzyme is covalently linked to DHFR in many protozoan parasites, existing as a DHFR–TS complex. Pyrimethamine, an inhibitor of microbial DHFRs, is used in the treatment of *Plasmodium* and *Toxoplasma* infections (Chapter 17).

With three exceptions, all the parasites discussed here possess pyrimidine biosynthetic, CTPS, RR, and TS enzymes. *G. lamblia* and *T. vaginalis* lack both RR and TS and must salvage pyrimidines and their deoxynucleosides. *T. foetus* possesses RR but not TS and must salvage pyrimidines and thymidine but not other deoxynucleotides.

Mammalian cells also can salvage pyrimidines, but they do so mostly at the nucleoside level. Salvage of pyrimidine bases is minimal, and UPRT is missing. Mammalian cells have cytosolic UK, TK1, and dCK enzymes, as well as a mitochondrial TK2. UK recognizes uridine and cytidine, TK1 is specific for thymidine, dCK, although it prefers deoxycytidine, can also phosphorylate deoxyadenosine and deoxyguanosine, and TK2 is less selective than TK1. There are also pyrimidine interconversion enzymes including dCD and dCMPD.

Apicomplexa

Plasmodium *spp.*

Uninfected red blood cells have virtually undetectable pyrimidine nucleotide pools and are deficient in pyrimidine nucleotide biosynthesis. The enzymes for UMP, TMP, and CTP synthesis, however, all can be found in *Plasmodium*-infected erythrocytes. Thus, *Plasmodium* is capable of making pyrimidines *de novo*. The enzymes involved in the pathway have not been well studied but seem functionally akin to their mammalian counterparts. The six enzymes in the *de novo* UMP pathway, however, are all distinct and not components of multifunctional enzymes; and PfDHODH appears to be mitochondrial. Metabolic studies have confirmed the presence of CTPSII, RR, and TS activities. TS is part of a DHFR–TS complex, as it is for most protozoan parasites. Pyrimidine salvage by *P. falciparum* and other *Plasmodium* species is minimal or non-existent. Any uracil salvage that is observed can theoretically be ascribed to PfOPRT of the *de novo* pathway, an enzyme that recognizes uracil poorly.

Toxoplasma gondii

T. gondii also synthesizes pyrimidine nucleotides *de novo*. The *T. gondii CPSII* gene has been cloned and disrupted, and mutants at the *CPSII* locus are auxotrophic for pyrimidines. These pyrimidine auxotrophs are completely avirulent in mice and are able to evoke long-term protective immunity to toxoplasmosis. Thymidylate nucleotides are synthesized from dUMP via DHFR–TS. All pyrimidine salvage in *T. gondii* proceeds through uracil, which is salvaged to the nucleotide level by uridine phosphoribosyltransferase (UPRT). Salvage is not essential for parasite viability, since UPRT-deficient mutants have been made and exhibit the same growth rate as wild-type parasites. These mutants, unlike wild-type parental cells, fail to take up uracil, uridine, and deoxyuridine. Neither wild-type nor UPRT-deficient *T. gondii* take up other pyrimidine bases or nucleosides, including cytosine, thymine, cytosine, deoxycytidine, and thymidine. Thus UPRT is the only enzyme in *T. gondii* that can salvage preformed pyrimidines. The *TgUPRT* gene has been cloned by insertional mutagenesis, and the recombinant protein has been produced in bulk and its crystal structure solved to 1.93 Å.

Eimeria tenella

E. tenella sporozoites can incorporate [^{14}C]aspartate and [^{14}C]orotate into pyrimidine nucleotides and thymidylate through DHFR–TS, so a pyrimidine biosynthetic pathway must be present. The parasite can also salvage certain pyrimidines including uracil, uridine, and cytidine, but not thymidine. The biochemistry of *E. tenella* pyrimidine enzymes has not been studied.

Kinetoplastida

Leishmania

The ability of *Leishmania* to carry out pyrimidine biosynthesis *de novo* was initially established from growth studies that indicated that exogenous pyrimidines were not needed for parasite proliferation. All of the pyrimidine biosynthetic enzymes were subsequently detected in a multiplicity of species. OPRT and ODC were shown to be biochemically separable enzymes, both of which are located in the glycosome, and DHODH is not associated with the parasite mitochondrion. Also, the DHFR and TS activities were shown to be part of a bifunctional DHFR–TS protein that has been well characterized at the biochemical and structural level.

Pyrimidine salvage and interconversion in *Leishmania* is far less complex than purine salvage. Orotate can be taken up through OPRT and distributed among UTP and CTP. Thymidine is incorporated inefficiently, a process that appears to be TK-mediated, since thymine is not taken up. Uridine cleavage, dCD, and UPRT activities have been detected in a multiplicity of strains. None of the enzymes has been well studied at the biochemical level.

Trypanosoma cruzi

Pyrimidine metabolism in *T. cruzi* is similar to that in *Leishmania*. All the biosynthetic enzymes have been found in extracts, and Aoki and coworkers have isolated all of the *T. cruzi* pyrimidine biosynthetic genes. The six enzymes are encoded by five genes, all of which are encompassed within a 25 kb DNA fragment, indicating that they are co-localized in the *T. cruzi* genome. The last two enzymes, OPRT and ODC, are encoded by a single gene and are covalently linked as a bifunctional protein in the order opposite to the mammalian OPRT–ODC bifunctional protein. The *T. cruzi* OPRT–ODC also possesses an archetypal glycosomal targeting signal. Thymidylate nucleotides are, as for all protozoan parasites, generated by a bifunctional DHFR–TS. Pyrimidine interconversion and salvage pathways in epimastigotes are like those of *Leishmania*. Pyrimidine metabolism appears similar in all stages of the *T. cruzi* life cycle, except that a UPRT activity has not been detected in amastigotes.

Trypanosoma brucei

African trypanosomes metabolize pyrimidines in a fashion akin to *T. cruzi* and *Leishmania*. All six biosynthetic enzymes have been detected in lysates of the bloodstream form of *T.b. brucei*, and UPRT, dCD, pyrimidine nucleoside cleavage activities, but not UK, have been detected in parasite homogenates.

In summary, pyrimidine metabolism in Kinetoplastida is like that of the mammalian host. The only discrepancies uncovered to date are the bifunctional and glycosomal nature of the last two UMP biosynthetic enzymes, the apparent cytosolic milieu of the DHODH, and the existence and absence of UPRT and UK enzymes, respectively.

Amitochondriates

Tritrichomonas foetus

T. foetus cannot synthesize pyrimidine nucleotides *de novo* and are obligatory scavengers of host pyrimidines. Uracil is the pyrimidine that is most efficiently salvaged. Uridine, cytidine, and thymidine are taken up at rates 10%, 10%, and 1% that of uracil, respectively, and cytosine and thymine are not metabolized. Uracil and uridine are only incorporated into RNA, and thymidine only into DNA. The lack of UMP synthesis and DHFR–TS enzymes is consistent with the metabolic data. Uracil is incorporated through UPRT. Other pyrimidine salvage enzymes detected include TP, CP, UP, and dCD. The lack of TS activity and the inability to incorporate uracil into DNA indicates that *T. foetus* is dependent upon thymidine salvage. Yet the parasite apparently possesses a hydroxyurea-refractory RR activity that is capable of synthesizing deoxycytidylate nucleotides.

Trichomonas vaginalis

T. vaginalis cannot synthesize pyrimidines *de novo*, a conclusion based on experiments in which the parasite failed to incorporate radiolabeled precursors into pyrimidine nucleotides. Pyrimidine salvage in *T. vaginalis* differs quantitatively but not qualitatively from that in *T. foetus*. Unlike *T. foetus*, which salvages uracil most efficiently, cytidine is the preferred pyrimidine, although uracil, uridine, and thymidine are also taken up. Cytidine is incorporated into both cytidylate and uridylate nucleotides, whereas uracil and uridine are preferentially taken up into uridylates.

Salvage enzymes that have been detected in *T. vaginalis* are also different from those found in *T. foetus*. UPRT levels are low, and uracil is converted to uridine by UP. Uridine, cytidine, and thymidine are all converted to the monophosphate by pyrimidine NPTs. No pyrimidine nucleoside kinases have been detected in *T. vaginalis*. *T. vaginalis* also lacks RR. Pyrimidine deoxynucleotides are obtained by salvage of deoxynucleosides via a deoxynucleoside phosphotransferase activity, which recognizes both purine and pyrimidine substrates.

Giardia lamblia

G. lamblia is incapable of pyrimidine biosynthesis but has extensive pyrimidine salvage and interconversion capacities. The parasite also lacks an RR activity and must therefore salvage both deoxycytidine and thymidine. A thymidine NPT has been analyzed, but the route of deoxycytidine incorporation is unknown.

Entamoeba histolytica

E. histolytica can incorporate orotate into nucleic acids, possesses ATC activity, and can grow in pyrimidine-depleted medium. Thus, the parasite appears to be capable of pyrimidine biosynthesis *de novo*. The ability to proliferate axenically in pyrimidine-deficient medium also implies that the parasite expresses both RR and TS activities. *E. histolytica* can also salvage pyrimidines, but the enzymes have not been studied in detail.

Helminths

Although investigations have been limited to representative organisms, all parasitic helminths appear to be capable of pyrimidine biosynthesis *de novo*. The most detailed biochemical investigations have been accomplished with *S. mansoni*, an organism in which all of the pyrimidine biosynthetic enzymes have been detected. Like mammalian cells, the first three pyrimidine biosynthetic enzymes appear to be components of a multifunctional complex, the fourth appears to be membrane-bound, and the last two part of a bifunctional complex. *S. mansoni* also appears to salvage pyrimidines, including cytidine, uridine, thymidine, deoxycytidine, orotate, and uracil. Enzyme activities that have been described include phosphorylases, phosphotransferases, and kinases for pyrimidine nucleosides and PRTs for orotate and uracil. The existence of RR activity is supported by the conversion of radiolabeled uracil into deoxynucleotides and the ability of hydroxyurea, the RR inhibitor, to interfere with DNA synthesis. Several lines of evidence also support the existence of pyrimidine biosynthesis and salvage in other trematodes, including *Fasciola gigantica*, *Clonorchis sinensis*, and *Paragonimus ohirai*. These include detection of biosynthetic and salvage activities in worm lysates and metabolic labeling studies.

There is also good evidence for pyrimidine biosynthesis in cestodes and nematodes. Five of the six *de novo* enzymes have been measured in the cestode *Hymenolepis diminuta* and ATC was found in *Moniezia benedeni*. Cestodes can also salvage pyrimidines, but only a TK activity from *H. diminuta* has been characterized. Five pyrimidine biosynthesis enzymes have also been detected in the nematodes *Nippostrongylus brasiliensis* and *Trichuris muris*, and CPSII and ATC have been found in *Ascaris*. TS and pyrimidine salvage activities have also been observed in other nematode species.

ACKNOWLEDGMENTS

The authors thank the National Institute of Allergy and Infectious Disease for their support through grants AI23682, AI46416, and AI44138.

FURTHER READING

Aronov, A.M., Munagala, N.R., Ortiz De Montellano, P.R., Kuntz, I.D. and Wang, C.C. (2000). Rational design of selective submicromolar inhibitors of *Tritrichomonas foetus* hypoxanthine–guanine–xanthine phosphoribosyltransferase. *Biochemistry* **39**, 4684–4691.

Berens, R.L., Krug, E.C. and Marr, J. J. (1995). Purine and pyrimidine metabolism. In: Marr, J.J. and Muller, M. (eds). *Biochemistry and Molecular Biology of Parasites*, London: Academic Press, pp. 323–336.

Carter, N.S., Ben Mamoun, C., Liu, W. *et al.* (2000). Isolation and functional characterization of the PfNT1 nucleoside transporter gene from *Plasmodium falciparum*. *J. Biol. Chem.* **275**, 10683–10691.

Carter, N.S., Landfear, S.M. and Ullman, B. (2001). Nucleoside transporters of parasitic protozoa. *Trends Parasitol.* **17**, 142–145.

Donald, R.G., Carter, D., Ullman, B. and Roos, D.S. (1996). Insertional tagging, cloning, and expression of the *Toxoplasma gondii* hypoxanthine-xanthine-guanine phosphoribosyltransferase gene. Use as a selectable marker for stable transformation. *J. Biol. Chem.* **271**, 14010–14019.

European Bioinformatics Institute (2002). Parasite genomes Wu-Blast2 (Submission Form). http://www.ebi.ac.uk/blast2/parasites.html

Gao, G., Nara, T., Nakajima-Shimada, J. and Aoki, T. (1999). Novel organization and sequences of five genes encoding all six enzymes for *de novo* pyrimidine biosynthesis in *Trypanosoma cruzi*. *J. Mol. Biol.* **285**, 149–161.

Hall, S.T., Hillier, C.J. and Gero, A.M. (1996). *Crithidia luciliae*: regulation of purine nucleoside transport by extracellular purine concentrations. *Exp. Parasitol.* **83**, 314–321.

Hwang, H.Y. and Ullman, B. (1997). Genetic analysis of purine metabolism in *Leishmania donovani*. *J. Biol. Chem.* **272**, 19488–19496.

Kicska, G.A., Tyler, P.C., Evans, G.B., Furneaux, R.H., Schramm, V.L. and Kim, K. (2002). Purine-less death in *Plasmodium falciparum* induced by immucillin-H, a transition state analogue of purine nucleoside phosphorylase. *J. Biol. Chem.* **277**, 3226–3231.

Krug, E.C., Marr, J.J. and Berens, R.L. (1989). Purine metabolism in *Toxoplasma gondii*. *J. Biol. Chem.* **264**, 10601–10607.

Landfear, S.M. (2001). Molecular genetics of nucleoside transporters in *Leishmania* and African trypanosomes. *Biochem. Pharmacol.* **62**, 149–155.

Marr, J.J. and Berens, R.L. (1983). Pyrazolopyrimidine metabolism in the pathogenic Trypanosomatidae. *Mol. Biochem. Parasitol.* **7**, 339–356.

Marr, J. J. and Berens, R.L. (1986). Purine and pyrimidine metabolism in Leishmania. In: Chang, K.-P. and Bry, R.S. (eds). *Leishmaniasis*, New York: Academic Press, pp. 65–79.

Plasmodium falciparum Genome Project (2002). Plasmodb: the *Plasmodium* genome resource. http://plasmodb.org/

Schumacher, M.A., Scott, D.M., Mathews, H. *et al.* (2000). Crystal structures of *Toxoplasma gondii* adenosine kinase reveal a novel catalytic mechanism and prodrug binding. *J. Mol. Biol.* **298**, 875–893.

Schwartzman, J.D. and Pfefferkorn, E.R. (1981). Pyrimidine synthesis by intracellular *Toxoplasma gondii*. *J. Parasitol.* **67**, 150–158.

Shi, W., Munagala, N.R., Wang, C.C. *et al.* (2000). Crystal structures of *Giardia lamblia* guanine phosphoribosyltransferase at 1.75 Å. *Biochemistry* **39**, 6781–6790.

Shih, S., Hwang, H.Y., Carter, D., Stenberg, P. and Ullman, B. (1998). Localization and targeting of the *Leishmania donovani* hypoxanthine–guanine phosphoribosyltransferase to the glycosome. *J. Biol. Chem.* **273**, 1534–1541.

C H A P T E R

10

Trypanosomatid surface and secreted carbohydrates

Salvatore J. Turco
Department of Biochemistry, University of Kentucky Medical Center, Lexington, KY, USA

INTRODUCTION

Trypanosomatid parasites have evolved unique lifestyles, shuttling between their insect vectors and vertebrate hosts, encountering extremely inhospitable environments specifically designed to keep such microbial intruders in check. Their survival tactics often involve glycoconjugates that form a protective obstacle against unwelcoming forces. A general trait of the parasite's cell surface architecture is an intricate and often highly structured glycocalyx that allows the parasite to interact with and respond to its external environment.

From the variant surface glycoprotein (VSG) of *Trypanosoma brucei* and the surface mucins of *Trypanosoma cruzi* to the various phosphoglycans of *Leishmania*, these glycoconjugates are obligatory for parasite survival and virulence. The variety of the glycoconjugate structures and, accordingly, the array of functions that have been attributed to these molecules, from host cell invasion to deception of the host's immune system, are simply amazing. Also remarkable is the observation of parallels and similarities in structure that underscore evolutionary relationships between the various parasites. For example, while the main mechanism used for membrane protein and polysaccharide anchorage in the trypanosomatids is the glycosylphosphatidyl inositol (GPI) anchor, the roles of the anchor are diverse and parasite-specific.

In this chapter the structures and functions of the major glycoconjugates of trypanosomatid parasites are highlighted. However, the structures of GPI anchors are only briefly detailed, since several excellent reviews have been published on the subject, nor will the biosynthetic pathways be extensively outlined.

Molecular Medical Parasitology
ISBN 0–12–473346–8

AFRICAN TRYPANOSOMES

Variant surface glycoprotein

The cell surface of the metacyclic *T. brucei* trypomastigotes is covered with a dense glycocalyx composed of about 10 million molecules of VSG. The parasite genome harbors several hundred VSG genes that are expressed sequentially and allow the parasite to evade recognition by the humoral immune system (Chapter 5). Each VSG is encoded by a single VSG gene. As the parasites multiply in the host bloodstream and tissue spaces, the host immune system mounts a response that is effective against only a certain population of parasites, those expressing the antigenic VSG on their coat. Those that have 'switched' to an alternative VSG coat escape the immune system. The glycocalyx also provides a diffusion barrier, thus preventing complement-mediated lysis.

The VSGs are dimeric proteins, consisting of two 55 kDa monomers. Despite very low sequence identities, VSG homodimers assume almost indistinguishable tertiary structures, which accounts for how antigenically diverse VSGs amass into functionally identical coat arrays. Each monomer carries at least one occupied *N*-linked glycosylation site. The amino terminal domains comprise about 75% of the mature protein and display the most sequence diversity, while the carboxy terminal domains are more similar. The diversity of the amino termini gives the VSGs their unique immunological properties. VSGs are classified as class I, II, III or IV based on peptide homology of the C-terminal domains. Most VSG variants belong to Class I or II. Type I VSGs carry one conserved *N*-glycosylation site, 50 residues from the C-terminus, which is occupied by high mannose structures ($Man_{5-9}GlcNAc_2$, predominantly $Man_7GlcNAc_2$). The class II glycosylation sites are 5–6 residues and 170 residues upstream of the C-terminus. The C-terminal site is usually occupied by high mannose structures or by polylactosamine-rich structures, while the inner site is modified by smaller oligosaccharides such as $Man_{3-4}Glc$ NAc_2 or $GlcNAcMan_3GlcNAc_2$. For example, one well characterized class III VSG from *T. brucei* contains three putative *N*-glycosylation sites modified with either high mannose or biantennary structures. Both the complex and polylactosamine glycan chains may be terminated with an α1,3-linked galactose. Each VSG monomer is GPI-anchored to the plasma membrane. The GPI anchor is pre-assembled as a precursor structure that is then transferred to the mature C-terminal amino acid (aspartate for class I, serine for class II, asparagine for class III) with the concomitant loss of the C-terminal GPI anchor signal sequence.

The mature anchor consists of the classical GPI core ethanolamine-HPO_4-6Man(α1,2) Man(α1,6)Man(α1,4)GlcN(α1,6)-inositoldimyristoylglycerol; however in class I and II, the anchor is uniquely modified by galactose side chains. Class I VSG GPIs contain an average of three Gal residues, while class II VSGs contain up to six Gal residues. Class II VSGs usually contain an additional α-Gal attached to the 2-position of the non-reducing α-Man residue and one-third of the structures contain an additional β-Gal attached to the 3-position of the middle Man.

Procyclins (Procyclic Acidic Repetitive Proteins, PARPS)

For many years, procyclic forms of *T. brucei* (the stage found in the tsetse flies) were believed to be 'uncoated' during transformation of bloodstream forms to procyclics as the VSG coat was shed. This view was supported by radioiodination experiments that labeled

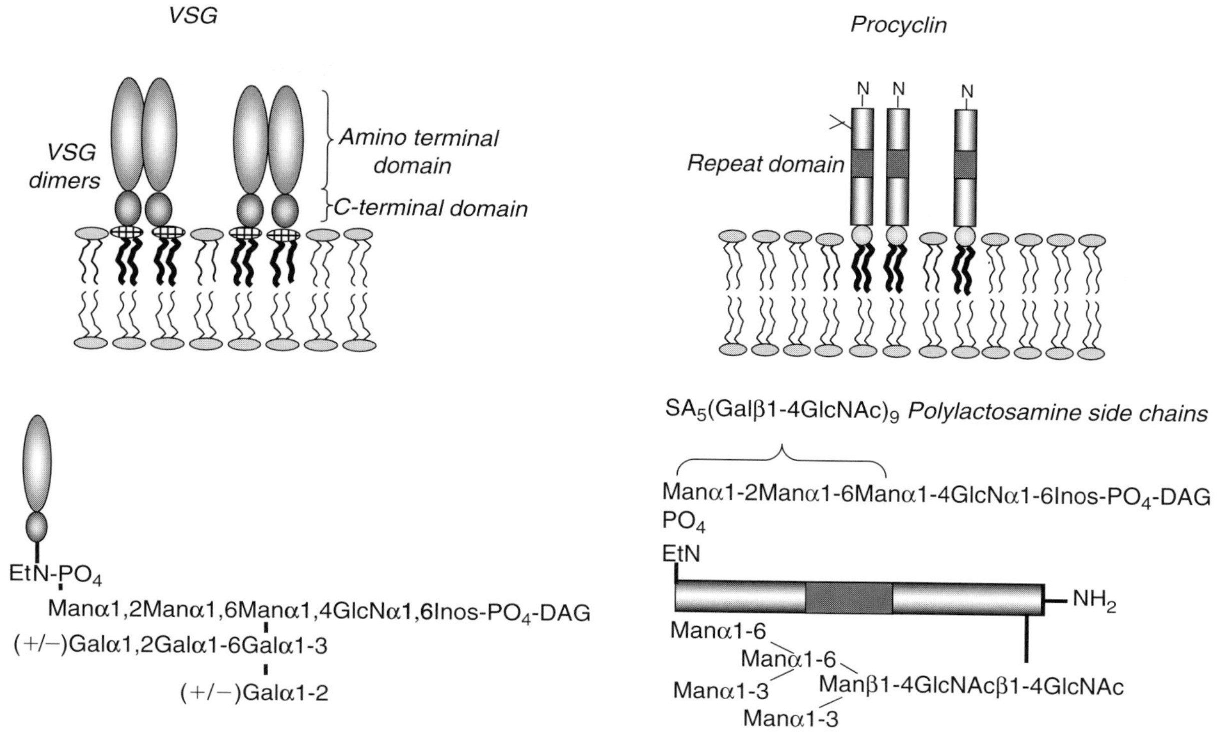

FIGURE 10.1 A schematic representation of the major surface glycoconjugates of procyclic and metacyclic *T. brucei*. VSG (variant surface glycoprotein) is the major component of the metacyclic form, each molecule consisting of two GPI-anchored *N*-glycosylated monomers. The shaded ovals represent the protein component. The surface of the procyclic form is densely covered with procyclins (PARPs or procyclic acidic repetitive proteins). These are GPI-anchored polypeptides with polyanionic repeat domains (shaded in diagram). The anchor structures and the *N*-linked oligosaccharide of procyclin are detailed below the schematic. DAG, diacylglycerol; EtN, ethanolamine; GlcN, glucosamine; Inos, inositol; SA, sialic acid.

over 25 different surface proteins in procyclics. However, procyclics contain a surface coat composed largely of acidic glycoproteins called the procyclins, also called PARPs or Procyclic Acidic Repetitive Proteins (Figure 10.1). Each cell expresses about 5 million copies of procyclins. The procyclins form a dense glycocalyx of GPI anchors, with the polyanionic polypeptide repeat domains projecting above the membrane. There are two families of procyclins: the EP-procyclins have glutamate–proline repeats while the GPEET-procyclins bear glycine–proline–glutamate–glutamate–threonine repeat sequences. Some members of the EP-procyclin family are *N*-glycosylated adjacent to the repeat domain. An unusual feature of this glycosylation site is that only a single type of *N*-glycan is found, namely a $Man_5GlcNAc_2$ oligosaccharide. Properly glycosylated forms of EP are required for concanavalin A-induction of a unique form of cell death in *T. brucei*, defined as proto-apoptosis. The GPI anchors are modified with unusual branched polylactosamine {Gal(β1,4)GlcNAc} glycans.

Interestingly, when the parasites are grown in culture, the β-galactose termini of these side chains are terminated with sialic acid by the action of a trans-sialidase enzyme. Recent work has shown that GPEET-procyclin is phosphorylated.

The various procyclins appear to be expressed in a cell cycle-specific manner. Early in infection GPEET2 is the only procyclin detected. However, GPEET2 disappears in about a week and is replaced by several isoforms of glycosylated EP, but not the unglycosylated isoform EP2. Unexpectedly, the N-terminal domains of all procyclins appear to be quantitatively removed by proteolysis in the fly, but not in culture. These findings suggest that one function of the protease-resistant C-terminal domain, containing the amino acid repeats, is to protect the parasite surface from digestive enzymes in the tsetse fly gut. Moreover, it has been hypothesized that the oligosaccharide chains may serve as lectin-binding ligands within the tsetse fly midgut. Based on their polyanionic nature and extended conformation, a likely function of procyclins is to afford protection of the parasite in the digestive confines of the tsetse fly midgut.

Intriguingly, while compelling evidence has been obtained that GPI biosynthesis is essential to the bloodstream forms of *T. brucei*, GPI biosynthesis in procyclic forms is non-essential, even though procyclics are rich in the GPI-anchored procyclins. Upon deletion of individual genes of the procyclin family, all but one could be deleted before producing non-viable parasites. To study the function of procyclins, mutants have been generated in the Roditi laboratory that have no EP genes and only one copy of GPEET. This last gene could not be replaced by EP procyclins, and could only be deleted once a second GPEET copy was introduced into another locus. The EP knockouts are morphologically indistinguishable from the parental strain, but their ability to establish a heavy infection in the insect midgut is severely compromised; this phenotype can be reversed by the reintroduction of a single, highly expressed EP gene. These results suggest that the two types of procyclin have different roles, and that the EP form, while not required in culture, is important for survival in the fly.

TRYPANOSOMA CRUZI

Mucins

T. cruzi has a dense and continuous coat composed of a layer of type-I glycosylinositolphospholipids or GIPLs, and a family of small mucins that project above the GIPL layer. The GIPLs have the same anchor structure as the *T. brucei* procyclins but are heavily substituted with Gal, GlcNAc and sialic acid. The mucins are rich in threonine, serine and proline residues that are heavily glycosylated with *O*-linked GlcNAc as the internal residue, and further modified with 1–5 galactosyl residues. Two groups of mucin-like molecules have been identified and are encoded by the diverse MUC gene family. The glycoproteins in the first group are 35–50 kDa in size and are expressed on the parasitic forms found in the insect (epimastigote and metacyclic trypomastigote). The second group consists of 80–200 kDa glycoproteins that are expressed in the cell culture-derived trypomastigotes. These glycoconjugates express the Ssp-3 epitope, which were later called F2/3 glycoproteins.

The glycan structures vary with the developmental stages of the parasite, a notable modification being the trans-sialylation of the glycan in the metacyclic stages. Trypanosomes are unable to synthesize sialic acid *de novo*, but do have the unusual ability to acquire it from

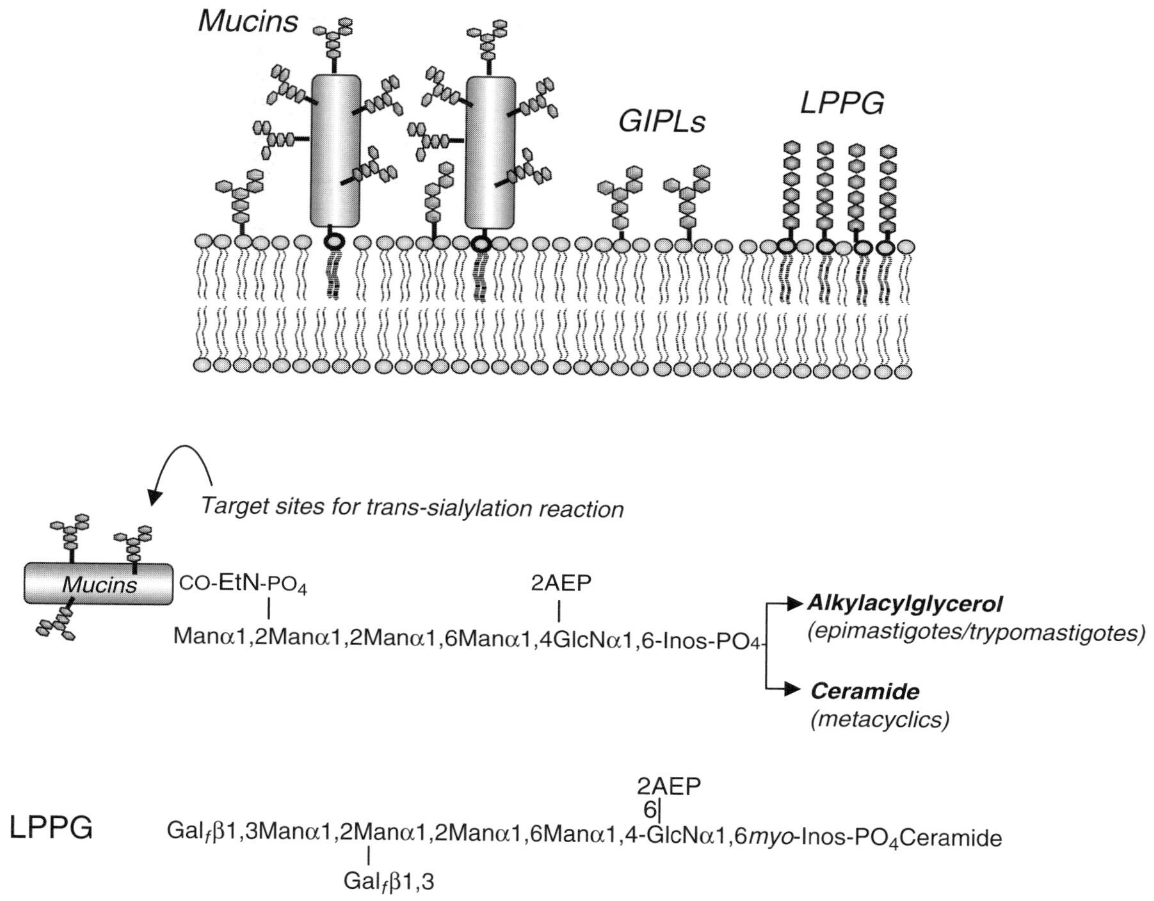

FIGURE 10.2 Schematic representation of the surface coat of *Trypanosoma cruzi*. The cell surface of *T. cruzi* is covered with a dense layer of mucins, GIPLs (glycosylinositolphospholipids), and LPPG (lipopeptidophosphoglycan). The structures of the mucin anchors and the predominant LPPG species are outlined. AEP, aminoethylphosphonate; EtN, ethanolamine; GlcN, glucosamine; Inos, inositol.

host glycoconjugates, thus accounting for the requirement for a trans-sialidase. The galactose-rich mucin is the acceptor for the trans-sialylation reaction in which α2,3-linked sialic acid residues are transferred from host glycoconjugates by the trans-sialidase enzyme to the terminal β-galactose residues of the *T. cruzi* surface mucin. Interestingly, *Endotrypanum,* a trypanosomatid parasite of sloths, also has the ability to incorporate host-derived sialic acid into molecules on its own surface.

The complete structures of the sialic acid acceptor *O*-linked oligosaccharide have been elucidated for epimastigotes of G strain and Y strain, and metacyclic forms of G strain. The structures of the *O*-linked oligosaccharides are conserved between the epimastigotes and the metacyclic trypomastigotes. However, polymorphisms have been identified among the different strains (e.g. the presence of galactofuranose rather than galactopyranose) and only the G strain expresses oligosaccharides containing a β-galactofuranose residue. The structure of the *O*-linked glycans from mucins of a CL-Brener myotropic strain of *T. cruzi*, the strain chosen for the *Trypanosoma* genome project, has been recently reported. The glycan structures are similar to those from the reticulotropic Y strain. The differences in structures of *O*-linked oligosaccharides from mucins of various *T. cruzi* strains and developmental stages may ultimately be related to distinctions in infectivity and tissue tropism, or to different growth conditions. The detailed characterization of these differences may eventually permit the identification of *T. cruzi* ligands for receptors in the vertebrate host.

Both parasite and host-cell molecules are believed to be involved in the complex process of host-cell invasion. The surface coat of *T. cruzi* seems to have a primarily protective function, and the sialylation of the mucins provides the parasite with the ability to survive in different environments. Sialylation is proposed to reduce the susceptibility of the parasite to anti-α-galactose antibodies present in the mammalian bloodstream. It has been proposed that the heavily sialylated coat may also provide a structural barrier to complement-mediated lysis, in escape from the endosomes, in developmental stage transition, in promoting adherence to the macrophage, in induction of protective lytic antibodies, and in the production of NO and cytokines.

The structures of the GPI anchors of the mucins undergo modifications with parasite differentiation. The epimastigote mucin GPIs contain *sn*-1-alkyl-2-acylglycerol, while over 70% of the metacyclic mucins contain ceramide lipids; reviewed by Ferguson (1997, 1999; see Further Reading). Moreover, the ceramide lipid species is found on LPPG (lipopeptidophosphoglycan, see below), which is expressed in large amounts on the epimastigote surface. GPI anchors from the trypomastigote stage of *T. cruzi* have been implicated in the induction of NO, IL-12 and TNF-α. Trypomastigote GPIs contain unsaturated fatty acids in the *sn-2* position of the glycolipid component. The observations that in non-infective epimastigotes, the GPI-lipid anchor of mucins is a 1-alkyl-2-acyl-PI, and that in infective metacyclic forms the mucins possess mostly ceramide-phosphoinositol suggest the possibility that they may play a key role in the release process during infection. Conceivably, further clarification of this distinction might lead to novel targets for chemotherapeutic intervention.

Lipopeptidophosphoglycan

Lipopeptidophosphoglycan (LPPG) is the major cell surface glycan of the *T. cruzi* epimastigote, with about 1.5×10^7 copies per cell. The expression of LPPG appears to be developmentally regulated, as it is present in very low levels in the stages that infect the mammalian host. LPPG consists of a glycan linked to an inositol-phosphoceramide via a non-acetylated glucosamine (Figure 10.2). The glycan structure contains mannose, terminal galactofuranose and 2-aminoethylphosphonate. The LPPG fraction extracted from *T. cruzi* epimastigotes contains three species that have slight variations in the glycan structure, with respect to the position of the Gal_f. The major species (65%) contains two terminal Gal_f residues linked β1,3 to Man(α1,2)Man. The lipid component is an inositol phosphoceramide containing mainly palmitoylsphinganine, palmitoylsphingosine and lignoceroylsphinganine. LPPG is considered a member of the GPI family based on the presence of the Man(α1,4)GlcN (α1,6)*myo*-inositol-1-PO_4-lipid motif, the hallmark of all GPI anchors. LPPG differs from the *Leishmania* GIPLs by the presence of the phosphoceramide moiety, a feature found in the LPPG of *Acanthamoeba castellanii* and in a lipid anchor found in *Dictyostelium discoideum*. Moreover *T. cruzi* LPPG contains 2-aminoethylphosphonate as does the *Acanthamoeba castellanii* phosphoglycan.

LEISHMANIA

Leishmania exist as flagellated, extracellular promastigotes in the midgut of their sand fly vector, and as intracellular, aflagellar amastigotes in the phagolysosomes of the mammalian macrophage. The parasites are transmitted when a sand fly takes a blood meal from an infected individual. Within the sand fly midgut the promastigotes develop from the avirulent procyclic form to the virulent metacyclic form in a process termed metacyclogenesis. Throughout their life cycle, *Leishmania* survive and proliferate in highly hostile environments and have evolved special mechanisms which enable them to endure these adverse conditions, including a dense cell surface glycocalyx composed of lipophosphoglycan, glycosylinositolphospholipids and secreted glycoconjugates, proteophosphoglycan (PPG) and secreted acid phosphatase (sAP).

Lipophosphoglycan

Lipophosphoglycan (LPG), the predominant cell surface glycoconjugate of *Leishmania* promastigotes, is localized over the entire parasite surface including the flagellum. Found in all species of *Leishmania* that infect humans,

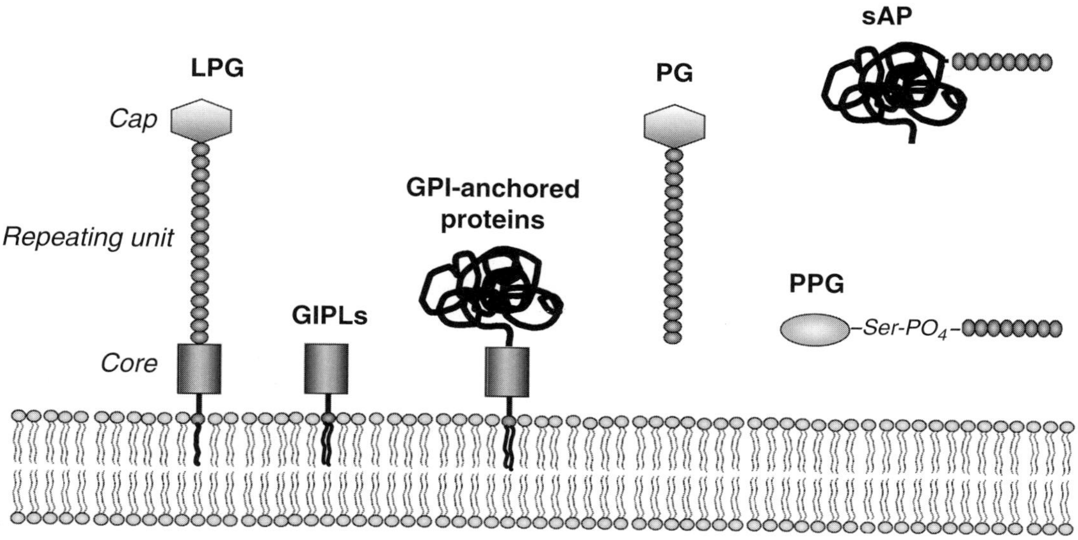

FIGURE 10.3 Schematic representation of the surface coat and secreted glycoconjugates of *Leishmania*. The small oval 'repeat unit' represents Gal-Man-PO_4 units. In PPG and sAP, the repeat unit chain is attached to the polypeptide via Man-PO_4-serine linkage. GIPLs, glycosylinositolphospholipids; LPG, lipophosphoglycan; PG, phosphoglycan; PPG, proteophosphoglycan; sAP, secreted acid phosphatase.

LPG is composed of four domains, (i) a 1-*O*-alkyl-2-*lyso*-phosphatidyl(*myo*)inositol lipid anchor, (ii) a glycan core, (iii) Gal(β1,4)Man (α1)-PO_4 backbone repeat units, and (iv) an oligosaccharide cap structure. The LPG backbone contains the 4-*O*-substituted mannose residue, which is not present in any other known eukaryotic glycoconjugate. The structure of the LPG from *Leishmania donovani* is shown in Figure 10.4A and is the prototype of leishmanial LPGs. Structural analysis of LPG from different species has revealed complete conservation of the lipid anchor, the glycan core and the Gal(β1,4)Man(α1)-PO_4 backbone of repeat units. The distinguishing features of LPGs are variability of sugar composition and sequence of branching sugars attached to the repeat Gal-Man-phosphate backbone and of the cap structure. The C3 hydroxyl of the repeat unit galactose is the site of most side-chain modifications. For example, the LPG of *L. donovani* from Sudan has no side chains, the *L. donovani* LPG from India has 1–2 β-Glc every 4–5 repeat units, the *L. major* LPG has 1–4 residues of β-Gal and is often terminated with arabinose, the *L. mexicana* LPG contains

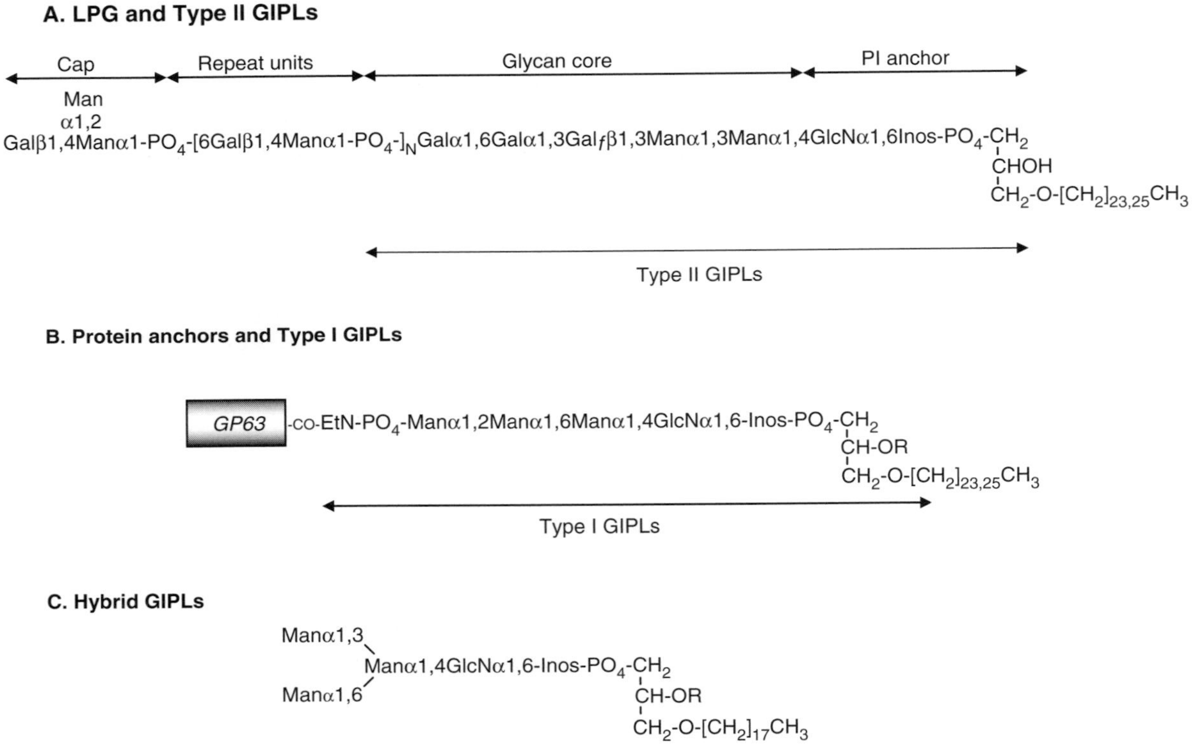

FIGURE 10.4 Structures of *Leishmania* surface glycoconjugates. (A) LPG (lipophosphoglycan), the predominant glycoconjugate of promastigotes, consists of a Gal-Man-PO_4 repeat unit backbone attached to a lipid anchor via a glycan core. The structure of *L. donovani* promastigote LPG is shown. Type II GIPLs share the LPG anchor and glycan core structure. (B) Structure of GP63/glycoprotein anchors, also shared by Type I GIPLs. (C) Structure of hybrid GIPLs. 'R' represents an ester-linked fatty acid.

β-Glc, and the *L. tropica* LPG has the most complex side-chain modifications with over 19 different types of glycans so far identified. *L. major* has the simplest cap structure, Man (α1–2)Man(α1), while the most common *L. donovani* cap is the branched trisaccharide Gal(β1–4)[Man(α1–2)] Man(α1).

LPG is anchored by the unusual phospholipid 1-*O*-alkyl-2-*lyso*-phosphatidyl (*myo*)inositol. In all species of *Leishmania* examined thus far, the aliphatic chain consists of either a C_{24} or C_{26} saturated, unbranched hydrocarbon. Similar to many GPI-anchored proteins, LPG can be hydrolyzed by bacterial phosphatidylinositol-specific phospholipase C producing 1-*O*-alkylglycerol and the entire polysaccharide chain. Attached to the inositol of the lipid anchor of LPG is the glycan core region. The glycan core consists of an unacetylated glucosamine, two mannoses, two galactopyranoses, and a galactofuranose. The presence of the last sugar is extremely unusual in eukaryotic glycoconjugates, especially since the furanoside is internal in a carbohydrate chain. As with all other GPI-anchored proteins known, LPG possesses the Man(α1–4)GlcN(α1–6)*myo*-inositol-1-phosphate motif. The LPG cores of *L. donovani* and *L. mexicana* have a glucosyl-α1-phosphate attached through a phosphodiester linkage to the C-6 hydroxyl of the proximal mannose residue. A substantial percentage of the *L. major* LPG also contains an identical glucosyl-α1-phosphate substitution. Another interesting sequence in the core region is the Gal(α1–3)Gal unit, which is believed to be the epitope for circulating antibodies in patients with leishmaniasis.

The three-dimensional structure of the basic repeating phosphate-6Gal(β1–4)Man (α1) disaccharide units of LPG has been determined. The repeating units, as expected, exist in solution with limited mobility about the Gal(β1–4)Man linkages. In contrast, a variety of stable rotamers exist about the Man(α1)-PO_4-6Gal linkages. An important feature of each of these low energy conformers is that the C-3 hydroxyl of each galactose residue is exposed and freely accessible. The C-3 hydroxyl of the Gal is the particular position that is substituted with glucose in the LPG from *L. mexicana*, or with galactose, glucose, and arabinose in the LPG from *L. major*. Thus, these additional units could be accommodated without major conformational changes of the repeat backbone. Another intriguing finding from the molecular modeling studies is that each of the stable conformers of the Man(α1)-PO_4-6Gal linkages may exist in a different configuration within the same LPG molecule. These torsional oscillations confer an ability upon the LPG molecule to contract or expand in a manner reminiscent of a 'slinky spring', resulting in a molecule whose length is potentially 90 Å when fully contracted to a length of 160 Å when fully extended, assuming an average of 16 repeat units.

A peculiar and significant observation is that extensive modifications in the structure of LPG accompany the process of metacyclogenesis, in which the promastigote converts from an avirulent into a virulent form. However, the 1-*O*-alkyl-2-*lyso*-phosphatidylinositol lipid anchor and the glycan core are conserved in LPGs regardless of the developmental stage of the parasite. During metacyclogenesis of *L. major* promastigotes, LPG undergoes elongation due to an approximate doubling in the number of repeating oligosaccharide-phosphate units, and a downregulation in the number of side-chain substitutions expressing a terminal β-linked galactose. Similarly, the *L. donovani* LPG also undergoes an approximate doubling in size when compared to the procyclic form of LPG, in this case due to an approximate doubling in the number of repeating Gal-Man-phosphate units. Unlike the situation in *L. major*, however, in *L. donovani* terminally exposed capping sugars are lost, which appears to be due not to a downregulation of their expression, but to a conformational change affecting their availability for binding. Such changes in LPG structure have profound implications on function, and suggest important points of regulation of the glycosyltransferases involved in LPG biosynthesis. These structural changes are believed to obscure epitopes that promote adherence to the midgut and allow passage of the parasites to the mouthparts of the sand fly. Genetic results obtained with LPG-deficient mutants strongly support the conclusions that LPG is not essential for survival in the early blood-fed midgut but, along with other secreted phosphoglycan-containing glycoconjugates, can protect promastigotes from the digestive enzymes in the gut. Furthermore, the LPG-deficient mutants provided additional evidence that LPG is required to mediate midgut attachment and to maintain infection in the fly during excretion of the digested blood meal.

LPG expression is not only developmentally regulated within the promastigote during metacyclogenesis, but also during promastigote–amastigote conversions. The number of LPG molecules expressed by the intracellular amastigotes is substantially lower than in the promastigote. Promastigotes express about $1–5 \times 10^6$ copies per cell, while amastigotes express about 100 or 1000 copies per cell. The amastigote LPG from *L. major* is biochemically and antigenically distinct from promastigote LPG. The glycan core and PI anchor structures are conserved among promastigotes and amastigotes with slight variations in the level of glucosylation. The amastigote LPG contains an average of 36 repeat units, 70% of which are unsubstituted, with the rest containing glucopyranose and galactopyranose, but not arabinose. Also, side chains of up to 11 Gal residues have been identified, compared to the typical 1–3 Gal residues in the LPG side chains of promastigotes. The cap structure is predominately Gal(β1,4)Man(α1-) whereas the major promastigote LPG cap is Man(α1,2)Man(α1). No functions have yet been attributed to the amastigote version of LPG.

LPG has been implicated in a number of functions within the mammalian host. Within the bloodstream, LPG prevents complement-mediated lysis by preventing insertion of the C5b-9 membrane attack complex into the promastigote membrane. In addition, LPG serves as a ligand for receptor-mediated endocytosis by the macrophage via complement receptors as well as the mannose receptor. Inside the macrophage, LPG inhibits PKC and the microbicidal oxidative burst as well as inhibiting phagosome–endosome fusion. These *in vitro* data suggest that LPG is a virulence factor; however, the molecular tools to unequivocally substantiate this claim have been lacking. Furthermore, there are complications with the structural relatedness of LPG with other glycoconjugates, such as the PPGs, which present difficulties in assigning the proper structure–function relationships. Recent advances in the knockout of genes involved in phosphoglycan biosynthesis and the isolation of the corresponding genes have now permitted the direct testing of LPG as a virulence factor in accord with Koch's postulates. Surprisingly, while LPG is clearly a virulence factor in *Leishmania*–sand fly interactions, two independent approaches have resulted in apparently conflicting conclusions in mice. In *L. major*, use of a knockout line that lacks an LPG biosynthetic gene (*LPG1*, a putative galactofuranosyltransferase gene), resulted in a dramatic reduction in virulence. Importantly, the re-introduction of *LPG1* restored virulence. In contrast, similar approaches using two distinct *L. mexicana* knockouts (either *LPG1* or *LPG2*, the GDP-Man transporter gene) failed to implicate LPG, or, surprisingly, any phosphoglycan-containing substances as virulence determinants. One possible explanation for these differences may be related to distinctions in immune responses to *L. major* and *L. mexicana*. Alternatively, the biology of infection by *L. mexicana* might place less emphasis on phosphoglycan-containing molecules separate from the immune response; perhaps it arises from differences in the rate of replication in the host, for example.

Glycosylinositolphospholipids

The glycosylinositolphospholipids (GIPLs) are a major family of low molecular weight glycolipids synthesized by *Leishmania* that are not attached to either protein or polysaccharides. GIPLs are expressed in very high copy numbers, about 10^7 copies per cell on both promastigote and amastigote surfaces. Three major lineages of GIPLs have been identified that are expressed to different levels in different species or developmental stages. Based on the pattern of their glycan headgroups, GIPLs are classified as Type I (analogous to protein GPI anchors), Type II GIPLs (analogous to LPG anchors) or hybrid (containing features of both). The lipid components of the hybrid (Figure 10.4C) and Type I GIPLs (Figure 10.4B) are distinct from those of protein or LPG anchors, in that they are rich in alkyl-acyl-PI with shorter, namely C18:0 alkyl chains. The Type II GIPLs (Figure 10.4A) are more heterogeneous and contain longer alkyl chains, namely C24:0 or C26:0. Not much is known about the functions of the GIPLs. The use of the mannose receptor in parasite attachment to the macrophage suggests that the mannose-rich GIPLs may play a role in macrophage invasion. Since the levels of LPG and GP63 are dramatically downregulated, the GIPLs are the major constituents of the amastigote surface and are presumably involved in protecting the parasite from environmental hazards, as well as playing some role in parasite–host interactions, especially in the mammalian stage. In fact, GIPLs may be involved in modulating signaling events in the macrophage such as NO synthesis and the oxidative burst. Enzymes involved in GPI biosynthesis are essential for parasite virulence, emphasizing the importance of protein-free GPI glycolipids in parasite viability.

Unlike *L. major*, whose GIPLs are largely galactosylated, promastigotes of *L. donovani* synthesize abundant GIPLs which contain one to four mannose residues and are not galactosylated. Although *L. donovani* amastigotes do not synthesize LPG, they continue to synthesize GIPLs in quantities comparable to those present in promastigotes. The *L. donovani* amastigote GIPLs, containing one to three mannose residues, are structurally different from promastigote GIPLs and appear to be precursors to glycolipid anchors of proteins. GIPLs have also been isolated and characterized from *L. mexicana*, *L. aethiopica* and *L. adleri* and show additional features, including the presence of phosphoethanolamine. The major GIPLs have lower turnover rates than LPG. Compartmentalization of different GPI pathways may be important in regulating the species and stage-specific expression of different GPI structures. Of interest, ceramidephosphoinositides also have been found in *L. donovani*, and appear to be further substituted with carbohydrate residues as observed in *Trypanosoma cruzi*, yeast, fungi, and plants. However, the structures have not yet been elucidated.

GIPLs that are structurally related to the *Leishmania* Type II and/or hybrid GIPLs and the *T. cruzi* LPPG have been identified in *Leptomonas samueli*, *Endotrypanum schaudinni, Herpetomonas samuelpessoai* and *Crithidia fasciculata*. The *Leptomonas* glycophosphosphingolipid resembles that of *Leishmania* in the α-linked Man in the GPI core, and resembles that of *T. cruzi* in the presence of aminoethylphosphonate and ceramide lipids. The *Endotrypanum* GIPLs are similar to *Leishmania* Type II and hybrid GIPLs but are sphingolipids rather than glycerol lipids. Several *Endotrypanum* GIPLs contain D-arabinose, and the terminal saccharide chains resemble the phosphoglycan side chains of *L. major* LPG. The *Leishmania* GIPLs contain predominantly α-linked Gal while *Endotrypanum* contains β-linked Gal. Such similarities are consistent with the close evolutionary relationship between *Leishmania*, trypanosomes, *Leptomonas* and *Endotrypanum*.

GP63

GP63 (Figure 10.4B) is the major cell surface glycoprotein of *Leishmania* promastigotes with 500 000 copies per cell, accounting for 1% of all protein. In amastigotes, GP63 is expressed to a lower level and is found in the flagellar pocket as opposed to covering the entire surface, as in promastigotes. GP63 is a 63 kDa zinc metalloprotease and is GPI-anchored to the cell surface via a myristic acid-containing GPI anchor. The amastigote GP63 subpopulation found in the flagellar pocket lacks a membrane anchor. An active site structural motif found in other zinc proteases has been identified in structural studies that may aid design of specific inhibitors. GP63 contains three potential glycosylation sites, and the *N*-linked glycans have been characterized in *L. mexicana* and *L. major*. The glycans are biantennary high-mannose type, and some bear a terminal glucose in α1,3 linkage. $Man_6GlcNAc_2$ and Glc $Man_6GlcNAc_2$ are present in all promastigotes species examined. In amastigotes the structures are more variable, while in *L. donovani* no *N*-linked glycans appear to be present. The terminal glucose in the GP63 glycan is highly unusual with respect to oligomannose structures found in glycoproteins. Whether the stage-specific changes in glycan structure affect parasite infectivity and development is unknown.

The abundance and location of GP63 suggest that it may play a role in the infection process. GP63 expression is decreased in avirulent compared to virulent promastigotes, and GP63 is found in increased amounts in stationary phase *L. braziliensis*. By virtue of its activity, GP63 may interact with and proteolytically degrade host macromolecules. Indeed, solubilized GP63 molecules bind and cleave the human plasma protease inhibitor α2-macroglobulin. The formation of α2- macroglobulin-GP63 complexes results in a dose-dependent inhibition of the proteolytic activity of soluble GP63 on azocasein. However, α2-macroglobulin may not interact with promastigotes in the bloodstream, since GP63 at the surface of live promastigotes does not recognize α2-macroglobulin. GP63 may also participate in the attachment of promastigotes to macrophage surface receptors via complement components and protect the parasite against complement-mediated lysis. Attempts to obtain mutants that are defective in GP63 by targeted gene deletion have been hindered by the fact that GP63 is encoded by a multigene family. Targeted deletion of six out of seven GP63 genes did not affect parasite growth *in vitro* or prevent disease in mice. Knockouts of GPI8, the GPI:protein transaminidase which eliminates the expression of GP63 along with other GPI-anchored proteins, grew normally in culture and macrophage infectivity *in vitro* was unaffected. More importantly, the ΔGPI8 mutant was able to establish infection in mice, suggesting that GP63 is not essential for growth or infectivity in mammals.

Membrane-bound acid phosphatase

Leishmania parasites have two distinct types of acid phosphatase: a surface membrane-bound and a secreted form of the enzyme (described below). The membrane-bound acid phosphatase has been purified from *L. donovani* and could be resolved into one major and two minor isoenzymes. The native enzyme appears to be a homodimer, has a pH optimum between 5.0 and 5.5, and possesses an extracellularly oriented active site. Some of the preferred natural phosphomonoester substrates for the enzyme include phosphotyrosine, fructose 1-6-bisphosphate, and AMP. Several biological roles of the membrane-bound acid phosphatase have been proposed. The enzyme may provide the parasite with a source of inorganic phosphate by hydrolyzing phosphomonoesters of metabolites. Another possible function for the acid phosphatase is in suppressing toxic oxygen metabolite production (the oxidative burst) normally induced upon entry of the microbes into macrophages, by dephosphorylating protein substrates that are critical for macrophage activation. One such example is the NADPH oxidase which becomes activated by phosphorylation by protein kinase C during microbial invasion of macrophages. However, these phosphoproteins are not especially good substrates for the acid phosphatase and, therefore, it is unlikely that the enzyme acts as a phosphoprotein phosphatase.

Mannan-like carbohydrate

An intracellular polysaccharide that appears to be composed entirely of mannose residues has been partially characterized from *L. donovani*. This mannan-like material appears to be a storage compound analogous to glycogen.

Secreted glycoconjugates

In addition to cell-surface LPG and GIPLs, *Leishmania* secrete a family of heavily glycosylated proteins and proteoglycans that are important for parasite virulence. Most of these glycans are similar in structure to those found on LPG, notably the Gal-Man PO_4 repeat unit motif. The structural features of secreted acid phosphatase, phosphoglycan and proteophosphoglycan are briefly outlined below.

Secreted acid phosphatase

With the exception of *L. major*, all *Leishmania* promastigotes abundantly secrete sAP (Figure 10.5A) from the flagellar pocket. The secreted glycoproteins and proteoglycans tend to form distinct macromolecular complexes which are found both in the flagellar pocket and in the culture media. Old world species such as *L. donovani, L. tropica* and *L. aethiopica* secrete mono- or oligomeric sAPs, whereas South American species such as *L. mexicana, L. braziliensis* and *L. amazonensis* secrete sAPs that aggregate into large pearl-like filamentous polymers. The sAPs are encoded by multiple genes that have high levels of sequence identity. The *L. donovani* sAP peptides are heavily glycosylated on C-terminal serine/threonine-rich domains. The glycans are phosphodiesters linked to serine residues and commonly consist of the 6Gal(β1,4) Man(α1-)PO_4 repeat units found on LPG. The target sites of phosphoglycosylation are not random, but rather are composed of repetitive motifs, with modifications on select serine residues. In *L. mexicana*, sAP is extensively modified with Man (α1-)PO_4 residues that may be further elaborated with α1,2-linked mannose oligosaccharides or short PO_4-Gal-Man repeats. Often the phosphoglycan chains are terminated with a diverse set of oligomannose

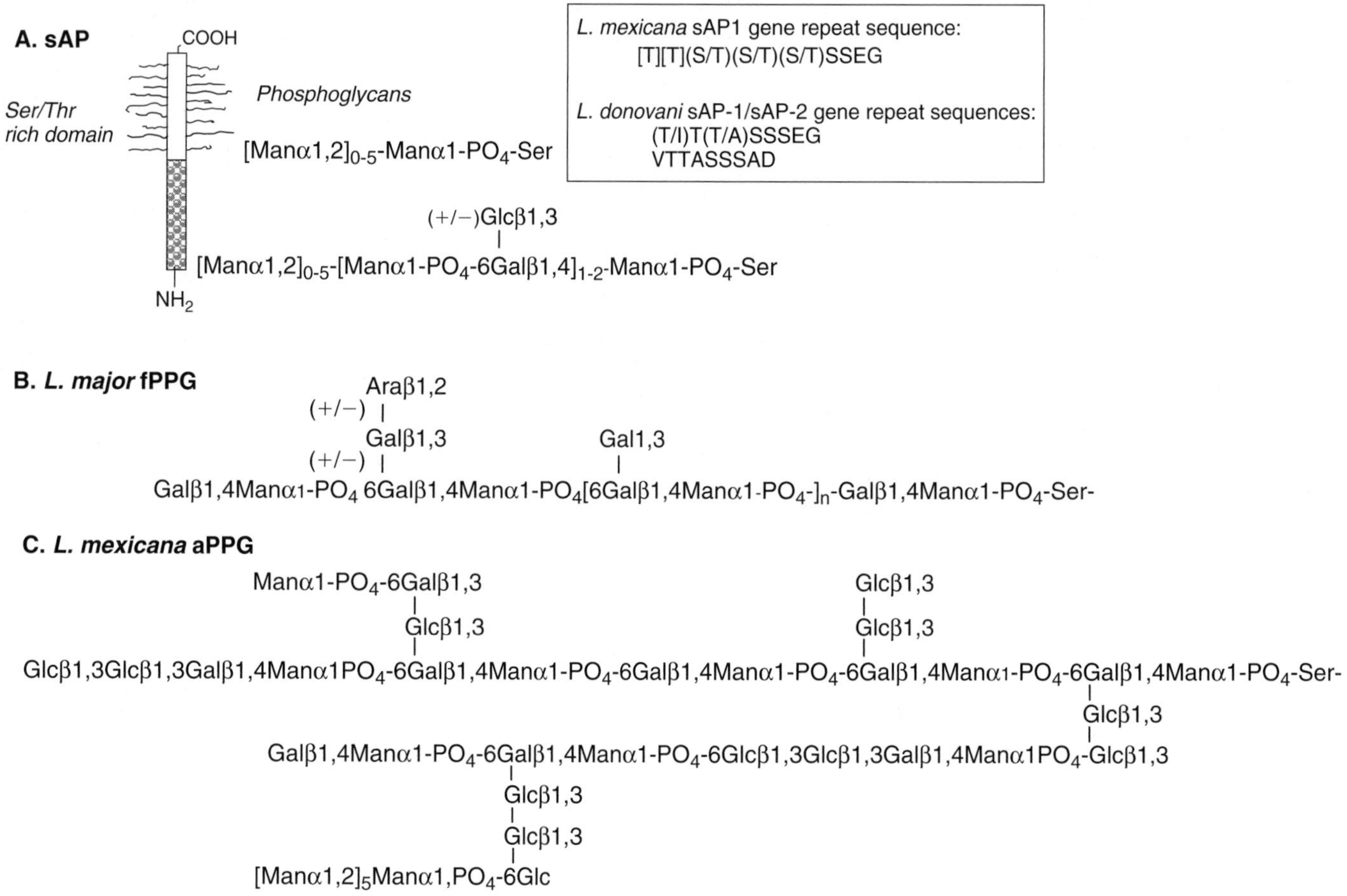

FIGURE 10.5 Structures of secreted *Leishmania* glycoconjugates. (A) Acid phosphatase (sAP). The acid phosphatases are composed of Ser/Thr-rich peptides that are heavily glycosylated on serine with phosphoglycan chains, similar to those found on LPG. The consensus sequences of the repeat domains where phosphoglycosylation occurs are shown in the box. (B) *L. major* fPPG (filamentous proteophosphoglycan) is a mucin-like molecule with extensive phosphoglycosylation on serine residues. (C) *L. mexicana* aPPG (amastigote-specific PPG) is distinct from fPPG, and although it shares phosphoglycan structures with LPG and sAP, it is also modified with unique branched glycans.

cap structures. Interestingly, the secreted enzyme also does not appear to inhibit the neutrophil oxidative burst, unlike its membranous counterpart.

Proteophosphoglycan

Proteophosphoglycans (PPG) in promastigotes include filamentous PPG or fPPG (Figure 10.5B) and a putative GPI-anchored cell-associated form or mPPG. Amastigotes secrete their own non-filamentous and stage-specific form termed aPPG. The filamentous form, fPPG, is secreted by promastigotes of all *Leishmania* species, and forms a highly viscous mesh within which the parasites lie embedded. fPPG consists of 95% phosphoglycans, with an abundance of serine, alanine and proline in the peptide component. Over 80–90% of the serine residues are phosphoglycosylated with short PO_4-Gal-Man repeats attached via phosphodiester bonds, which are terminated by small oligosaccharide cap structures. Although the function of fPPG is not yet defined, it is believed that the gel-like matrix formed by the interlocking filaments trap the parasites in the sand fly anterior gut. Furthermore, it has been hypothesized that the presence of the parasite plug deters the ingestion of a second bloodmeal, thereby encouraging the sand fly to probe several hosts and, in the process, improve the chances of transmission.

L. mexicana amastigotes secrete copious amounts of aPPG (amastigote-specific PPG) (Figure 10.5C) into the phagolysosomal compartment. PPG consists of a defined polypeptide backbone modified with phosphoglycans on serine residues. The glycans include previously identified structures common to LPG and sAP, but also include a large number of novel highly branched structures, unlike fPPG. On average, each molecule contains six mono- or multiphospho-oligosaccharide chains, capped by four neutral caps. The glycans include neutral structures like $[Glc(\beta1,3)]_{1\text{-}2}$ Gal(β1,4)Man, monophosphorylated glycans containing the conserved PO_4-Gal-Man backbone but incorporating unusual stage-specific modifications such as Gal(β1,3) or $[Glc(\beta1,3)]_{1\text{-}2}$ Glcβ1, and monophosphorylated tri- and tetrasaccharides that are monophosphorylated on the terminal hexose, as well as tri- and tetraphosphorylated glycans. aPPG is found in mg/ml concentrations in parasite-containing vacuoles. Within the macrophage, aPPG is believed to contribute to the formation of the parasitophorous vacuole, thus participating in the maintenance of infection in the mammalian host. PPG appear to activate the complement system via the mannose-binding pathway by virtue of their large numbers of potential mannose-binding lectin-binding sites. PPG also may contribute to the binding of *Leishmania* to host cells and may play a role in modulating the biology of the infected macrophage at the early stage of infection. A membrane-associated form of PPG (mPPG) has been identified in *L. major* promastigotes.

CRITHIDIA

Crithidia are monogenetic members of the trypanosomatid family that colonize the digestive tract of flies. Two major types of glycans have been identified and characterized in *Crithidia*; a β1,2-linked D-mannan, which is a polymer of 50 or more Man residues, and a lipoarabinogalactan that consists of a lipid-anchored β1,3-linked D-galactan modified with Ara residues (Figure 10.6). The lipid anchor is similar but not identical to the *Leishmania* LPG anchor, and consists of a glucosaminyl-inositol phosphoceramide. The deduced structure of the arabinogalactan of *Crithidia*

Crithidia fasciculata **lipoarabinogalactan (LAG)**

$[\text{Ara}1,2]_{\sim 8}$-$[\text{Gal}\beta 1\text{-}3]_{12}$-Gal$\beta$1-3Gal$\beta$1-3Gal$_f\beta$1-3Man$\alpha$1-3Man$\alpha$1-4GlcN$\alpha$1-3-Inos-$PO_4$-Ceramide
PO_4
αGlc

FIGURE 10.6 Structure of lipoarabinogalactan (LAG) from *Crithidia fasciculata.*

is $[\text{Ara}_p1\text{-}2]_x$-$[\text{Gal}\beta 1,3]_n$-Gal($\beta$1,3) Gal($\alpha$1,3)Gal$_f$ (β1,3)Man(α1,3)Man(α1,4)GlcN(α1,6)-inositol phosphoceramide. The unusual D-Ara$_p$ has been found in *E. schaudinni* and the LPG of *Leishmania*. Earlier studies reported the presence of soluble arabinogalactan (Gorin *et al.*, 1979), but purification procedures used at the time probably resulted in hydrolysis of the molecule at the galactofuranose residue. Thus, the soluble glycan probably corresponds to the lipid-anchored lipoarabinogalactan reported by Schneider and colleagues.

Interestingly, sialic acid-bearing glycoconjugates have been identified on the surface of *Crithidia*, although the structures of these molecules are unknown. The presence of sialoglycoconjugates in *Crithidia* growing in sialic acid-free chemically defined medium suggests that the residues are synthesized *de novo* and not transferred from existing glycoconjugates by a trans-sialidase, as observed in other trypanosomatids.

FURTHER READING

Ferguson, M.A.J. (1997). The surface glycoconjugates of trypanosomatid parasites. *Philos. Trans. R. Soc. Lond.* **352**, 1295–1302.

Ferguson, M.A.J. (1999). The structure, biosynthesis and functions of glycosylphosphatidylinositol anchors, and the contributions of trypanosome research. *J. Cell Sci.* **112**, 2799–2809.

Ilg, T. (2000). Proteophosphoglycans of *Leishmania. Parasitol. Today* **16**, 489–497.

Ilg, T., Handman, E. and Stierhof, Y.D. (1999). Proteophosphoglycans from *Leishmania* promastigotes and amastigotes. *Biochem. Soc. Trans.* **27**, 518–525.

Ilgoutz, S.C. and McConville, M.J. (2001). Function and assembly of the *Leishmania* surface coat. *Int. J. Parasitol.* **31**, 899–908.

Mengeling, B.J., Beverley, S.M. and Turco, S.J. (1997). Designing glycoconjugate biosynthesis for an insidious intent: phosphoglycan assembly in *Leishmania* parasites. *Glycobiology* **7**, 873–880.

Sacks, D.L. (2001). *Leishmania*–sand fly interactions controlling species-specific vector competence. *Cell. Micro.* **3**, 189–196.

Sacks, D. and Kamhawi, S. (2001). Molecular aspects of parasite–vector and vector–host interactions in leishmaniasis. *Annu. Rev. Microbiol.* **55**, 453–483.

Schenkman, S., Eichinger, D., Pereira, M.E.A. and Nussenzweig, V. (1994). Structural and functional properties of *Trypanosoma trans*-sialidase. *Annu. Rev. Microbiol.* **48**, 499–523.

Turco, S.J., Spath, G.F. and Beverley, S.M. (2001). Is lipophosphoglycan a virulence factor? A surprising diversity between *Leishmania* species. *Trends Parasitol.* **17**, 223–226.

C H A P T E R

11

Intracellular signaling

Larry Ruben[1], *John M. Kelly*[2] *and Debopam Chakrabarti*[3]

[1]Department of Biological Sciences, Southern Methodist University, Dallas, TX, USA;

[2]Department of Infectious and Tropical Diseases, London School of Hygiene and Tropical Medicine, London, UK; and

[3]Department of Molecular Biology & Microbiology, University of Central Florida, Orlando, FL, USA

OVERVIEW

Signal pathways are used by cells to elicit coordinated changes in activity. Components of motility, cell division, gene expression, energy metabolism and secretion are potential targets of regulatory control. Signal processes are of interest because when they go awry, the result is inappropriate cell behavior that can be lethal. Consequently, some of the most heavily prescribed pharmaceutical agents are directed towards ablation or activation of signal cascades.

A variety of molecules have been adopted for use in signal production. These molecules are typically allosteric regulators of specific enzyme targets, and are neither substrates nor products of the metabolic pathways they regulate. Among the most widely studied signal molecules are divalent cations, nucleotides, phosphoryl transfer and phospholipid metabolites. Phospholipids provide an especially diverse source of signal molecules where polar head groups (in the case of inositol phosphates), fatty acyl chains (such as arachidonic acid and its metabolites) and diacylglycerol can function to regulate cell activity. The following review focuses on intracellular signals within protozoan parasites. Emphasis is placed upon the ability of Ca^{2+}, cyclic nucleotides or phosphoryl transfer to regulate cellular events. Comparisons are made with metazoans, whose broad range of interactions creates a rich tapestry of signaling networks. By comparison, the genomes of unicellular organisms reveal a somewhat reduced signal capacity, with diminished numbers of plasma membrane receptors, hormone responsive transcription factors, and

Molecular Medical Parasitology
ISBN 0-12-473346-8

protein recruitment domains. Nonetheless, it is clear that protozoan parasites contain the basic biochemical components needed to generate a variety of signals. Indeed, some signal components that are conserved between metazoans and protozoan parasites have been identified by genome analysis. Future work is expected to identify mechanisms that affect the timing and spatial organization of signal production, identify the variety of cell activities subject to signal control, and demonstrate the utility of signal processes as therapeutic targets against protozoan parasites.

CALCIUM

Signal properties of Ca^{2+}

Ca^{2+} has several features that make it suitable to interact with proteins and serve as a signal. In particular, Ca^{2+} can be readily distinguished from other abundant cations and anions due to its unique combination of ionic radius and charge density. Ca^{2+}-binding proteins create selective pockets that are large enough for Ca^{2+} with an ionic radius of 1.14 Å but too large for Mg^{2+} with an ionic radius of 0.86 Å. Although Na^{+} has an ionic radius of 1.16 Å, it is distinguished from Ca^{2+} by its different charge density. The tightly fitting Ca^{2+} is then in position to have its water shell displaced. Because Ca^{2+} has a lower energy of hydration than other divalent cations, it exhibits faster association kinetics when it binds to proteins. The Ca^{2+} signal is further aided by a steep concentration gradient that is maintained across the plasma membrane. The ability of Ca^{2+} and phosphate to form a precipitate suggests that early in the evolution of life, Ca^{2+} was expelled from the cell. Intracellular free Ca^{2+} concentrations ($[Ca^{2+}]_i$) are typically maintained around 100 nM while the extracellular environment is maintained at values around 2 mM. Therefore, small fluxes in Ca^{2+} are able to significantly raise $[Ca^{2+}]_i$ above the background value. The resultant high signal-to-background ratio is a critical feature of the Ca^{2+} signal.

A Ca^{2+} signal occurs when the free concentration of Ca^{2+} is raised to an extent that specific Ca^{2+}-binding proteins become activated. However, this phenomenon of amplitude regulation does not explain all of the effects of Ca^{2+} signal propagation. In particular, oscillations in amplitude occur. Generally, oscillations result when Ca^{2+} is released from an internal compartment followed by re-filling of the compartment as $[Ca^{2+}]_i$ is returned to baseline values. The oscillations can be duplicated experimentally with pulsed release of caged Ca^{2+} or with pulsed shifts in medium Ca^{2+} under conditions where an influx channel remains constantly open. Both methods activate specific cell responses including gene transcription and protein kinase activity when the oscillation frequency is appropriate. Therefore the Ca^{2+} signal is subject to amplitude and frequency modulation. Spatial confinement of the Ca^{2+} signal is also important. Free diffusion of Ca^{2+} within cells is limited, in part because of the activity of specific organelles. When the Ca^{2+} sensitive luminescent protein aequorin is targeted to various cellular compartments, a disproportionate accumulation of Ca^{2+} into mitochondria is observed. Proximity of mitochondria to the plasma membrane and to Ca^{2+}-releasing organelles such as endoplasmic reticulum (ER) allow them to sequester Ca^{2+} even when the average $[Ca^{2+}]_i$ is below the dissociation constant of the mitochondrial transport systems. Although Ca^{2+} oscillations have been detected in *Entamoeba*, further work is needed to learn if other protozoan parasites share this ability. Nonetheless, stored Ca^{2+} in intracellular organelles has been detected. Studies with targeted aequorin in procyclic forms of *Trypanosoma brucei* reveal that, as in the mammalian host, the mitochondrion can limit the free diffusion of Ca^{2+}. Selective accumulation of Ca^{2+} within the mitochondrion has been observed following influx across the plasma membrane or following release from the acidocalcisome. Only small amounts of the remaining Ca^{2+} are free to contribute to the signal process.

Ca^{2+} homeostatic pathways

A combination of Ca^{2+} transport into organelles, polyanionic sites along plasma membranes, and high capacity Ca^{2+}-binding proteins help keep the level of $[Ca^{2+}]_i$ low (Figure 11.1A). The same homeostatic processes that maintain a low level of $[Ca^{2+}]_i$ can also release Ca^{2+} during the signal process (Figure 11.1B). In parasitic protozoa, the role of Ca^{2+} homeostasis has been investigated following disruption with ionophores, or conversely, following stabilization of $[Ca^{2+}]_i$ with intracellular Ca^{2+} chelators. Cell processes as diverse as secretion in *Entamoeba* and cell invasion of *T. cruzi*, *P. falciparum* and *T. gondii* are affected by these treatments. In mammalian cells, it is difficult to find an organelle that is not capable of Ca^{2+} transport. Along with active transport across the plasma membrane, Ca^{2+} sequestration occurs in mitochondria, ER, Golgi apparatus, lysosomes, secretory vesicles and within the nuclear envelope. Organelles are necessary to maintain Ca^{2+} homeostasis and to sequester Ca^{2+} when the signal is terminated. Parasitic protozoa vary in organelle content from the relatively simple *Giardia* and *Entamoeba* to the more elaborate kinetoplastids. Parasites also encounter environments that vary in Ca^{2+} content, such as blood, the gut lumen of insects or mammals, or the parasitophorous vacuole of intracellular parasites. It is not yet known whether reliance upon extracellular versus stored Ca^{2+} varies with these different situations.

Mitochondrial Ca^{2+} transport

A variety of Ca^{2+} transporting organelles have been detected in different protozoan parasites. A permeabilized cell system has proven to be generally useful in this regard. Cell membranes are disrupted with cholesterol agents such as digitonin, saponin or filipin. After permeabilization, specific organelles can be targeted for study with selective energy sources or inhibitors. The azo dye arsenazo III changes its absorption spectrum upon binding Ca^{2+} and consequently can be used to monitor Ca^{2+} sequestration or release. Mitochondrial Ca^{2+} transport requires an oxidizable substrate or ATP hydrolysis, is sensitive to electron transport inhibitors and is not sensitive to vanadate. Ca^{2+} transport into energized mitochondria of apicomplexans and kinetoplastids has been reported. Generally, mitochondria serve as a high capacity and low affinity Ca^{2+} reservoir. In permeabilized cells, the kinetoplastid mitochondrion can buffer Ca^{2+} in the medium to a fixed value around 700 nM. The remaining Ca^{2+} buffering capacity of the cell depends upon vanadate-sensitive ATPase pumps and is sufficient to lower $[Ca^{2+}]$ in the medium to a value around 50–100 nM. This value corresponds to the resting level of Ca^{2+} in cytoplasm. Within the mitochondrion of procyclic trypanosomes, the resting level of Ca^{2+} is around 400 nM (measured with targeted aequorin). When $[Ca^{2+}]_i$ is perturbed with agents that induce influx across the plasma membrane or release from acidocalcisomes, a rapid and transient accumulation of Ca^{2+} within the mitochondrion is observed. As a consequence, the cytosolic Ca^{2+} signal is attenuated, while the intramitochondrial free Ca^{2+} concentration reaches around 8–10 μM. The rapid uptake of

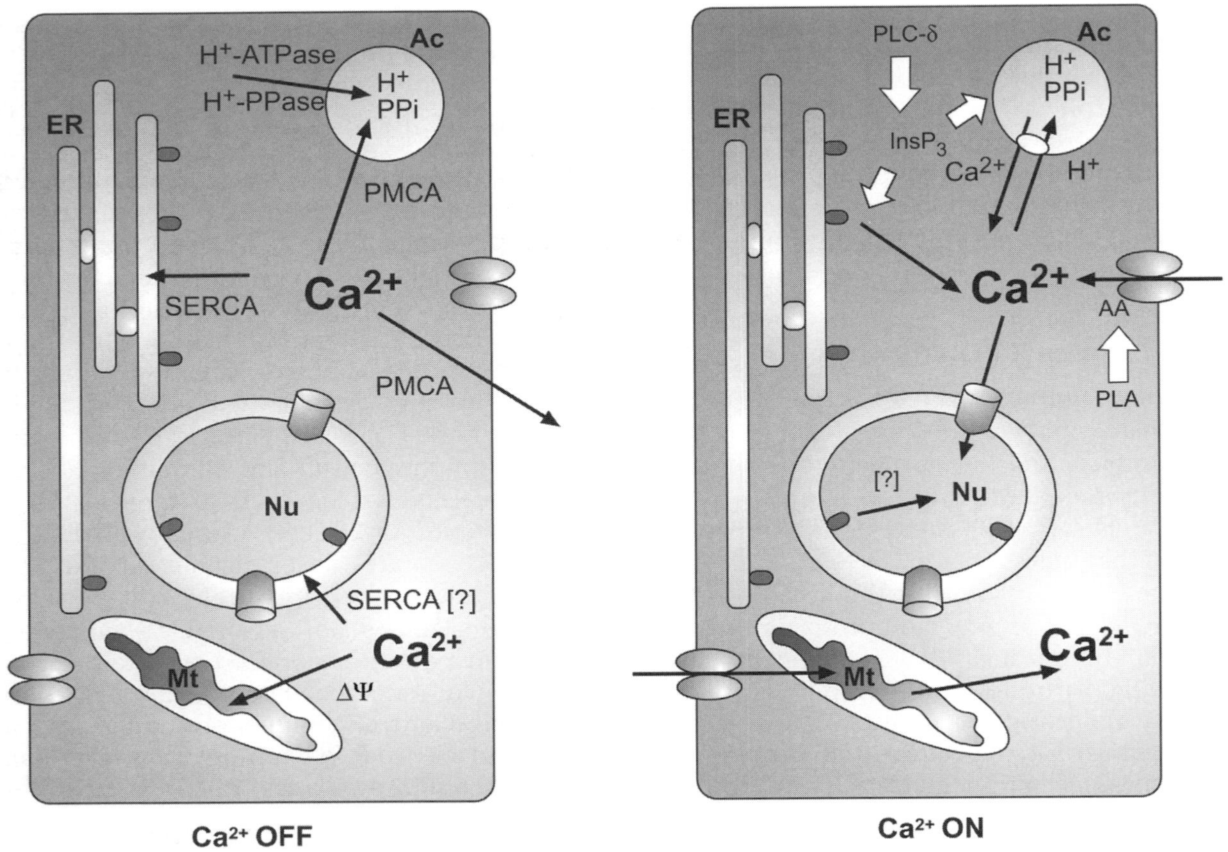

FIGURE 11.1 (See also Color Plate 4) Regulation of intracellular free Ca^{2+} concentrations during the signal process. Cells utilize redundant energy-dependent processes to store and release Ca^{2+} during the signal process. In various protozoan parasites Ca^{2+} sequestration has been detected in the acidocalcisome (Ac), mitochondrion (Mt) and endoplasmic reticulum (ER). Whether the nuclear envelope stores Ca^{2+} in protozoan parasites as occurs in mammalian cells is not known. When a Ca^{2+} signal is terminated (Ca^{2+} OFF) the mitochondrial electrochemical gradient ($\Delta\Psi$) is used to sequester Ca^{2+} in the matrix space. Additionally, energy-dependent P-ATPase pumps transport Ca^{2+} out of the cytoplasm. Vacuolar type Ca^{2+}-ATPase pumps (PMCA) have been identified in association with acidocalcisomes and the plasma membrane. SERCA type pumps of the ER have also been identified. During the signal process (Ca^{2+} ON), $InsP_3$-dependent Ca^{2+} release from the ER and acidocalcisome has been reported from some organisms, but not others. Ca^{2+} influx across the plasma membrane can be initiated with arachidonic acid (AA) in some organisms.

Ca^{2+} and its gradual release may also extend the Ca^{2+} signal, as has been proposed for mammalian cells. In mammalian cells, Ca^{2+} within mitochondria stimulates a variety of dehydrogenases and primes the mitochondria to increase their energy output. However, Ca^{2+}-dependent dehydrogenases have not yet been identified in kinetoplastid or apicomplexan parasites. Organisms such as *Entamoeba* are capable of storing intracellular Ca^{2+} in the absence of mitochondria.

Endoplasmic reticulum

Functions of stored Ca^{2+}

In mammalian cells, exchangeable Ca^{2+} in the endoplasmic reticulum (ER) has been extensively studied. The ER Ca^{2+} pool is sufficient to trigger cell events, while the stored Ca^{2+} appears to be essential for translation and proper folding of proteins. In contrast with the mammalian ER, much less is known about the corresponding Ca^{2+} pool in protozoan parasites. In mammalian cells, the level of Ca^{2+} within the ER has been quantified with mutated aequorins, aequorin-Sr^{2+} reconstitution systems, sequestered fura-2, and chameleons. The average concentration of Ca^{2+} in the ER vastly exceeds the exchangeable pool in mitochondria. Concentrations in the millimolar range have been measured. Interestingly, Ca^{2+} is not uniformly dispersed throughout the ER, and visualization of the different subcompartments with sequestered fura-2 reveals that variable amounts of the stored Ca^{2+} can be released in response to inhibitors of the Ca^{2+}-ATPase or agonists of ER Ca^{2+} channels. High levels of stored Ca^{2+} appear to be necessary for protein translation, since depletion of the ER pool decreases the rate of translation. Much of the stored Ca^{2+} in ER is bound to high capacity, low affinity binding proteins such as calsequestrin, calreticulin, calnexin and BiP. When the expression level of BiP is increased by changes in copy number or upon induction of the unfolded protein response, the level of stored ER Ca^{2+} also increases. BiP, calreticulin and calnexin serve as Ca^{2+}-dependent chaperones. In *T. cruzi*, the calreticulin gene has been cloned and its ability to modulate glycoprotein folding has been demonstrated. The *T. cruzi* calreticulin is contained in a microsomal fraction where it associates with the monoglucosylated carbohydrate moiety of cruzipain. The calreticulin from *L. donovani* has also been cloned and the recombinant protein binds Ca^{2+}. Although the recombinant calreticulin can be phosphorylated *in vitro*, it binds to leishmanial RNAs independent of its phosphorylation state. Along with the low affinity, high capacity Ca^{2+}-binding chaperones, an unusual protein which appears to be a fusion of sarcalumenin and an Eps15-like domain has recently been described in *P. falciparum*. In mammalian cells, Eps-15 functions as a tyrosine kinase substrate. It associates with AP-2 of coated vesicles, and plays a role in vesicle sorting within the endomembrane system. Sarcalumenin from mammalian cells resides within the ER lumen and modulates sensitivity of the ryanodine-sensitive Ca^{2+} release channel. The role of this unusual fusion protein in *P. falciparum* is not known. A separate family of Ca^{2+}-binding proteins referred to as the CREC family resides in the secretory pathway. Proteins in this family contain multiple repeats of the EF-hand Ca^{2+}-binding motif (see below) and may be involved in protein trafficking. Unlike cytoplasmic EF-hand proteins, members of the CREC family have a low affinity for Ca^{2+}. Reticulocalbin, a member of this family, is essential, based on homozygous deletion studies in mice. The reticulocalbin homolog of *P. falciparum* has been cloned and is referred to as Pfs40/PfERC. The protein contains six EF-hand Ca^{2+}-binding motifs and an ER retrieval sequence at its carboxy terminus. The protein localizes to a reticular network within trophozoites, and its expression level increases during trophozoite maturation. Overall, these observations indicate that the ER of protozoan parasites contains Ca^{2+}-binding proteins capable of mediating lumenal activities including translation, protein folding and protein trafficking.

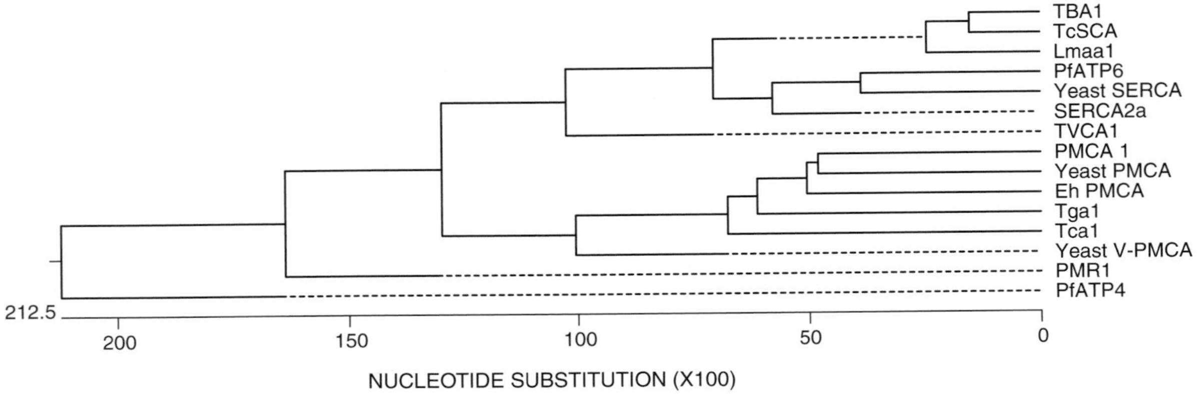

FIGURE 11.2 Phylogenetic relationships between Ca^{2+}-ATPase pumps from parasitic protozoa. Dendogram obtained from the multiple sequence alignments generated by the ClustalV method. The units at the bottom of the tree indicate the number of substitution events. The individual sequences are *T. brucei* SERCA (TBA1, accession #P35315); *T. cruzi* SERCA (TcSCA, accession #AAD08694); *L. amazonensis* SERCA (Lmaa1, accession #AAC47505), *P. falciparum* SERCA (PfATP6, accession #737940); Yeast SERCA pump (accession #P92939); chicken cardiac SERCA 2a (accession #A40812); *T. vaginalis* SERCA (TVCA1, accession #AA817958); rabbit PMCA1 (accession #J03754); yeast PMCA (accession #022218); *E. histolytica* PMCA (accession # T18294); *T. gondii* PMCA (Tga1, accession #AAF72330); *T. cruzi* PMCA (Tca1, accession #T30303); yeast vacuolar-PMCA (accession #P38929); yeast Golgi Ca^{2+}-ATPase (PMR1, accession #P13586) and *P. falciparum* Ca^{2+}-ATPase (PfATP4, accession #AF203980).

Ca^{2+} sequestration mechanisms

Ca^{2+} is sequestered in the ER by means of P-type sarcoplasmic-endoplasmic reticulum ATPase (SERCA) pumps (Figure 11.2). The SERCA pumps are a family of proteins with ten membrane-spanning domains (M1–M10) and high affinity Ca^{2+} binding residues within M4–M6 and M8. Inhibitors of these pumps, such as thapsigargin, cyclopiazonic acid or 2,5-di-(*t*-butyl)-1,4-hydroquinone cause Ca^{2+} to leak out of the ER. Since the release of Ca^{2+} from the ER is often sufficient to elicit a cell response, these inhibitors are useful tools for assessing the role of ER Ca^{2+} in cell regulation. Based upon homology with mammalian proteins, SERCA pumps have been cloned from *T. brucei* (TBA1), *T. cruzi* (TcSCA), *L. m. amazonensis* (Lmaa 1), *T. vaginalis* (TVCA1), and *P. falciparum* (PfATP 6). The proteins are closely related to each other and to mammalian SERCA-type pumps. However, some differences from the mammalian proteins have been observed. In particular, the thapsigargin binding site from kinetoplastids exhibits sequence differences. Additionally, the phospholamban binding site of cardiac SERCA pumps is missing in kinetoplastids. Instead, putative protein kinase A phosphorylation sites are present. Similar sites are absent in mammalian SERCA pumps. Whether phosphorylation of the kinetoplastid SERCA is used to regulate ATPase activity is not known. Interestingly, a novel Ca^{2+}-ATPase from *P. falciparum* (PfATP 4) has a structure that is distinct from known P-ATPase families including SERCA pumps, the plasma membrane Ca^{2+}-ATPases (PMCA), and the yeast Golgi-type (PMR1). Sequence differences make PfATP4 refractory to thapsigargin but allow it to remain sensitive to cyclopiazonic acid. Only the TBA1, TcSCA and PfATP 4 have been functionally characterized. The other proteins are presumed to function as Ca^{2+} pumps based on sequence homologies. Susceptibility of PfATP 4 and TcSCA to cyclopiazonic acid suggest that this inhibitor might be a useful tool for disrupting Ca^{2+} homeostasis in these organisms. Whether or not overexpression of the SERCA pumps changes the level of Ca^{2+} stored in the ER and alters cell function is not known. Nonetheless, the expression level of PfATP 4 varies during the *Plasmodium* life cycle, with the maximum level reached in merozoites compared to ring stages. In *Leishmania*, changes in the expression level of Lmaa 1 have also been detected. Here, overexpression correlates with increased virulence. However, Ca^{2+} storage within the ER could not be measured. In procyclic *T. brucei*, a tenfold overexpression of TBA1 was achieved with no effect on parasite growth rate or basal levels of $[Ca^{2+}]_i$. The expression levels of TBA1 and TcSCA are constant throughout the life cycle.

Ca^{2+} release channels of the ER and regulation by phospholipid metabolites

During the signal process, stored Ca^{2+} in the ER is released through specific receptors that also function as channels. The receptors are divided into two related families, depending upon whether release is activated by binding to inositol 1,4,5 trisphosphate ($InsP_3$) or to the fungal toxin ryanodine. In addition, both receptor families open in response to an elevation in cytoplasmic Ca^{2+} concentration. This phenomenon is known as Ca^{2+}-induced Ca^{2+} release (CICR). The process is biphasic since too much external Ca^{2+} can close the channel and prevent further release. Ca^{2+} release through the $InsP_3$ receptor is also quantal. In other words, discrete amounts of Ca^{2+} are released in response to increasing amounts of agonists. Structurally, the receptor families are highly related in their carboxyl termini, the region that spans the membrane and forms the functional channel. The amino terminus of each is variable and binds to the different channel agonists. Along with its distribution on the cytoplasmic face of the ER, the $InsP_3$ receptor is also found on the nuclear matrix side of the nuclear envelope. In mammalian cells, stored Ca^{2+} within the nuclear envelope has been detected with sequestered fura-2. Presence of the $InsP_3$ receptor might represent a regulated route for the direct movement of Ca^{2+} into the nucleus. Ca^{2+} movement into the nucleus has been measured with targeted aequorin in *T. brucei* procyclic forms. However, the Ca^{2+} appeared to pass through nuclear pores. Whether the nuclear envelope of parasites can modulate nuclear Ca^{2+} levels is not known. In cells that are not electrically coupled (a model that may approximate the situation in pathogenic protozoa), depletion of the ER Ca^{2+} pool is associated with regulation of store-operated Ca^{2+} channels (SOCs) in the plasma membrane. This phenomenon of Ca^{2+} gating is referred to as capacitative Ca^{2+} entry.

Because the ER Ca^{2+} channels can be regulated by $InsP_3$, the metabolism of phosphatidylinositol (4,5) bisphosphate is of broad interest. Additionally, phosphorylated forms of phosphatidylinositol can serve to recruit proteins that contain pleckstrin homology (PH) domains. Interestingly, two proteins that are thought to utilize PH domains as membrane anchors lack these motifs in the parasite homologs. These proteins include phospholipase C-δ from *T. cruzi* and protein kinase B from *Plasmodium*.

In *T. cruzi*, phosphatidylinositol and its metabolites have been detected by TLC and HPLC. The amount of $InsP_3$ increases upon exposure of permeabilized cells to Ca^{2+}, or upon treatment of intact cells with the cholinergic agonist carbachol. *Plasmodium* also varies the amount of phosphatidylinositol (4,5) bisphosphate in the membranes, with the highest amounts detected in later stages of the asexual cycle. Exogenous Ca^{2+} potentiates the release of $InsP_3$ from membranes of *Plasmodium*. The level of $InsP_3$ becomes elevated early during gametogenesis, at the onset of flagellum development.

$InsP_3$ release from the membrane is mediated by members of the phosphatidylinositol-specific phospholipase C family of proteins (PLC). Three broad groups of PLC have been described. PLC-β is activated by the alpha subunit of trimeric G proteins (G_q). PLC-γ has an SH2 domain and is recruited to phosphorylated tyrosine residues on catalytic receptors. Finally, PLC-δ lacks binding sites for G_q or SH2 domains and may be regulated by association with small G proteins. To date, *T. cruzi* is the only protozoan parasite from which the PLC has been cloned. Similar to other single-celled organisms, the *T. cruzi* enzyme is of the δ-type. However, it lacks the PH domain usually found at the amino terminus of other δ-type enzymes. Instead, it is acylated by myristate and palmitate, and this modification may help target the enzyme to appropriate membranes. The enzyme also has an unusual bridge of charged amino acids (mostly acidic) between its two catalytic X and Y domains. Enzyme activity is stimulated with Ca^{2+}, consistent with the Ca^{2+}-dependent rise in $InsP_3$ observed in permeabilized cells. Conditions that stimulate the *in vitro* transition from trypomastigotes to amastigote forms simultaneously elevate the expression level of PLC-δ and increase the cellular pool of $InsP_3$. Despite the ability of *T. cruzi* to regulate $InsP_3$ levels in the cell, to date it has not been possible to show $InsP_3$-dependent Ca^{2+} release from the ER of this organism. Unlike the situation in *T. cruzi*, the addition of $InsP_3$ to isolated permeabilized *Plasmodium* trophozoites causes Ca^{2+} release from what appear to be two distinct compartments. One compartment is depleted by inhibitors of the SERCA pumps, including thapsigargin and 2,5-di(*t*-butyl)-1,4 benzohydroquinone, and thus is likely to derive from the ER. The other compartment is discharged with the ion exchanger nigericin or neutralizing agent chloroquine. These reagents disrupt pH gradients within acidic organelles, suggesting that $InsP_3$ might also release Ca^{2+} from acidocalcisomes. Interestingly, in *E. histolytica*, where a well defined ER is not present, permeabilized cells still release Ca^{2+} from an intracellular compartment in response to $InsP_3$ as well as Ins(2,4,5)P_3. Depletion of the Ca^{2+} pool with $InsP_3$ does not prevent further Ca^{2+} release with inositol (1,3,4,5) tetrakisphosphate ($InsP_4$), and the association of $InsP_3$ to membranes could not be removed with excess unlabeled $InsP_4$ or vice versa. Taken together, these data suggest that two Ca^{2+} pools exist in these cells, with differing sensitivities to inositol phosphates. Within *Entamoeba*, the stored Ca^{2+} is likely contained in vacuoles. A plasma membrane-type PMCA Ca^{2+}-ATPase has been localized to vacuoles in this organism. Whether these vacuoles share other properties with acidocalcisomes has yet to be determined.

In summary, much of the nuance involved in ER Ca^{2+} storage and release has not been determined for protozoan parasites. Unlike the mammalian system where the ER Ca^{2+} pool is of central importance to the regulation of cell response, the parasite ER pool is of comparatively unknown significance. Nonetheless, it is likely that parasites will use the stored Ca^{2+} to mediate ER activities associated with translation and to modulate $[Ca^{2+}]_i$ concentrations during signal production.

Acidocalcisomes

Perphaps the most remarkable aspect of Ca^{2+} homeostasis in pathogenic protozoa has been the discovery of exchangeable Ca^{2+} in an acidic compartment, referred to as the acidocalcisome. The acidocalcisome is of interest because the Ca^{2+} reservoir is very large compared with other intracellular pools, and because the acidocalcisome does not appear to have a counterpart in the mammalian host. Acidocalcisomes have been identified in a wide range of kinetoplastid and apicomplexan parasites. The low pH environment of the acidocalcisome can be identified with fluorescent probes, such as acridine orange, that shift their fluorescence and absorption spectra when they accumulate in acidic compartments. Interestingly, pyrophosphate serves as an energy source for acidification of the compartments in *T. cruzi, T. brucei, L. donovani, P. knowelsi* and *T. gondii.* A plant-like vacuolar H^+-pyrophosphatase has been cloned from *T. cruzi* and *T. gondii.* This enzyme is sensitive to the pyrophosphate analogs imidodiphosphate or aminomethylenediphosphate (AMDP). In addition to the pyrophosphatase, H^+ is pumped into the acidocalcisome with a vacuolar type H^+-ATPase. This enzyme is sensitive to bafilomycin A_1, and bafilomycin-sensitive acidification of acidocalcisomes has been detected in *T. cruzi, T. brucei, L. amazonensis, T. gondii* and *P. berghei.* A Na^+/H^+ exchanger has also been identified, and pH within the organelle can be neutralized in the presence of elevated Na^+. When present, the Na^+/H^+ exchanger is sensitive to 3,5-dibutyl-4-hydroxytoluene, but not to the more standard inhibitor ameloride.

The H^+ gradient is required to help retain Ca^{2+} in the acidocalcisome. Inhibitors of vacuolar ATPase (bafilomycin A_1) and pyrophosphate-dependent proton pumps (AMDP) can disrupt acidification. These agents, along with ion exchangers that neutralize the compartment, can cause the release of stored Ca^{2+}. Ca^{2+} release occurs as the acidocalcisome attempts to restore the pH gradient by Ca^{2+}/nH^+ exchange. In *Plasmodium,* but not kinetoplastids, Ca^{2+} efflux also results from exposure to $InsP_3$. By contrast, arachidonic acid releases Ca^{2+} from the acidocalcisomes of procyclic *T. brucei.* The transport of Ca^{2+} into the acidocalcisome occurs by means of a vacuolar PMCA-type Ca^{2+}-ATPase pump. The gene for the acidocalcisome PMCA has been cloned from *T. cruzi* (Tca1) and from *T. gondii* (TgA1). The protein product is distinct from other PMCA enzymes, but similar to vacuolar PMCAs of *Entamoeba* and yeast in its lack of a calmodulin-binding site. The pump is localized along the plasma membrane as well as the acidocalcisome in both *T. cruzi* and *T. gondii.* The amount of Ca^{2+}-ATPase varies during the life cycle of *T. cruzi,* with an increase detected in amastigote forms. A corresponding increase in nigericin-releasable Ca^{2+} from amastigote forms has also been observed.

X-ray microprobe analysis of acidocalcisomes reveals that, in addition to Ca^{2+}, other metals, including Zn^{2+}, Mg^{2+}, Na^+ and K^+ and phosphorus are present. Electron energy loss studies suggest that the phosphorus is bound to oxygen in phosphate chains. Large quantities of long-chain phosphates within the acidocalcisomes of *T. cruzi, T. brucei, L. major* and *T. gondii* have been identified with ^{31}P-NMR. The long-chain polyphosphates of *T. cruzi* accumulate during the transition from trypomastigote to amastigote, and are released when cells are subjected to osmotic stress. Pyrophosphate analogs are able to inhibit growth of *T. cruzi, P. falciparum, T. gondii* and *Cryptosporidium parvum.* Whether the deleterious effects of these analogs result from disruption of acidocalcisomes or other primary targets remains to be determined.

Although acidocalcisomes are ubiquitously distributed in kinetoplastids and apicomplexans, variations in this organelle have been observed. Within the same organism, such as *L. donovani,* two distinct acidic compartments have been identified that contain either the

V-H^+-ATPase or the V-H^+-pyrophosphatase. Moreover, the Na^+/H^+ exchanger was only found in acidocalcisomes with V-H^+-ATPase activity. In *T. cruzi,* only a subset of the acidocalcisomes that can be labeled with antibodies against the vacuolar PMCA pump also label with antibodies against the V-H^+-ATPase. Different developmental stages also exhibit biochemical changes in acidocalcisome activities. The V-H^+-pyrophosphatase and Na^+/H^+ exchanger are each present in procyclic forms of *T. brucei,* but are absent from bloodstream forms. In general, the need for different types of acidocalcisomes is not clear. Understanding the role of stored Ca^{2+} within these organelles and the physiological mechanisms of Ca^{2+} release remain important questions for future research.

Plasma membrane transport systems

Along with the sequestration of Ca^{2+} into organelle compartments, the $[Ca^{2+}]_i$ can be lowered by transport out of the cell. Ca^{2+}-ATPase pumps of the PMCA family are involved (Figure 11.2). Ca^{2+}-dependent ATPase activity has been detected in plasma membrane fractions from *L. donovani, T. brucei, T. cruzi, Entamoeba invadens* and *T. gondii*. The enzyme activity from *L. donovani* and *T. cruzi* is sensitive to calmodulin, while the activity from *E. invadens* or *T. brucei* is not. As indicated above, members of the PMCA family have also been cloned from *T. cruzi, T. gondii, P. falciparum* and *E. histolytica*. These enzymes generally localize to multiple cellular compartments.

Ultimately cells rely on extracellular Ca^{2+} to initiate signals, or to refill internal pools when they are emptied. Ca^{2+} entry across the plasma membrane is mediated by channels. These include receptor-operated channels (ROCs), voltage-operated channels (VOCs), store-operated channels (SOCs) and arachidonic acid-regulated channels (ARCs). In mammalian cells, VOCs are the best characterized. The $\alpha 1$ subunit from skeletal muscle is capable of conducting Ca^{2+} currents, although expression level and gating properties are altered unless co-expressed with other channel subunits. Multiple isoforms of the $\alpha 1$ subunits have been identified, and these help divide voltage-gated channels into three distinct families. The $\alpha 1$ subunits each contain four sets of six transmembrane helices (S1–6). The membrane-associated loop between helices 5 and 6 contains glutamate residues that provide selectivity for Ca^{2+}, while the S4 segment contains the voltage sensor. Sensitivity to dihydropyridines is derived from sequences in S6. Other associated proteins include the $\alpha_2\delta$ complexes, cytosolic β subunit and transmembrane γ subunit. Currently, electrophysiological studies are needed to evaluate channel activity in parasites. While VOCs have not been described in protozoan parasites, it is interesting to note that the flagellum of *Chlamydomonas* contains VOC activity. In contrast with VOCs, the distribution of SOCs is more pervasive. Nonetheless, the structure is less well understood. SOC activation is secondary to emptying of the ER Ca^{2+} pool, by a process that is referred to as capacitative Ca^{2+} entry. A conformational coupling mechanism that mechanically connects the ER Ca^{2+} release channel with the plasma membrane Ca^{2+} influx channel has been proposed. In addition, a soluble influx factor that is formed on the surface of the ER and released may also be involved. In protozoan parasites it is not known whether cells utilize SOCs and respond to Ca^{2+} release from internal pools. The best characterized Ca^{2+} influx process is in *T. brucei,* and implicates ARCs. In bloodstream forms of *T. brucei,* Ca^{2+} influx across the plasma membrane can be induced by exposure to low concentrations of amphiphilic peptides such as melittin or mastoparan.

Similar results are obtained with insect forms of *T. brucei*, intracellular forms of *T. cruzi* and promastigote forms of *L. donovani*. The Ca^{2+} influx exhibits some of the hallmarks of channel activity, including selectivity for Ca^{2+} over other divalent cations, and inhibition by heavy metals. The process does not require emptying of stored Ca^{2+} from the ER or the acidocalcisome. Conversely, release of Ca^{2+} from the acidocalcisome with nigericin or monensin in intact cells does not result in Ca^{2+} influx across the plasma membrane. Therefore, Ca^{2+} influx in *T. brucei* does not appear to involve the capacitative process. Instead, phospholipase activity is required for Ca^{2+} influx. Phospholipase A_2 inhibitors block the effect of melittin, while free arachidonic acid mimics the effect of melittin. In addition, *P. falciparum* and *T. brucei* are each capable of producing further lipid signals in the form of prostaglandins $F_{2\alpha}$ and D_2, as well as arachidonic acid. The prospect of lipid-based signals generated by these parasites opens up new areas for future research.

Ca^{2+}-binding proteins

Low affinity proteins

Cells respond to Ca^{2+} signals in different ways. The complement of Ca^{2+}-binding proteins determines the repertoire of cellular responses. In general, Ca^{2+}-binding involves oxygen atoms that are provided by acidic side chains in amino acids, the carbonyl of the peptide bond, or water within the Ca^{2+}-binding site. Low affinity Ca^{2+}-binding proteins typically use repeats of acidic residues to bind Ca^{2+}. Along with the low affinity Ca^{2+}-binding proteins of the ER, both *T. brucei* and *P. falciparum* contain a homolog of the translationally controlled tumor protein. Ca^{2+}-binding has been confirmed by ^{45}Ca-overlay procedures. The *P. falciparum* protein has been localized to the cytoplasm and to food vacuoles. Although the function of the translationally controlled tumor protein is not known, it has been identified as the primary binding target for the anti-malaria drug artemisinin, and therefore may be essential for cell survival.

High affinity Ca^{2+}-binding proteins

Annexins

In contrast with low affinity Ca^{2+}-binding proteins, the high affinity Ca^{2+}-binding proteins utilize well defined motifs to coordinate with Ca^{2+}. Annexins are an extensive family of Ca^{2+}- and phospholipid-binding proteins that share conserved repeats and an annexin consensus sequence. In mammalian cells, annexins play essential roles in the mediation of cytoskeleton organization, secretion, and regulation of ion channels. Considering the broad range of annexin activities, it is surprising that only one member of the annexin family has been described in any protozoan parasite. The *Giardia* protein alpha 1-giardin has been classified as annexin XIX. Its role is not well understood, but it presumably mediates microtubule organization in the ventral disk. Limited numbers of annexin family members appear to be present in the genomes of *Plasmodium*, *T. gondii* and *T. cruzi*, but not *T. brucei*. Procedures that biochemically detect annexins in mammalian cells are not successful in detecting annexins in *T. brucei*. Why the protozoan parasites appear to have limited numbers of annexin family members is not known.

Protein kinase C

Protein kinase C (PKC) is another extensive family of phospholipid- and Ca^{2+}-binding proteins. Various isoforms are classified as 'classical' (bind Ca^{2+}, phosphatidylserine and diacylglycerol), 'novel' (lack the Ca^{2+}-binding domain), or 'atypical' (are insensitive to Ca^{2+} or diacylglycerol). Ca^{2+} binding is mediated by a cleft structure referred to as the C2 domain. Other proteins, such as PLC-δ_α and synaptotagmin, also utilize a similar structure to bind Ca^{2+}. In mammalian cells, PKC isoforms have a wide range of targets, and these are generally responsible for the sustained phase of the Ca^{2+} response. Artificial and endogenous substrates have been used to detect PKC-like activity in cell fractions from *T. brucei*, *T. cruzi*, *Giardia* and *Entamoeba*. Surprisingly, no member of the PKC family has been cloned from parasitic protozoa. The activity of PKC is determined in part by its distribution in the cell. Mammalian cells utilize a signal anchor protein (the Receptor for Activated C-Kinase, or RACK1) as a scaffold for signal complex formation. RACK1 is a beta-propeller protein with a structure that is similar to G_β of heterotrimeric G proteins. Homologs of RACK1 have been identified in *T. brucei*, *Leishmania* and *Crithidia*. The binding partners for parasite RACK1 are not known, but in *T. brucei*, mRNA levels for RACK1 are elevated in G_0 arrested stumpy cells and in cells induced to undergo an apoptosis-like cell death.

EF-hand Ca^{2+}-binding proteins

Within parasitic protozoa a great diversity in EF-hand Ca^{2+}-binding proteins has been described (Figure 11.3A, B). The EF-hand was originally identified by X-ray crystallography of muscle parvalbumin. The motif involves two alpha helices, separated by a Ca^{2+}-binding loop (helix-loop-helix motif). The X, Y, Z, −X, −Y, −Z coordination sites are residues 1, 3, 5, 7, 9 and 12 within the loop. Water within the loop contributes a seventh coordination site, although in parvalbumin this is not the case and instead, the amino acid in the −Z position (Glu) uses both of its oxygens to bind Ca^{2+}. The result is a pentagonal bipyramidal organization of Ca^{2+} coordination within the loop.

Calmodulin is the most widely distributed of the EF-hand motif proteins (Figure 11.3A). Calmodulin contains four EF-hand motifs that change conformation upon binding Ca^{2+} and expose a central helical region that serves as an interaction site for target proteins. Calmodulin is present in all eukaryotic cells and has been cloned from *T. brucei*, *T. cruzi*, and *P. falciparum*. Inhibitors of calmodulin, including the naphtholenesulfonamides (W7) along with the phenothiazines (trifluoperazine and chlorpromazine) have deleterious effects on *T. brucei*, *T. cruzi*, *T. gondii*, *Plasmodium*, *Giardia* and *Entamoeba*. In *T. brucei*, the calmodulin gene is contained within a locus of four tandem repeats. Knockout of one locus slows trypanosome growth, while knockout of both loci cannot be achieved, suggesting that calmodulin is essential to these cells. Calmodulin has no enzymatic activity of its own, and instead modulates the activity of target proteins in response to Ca^{2+}. As a consequence, the identification of calmodulin-binding proteins is an important step towards understanding Ca^{2+} regulation in parasitic protozoa. Slight differences in gene sequence between the various parasite calmodulins may allow interactions with unique parasite-specific target proteins. However, this hypothesis has yet to be verified. In *T. brucei*, elongation factor-1α (EF-1α) has been identified as a calmodulin-binding protein. The interaction between calmodulin and EF-1α has been confirmed in plants and mammals. Calmodulin-binding may coordinate the many activities of EF-1α including its ability to regulate protein translation, cytoskeletal organization, inositide metabolism and proteasome activity. Calmodulin also appears to associate with the paraflagellar rod. Whether calmodulin contributes to bending of the rod is not known. However, in *T. cruzi*, a calmodulin-sensitive

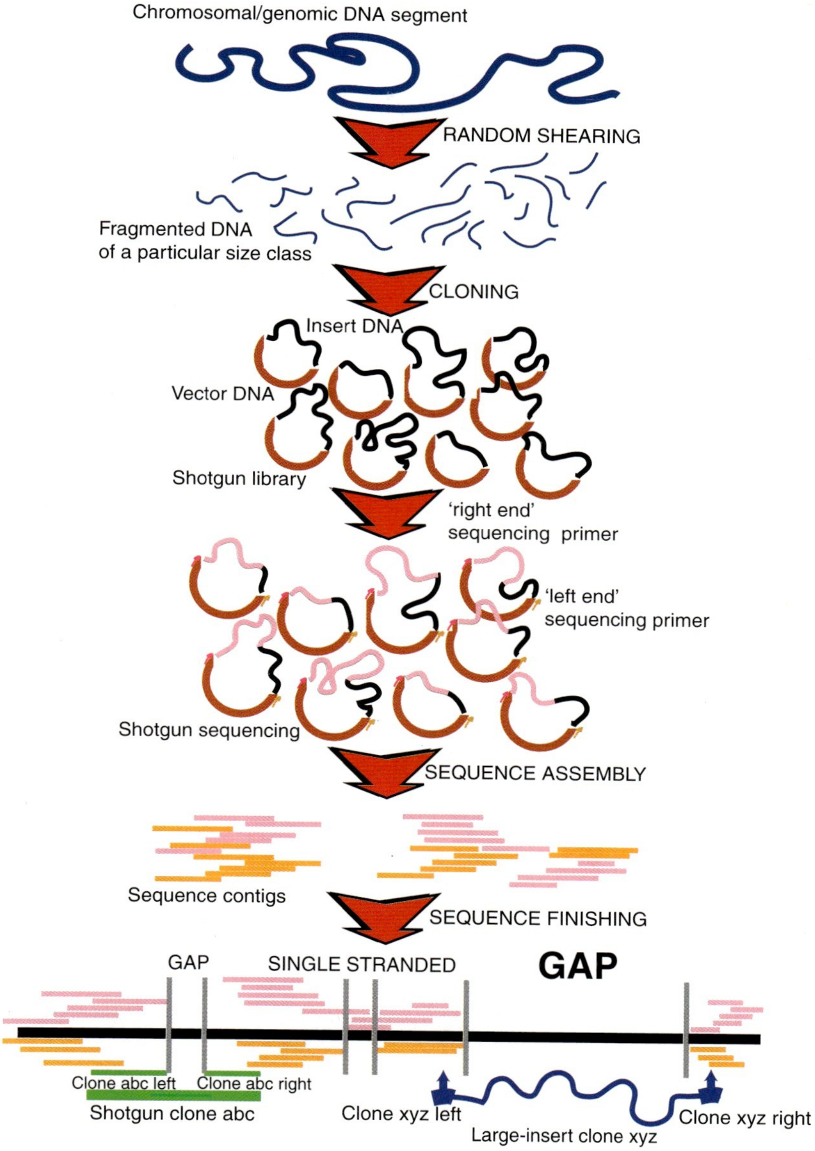

COLOR PLATE 1 (See also Figure 1.1) Sequencing a genome. This figure illustrates the steps involved in determining the sequence of a genome (or genomic segment, such as a chromosome or large-insert clone). The DNA is first sheared and cloned to make a shotgun library, which is then sequenced using universal 'left' and 'right' primers to generate a shotgun sequence set for the segment. This shotgun sequence set is then assembled into contigs of sequence reads that overlap each other. These contigs are turned into a finished product through the use of linking clone data: either matching left and right reads from individual clones within the shotgun sequence dataset, or end-sequences from larger-insert clones. It is usual to finish the sequence by obtaining sequence from both strands, resolving all ambiguities in the predicted consensus, and linking all contigs.

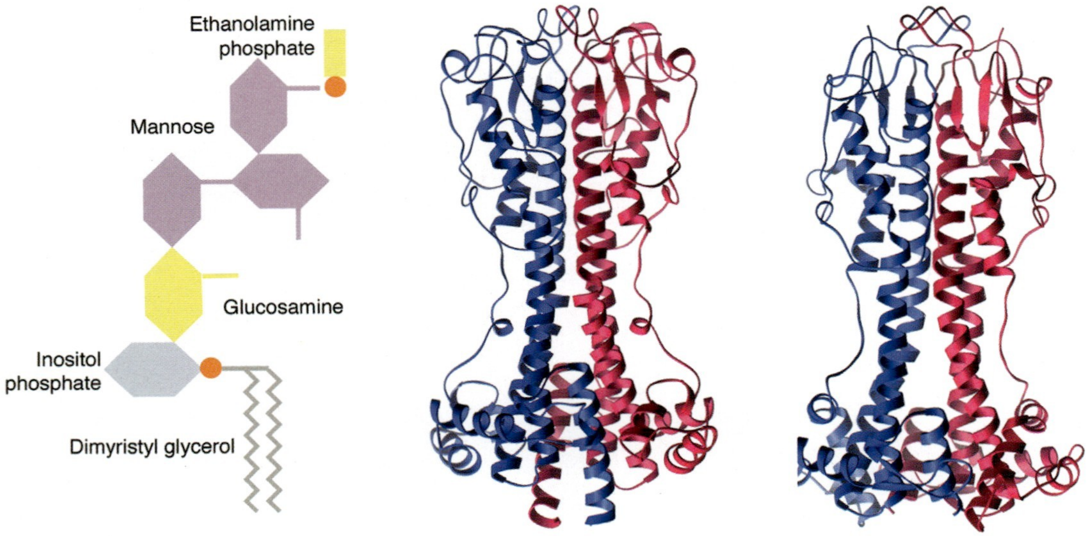

COLOR PLATE 2 (See also Figure 5.2) (Left) Structure of the minimal ('core') VSG GPI anchor and (right) three-dimensional structures of the amino-terminal domains of VSG MIT at 1.2 (left) and ILT at 1.24 (right). The dark and light ribbons indicate the atomic traces of the two identical units comprising the dimeric domain. The amino termini are at the top of the molecule, which forms the outer face of the coat. The bottom of this domain is linked to the carboxy-terminal domain, for which a structure is not available, which is then linked to the GPI anchor. The image was kindly provided by Ms Lore Leighton.

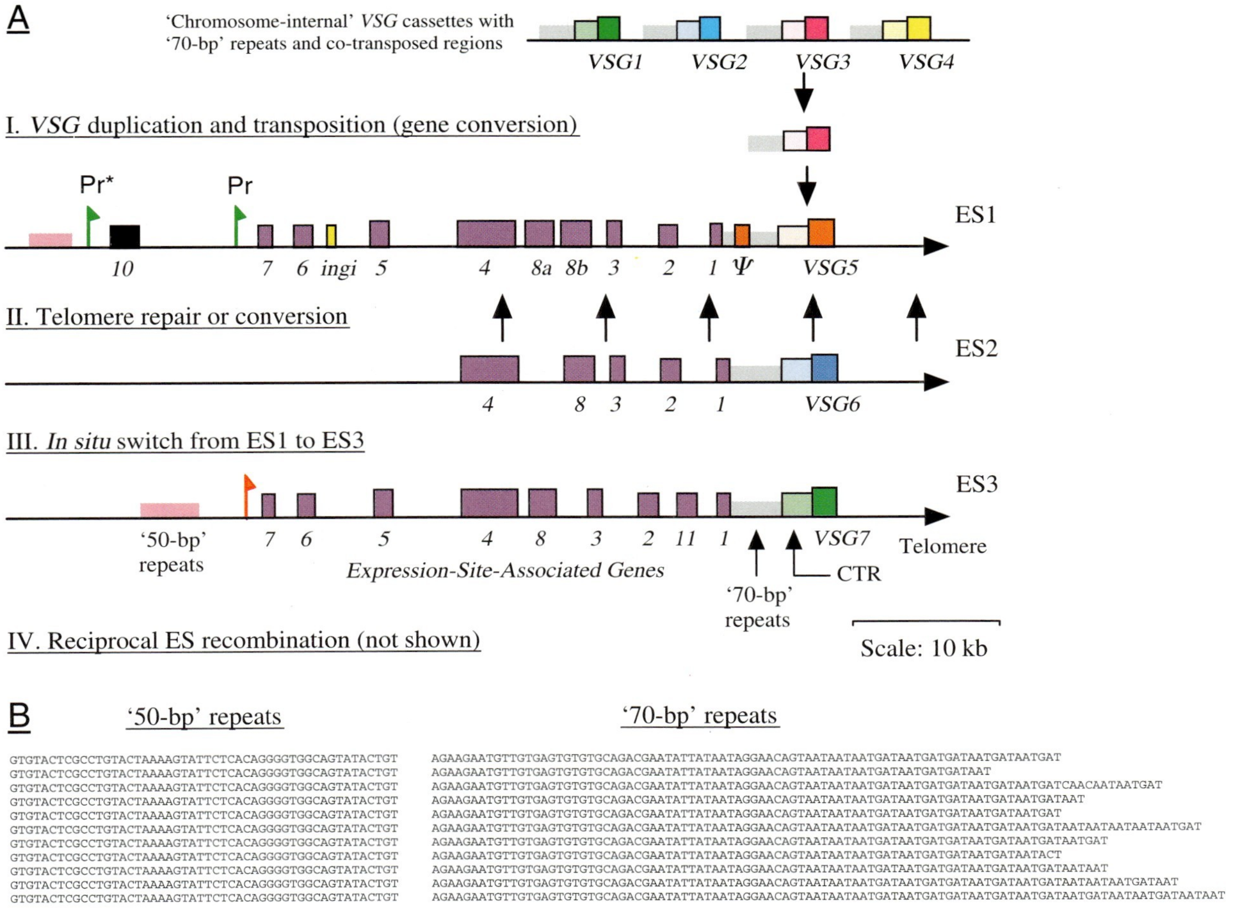

COLOR PLATE 3 (See also Figure 5.4) *VSG* expression site organization and alternative mechanisms for VSG switching. *ESAG*s are generally arranged in the order shown, but there can be duplications or triplications of some *ESAG*s and a duplicated promoter (Pr*) in some ESs. CTR is a region upstream of the VSG known as the co-transposed region: ψ a *VSG* pseudogene. (B) Examples of the '50-bp' and '70-bp' repeat sequences found upstream of bloodstream-form ES promoters and *VSG*s, respectively.

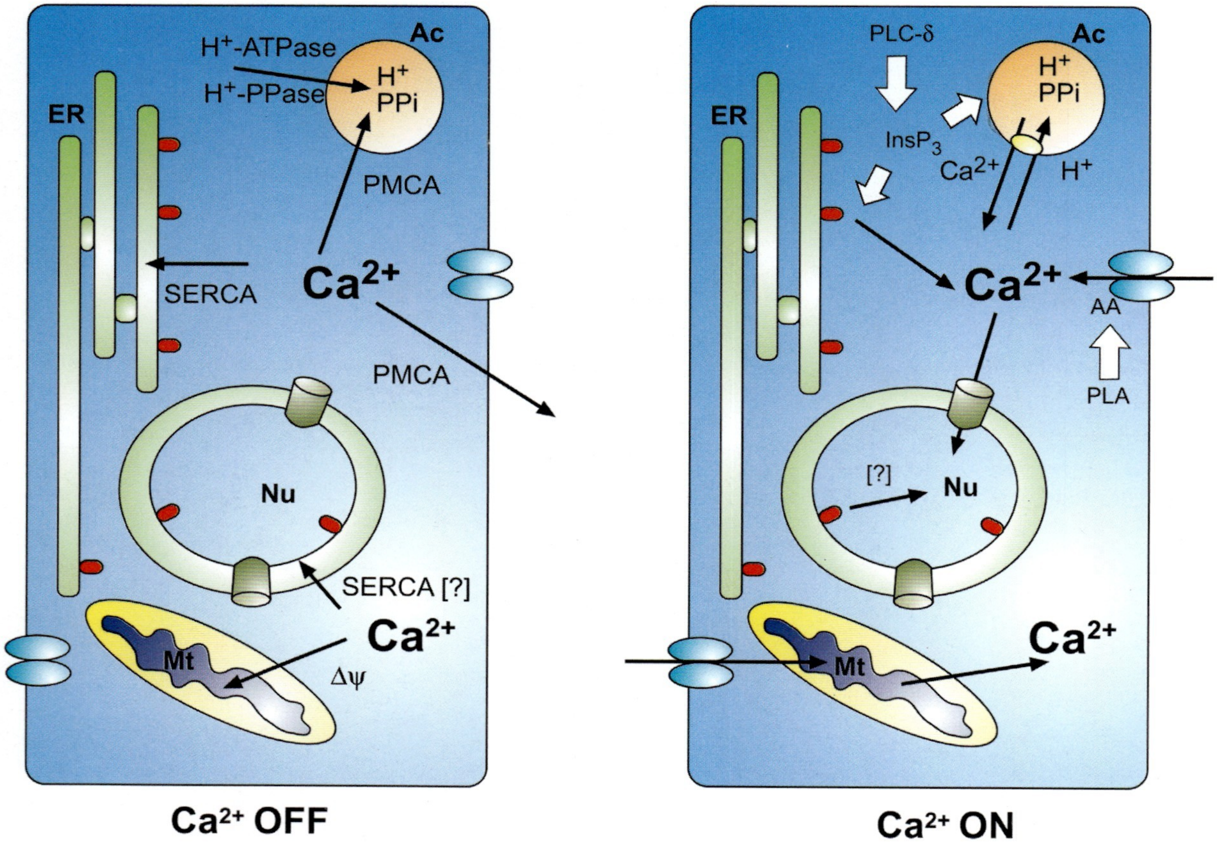

COLOR PLATE 4 (See also Figure 11.1) Regulation of intracellular free Ca^{2+} concentrations during the signal process. Cells utilize redundant energy-dependent processes to store and release Ca^{2+} during the signal process. In various protozoan parasites Ca^{2+} sequestration has been detected in the acidocalcisome (Ac), mitochondrion (Mt) and endoplasmic reticulum (ER). Whether the nuclear envelope stores Ca^{2+} in protozoan parasites as occurs in mammalian cells is not known. When a Ca^{2+} signal is terminated (Ca^{2+} OFF) the mitochondrial electrochemical gradient ($\Delta\Psi$) is used to sequester Ca^{2+} in the matrix space. Additionally, energy-dependent P-ATPase pumps transport Ca^{2+} out of the cytoplasm. Vacuolar type Ca^{2+}-ATPase pumps (PMCA) have been identified in association with acidocalcisomes and the plasma membrane. SERCA type pumps of the ER have also been identified. During the signal process (Ca^{2+} ON), $InsP_3$-dependent Ca^{2+} release from the ER and acidocalcisome has been reported from some organisms, but not others. Ca^{2+} influx across the plasma membrane can be initiated with arachidonic acid (AA) in some organisms.

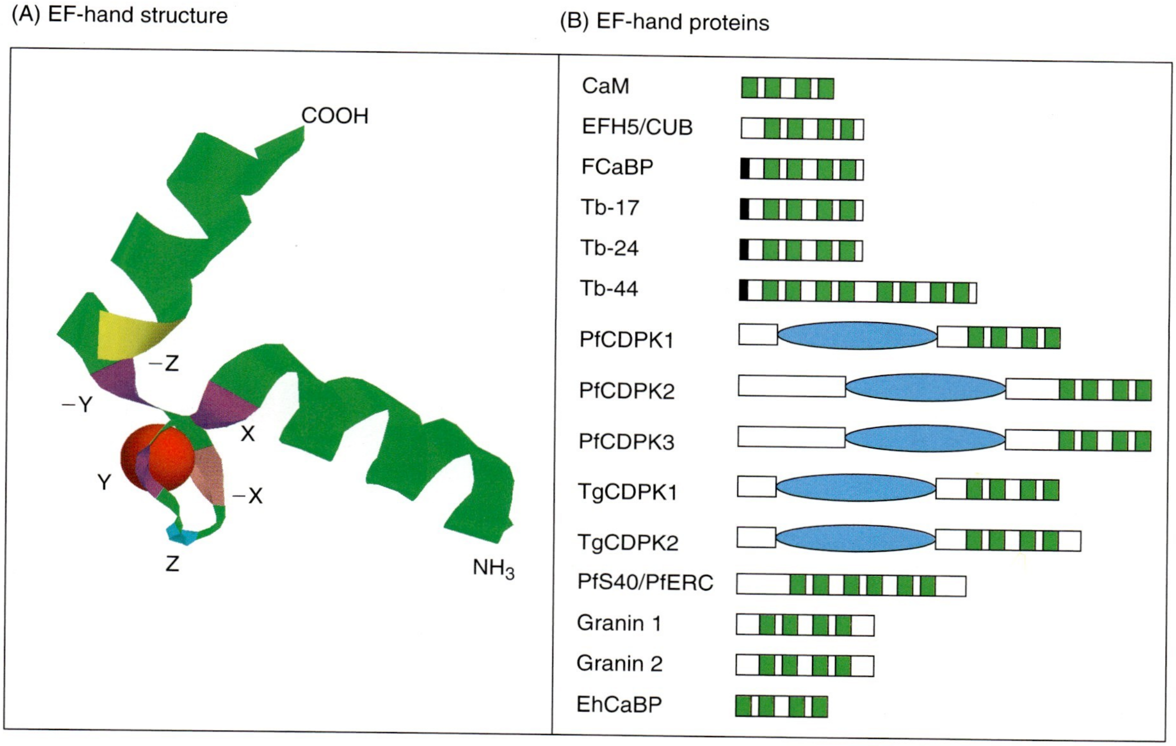

COLOR PLATE 5 (See also Figure 11.3) EF-hand structures in parasitic protozoa. (A) EF-hand #2 from vertebrate calmodulin was modeled to show its helix-loop-helix structure. The bound Ca^{2+} is represented as a red ball and is nestled in the loop between the two alpha helices. Residues 45–75 are shown. The Ca^{2+} coordination sites are colored and represent X (D56, purple); Y (D58, purple); Z (N60, blue); −X (T62, red); −Y (D64, purple) and −Z (E67, yellow). The model was constructed with Rasmol V. 2.6.1 and used calmodulin file PDB. 1CLL. (B) The variety of EF-hand proteins from different protozoan parasites is shown. Each EF-hand is represented by a green bar. The majority of proteins consist of little more than EF-hands and have no enzymatic activity. Not all of the EF-hand structures are functional Ca^{2+}-binding sites. In addition to the EF-hands, some of the proteins are acylated at the amino terminus (black bar) or contain a protein kinase catalytic domain (blue oval). In addition to calmodulin (CaM), which is found in all of the parasitic protozoa, *T. cruzi* also contains CUB and FCaBP. *T. brucei* contains EFH5, Tb-17, Tb-24 and Tb-44. *P. falciparum* contains PfCDPK1–3 and PfS40/PfERC. *T. gondii* contains TgCDPK1–2. *E. histolytica* contains granins 1–2 and EhCaBP.

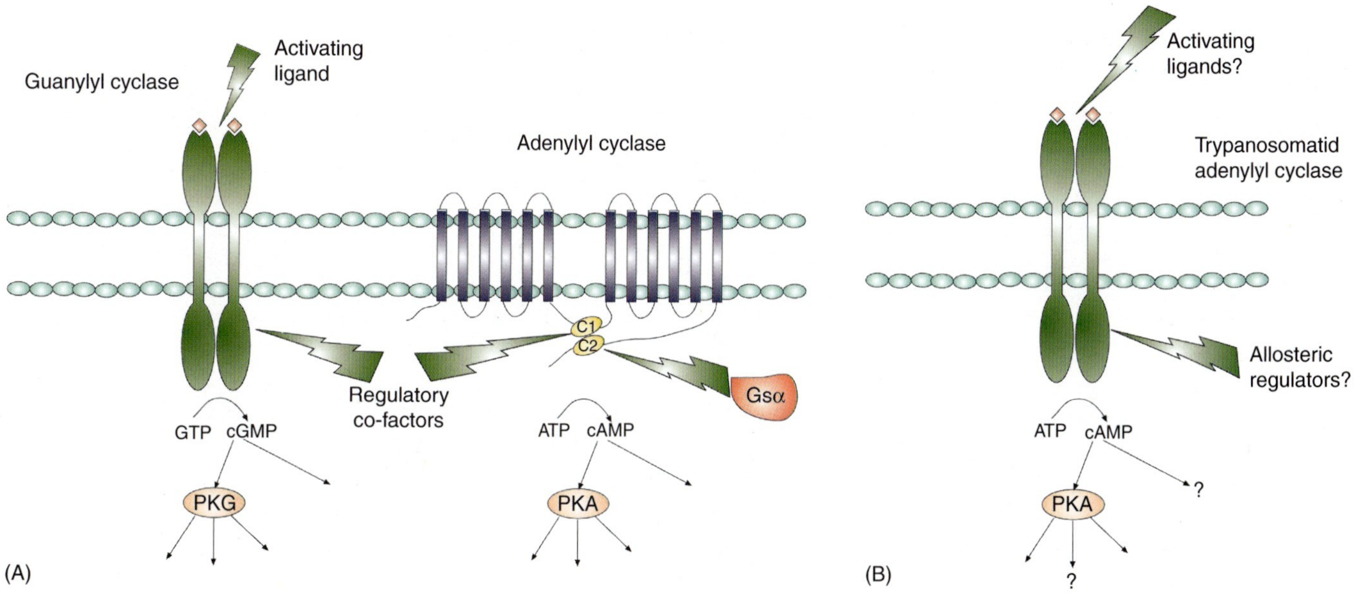

COLOR PLATE 6 (See also Figure 11.4) (A) Structural representation of mammalian membrane-localized guanylyl cyclase (GC) and adenylyl cyclase (AC). The homodimeric catalytic regions of GC are activated by ligands that bind to the extracellular receptor domain. Signal transduction can then occur via cGMP-dependent protein kinase (PKG), or by alternative pathways. With AC, the activating ligands bind to G-protein-coupled receptors. This leads to the dissociation of the linked heterotrimeric G-protein complex and activation of the catalytic site of AC, by $G_{s\alpha}$ in the example shown. cAMP can act directly on proteins, or via cAMP-dependent protein kinase (PKA). The activity of both GC and AC can be fine-tuned by other cytosolic regulatory cofactors. (B) Structural representation of trypanosomatid AC. It is presumed, although not proven, that the intracellular homodimer is activated by ligands that bind to the receptor domain. In trypanosomes, the extensive repertoire of AC genes provides the potential to respond to a wide diversity of activating ligands. The precise nature of the downstream signal-transduction pathways are not known, but it is likely that PKA plays a role similar to that in mammalian cells. Structural analysis of *T. brucei* AC has indicated the possibility of allosteric regulatory sites in the catalytic domains.

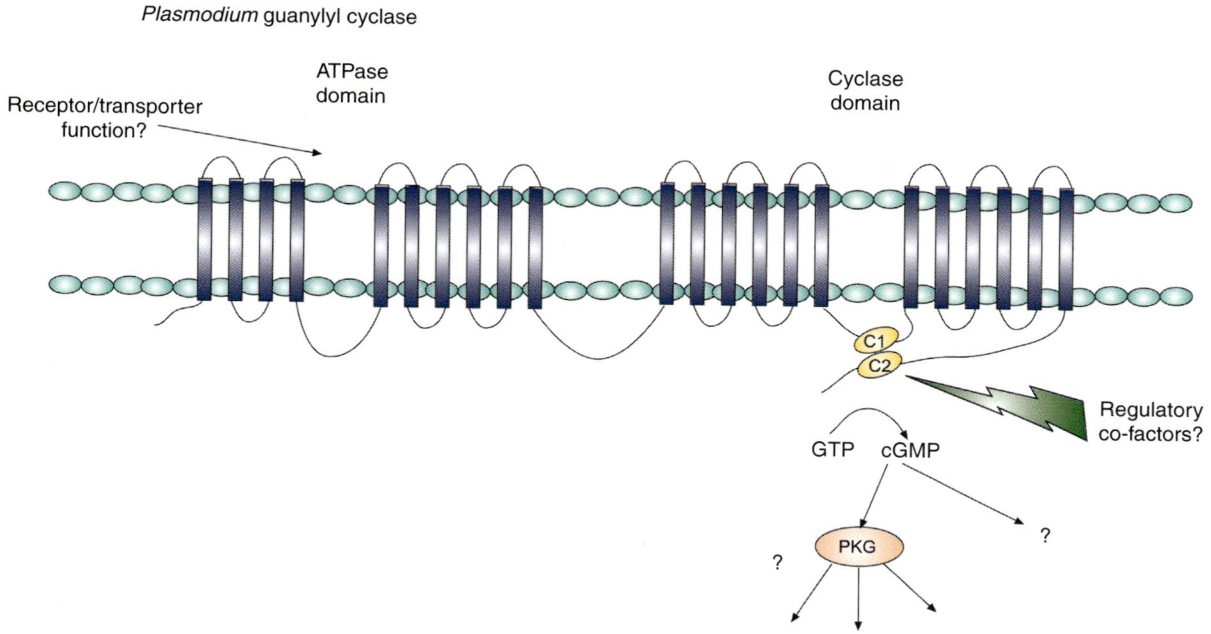

COLOR PLATE 7 (See also Figure 11.5) Model for the structure of *Plasmodium* guanylyl cyclase (GC). The amino acid sequences of the *P. falciparum* GCs (PfGCα and PfGCβ) suggest that they are bifunctional integral membrane proteins. The amino terminal regions have similarities to P-type ATPases of other organisms and the topology of the carboxyl terminal regions conforms to that of mammalian G-protein-dependent adenylyl cyclases (Figure 11.4A). The function of the ATPase domain remains to be determined, but the catalytic regions (C1 and C2) exhibit GC activity. cGMP-dependent protein kinase (PKG) is thought to have a role in signal transduction, although the mechanisms involved have not been characterized. Similarly, little is known about regulatory cofactors that could act directly on the GC catalytic regions.

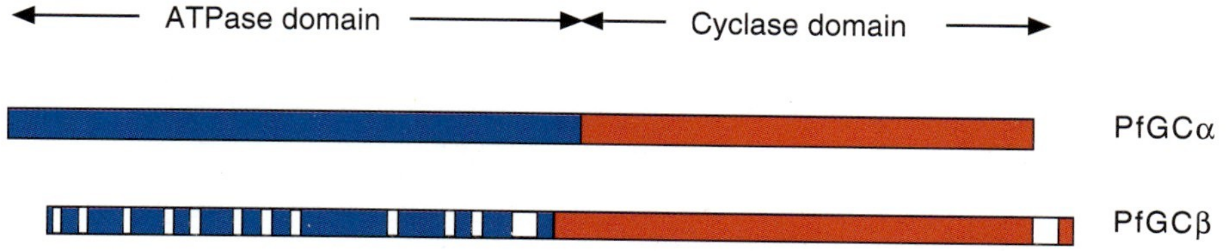

COLOR PLATE 8 (See also Figure 11.6) Structural organization of the bifunctional genes encoding guanylyl cyclase in *P. falciparum*. *PfGCα* is uninterrupted by introns, but there are 13 (shown in white) in *PfGCβ* and all but one of these are in the ATPase domain.

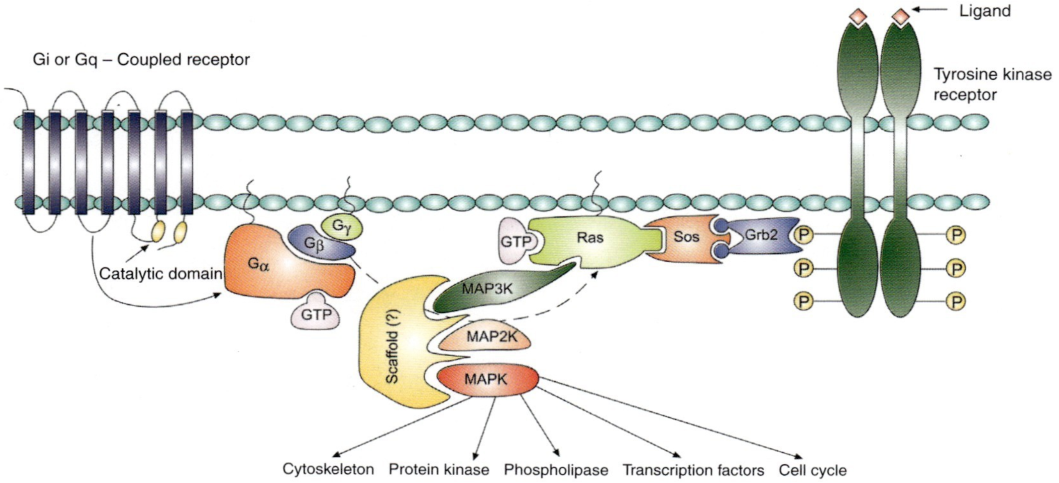

COLOR PLATE 9 (See also Figure 11.8) Activation of the MAPK pathway by receptor tyrosine kinases and seven transmembrane or serpentine receptors. Binding of a ligand to receptor activates guanine nucleotide exchange factor (GEF) Sos through the adaptor protein Grb2. Sos then activates membrane-bound Ras by exchanging GDP with GTP. The activated Ras, through an unknown mechanism, recruits and activates MAP3K and thereby the MAPK module. A scaffold protein, such as Ste5 from the *S. cerevisiae* mating response pathway, often brings the individual members of the MAPK module together. Recently, similar scaffold proteins, MP1 and JIP-1, have been discovered in mammalian cell stress response pathways. The MAPK module can also be activated by signals received through serpentine receptors. Binding of a ligand to a serpentine receptor activates membrane-anchored heterotrimeric G-protein. The exchange of GDP with GTP separates G_α and G_β subunits. The MAPK module is then recruited to the free G_β subunit through additional intermediary proteins. For example, in *S. cerevisiae* Ste20, a MAP4K acts to recruit the MAPK module to G_β. Activated MAPKs then phosphorylate a variety of cellular substrates.

COLOR PLATE 10 (See also Figure 15.5) Photograph of *Ascaris suum*. This photograph shows *Ascaris suum* within 24 hours of extraction from the host's gut. The pinkish color and the vitality of the nematodes indicate that they are in good body condition. The photograph also illustrates the large size of *A. suum*; adults are 20–30 cm long and about 5 mm in diameter.

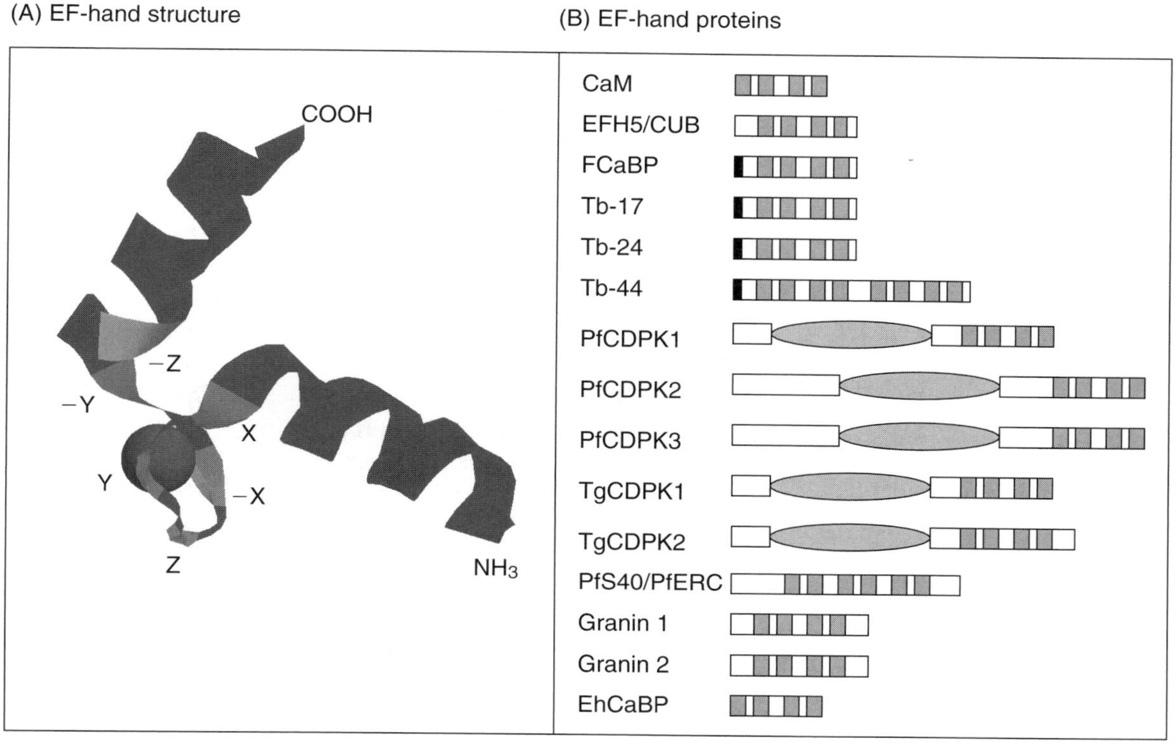

FIGURE 11.3 (See also Color Plate 5) EF-hand structures in parasitic protozoa. (A) EF-hand #2 from vertebrate calmodulin was modeled to show its helix-loop-helix structure. The bound Ca^{2+} is represented as a ball and is nestled in the loop between the two alpha helices. Residues 45–75 are shown. The Ca^{2+} coordination sites are X (D56); Y (D58); Z (N60); −X (T62); −Y (D64) and −Z (E67). The model was constructed with Rasmol V. 2.6.1 and used calmodulin file PDB.1CLL. (B) The variety of EF-hand proteins from different protozoan parasites is shown. Each EF-hand is represented by a gray bar. The majority of proteins consist of little more than EF-hands and have no enzymatic activity. Not all of the EF-hand structures are functional Ca^{2+}-binding sites. In addition to the EF-hands, some of the proteins are acylated at the amino terminus (black bar) or contain a protein kinase catalytic domain (gray oval). In addition to calmodulin (CaM), which is found in all of the parasitic protozoa, *T. cruzi* also contains CUB and FCaBP. *T. brucei* contains EFH5, Tb-17, Tb-24 and Tb-44. *P. falciparum* contains PfCDPK1–3 and PfS40/PfERC. *T. gondii* contains TgCDPK1–2. *E. histolytica* contains granins 1–2 and EhCaBP.

kinase (CaM kinase II) is also localized to the flagellum, suggesting a mechanism by which flagella-localized calmodulin might coordinate bending of the axoneme and paraflagellar rod. Various *T. cruzi* enzymes also exhibit a sensitivity to calmodulin, including phosphodiesterase and nitric oxide synthase. Less is known about calmodulin-binding partners from *Plasmodium, Toxoplasma, Giardia* and *Entamoeba*.

In kinetoplastids, the calmodulin gene is adjacent to the polyubiquitin gene locus and separated by yet another EF-hand encoding gene. In *T. brucei*, this gene is referred to as EFH5 while in *T. cruzi* it is referred to as CUB. Although a role for EFH5/CUB is not known, the knockout of both CUB alleles only occurred in cells that were simultaneously transformed with a functional copy of the gene. The inability to knock out both alleles in the absence of gene replacement demonstrates that CUB is essential to parasite survival. Its function is unknown.

Kinetoplastids also contain an EF-hand protein that is associated with the flagellum, that was initially described in *T. cruzi* as FCaBP, and in *T. brucei* as Tb17. FCaBP is tandemly repeated within the *T. cruzi* genome and exhibits slight variations between different strains. By contrast, *T. brucei* contains four distinct members of the FCaBP group (including Tb17) that have been collectively referred to as calflagins. FCaBP has four EF-hands of which two at the carboxyl terminus appear to bind Ca^{2+} with high affinity. FCaBP is of unknown function but is acylated at the amino terminus by myristic acid and palmitic acid. These post-translational modifications appear to be important for its flagellar location. A Ca^{2+}-acyl switch may allow FCaBP to reversibly associate with membranes and regulate enzyme activities. To date, binding partners for FCaBP/calflagins are not known. However, FCaBP resembles the acyl-switch protein recoverin of sensory flagella in mammalian cells. In the presence of Ca^{2+}, recoverin binds to and inhibits rhodopsin kinase. It is not yet known why the flagellum of kinetoplastids contains different EF-hand family members.

Apicomplexans provide further evidence of the diverse structures that incorporate the EF-hand motif. A family of calmodulin-like domain protein kinases (CDPKs) is found in *P. falciparum* and *T. gondii*. While typical Ca^{2+}-sensitive kinases associate reversibly with calmodulin, the CDPKs possess their own EF-hand motifs. Similar kinases are found in plants. In *P. falciparum*, CDPK1 associates with sites of membrane fusion, and may be associated with the invasion process. By contrast, PfCDPK3 is expressed specifically in the sexual erythrocytic stage where it may mediate Ca^{2+}-sensitive steps of gametogenesis. In *T. gondii*, a similar situation exists, except that only two members of the CDPK family have been identified in the genome (TgCDPK1 and TgCDPK2). Only TgCDPK2 is expressed in tachyzoites. An inhibitor of Ca^{2+}-sensitive protein kinases KT5926 blocks the activity of TgCDPK2, prevents phosphorylation of specific parasite proteins, and also blocks motility and attachment to host cells.

In *E. histolytica*, a novel protein with EF-hands has been identified along with calmodulin; the protein is called EhCaBP. It is not found in *E. invadens*. Binding partners for EhCaPB have been detected by affinity procedures, and the subset of proteins that associate with EhCaBP in trophozoites is different from the binding partners for calmodulin. In keeping with that observation, EhCaBP is able to activate a novel plant protein kinase, while calmodulin cannot. Additionally, two EF-hand proteins have been found associated with cytoplasmic granules. Each of the proteins, called granins 1 and 2, is predicted to contain three functional EF-hands. Since granule discharge may be triggered by Ca^{2+}, the granins, along with calmodulin and EhCaBP, may help regulate exocytosis in these organisms. Ca^{2+}-sensitive discharge of dense granules from *Entamoeba* has been reported.

Taken as a whole, it is evident that along with calmodulin, parasites contain a wide range of EF-hand proteins. Many of these proteins likely interact with targets that are not present in the mammalian host. *Plasmodium* and *Toxoplasma* contain a variation on that theme by utilizing proteins whose catalytic domain is fused to the EF-hand Ca^{2+} sensing domain. In addition, other calcium-binding proteins exist in these cells. Inducible RNAi systems, DNA microarrays and proteome analyses will go a long way towards establishing the function of the novel parasite proteins.

Ca^{2+} and cell function

Ca^{2+} signals are designed to trigger rapid changes in cell activity. In parasitic protozoa, a role for Ca^{2+} is most apparent in organisms that attach to the host or become internalized during the invasion process. For *E. histolytica*, the invasion process involves a combination of adhesion, secretion, phagocytosis and growth. Aspects of each of these steps are sensitive to Ca^{2+}. When *E. histolytica* makes contact with the extracellular matrix (ECM), surface proteins related to β-integrins are employed. The integrins associate with fibronectin in the ECM and when this occurs, fura-2-loaded *E. histolytica* exhibit a rise in Ca^{2+}. If ionophores are added, the increase in $[Ca^{2+}]_i$ is by itself sufficient to reorganize actin filaments and increase adhesion. Ameba can also attach to other cells by interactions involving surface lectins. When *E. histolytica* makes contact with CHO cells, Ca^{2+} oscillations occur. Ca^{2+} antagonists block cytoadherence. Carbohydrates inhibit the lectin interactions and prevent the elevation in $[Ca^{2+}]_i$. In addition to the change in parasite $[Ca^{2+}]_i$, the host cell also elevates $[Ca^{2+}]_i$ as a prelude to cell death. The rise in parasite $[Ca^{2+}]_i$ probably mediates dense granule secretion where a variety of EF-hand Ca^{2+}-binding proteins have been localized, including calmodulin and granins. Calmodulin antagonists prevent secretion. Other mediators of the Ca^{2+} signal may also be involved. For example, killing of CHO cells is enhanced by TPA (12-*o*-tetradecanoylphorbol 13-acetate), and blocked by sphingosine, an inhibitor of PKC. Overall, the results suggest that both calmodulin and PKC may be involved in parasite invasiveness. Further work is needed to clarify the role of specific signal pathways during invasion.

In contrast with *Entamoeba*, the apicomplexans are internalized by the host cell and reside within the parasitophorous vacuole. The process, typified by *T. gondii*, involves combinations of apical attachment, actin/myosin-dependent gliding motility, conoid extrusion, release of micronemal proteins, formation of the moving junction and internalization. Replication of the parasite occurs within the parasitophorous vacuole and egress must occur before another round of cell infection is initiated. Changes in $[Ca^{2+}]_i$ are needed for secretion of microneme contents, including MIC2 and the *T. gondii* AMA-1. The MIC2 is required for parasite adhesion in *Toxoplasma* and *Plasmodium*. The secretion is triggered by ionophore and prevented by intracellular Ca^{2+} buffers such as BAPTA. The release of stored Ca^{2+} is sufficient for conoid extrusion and for microneme secretion, since both processes can occur in the presence of thapsigargin or upon neutralization of acidic compartments with NH_4Cl. Additionally, a staurosporine-sensitive protein kinase appears to stimulate microneme secretion. The Ca^{2+}-sensitive kinase CDPK may help transduce the Ca^{2+} signal. When CDPK is inhibited with KT5926, invasion is also blocked.

The converse of the invasion process is parasite egress. Ionophores can stimulate egress, implicating Ca^{2+}. Thiol reduction also appears to be important, perhaps by activating an oxidized, inactive nucleotide triphosphate hydrolase of parasite origin. This enzyme can be activated by dithiothreitol and results in a concomitant Ca^{2+} flux from tubulovesicular Ca^{2+} stores. Prior to egress, the host plasma membrane becomes permeabilized. As a consequence, extraparasitic K^+ concentrations decrease. The decrease in $[K^+]$ is sufficient to stimulate PLC activity and release Ca^{2+}

from intracellular stores. Mutation analysis of *T. gondii* has identified cells that are resistant to the effects of ionophore A23187. Further characterization of the mutant cells should reveal gene products that help sensitize the egress process to Ca^{2+}. Invasion of host red blood cells (RBCs) by *Plasmodium* also has a Ca^{2+} sensitive component. Moreover, the infected RBC also appears to be modified in terms of its Ca^{2+} transport properties. Patch clamp studies reveal an inwardly directed Ca^{2+} current that is 18-fold greater than in the uninfected RBC. A consequence of this change is an elevation in $[Ca^{2+}]_i$ in the infected RBC.

Kinetoplastids also are invasive to host cells. *T. cruzi* makes contact by means of cell-surface glycoconjugates. Contact is associated with Ca^{2+} flux in the parasite cytoplasm. The elevated $[Ca^{2+}]_i$ appears to be important for invasion since stabilization of $[Ca^{2+}]_i$ with BAPTA or Quin-2 inhibits this process. Calmodulin antagonists are also inhibitory, but do not have enough specificity to directly implicate calmodulin in the invasion process. During invasion, contact with parasites creates a bidirectional signal. Association of host cells with specific metacyclic glycoproteins, such as gp82, is sufficient to initiate a Ca^{2+} flux. Additionally, a parasite-derived serine endoprotease (oligopeptidase B) generates a product that initiates Ca^{2+} flux in the host cells. The product appears to be recognized by a host-cell receptor that is coupled to a pertussis toxin-sensitive trimeric G protein. The subsequent production of $InsP_3$ releases Ca^{2+} from the host ER through the $InsP_3$ receptor. A consequence of this activity is disruption of the cortical actin network, and fusion of lysosomes with the plasma membrane at the site of parasite entry. Buffering of the host cell $[Ca^{2+}]_i$ with BAPTA, or depletion of intracellular pools with thapsigargin, blocks parasite entry. The fusion process involves the vesicle-associated Ca^{2+} sensor protein synaptotagmin VII. Cell entry can be blocked with antibodies against its C(2)A domain. Kinase cascades initiated by glycosylphosphatidylinositols or by the transforming growth factor β (TGFβ) receptor have also been implicated in *T. cruzi* invasion. How the various Ca^{2+} dependent and independent pathways work together to allow invasion to occur is not known.

CYCLIC NUCLEOTIDES

Cyclic nucleotides (cAMP and/or cGMP) are important second messengers in signal transduction pathways, and can be involved in a variety of functions ranging from simple environmental responses in prokaryotes and lower eukaryotes to more complex processes, such as memory storage in *Drosophila* and hormonal control of metabolism in mammals. In parasitic protozoa, particularly trypanosomes and *Plasmodium*, cyclic nucleotides have been implicated in the control of differentiation. Furthermore, many of the enzymatic components of cyclic nucleotide pathways are structurally distinct from their mammalian counterparts. Multiple isoforms of purine nucleotide cyclases, cyclic nucleotide-dependent protein kinases and phosphodiesterases contribute to further structural diversity. This arrangement facilitates the fine tuning of cellular responses, and also provides opportunities for the development of highly specific inhibitors. For example Viagra, the drug that is used to treat male impotency, is an inhibitor of a type 5 phosphodiesterase with cGMP-specific activity. The major differences that have now been identified between host and parasite enzymes involved in cyclic nucleotide signaling could therefore be exploitable in terms of drug development.

The structure and activity of mammalian adenylyl and guanylyl cyclases

Nine different mammalian adenylyl cyclase (AC) cell-surface isoforms and one soluble isoform have now been characterized, each of which has a distinct tissue distribution and regulatory specificity. All membrane forms conform to the same basic structure (Figure 11.4A); they consist of two catalytic pseuodsymmetrical domains (C1 and C2) separated by six transmembrane helices, with a further six transmembrane helices located towards the amino terminus. The two catalytic domains dimerize to form the active enzyme. Mammalian ACs have no known receptor capability. Instead the activating ligand binds to a G-protein-coupled receptor, leading to the dissociation of the G_{α} from the $G_{\beta\gamma}$ subunits in the associated heterotrimeric G-protein complex. The binding of either G_{α} or G_{β} results in the activation or inhibition of AC depending on the AC and G protein isoforms involved (Figure 11.4A). It has been estimated that each receptor may activate up to 100 heterotrimeric G-protein molecules.

In mammalian cells, cAMP has a major role in controlling gene expression and metabolism. One of the major signal transduction pathways involves the binding of cAMP to the regulatory subunits of specific cAMP-dependent protein kinases (PKA). This causes dissociation

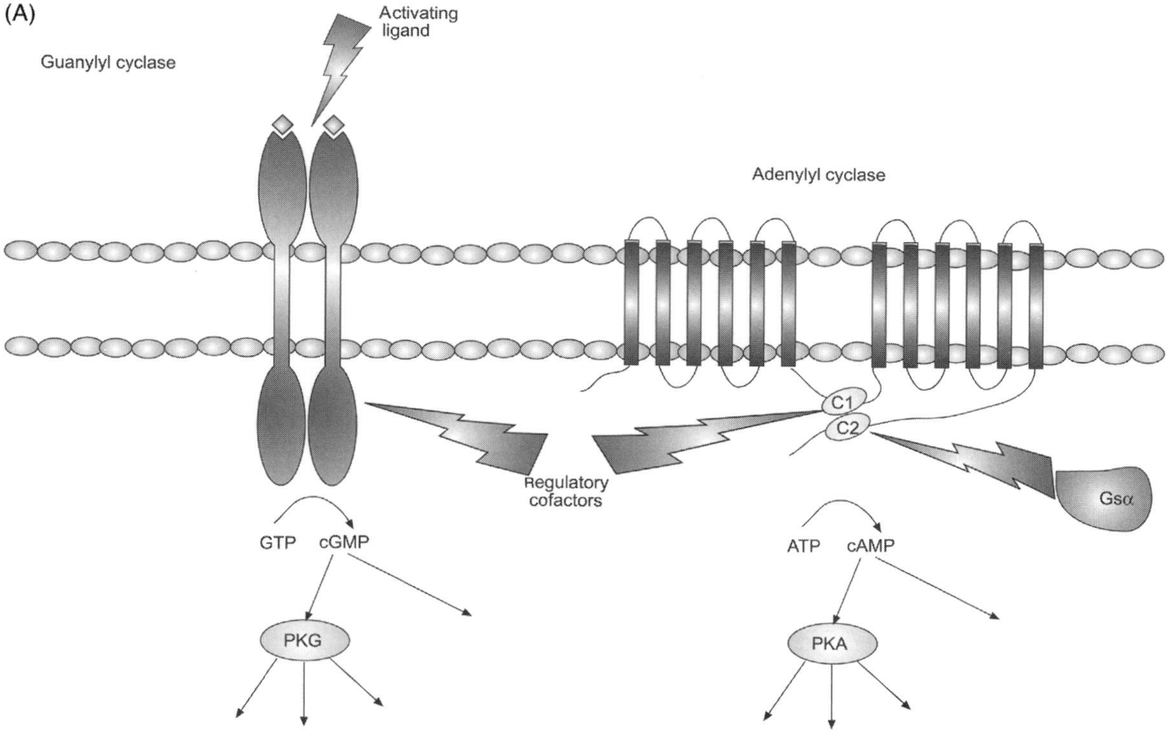

FIGURE 11.4 *(Continued on next page)*

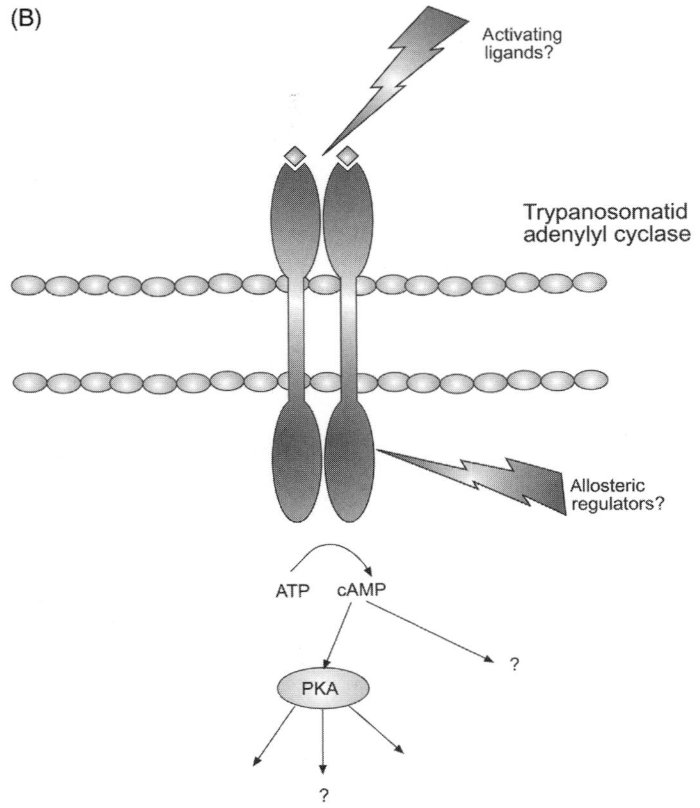

FIGURE 11.4 (See also Color Plate 6) (A) Structural representation of mammalian membrane-localized guanylyl cyclase (GC) and adenylyl cyclase (AC). The homodimeric catalytic regions of GC are activated by ligands that bind to the extracellular receptor domain. Signal transduction can then occur via cGMP-dependent protein kinase (PKG), or by alternative pathways. With AC, the activating ligands bind to G-protein-coupled receptors. This leads to the dissociation of the linked heterotrimeric G-protein complex and activation of the catalytic site of AC, by $G_{S\alpha}$ in the example shown. cAMP can act directly on proteins, or via cAMP-dependent protein kinase (PKA). The activity of both GC and AC can be fine-tuned by other cytosolic regulatory cofactors. (B) Structural representation of trypanosomatid AC. It is presumed, although not proven, that the intracellular homodimer is activated by ligands that bind to the receptor domain. In trypanosomes, the extensive repertoire of AC genes provides the potential to respond to a wide diversity of activating ligands. The precise nature of the downstream signal-transduction pathways are not known, but it is likely that PKA plays a role similar to that in mammalian cells. Structural analysis of *T.brucei* AC has indicated the possibility of allosteric regulatory sites in the catalytic domains.

of the catalytic subunits, which become active and can phosphorylate a diverse group of proteins, ranging from transcription factors to cyclic nucleotide-gated ion channels. As an example, activated PKA can be translocated to the nucleus where it phosphorylates the transcriptional activator CREB. Phosphorylated CREB then binds to the cAMP response element upstream of specific genes and enhances their transcription. cAMP-mediated activation can also occur independently of the PKA pathway as occurs, for example, with the cAMP-gated cation channels in olfactory neurons. AC activity is closely coordinated with that of phosphodiesterases, which breakdown cAMP by hydrolysis of the 3′-ester bond.

Guanylyl cyclases (GCs) exist as two main forms that are expressed in almost all types of mammalian cells. The receptor forms have a single membrane-spanning helix and an extracellular domain that is able to bind directly to the corresponding activator (Figure 11.4A). The activators can include peptide hormones or bacterial toxins, depending on the isoform. Binding of the activator to the cyclase leads to activation of the single intracellular catalytic domain that functions as a homodimer. The soluble, cytosolic form of GC exists as heterodimers of the α and β subunits, and is activated by nitric oxide in the presence of heme. cGMP signaling has a central role in numerous diverse physiological processes, including photoreceptor signal transduction, neurotransmission, and electrolyte homeostasis. In sea urchins, a membrane-bound GC found on the surface of sperm can act as a receptor for peptides released by the egg and regulate sperm chemotaxis. cGMP binds to and activates cGMP-dependent protein kinases, which then phosphorylate and regulate the activity of a number of specific proteins. In other circumstances cGMP can act directly, as in the case of the cGMP-gated ion channels in retinal cone cells. The intracellular level of cGMP is tightly regulated by the interplay between the activities of GC and cGMP-specific phosphodiesterases.

The catalytic domains of GC and AC have several shared features that reflect their common evolutionary origin. A small number of residues in the active site, located within hydrophobic pockets, are involved in conferring purine specificity; Lys and Asp in AC and Glu, Arg and Cys in the case of GC. The purine specificity of GC can be altered by changing these key residues (Glu and Cys) to their counterparts (Lys and Asp) in AC, resulting in an enzyme with AC activity. In homodimeric cyclases the polypeptide chains form two symmetrical active sites. However, with heterodimeric cyclases, such as mammalian AC, one of the adenosine binding sites interacts with allosteric activators such as the diterpene forskolin, and presumably an endogenous structurally related molecule.

Trypanosomatid adenylyl cyclases

In *Leishmania, T. brucei* and *T. cruzi,* proteins with AC activity have been predicted to conform to receptor-type cyclases, and have structures more typical of the mammalian membrane-bound GCs (Figure 11.4B). The trypanosomatid ACs have a large extracellular domain, a single transmembrane helix and a cytosolic catalytic domain. In *T.brucei* at least, ACs appear to be confined to the flagellum. No *GC* genes have been identified in the trypanosomatids.

The ACs are expressed by gene families that vary in complexity among trypanosomatid parasites. In the simplest case, *L. donovani,* the *AC* genes are organized as a cluster of five. Two of these (designated *Rac-A* and *B*) are developmentally regulated and are expressed only in the promastigote (insect) stage of the life cycle. The Rac-A protein functions as an AC, but Rac-B exhibits no cyclase activity. Interestingly, the activity of Rac-A was downregulated when co-expressed with Rac-B in *Xenopus* oocytes. The functional relevance of this observation remains to be determined. Preliminary data from the *T. brucei* Genome Project suggest that several hundred *AC*-like genes may be present, and that they fall into two main categories. Genes belonging to one group, designated

ESAG4 (expression site associated gene 4), are localized close to the ends of chromosomes in the telomeric variant surface glycoprotein (VSG) expression sites (for review, see Chapter 5). These genes are expressed specifically in the mammalian bloodstream stage of the life cycle. The second group of genes, termed *GRESAG4s* (genes related to *ESAG4*), are widely dispersed in the genome and can occur in clusters, or in single copies. These genes appear to be constitutively expressed. In *T. cruzi*, sequencing data indicate the presence of a polymorphic family of more than 30 *AC* genes. These genes are localized on at least six chromosomes and are scattered rather than clustered. *AC* pseudogenes have also been identified in *T. cruzi*.

It has been suggested that trypanosomatid ACs are activated directly by external ligands, with the extracellular domain acting as the receptor, similar to mammalian membrane-bound GCs (Figure 11.4A, B). This direct activation would negate the requirement for G-protein-dependent activation. Consistent with this, heterotrimeric G-protein subunit homologs have not yet been identified by any of the three trypanosomatid Genome Projects, and characteristic G-protein-binding sequences are absent from the trypanosomatid ACs. In contrast to the conserved catalytic domains of trypanosome ACs, the extracellular regions are extremely divergent at the sequence level. This heterogeneity may play a role in immune evasion, or may reflect functional differences between isoforms, such as ligand-binding specificity. It is also possible that both of these forces have acted during evolution to generate the diversity of the AC repertoire within the constraints necessary for maintaining receptor function.

Trypanosome ACs are inactive as monomers, and activity is dependent on dimer formation. With members of the *T. brucei* GRESAG4 family, recombinant forms of the catalytic domains can dimerize spontaneously; however this does not increase activity above basal levels. Activity is considerably enhanced, though, when a yeast leucine zipper sequence is added to the amino terminus of these catalytic domains. The leucine zipper may promote a dimer conformation that is more favorable for activity, perhaps mimicking the effects of ligand binding to the extracellular domain of the native protein. Since the catalytic domain is formed from two homologous polypeptides, the trypanosomal enzyme must have two symmetrical active sites. At each active site, binding of the ATP-complexed metal ion is contributed by one chain of the homodimer, while purine-binding, Mg^{2+}-binding and catalytic residues are contributed by the other. In this type of enzyme, the second active site corresponds to the forskolin-binding site of the mammalian heterodimeric AC. This model has been confirmed following the determination of the structure of a *T. brucei* AC catalytic domain by X-ray crystallography and by further biochemical and mutagenesis analyses. In terms of structure, the catalytic domain of the *T. brucei* AC closely resembles that of the mammalian enzyme, although it does contain an additional insertion sequence that forms a unique motif called the Δ-subdomain. The sequence of the Δ-subdomain is very highly conserved among trypanosomatid ACs. It is located in a position that corresponds to a site where regulatory cofactors interact with the mammalian AC, and with membrane-bound GCs. Most interestingly, the observation that D-DTT can bind in a stereospecific manner to a pocket below the Δ-subdomain has led to the tentative suggestion that *in vivo* this region may act as an allosteric control site, specific for other small molecules that have a regulatory role.

Other enzymic components of the trypanosomatid cAMP-signaling pathways

Few attempts have been made so far to functionally characterize other components of the trypanosomatid cAMP signal transduction pathways. This should change rapidly in the near future, stimulated by the rapid progress of the trypanosomatid Genome Projects. For example, at the time of writing, sequences corresponding to putative catalytic (*L. major* and *T. brucei*) and regulatory (*L. major*) subunits of PKA have been deposited in the database. Two types of *T. brucei* cAMP-specific phosphodiesterase have also been identified. One, which belongs to the class 1 phosphodiesterases, was isolated after complementation of yeast mutants. TbPDE1 is encoded by a single-copy gene and is expressed constitutively. The second type corresponds to the class 2 phosphodiesterases, and is encoded by a family of at least five genes, some of which are clustered. The activity of the expressed product of one of these genes (TbPDE2A) was reduced by some, but not all, of a group of phosphodiesterase inhibitors. Interestingly, the active inhibitors trequinsin, dipyridamole, sildenafil citrate (Viagra) and ethaverine also block the proliferation of bloodstream parasites, suggesting that this phosphodiesterase activity could be essential.

Possible roles for cAMP signaling in trypanosomatids

Given the abundance of *AC* genes in the *T. brucei* genome, it would seem apparent that cAMP-signaling is of major regulatory importance during the parasite life cycle. However, surprisingly little is known about the role(s) of this signaling system, the nature of the external activators, and why *T. brucei* requires such an extensive repertoire of genes. Most of the available evidence suggests a role in differentiation.

When bloodstream *T. brucei* are triggered to differentiate to the procyclic form by the addition of citrate/cis-aconitate and reduction of the ambient temperature to 27°C, two phases of transient AC activation can be detected. The first occurs after 6–10 h, and is immediately followed by the release of the surface VSGs. This pulse of AC activity also precedes the first cell division, and the loss of the surface AC isoforms encoded by the *ESAG4* genes. The second pulse of activity, after 20–40 h, coincides with the beginning of cell proliferation. Treatment of bloodstream forms of *T. brucei* to induce shedding of the VSG (low pH and trypsin digestion) also leads to activation of AC. In addition, activation of AC and release of VSG can be induced by specific protein kinase C (PKC) inhibitors, suggesting a possible inhibitory role for this regulatory kinase. Despite these findings, it appears that the simultaneous release of VSG and the activation of the cAMP pathway in response to cellular stress occur independently and are not mechanistically linked. Further evidence for the antagonistic action of PKC and the cAMP signal transduction pathway has come from a study of disaggregation of *T. brucei* bloodstream forms following exposure to anti-VSG antibodies or immune serum. Disaggregation is a regulated PKC-dependent process that can be inhibited by increased levels of cAMP.

Compelling evidence of a role for cAMP signaling in differentiation has come from studies on the development of the non-dividing stumpy forms of the parasite, that are pre-adapted for transmission to the tsetse fly. Stumpy cell induction is a response to cell density, and appears to be mediated by the release of a low molecular weight 'stumpy induction factor' (SIF). SIF causes cell-cycle arrest in the G_1/G_0 phase of the slender bloodstream forms and induces differentiation, a process that is accompanied by a 2–3-fold increase in the level of intracellular cAMP. Addition of membrane permeable cAMP analogs or phosphodiesterase inhibitors to parasite cultures also mimics the effects of SIF, suggesting that differentiation is mediated by the cAMP signaling pathway.

In *T. cruzi* cAMP also has a role in differentiation. Addition of cAMP analogs to epimastigote cultures promotes metacyclogenesis. In addition, two putative activating ligands that result in both increased cAMP levels and enhanced differentiation have been identified. One is the globin-derived factor (GDF), a small peptide which arises from proteolytic cleavage of α^D-globin within the hindgut of the insect vector. Administration of this peptide *in vitro* results in activation of parasite AC in membrane fractions and enhanced differentiation from epimastigotes to metacyclic trypomastigotes. When infected triatomine bugs are fed on plasma rather than blood, parasite differentiation is inhibited unless the GDF peptide is included. In the same way, metacyclic trypomastigotes proteolytically degrade fibronectin into several peptides, two of which appear to activate AC and initiate transition to the amastigote stage.

Although putative activators have now been identified, more detailed dissection of this signal transduction pathway has not been reported. In addition, many of the experiments used to investigate the role of the AC pathway in *T. cruzi* were designed on the assumption that the parasite enzyme would conform to the structure of mammalian G-protein-dependent ACs, and respond similarly to activators and inhibitors. As outlined above, so far this has not proved to be the case. For example, the *T. cruzi* enzyme lacks the sequences necessary for G_α and $G_{\beta\gamma}$ binding, and in biochemical assays does not respond to forskolin. The interpretations drawn from these earlier reports should now be treated with caution, and a re-evaluation of the molecular basis of cAMP-mediated signaling in *T. cruzi* is warranted.

Guanylyl cyclase activity in *Plasmodium*

In the human malaria parasite *P. falciparum*, GC activity is associated with two large proteins (PfGCα and PfGCβ) that have topological features reminiscent of mammalian membrane-localized ACs (Figure 11.5). Unusually, these proteins both appear to be bifunctional in that their amino-terminal domains have strong similarities to P-type ATPases, including a region predicted to contain ten membrane-spanning helices. The sequence and structure of the carboxy terminal regions of these proteins conforms to those expected of a mammalian AC, with two sets of six transmembrane sequences, each followed by the C1 and C2 catalytic domains. However, the amino acids that are thought to be enzymatically important, and are normally present in the C2 domain of mammalian AC, are located in the C1 domain of the *P. falciparum* proteins. Likewise, amino acids that are present in the C1 domain of the mammalian enzyme, and that have a mechanistic role, are located in the C2 domain of the malarial cyclases. In addition, the key residues in the *P. falciparum* enzyme that determine purine specificity are characteristic of GC.

Biochemical analysis of the C1 and C2 domains has confirmed that they have GC, but not AC, activity. In addition, some of the residues in the cation-binding helices of the ATPase domain, including an aspartate essential for ion transport, are not present. As a consequence, it has been postulated that these proteins may associate with organic molecules rather than inorganic ions. Therefore the ATPase

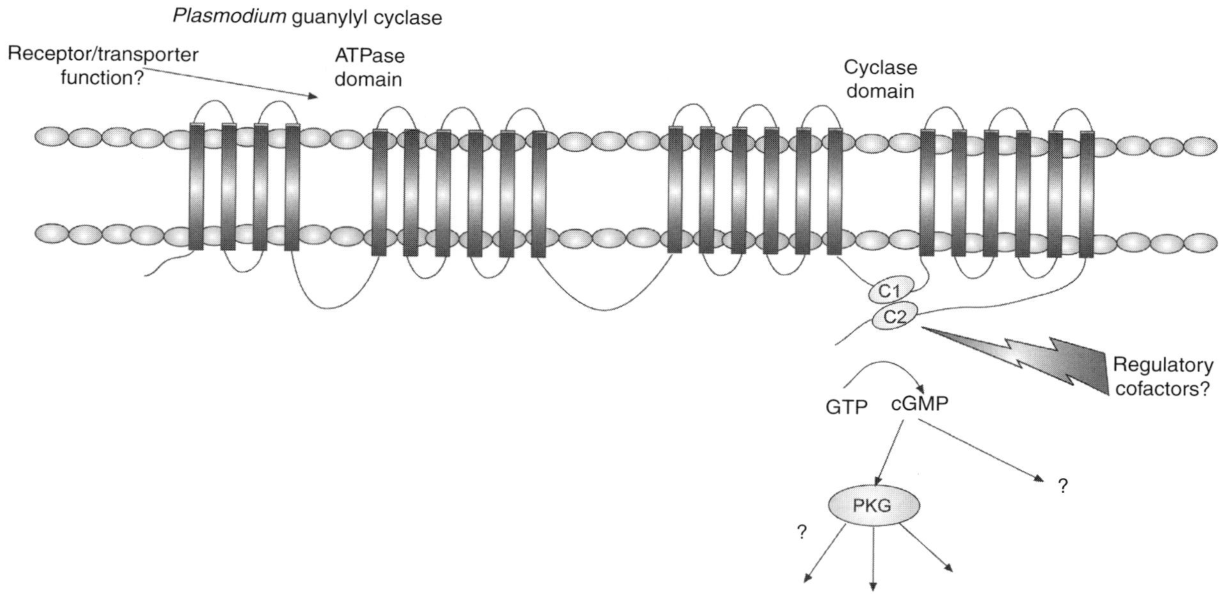

FIGURE 11.5 (See also Color Plate 7) Model for the structure of *Plasmodium* guanylyl cyclase (GC). The amino acid sequences of the *P. falciparum* GCs (PfGCα and PfGCβ) suggest that they are bifunctional integral membrane proteins. The amino terminal regions have similarities to P-type ATPases of other organisms and the topology of the carboxy terminal regions conforms to that of mammalian G-protein-dependent adenylyl cyclases (Figure 11.4A). The function of the ATPase domain remains to be determined, but the catalytic regions (C1 and C2) exhibit GC activity. cGMP-dependent protein kinase (PKG) is thought to have a role in signal transduction, although the mechanisms involved have not been characterized. Similarly, little is known about regulatory cofactors that could act directly on the GC catalytic regions.

domain may act as a receptor and be functionally, as well as physically, linked to the GC protein. Cyclases with the unusual topology of the *Plasmodium* GC have also been identified in the taxonomically related ciliates *Paramecium* and *Tetrahymena.*

In *P. falciparum,* expression of both GC mRNAs was detectable in the sexual, but not the asexual, stages of the life cycle, and PfGCα has been localized to the parasite/parasitophorous vacuole membrane region. In agreement with this observation, GC activity is associated predominantly with the membrane fractions of mature gametocytes. GC activity is Mg^{2+}/Mn^{2+} dependent, but can be inhibited by Ca^{2+}. From a functional perspective it is not immediately obvious why the malaria parasite has two proteins with the same activity that are both expressed during the sexual stages of the life cycle. One possibility is that they are expressed differentially during this complex life-cycle phase. Another is that they are co-expressed, but that they respond to different activators.

Both PfGCα and PfGCβ are encoded by large single-copy genes that differ in an interesting way. The *PfGCα* sequence is uninterrupted by introns, but *PfGCβ* contains 13, with all but one being confined to the ATPase domain (Figure 11.6). It can be presumed that the progenitor PfGC protein arose from the fusion of two

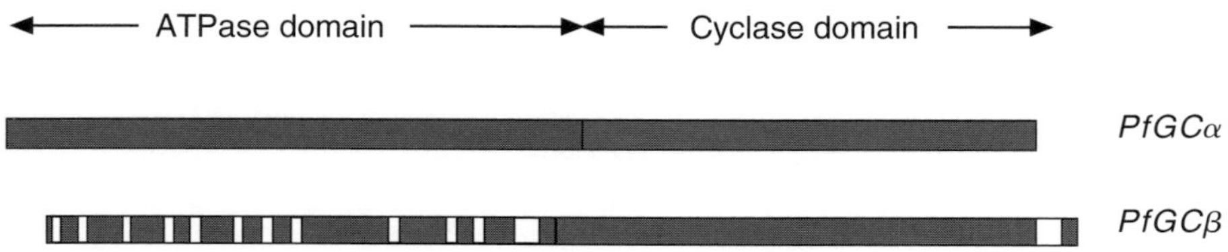

FIGURE 11.6 (See also Color Plate 8) Structural organization of the bifunctional genes encoding guanylyl cyclase in *P. falciparum*. *PfGCα* is uninterrupted by introns, but there are 13 (shown in white) in *PfGCβ* and all but one of these are in the ATPase domain.

genes, one encoding a P-type ATPase and the second encoding a cyclase. There are at least two other examples in *Plasmodium* of bifunctional proteins that occur as distinct molecules in other organisms; dihydrofolate reductase–thymidylate synthase and dihydropteroate synthetase–pyrophosphokinase. The fusion event that gave rise to the bifunctional proteins with GC activity almost certainly occurred in a common ancestor of *Plasmodium, Paramecium* and *Tetrahymena,* in agreement with the classification of the ciliates with the apicomplexans and dinoflagellates in the Alveolata. In *Plasmodium,* this initial fusion event appears to have been followed by gene duplication to create two copies that then diverged to become *PfGCα* and *PfGCβ*. Although it cannot be excluded that the progenitor gene contained introns, their absence from the *Paramecium* and *Tetrahymena* genes suggests that the most likely explanation for their presence in *PfGCβ* is that they have been added over time. Why they are found in only one of the *Plasmodium* genes, and why they are predominantly restricted to the ATPase domain, is a puzzle.

Other enzymatic components of the cyclic nucleotide signaling pathways in *Plasmodium* have been less well characterized. Two putative ACs have been identified in the genome database and G-protein-independent AC activity, distinct from that of red blood cells, has been detected in *P. falciparum* asexual blood stages. Putative homologs of both the catalytic and regulatory subunits of PKA have also been identified. Interestingly, the anti-malarial drug halofantrine is a potent and specific inhibitor of the catalytic subunit of mammalian PKA. The possibility that interaction between halofantrine and the parasite PKA contributes to anti-malarial activity should be worth investigating. Sequences corresponding to *P. falciparum* phosphodiesterases and a cGMP-dependent protein kinase (PKG) can also be found in the database. With the completion of a fully annotated *P. falciparum* genome sequence in the near future, it can be confidently expected that the entire repertoire of proteins involved in cyclic nucleotide-mediated signaling will be available. This should provide a framework for the detailed functional dissection of these signal transduction pathways.

Cyclic nucleotide signaling and differentiation of the malaria parasite

Cyclic nucleotides have been implicated in the triggering of differentiation in *Plasmodium,* although limited progress has been made in dissecting the precise mechanisms involved. For example, addition of membrane-permeable analogs of cAMP to cultures of *P. falciparum* that contain a high proportion of asexual erythrocytic parasites induces the development of male and female gametocytes (gametocytogenesis), an obligate step in the life cycle. In apparent contradiction of this observation, however, gametocyte producer and non-producer clones both have the same basal levels of cAMP, although PKA activity in the non-producer clone was significantly lower. cGMP has been implicated in exflagellation, a process that occurs in the mosquito midgut when eight flagellated male gametes erupt from a single gametocyte-infected red blood cell. Exflagellation is enhanced when either cGMP analogs or phosphodiesterase inhibitors are added to cultures of mature gametocytes.

Exflagellation can be triggered *in vitro* by a decrease in temperature and a bicarbonate ion-mediated increase in pH. Exflagellation can also be induced at a non-permissive pH in the absence of bicarbonate ions by a gametocyte-activating factor which has now been identified as xanthurenic acid, a product of tryptophan metabolism. Significantly, addition of xanthurenic acid to mature *P. falciparum* gametocyte membrane preparations enhances GC activity. These observations provide a link between an external stimulator of exflagellation and activation of a signal transduction pathway associated with this essential developmental process. Addition of xathurenic acid to the recombinant catalytic domains of PfGCβ, either individually or in combination, had no significant stimulatory effect on GC activity, suggesting that the mode of action probably does not involve direct interaction with the catalytic domains.

REVERSIBLE PROTEIN PHOSPHORYLATION

Reversible protein phosphorylation is widely used by eukaryotic cells to transmit signals. Phosphorylation–dephosphorylation of serine, threonine and tyrosine residues can trigger remarkable changes in protein conformation. The conformational changes, in turn, alter a variety of properties such as catalytic activity, intracellular localization, interaction with other proteins, and degradation. It is reasonable to predict that a wide range of protein kinases and phosphatases will function as molecular switches regulating various cellular activities. Recently, the complete sequencing of various genomes has shown the veracity of this prediction (Table 11.1). The yeast genome can be used as a gauge to roughly estimate the number of protein kinases and phosphatases that parasites might employ to regulate intracellular signaling.

TABLE 11.1 Number of protein kinases and phosphatases from different species

Type	Yeast (*S. cerevisiae*)	Fly (*Drosophila*)	Worm (*C. elegans*)	Human
Serine/threonine or dual-specificity kinase	114	198	315	395
Tyrosine kinase	5	47	100	106
Serine/threonine protein phosphatase	13	19	51	15
Tyrosine phosphatase	5	22	95	56
Dual-specificity protein phosphatase	4	8	10	29
Number of genes	6000	13 000	19 000	30 000

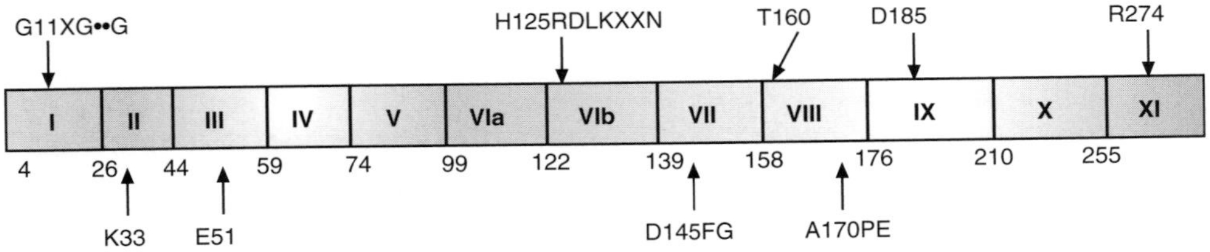

FIGURE 11.7 Structure of protein kinase catalytic domains. The 12 subdomains of the catalytic subunit are shown, with residues important for catalytic activity indicated by arrows.

Structure of protein kinases

All protein kinases contain a specific domain that catalyzes transfer of the γ-phosphate group from ATP or GTP to the acceptor hydroxyl group of serine, threonine or tyrosine residues. The catalytic domain and its key catalytic residues are well conserved amongst various kinases. The catalytic domain folds into a two-lobed structure that consists of twelve subdomains (Figure 11.7). The smaller amino terminal lobe contains subdomains I–IV and the larger carboxy terminal lobe contains subdomains V–XI. ATP or GTP binds deep inside the cleft between the two lobes, and is anchored by the GXGXXGXV (glycine loop) consensus sequence in subdomain I. The larger carboxy terminal lobe is involved in peptide substrate recognition and catalysis. Subdomain II contains an invariant Lys that is involved in orientation of the α and β phosphate groups of the purine nucleotide. A salt bridge is formed between this Lys and a conserved Glu in subdomain III. Subdomain VIB contains the HRDLKXXN consensus motif of the catalytic loop. The Asp within this consensus motif coordinates phosphate group transfer, while the Arg within this motif stabilizes the loop. Subdomain VII also folds into a β strand-loop-β strand structure similar to subdomain VIB. The conserved DFG motif is located in subdomain VII. The Asp residue in the DFG motif helps to coordinate with the Mg^{2+} or Mn^{2+} that connects with β- and γ-phosphates of the purine nucleotide. The DFG motif is also at the beginning of the activation domain for many kinases. The activation segment usually ends with the APE motif that faces the cleft region in subdomain VIII. The invariant Arg in subdomain XI stabilizes the large carboxy terminal lobe.

Mitogen-activated protein (MAP) kinase-mediated signaling

MAP kinases (MAPKs) are involved in signal cascades that typically activate gene transcription and various other cellular events in response to extracellular stimuli (Figure 11.8). Metazoans and yeast contain a family of MAP kinases for this purpose. External signals responsible for MAPK activation associate with catalytic receptor kinases, or with serpentine receptors coupled to heterotrimeric G proteins. Downstream of receptor activation, a variety of proteins can be used to stimulate the MAP kinase cascade, including Ras or Rac-like small GTP-binding proteins, or STE20 kinases. The catalytic and serpentine receptor pathways converge by each activating Ras.

MAP kinases become activated when they are phosphorylated on threonine and tyrosine residues within the activation region of subdomain VIII. These residues correspond to T183 and Y185 of the extracellular signal regulated protein kinases 2 or ERK2, and p38. In the

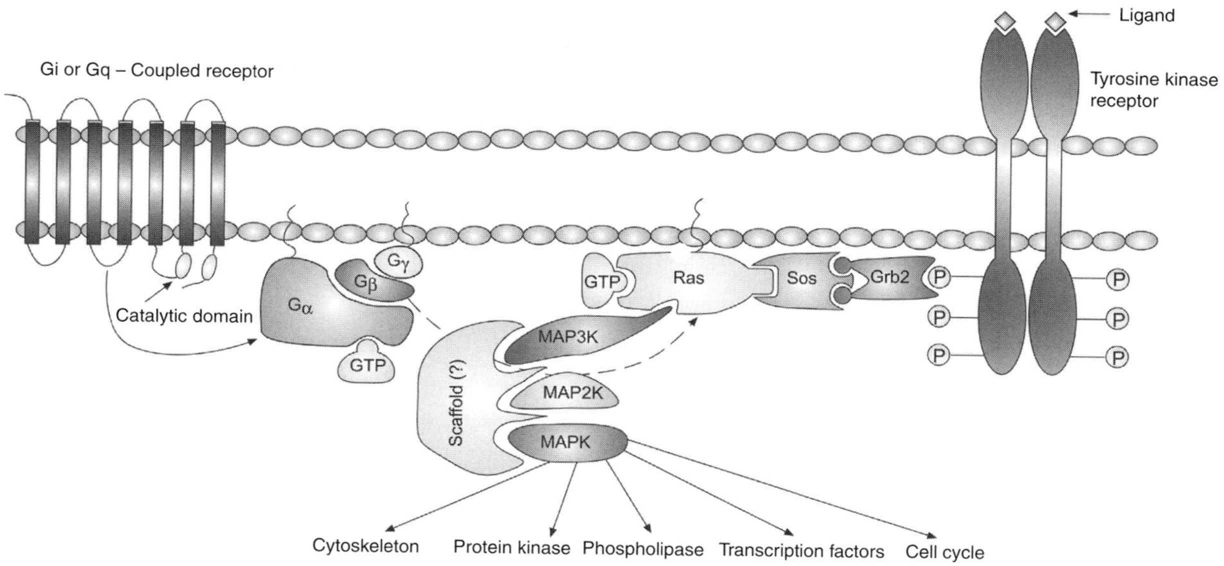

FIGURE 11.8 (See also Color Plate 9) Activation of the MAPK pathway by receptor tyrosine kinases and seven transmembrane or serpentine receptors. Binding of a ligand to receptor activates guanine nucleotide exchange factor (GEF) Sos through the adaptor protein Grb2. Sos then activates membrane-bound Ras by exchanging GDP with GTP. The activated Ras, through an unknown mechanism, recruits and activates MAP3K and thereby the MAPK module. A scaffold protein, such as Ste5 from the *S. cerevisiae* mating response pathway, often brings the individual members of the MAPK module together. Recently, similar scaffold proteins, MP1 and JIP-1, have been discovered in mammalian cell stress response pathways. The MAPK module can also be activated by signals received through serpentine receptors. Binding of a ligand to a serpentine receptor activates membrane-anchored heterotrimeric G-protein. The exchange of GDP with GTP separates G_α and G_β subunits. The MAPK module is then recruited to the free G_β subunit through additional intermediary proteins. For examples, in *S. cerevisiae* Ste20, a MAP4K acts to recruit the MAPK module to G_β. Activated MAPKs then phosphorylate a variety of cellular substrates.

unphosphorylated state the two lobes of MAPKs are at a distance that leads to misalignment of amino acid residues in the active site. Following dual phosphorylation by MAP kinase kinases (MAPKK), there is a conformational change in the activation domain that leads to proper alignment of residues involved in catalysis. Each MAPKK is in turn activated by serine/threonine phosphorylation in subdomain VIII by MAPKK kinases (MAPKKK).

It is now well established that the three-kinase module (MAPKKK-MAPKK-MAPK) is used to activate the mammalian and yeast MAPK signaling pathway. Different sets of stimuli activate different modules. Various isoforms have been identified for each member of these modules. Scaffold proteins that anchor various components present in a module provide the specificity of activation of a signaling pathway. For example, in yeast, the signaling pathway for mating response uses the scaffold protein Ste5 which binds the module proteins Ste11 (MKKK), Ste7 (MKK), and Fus3 (MAPK). However, in the osmosensing pathway, PBS2 acts both as the scaffold and MAPKK, anchoring the same MAPKKK (Ste11) but recruiting a different MAPK (Hog1). MAPKKKs are usually diverse in structure, allowing them to interact with different groups of proteins. The regulatory regions of MAPKKK may include pleckstrin homology (PH) domains, leucine-zipper dimerization sequences, src homology 3 domains (SH3), or association sites for GTP-binding proteins. This heterogeneity of MAPKKK regulatory domains allows MAPK modules to respond to an assortment of stimuli.

Downstream of MAPK activation is the regulation of gene expression brought about by phosphorylation of a variety of transcription factors such as Ets1, c-Jun, c-myc, Elk1, ATF-2, Max, p53 and NFAT4, among many. Although control of gene expression through phosphorylation is an important function of MAPKs, the pool of active MAPK does not completely translocate to the nucleus. Active MAPKs also regulate cytoplasmic proteins such as $p90^{srk}$ S6 kinase, MAPKAP kinase, or carbamoyl phosphate synthetase. The phosphorylation site of MAPK substrates usually contains a serine or threonine followed by proline at the P+1 position, hence MAPKs are known as proline-directed kinases. Sometimes a proline at the P−2 site is also present. Additional domains also are involved in substrate recognition by MAPKs: the D domain composed of basic and hydrophobic residues, and the FXFP sequence. It would be an important and interesting endeavor to characterize downstream processes of MAPK-mediated signaling pathways in protozoan parasites.

MAP kinase activation in protozoan parasites

MAPK homologs have been identified in many parasites including *Plasmodium, Trypanosoma, Leishmania,* and *Giardia.* How these structures might be activated is not known. While tyrosine kinase activity is present in these organisms, only indirect evidence is available to suggest catalytic receptors might also be present. Serpentine receptors have not been identified. Heterotrimeric G-protein activity has been hinted at with antibodies, labeled GTP analogs or ADP-ribosylation with cholera toxin or pertussis toxin. Indirect evidence for G_α exists in *T. brucei, T. cruzi, P. falciparum* and *T. gondii.* However, no genes corresponding to G_α have been cloned. Nonetheless, the treatment of asexual malaria parasites with cholera toxin induces gametocytogenesis, implicating G-protein-coupled signals in the sexual differentiation process. Ras-mediated signaling is an important component of MAPK activation. Genes coding for Ras have been identified in *Entamoeba.* Two isoforms of *Entamoeba* Ras, EhRas1 and EhRas2, are 47% similar to human Ras. Both EhRas1 and EhRas2 carry the signature cysteine-aliphatic-aliphatic-any amino acid (CAAX) prenylation sequence, and the prenylation of EhRas1 has been observed *in vitro.* Inhibitors of prenyl transferases are lethal to *T. brucei* and *P. falciparum.* Protein prenylation has also been demonstrated in *Giardia, T. brucei,* and *P. falciparum.*

MAP kinase pathways in apicomplexans

In *T. gondii,* studies with heterologous ERK antibodies and those using gel kinase assays suggest the presence of MAPK-like proteins in cell homogenates. In *P. falciparum* two putative MAPK homologs have been identified. PfMAP-1 contains MAPK conserved sequences, including the TXY activation motif and a large carboxy terminus extension of highly charged tetra- or octapeptide repeats. Recombinant PfMAP-1 autophosphorylates both the tyrosine and threonine residues in this motif, increasing catalytic activity to a small extent. The 826 amino acid residue PfMAP-1 has the potential to generate a 100 kDa protein. Interestingly, Western blot analysis indicated the presence of truncated isoforms of 40 and 80 kDa that are tyrosine phosphorylated. The 40 kDa isoform could be a proteolytic degradation product, and the 80 kDa form could be the full-length protein that migrates anomalously on gels. PfMAP-1 expression is detected in both asexual and sexual stages. PfMAP-2 appears to be a novel protein, although it contains significant homology to MAPKs. Recombinant PfMAP-2 exhibits typical MAPK properties such as autophosphorylation, phosphorylation of myelin basic proteins, and the reduction in activity following treatment with MAPK-specific phosphatases. However, the signature TXY activation sequence is substituted in PfMAP-2 with TSH. Replacement of the threonine in TSH with alanine abolishes autophosphorylation of PfMAP-2, suggesting a role for this residue in catalysis. PfMAP-2 expression is specific for gametocytes.

In a search for the MAPK-activating kinases MAPKK or MEK, the *P. falciparum* genome-sequence database has been searched using the conserved activation site sequences found in MEK1 and MEK2. A 996-amino acid protein with a long carboxy terminus extension and many conserved residues of the MEK family was identified. The putative Pfnek-1 catalytic domain however, shows highest homology to NIMA (never in mitosis)-like kinases, a family of cyclin-independent kinases required for progression into mitosis. Pfnek-1 phosphorylates PfMAP-2, but not PfMAP-1. Pfnek-1 also stimulates the ability of PfMAP-2 to phosphorylate myelin basic protein *in vitro*. Both PfMAP-2 and Pfnek-1 are expressed in gametocytes, suggesting that Pfnek-1 may be involved in PfMAP-2 activation during sexual stage differentiation.

MAP kinase pathways in kinetoplastids

More direct proof of a role for MAPKs in parasite differentiation comes from *Leishmania mexicana*. Deletion of the tandemly repeated secreted acid phosphatase (SAP) gene locus (*Δlmsap1/2*) curtails the transformation of promastigotes into amastigotes within macrophages. Analysis of the *lmsap1/2* locus revealed that a MAPK homolog (termed LMPK) is encoded by the intergenic region between the *sap* genes. LMPK contains signature MAPK sequences. Interestingly, the *lmpk* gene is sufficient to complement the *Δlmsap1/2* null mutation, as evidenced by progression of infection and appearance of amastigotes in transfected cells. Although the LMPK protein is expressed in both promastigotes and amastigotes, LMPK activity is found only in amastigotes, suggesting that appropriate activators, such as MAPKKs, are present only in amastigotes.

In *T. brucei*, a MAPK homolog, termed KFR1, with a strong homology to yeast MAPK homologs *KSS1/FUS3*, has been cloned. The *KFR1* gene is presumed to be essential to *T. brucei* procyclic forms, since viable cells could not be obtained when the gene was disrupted at both alleles. Recombinant KFR1 can utilize myelin basic protein as a substrate *in vitro*, in common with other MAPKs. Phosphorylation of KFR1 is important for activity, and dephosphorylation by MAPK (ERK1 and ERK2)-specific dual phosphatase abolishes activity. The specific activity of KFR1 varies during the life cycle, and is higher in bloodstream forms than procyclic forms. When an extra copy of KFR1 is introduced into procyclic forms, an increased abundance of *KFR1* transcripts is observed. However, the amount of KFR1 protein remains constant, indicating that the level of KFR1 is translationally controlled. Therefore, trypanosomes exert two levels of control over KFR1: regulating the amount of KFR1 by translational control, and the activity of KFR1 by phosphorylation. Serum starvation of bloodstream *T. brucei* causes rapid loss of KFR1 activity concomitant with a decrease in KFR1 phosphorylation. Activity and phosphorylation could both be restored gradually when cells were supplemented with serum or with interferon-γ (IFN-γ), but not when supplemented with platelet-derived growth factor or basic fibroblast growth factor. In mammalian cells, IFN-γ can activate receptor-associated tyrosine kinases such as Janus kinase (JAK). It is not known whether trypanosomes utilize JAK pathways to phosphorylate KFR1 in response to serum. However, IFN-γ levels rise in the mammalian host following infection with *T. brucei*, suggesting that the parasite may take advantage of its environment to foster proliferation.

MAP kinases from *Giardia*

Genes for ERK1 and ERK2 homologs (gERK1 and gERK2) have been recently cloned from *Giardia*, and the recombinant proteins phosphorylate myelin basic protein. gERK1 and gERK2 showed differential activity and subcellular distribution during the transition from trophozoites to cysts. This hints at a possible role of *Giardia* MAPKs in differentiation.

Overall, it is apparent that, as in mammalian cells and yeast, MAPKs have a major role in regulating parasite stage-specific growth and differentiation such as gametocytogenesis in *Plasmodium*, maturation into amastigotes following infection in *Leishmania*, proliferation of the bloodstream form of *Trypanosoma*, or encystation in *Giardia*.

Cyclin-dependent kinases (CDKs)

Cyclin-dependent kinases (CDKs) are key regulators of cell-cycle progression in all eukaryotes. These kinases serve as molecular switches that regulate cell-cycle transitions. CDKs are heterodimers of the catalytic kinase subunit complexed with the regulatory cyclin subunit. Cyclins exhibit periodicity of expression due to cell-cycle-dependent degradation. The monomeric catalytic subunit of CDK is inactive. The catalytic cleft is blocked by a 29-residue region known as the 'T-loop'. The T-loop resides within subdomain VII and VIII between Asp146 and Glu173. Binding of cyclin causes a number of critical changes in the CDK structure and function. A domain in cyclin known as the 'cyclin box' interacts with the T-loop and the canonical PSTAIRE motif of CDKs in subdomain III. This interaction causes reorientation of the T-loop so that the catalytic pocket is opened to the protein substrate and ATP. Additionally, the PSTAIRE helix reorients so that Glu55 forms a salt bridge with Lys33. At the same time, Asp145 moves into position so that it can coordinate with Mg^{2+} and γ-phosphate. Although cyclin binding alone restores catalytic activity of CDK, full activity requires phosphorylation at Thr160 by CDK-activating kinases (CAKs). Thr160 is buried in CDK until a complex with cyclin is formed. However, the binding of cyclin to CDK is independent of Thr160 phosphorylation. Following phosphorylation of Thr160, CDK shows increased flexibility in the glycine loop and the activation region. The inactivation of CDK is caused by the reversal of cyclin binding as well as by phosphorylation of Thr14 and Tyr15 in the glycine loop. In most eukaryotes there are several CDKs that associate with cyclins to regulate cell cycle transition. However, in yeast only one CDK homolog is present, and its specificity is governed by interaction with different cyclins. Timing of cyclin expression is therefore a critical component of cell-cycle regulation. For example, cyclin D is synthesized in G_1 and binds to CDK4 and CDK6 as cells are exiting the quiescent stage. Cyclin E–CDK2 and cyclin A–CDK2 complexes on the other hand are detected during transition of cells from G_1 to S phase.

CDKs and cyclins in kinetoplastids

A number of genes for CDK-like kinases have been identified in protozoan parasites. However, the exact physiological roles for many have yet to be determined. Since parasitic protozoa have complex life cycles, different CDKs might be used to regulate stage-specific events. For example, trypanosomatids contain a family of CDK-related kinases referred to as CRKs. *L. mexicana* CRK1 (LmmCRK1) has significant homology with mammalian and yeast CDK2s (CDC28 in budding yeast or cdc2 in fission yeast). However, the PSTAIRE cyclin-binding domain in LmmCRK1 is changed to PCTAIRE and LmmCRK1 fails to complement fission yeast temperature-sensitive cdc2 mutants. The activity of LmmCRK1 appears to be post-translationally regulated since the protein is detected at all stages, while the histone H1 kinase activity is present only in promastigotes and metacyclics, but not in amastigotes. Attempts to generate the *lmmcrk1* null phenotype have been unsuccessful, suggesting that LmmCRK1 has an essential function. The CDK–cyclin complex also contains additional proteins such as CKS (cyclin-dependent kinase subunit) that interact with substrates to stimulate kinase activity. The fission yeast homolog of CKS is known as $p13^{suc1}$. LmmCRK1 does not interact with the fission yeast $p13^{suc1}$. However, another CDK-related leishmanial kinase, LmmCRK3, which contains PQTALRE instead of the PSTAIRE sequence binds to fission yeast $p13^{suc1}$ and the *Leishmania* homolog $p12^{cks1}$. The association of CRK3 with $p13^{suc1}$ or $p12^{cks1}$ is reduced in extracts from metacyclic forms (cell cycle-arrested) compared with promastigote or amastigote forms. This reduced binding also correlates well with histone H1 kinase activity. LmmCRK3 also appears to be essential since attempts to disrupt its locus result in an increase in ploidy. The LmmCRK3 kinase activity of promastigotes appears to be required mainly during the G_2/M transition.

Three well studied *T. brucei* CRKs have substitutions in the canonical PSTAIRE sequence. TbCRKs 1–3 contain PCTAIRE, PSTAVRE, and PQTALRE motifs, respectively. An additional five CRK genes have been tentatively identified in *T. brucei*. TbCRK4, unlike other CRKs, contains large insertions in the catalytic domain and is expressed at all life-cycle stages. The kinase domain of TbCRK5 has strong similarity with MOK (MAPK/MAK/MRK overlapping kinase).

Three cyclin genes have been identified in *T. brucei*. *CYC1* was initially thought to be a mitotic cyclin. However, recent reports indicate that the original sequence was incorrect and CYC1 does not exhibit mitotic cyclin characteristics. Two additional *T. brucei* cyclin genes, *CYC2* and *CYC3*, have been isolated by complementation of yeast G1 cyclin mutants. *T. brucei* appears to contain eight cyclin homologs (*CYC2–9*). CYC2 is related to the budding yeast PHO80 cyclin. The ty-epitope tagged CYC2 associates with *L. mexicana* $p12^{cks1}$. Histone H1 kinase activity was detected in tyCYC immunoprecipitates, suggesting its role in the formation of an active CDK–cyclin complex. TbCYC3 shows homology to mitotic B-type cyclins. Both TbCYC2 and TbCYC3 are expressed in all stages, and an interaction of TbCRK3 with both TbCYC2 and TbCYC6 has been demonstrated by yeast two-hybrid screening. TbCYC6 showed significant sequence similarity with mitotic cyclins, and complements budding yeast *CLN123*$^-$ mutant. Functional knockout of TbCYC6 with RNA interference (RNAi) demonstrated that TbCYC6 is essential for mitosis in both procyclic and bloodstream forms of the parasite. Interestingly, in procyclic forms, the absence of TbCYC6 causes cytokinesis in the absence of mitosis. Consequently, cells with 1N2K (1 nucleus, 2 kinetoplasts) produce 1N1K daughter cells and a zoid (0N1K). However, in bloodstream forms the absence of TbCYC6 results in blocked mitosis and cytokinesis without affecting kinetoplast replication and segregation. These results suggest developmental stage-specific checkpoint control mechanisms in *T. brucei*.

CDKs and cyclins in Plasmodium

Similar to the trypanosomatids, several CDK homologs have been cloned from *P. falciparum*. However, the physiological functions of these kinases are not clear yet. PfPK5 contains a PSTTIRE motif instead of PSTAIRE and is the most likely CDK-like kinase candidate to drive the *Plasmodium* cell cycle. PfPK5 phosphorylates histone H1 *in vitro* and localizes to the nucleus at the beginning of nuclear division. Interestingly, PfPK5 shows robust activation by human cyclin H and p25 (CDK5-activating cyclin). The activation by cyclin H is surprising, since cyclin H is thought to be a partner of CDK-activating kinase CDK7 and not CDK1. Furthermore, p25 and cyclin H are equally effective in stimulating PfPK5 activity. It is widely known that p25 is quite specific about its preference for CDK5. In addition, cyclin H association also activates CDK5. These results provide strong evidence that PfPK5 is an ortholog of CDK5 and is promiscuous for its cyclin partner, in common with yeast Pho85 kinases. A *P. falciparum* cyclin homolog PfCYC1 also strongly stimulates PfPK5 activity. Similarly, a *Xenopus* protein RINGO (rapid inducer of G_2/M progression in oocytes) also strongly stimulates PfPK5 activity. RINGO has no sequence similarity to cyclins, but can bind and activate Cdc2. In contrast, human CDK inhibitor (CKI) $p21^{CIP1}$ inhibits PfPK5 activity *in vitro*.

Another CDK-related kinase in *P. falciparum*, PfPK6, contains a SKCILRE sequence in place of PSTAIRE. PfPK6 has characteristics of both CDKs and MAPKs, localizes mainly to the cytoplasm, and has a preference for Mn^{2+}. However, unlike PfPK5, PfPK6 appears to be a cyclin-independent kinase and shows strong autophosphorylation and histone H1 kinase activity in the absence of a cyclin. The IC_{50} values of PfPK6 to the CDK inhibitor rescovitine are close to MAPKs rather than CDKs. PfPK6 appears to interact with translational elongation factor-1α (EF-1α). EF-1α can interact with the cytoskeleton by binding and bundling actin filaments and microtubules, suggesting that PfPK6, rather than directly regulating cell-cycle transition, may have a secondary role in cell-cycle control through its involvement in translation regulation or cytoskeletal organization.

Additional CDK-related kinases in *Plasmodium* include Pfcrk1, which contains an AMTSLRE motif similar to PITSLRE in $p58^{GTA}$. PfCRK1 has been implicated in sexual stage development as the transcript accumulates in gametocytes. PfMRK uses NFVLLRE and is similar to NRTALRE from the CDK-activating kinase (CAK) CDK7. A complex between PfMRK and human cyclin H stimulates histone H1 kinase activity. Interestingly, PfMRK/human cyclin H does not activate PfPK5 suggesting that either PfMRK has a substrate preference different from PfPK5 or that it is not a CAK. All CDK-like kinases from *P. falciparum* autophosphorylate, unlike CDKs from other species.

In an effort to identify parasite intracellular targets of the purine CDK inhibitors, purvalanol was coupled to an agarose matrix. Surprisingly, casein kinase 1 (CK1) was the main protein that bound to the matrix in extracts of *P. falciparum, L. mexicana, T. cruzi,* and *T. gondii*, in contrast to results from metazoans. Predicted amino acid sequences from *P. falciparum* and *T. cruzi* CK1 genes show strong identity with known sequence isoforms, and recombinant proteins showed properties characteristic of

CK1. These intriguing results suggest that the parasite CK1s exhibit high affinities towards purvalanols. *L. donovani* CK1 in promastigotes is an ecto-protein kinase that is shed into the medium.

Protein phosphatases

Protein phosphatases are required to finely regulate the duration and magnitude of protein phosphorylation, and consequently to modulate the activation/deactivation of the signaling processes. Protein phosphatases are divided into three types: serine/threonine phosphatases specific for phosphorylated serine and threonine residues, tyrosine phosphatases specific for phosphotyrosine residues, and dual-specificity phosphatases that dephosphorylate phospho-serine/threonine and phosphotyrosine residues. All these phosphatases have unique structural characteristics and can be divided into different subfamilies. Phosphoserine/threonine-specific phosphatases can be subdivided into PPP and PPM families. Protein phosphatases PP1, PP2A, PP2B (calcineurin), PP4, PP5 and PP6 belong to the PPP subfamily. The catalytic domains of these phosphatases are quite diverse and often comprise the majority of the catalytic subunit. The PPM subfamily includes Mg^{2+}-dependent PP2C phosphatases. Substrate specificity of these phosphatases is controlled by a diverse group of regulatory subunits. All protein tyrosine phosphatases (PTPs) contain a conserved cysteine in the catalytic domain that is important for activity. Other regions in the PTPs are important for subcellular localization. PTPs can be integral membrane proteins, cytosolic type, or non-transmembrane type. The dual-specific phosphatases include vaccinia virus VH1 protein and its mammalian homolog VHR, and MAP kinase phosphatases (MKPs), among others.

Protein phosphatases from apicomplexans

Compared to efforts devoted to protein kinases, studies on protein phosphatases in protozoan parasites are not as extensive. Nonetheless, a wide variety of phosphatases appear to be used to regulate the growth and differentiation of medically important parasites. In *P. falciparum*, inhibitors of PP1 and PP2A (okadaic acid and calyculin, respectively) block intraerythrocytic growth. Since erythrocytes also have significant phosphatase activity, it is not clear whether this inhibition of parasite growth is due, to some extent, to inhibition of red cell metabolism. PfPP2A-like activity in *Plasmodium* is sensitive to low nanomolar concentrations of okadaic acid. In contrast, the PP2B-like activity is inhibited by the cyclosporin–cyclophilin complex, but is insensitive to okadaic acid. The peptide sequence derived from amino terminal microsequencing of a putative 35 kDa PfPP2A shows homology with PP2A subfamily sequences in the database. Using the amino terminal peptide sequence, a 309-residue PP2A ORF has been identified from the malaria genome database. The PfPP2A sequence contains all the signature motifs of serine/threonine phosphatases, including the GDXHGQ, GDXVDRG and RGNHE sequences important for catalysis, substrate binding, or metal-ion binding. PfPP2A also contains the SAPNYCYRCG motif essential for high-affinity binding of okadaic acid.

Another *P. falciparum* PP2A-like phosphatase, termed PP-β, has a catalytic domain simlar to PfPP2A. However, PfPPβ contains an amino terminal extension and is detected only in the sexual stages of the parasite. Recently, a novel protein phosphatase, PfPPJ, of the PPP superfamily has been characterized. The predicted primary amino acid sequence of PfPPJ contains signature sequences of the PPP superfamily, but has a unique carboxy terminal sequence and unique sensitivity to inhibitors.

PfPPJ is not sensitive to okadaic acid, microcystin-LR, and the PP1-specific mammalian heat-stable inhibitor-2. These observations suggest that PfPPJ does not belong to either the PP2A or PP1 class. Additionally, PfPPJ does not require calmodulin or Ca^{2+} for its catalytic activity and therefore does not belong to the PP2B class. PfPPJ is developmentally regulated and shows peak expression predominantly in the schizont stage.

Another phosphatase from *P. falciparum* is an atypical PP2C homolog. The predicted amino acid sequence of PfPP2C can be divided into two functional halves, PfPP2C-1 and PfPP2C-2. PfPP2C-1, PfPP2C-2, or the full-length PfPP2C complement *S. pombe* PP2C temperature-sensitive mutants. Recombinant protein derived from each half of PP2C is catalytically active. PfPP2C selectively dephosphorylates PKC-mediated phosphorylation of translation elongation factor-1β.

Protein phosphatases from kinetoplastids

Initial evidence for an important role of protein phosphatases in cell growth and differentiation of trypanosomatids came from the treatment of cells with okadaic acid. In *T. brucei*, okadaic acid treatment uncouples the nuclear and organellar division resulting in cells with multiple nuclei and a single kinetoplast. In *T. cruzi*, calyculin A-treated cells can undergo mitosis in the absence of cytokinesis. However, unlike in *T. brucei*, *T. cruzi* epimastigotes are relatively insensitive to okadaic acid. Two PP1 homologs, TcPP1α and TcPP1β have been cloned. Interestingly, TcPP1β is significantly more sensitive to calyculin A than okadaic acid (IC_{50} of 2 nM vs. 100 nM). A PP5-type *T. brucei* protein phosphatase (TbPP5) also has been characterized. It contains PP5 signature tetratricopeptide repeats at the amino terminus and the phosphatase catalytic domain at the carboxy terminus. The 52 kDa TbPP5 is expressed in both bloodstream and procyclic forms. TbPP5 is found predominantly in the cytoplasm although some nuclear localization has also been observed. The recombinant TbPP5 is catalytically active and shows modest stimulation by arachidonic acid. In *Leishmania chagasi* PP1 and PP2C-like activities have been detected, the *PP2C* (*LcPP2C*) gene has been cloned, and the recombinant protein is enzymatically active.

In summary, reversible phosphorylation mediated by an array of kinases and phosphatases is expected to play a key role in various cellular processes of protozoan parasites. Although a variety of kinases and phosphatases have been isolated from different species based on homology, their exact physiological roles are not clear. Ascertaining the role of these proteins will be a major challenge of the post-genome-project era. Many of these proteins may have evolved unusual functions as a means of controlling complex life-cycle events. These atypical characteristics will not be evident from primary amino acid sequences. Novel structure and properties of parasite kinases and phosphatases may be exploited in the design of new generation anti-parasitic therapeutics with a high degree of selectivity.

CONCLUSIONS

Overall it is evident that protozoan parasites contain a complex web of signaling components. Evolutionarily conserved enzymes involved in Ca^{2+} sensing, cyclic nucleotide metabolism, and phosphoryl transfer reactions have been identified. While many of these enzymes share homology with host proteins, others appear to be unique in structure or function. Inhibitor studies and the more limited gene knockout studies hint at the essential nature of the pathways regulated by these proteins. Critical processes such as parasite differentiation, growth, division or infectivity appear to be affected. The discovery of more specific inhibitors is predicted in the future as studies on signal pathways become more refined. Completion of genome sequences coupled with developments in microarray technologies, knockout methodologies and proteomics will speed this research. Key areas of future research include the identification of extracellular triggers that activate the intracellular signal pathways, characterization of signal anchor proteins that localize cell responses despite global production of signal molecules, and assignment of the many signal components into functional interacting pathways.

ACKNOWLEDGMENTS

The authors wish to thank Siglinde Quirk, University of Central Florida, for assistance with the graphics, and also David Baker and Martin Taylor, London School of Hygiene and Tropical Medicine, for their critical review of portions of the chapter. Research by the authors was supported by NIH grants AI24627 (L. R.) and AI48036 (D. C.), and by the Wellcome Trust and British Heart Foundation (J. M. K.).

FURTHER READING

Signal properties of Ca^{2+}

Docampo, R. and Moreno, S.N. (2001). The acidocalcisome. *Mol. Biochem. Parasitol.* **114**, 151–159.

Flawia, M.M., Tellez-Inon, M.T. and Torres, H.N. (1997). Signal transduction mechanisms in *Trypanosoma cruzi. Parasitol. Today* **13**, 30–33.

Garcia, C.R. (1999). Calcium homeostasis and signaling in the blood-stage malaria parasite. *Parasitol. Today* **15**, 488–491.

Godsel, L.M. and Engman, D.M. (1999). Flagellar protein localization mediated by a calcium-myristoyl/palmitoyl switch mechanism. *EMBO J.* **18**, 2057–2065.

Kieschnick, H., Wakefield, T., Narducci, C.A. and Beckers, C.J. (2001). *Toxoplasma gondii* attachment to host cells is regulated by a calmodulin-like domain protein kinase. *J. Biol. Chem.* **276**, 12369–12377.

Krishna, S., Woodrow, C., Webb, R. *et al.* (2001). Expression and functional characterization of a *Plasmodium falciparum* Ca^{2+}-ATPase (PfATP4) belonging to a subclass unique to apicomplexan organisms. *J. Biol. Chem.* **276**, 10782–10787.

Parsons, M. and Ruben, L. (2000). Pathways involved in environmental sensing in trypanosomatids. *Parasitol. Today* **16**, 56–62.

Sibley, L.D. and Andrews, N.W. (2000). Cell invasion by un-palatable parasites. *Traffic* **1**, 100–106.

Cyclic nucleotides

Bieger, B. and Essen, L.-O. (2001). Structural analysis of adenylate cyclases from *Trypanosoma brucei* in their monomeric state. *EMBO J.* **20**, 433–445.

Carrucci, D.J., Witney, A.A., Muhia, D.K. *et al.* (2000). Guanylyl cyclase activity associated with putative bifunctional integral membrane proteins in *Plasmodium falciparum. J. Biol. Chem.* **275**, 22147–22156.

Hurley, J.H. (1999). Structure, mechanism, and regulation of mammalian adenylyl cyclase. *J. Biol. Chem.* **274**, 7599–7602.

Linder, J.U., Engel, P., Reimer, A. *et al.* (1999). Guanylyl cyclases with the topology of mammalian adenylyl cyclases and an N-terminal P-type ATPase-like domain in *Paramecium, Tetrahymena* and *Plasmodium. EMBO J.* **18**, 4222–4232.

Sanchez, M.A., Zeoli, D., Klamo, E.M., Kavanaugh, M.P. and Landfear, S.M. (1995). A family of putative-adenylate cyclases from *Leishmania donovani. J. Biol. Chem.* **270**, 17551–17558.

Seebeck, T., Gong, K., Kunz, S., Schaub, R., Shalaby, T. and Zoraghi, R. (2001). cAMP signalling in *Trypanosoma brucei. Int. J. Parasitol.* **31**, 490–497.

Vassella, E., Reuner, B., Yutzy, B. and Boshart, M. (1997). Differentiation of African trypanosomes is controlled by a density sensing mechanism which signals cell cycle arrest via the cAMP pathway. *J. Cell Sci.* **110**, 2661–2671.

Protein phosphorylation

Cohen, P.T.W. (1997). Novel protein serine/threonine phosphatases: variety is the spice of life. *Trends Biochem. Sci.* **22**, 245–251.

Doerig, C., Chakrabarti, D., Kappes, B. and Mathews, K. (2000). The cell cycle in protozoan parasites. *Prog. Cell Cycle Res.* **4**, 163–183.

Engh, R.A. and Bossemeyer, D. (2001). The protein kinases activity modulation sites: mechanisms for cellular regulation-targets for therapeutic intervention. *Adv. Enzyme Regul.* **41**, 121–149.

Garcia, A., Cayla, X., Barik, S. and Langsley, G. (1999). A family of PP2 phosphatases in *Plasmodium falciparum* and parasitic protozoa. *Parasitol. Today* **15**, 90–92.

Kappes, B., Doerig, C.D. and Graeser, R. (1999). An overview of *Plasmodium* protein kinases. *Parasitol. Today* **15**, 449–454.

Le Roch, K., Sestier, C., Dorin, D. *et al.* (2000). Activation of a *Plasmodium falciparum* cdc2-related kinase by heterologous p25 and cyclin H. Functional characterization of a *P. falciparum* cyclin homologue. *J. Biol. Chem.* **275**, 8952–8958.

Mamoun, C.B., Sullivan, D.J. Jr., Banerjee, R. and Goldberg, D.E. (1998). Identification and characterization of an unusual double serine/threonine protein phosphatase 2C in the malaria parasite *Plasmodium falciparum. J. Biol. Chem.* **273**, 11241–11247.

Van Hellemond, J.J., Neuville, P., Schwarz, R.T., Matthews, K.R. and Mottram, J.C. (2000). Isolation of *Trypanosoma brucei* CYC2 and CYC3 cyclin genes by rescue of a yeast G(1) cyclin mutant. Functional characterization of CYC2. *J. Biol. Chem.* **275**, 8315–8323.

Wiese, M. (1998). A mitogen-activated protein (MAP) kinase homologue of *Leishmania mexicana* is essential for parasite survival in the infected host. *EMBO J.* **26**, 19–28.

C H A P T E R

12

Plastids, mitochondria, and hydrogenosomes

Geoffrey Ian McFadden
School of Botany, University of Melbourne, Victoria, Australia

INTRODUCTION

One of the key differences between eukaryotes and prokaryotes is the numerous subcellular compartments present in eukaryotic cells. In addition to the nucleus, eukaryotes possess compartments such as the endomembrane system, mitochondria, and in some cases plastids. None of these compartments occurs in prokaryotes. Indeed it is the presence of these compartments and the ability to compartmentalize processes that has allowed eukaryotic cells to expand in size and, more importantly, to differentiate into various cell types, each with a unique role, in a multicellular consortium. It is reasonably certain that prokaryotes preceded eukaryotes in the evolution of life on earth. Therefore, the simple prokaryotic cellular organization is considered ancestral to the more complex, compartmentalized eukaryotic cell organization. This chapter examines the origin of two eukaryotic compartments, the mitochondrion and plastid, their significance in the evolution of parasites, and the role of these organelles in treatment of several parasitic diseases.

ENDOSYMBIOSIS

The theory of endosymbiosis describes the origin of mitochondria and plastids from endosymbiotic bacteria. The theory is an old one, going back to early studies by light microscopists in the late 1800s, and has gained more and more support as the details of the inner workings of these organelles have emerged with the advent of new technologies such as biochemistry, electron microscopy and molecular biology. Indeed, as the evidence accumulates we are increasingly more certain that mitochondria and plastids are relics of once free-living prokaryotes now housed within eukaryotic cells. What are the main lines of evidence for this assertion? Firstly, these organelles are what is known as semi-autonomous. That

Molecular Medical Parasitology
ISBN 0–12–473346–8

is, they divide independently of the host cell, often existing as multicopy structures in each eukaryotic cell. Several of the same proteins responsible for the division process in bacteria (FtsZ, MinD, MinE) also participate in mitochondrial and plastid fission. Ultrastructurally, mitochondria and plastids resemble bacteria in select ways. The organelles are bounded by two membranes, which are likely homologous to the plasma membrane and outer membrane of the Gram-negative ancestors from which they derive. The inner membrane of both mitochondria and plastids is highly convoluted, forming cristae in mitochondria and thylakoids in photosynthetic plastids. These convolutions are undoubtedly adaptations to expanding the surface area for the major processes within the organelle, oxidative respiration in mitochondria and photosynthesis in plastids. No such folding of the inner membranes occurs in the prokaryotes thought to be ancestral to mitochondria, but inner membrane convolutions homologous to plastid thylakoids are seen in photosynthetic bacteria (the likely ancestors of plastids). It is abundantly clear that the metabolisms of the two organelles are descended from ancestral prokaryotic processes. The Krebs cycle and oxidative phosphorylation in mitochondria are virtually identical to the same pathways in bacteria, with most of the enzymes, cofactors, electron carriers and ATP synthases sharing the same evolutionary heritage. Similarly, the engines of photosynthesis in plant and algal plastids are plainly derived from the machinery of cyanobacteria-like prokaryotes.

Perhaps the most persuasive line of evidence for the endosymbiotic origin of mitochondria and plastids are the organelles' genomes. Both organelles have their own DNA, and the architecture of their genome is classically prokaryotic, being circular and having a single origin of replication and genes arranged in operons like prokaryotes. This organization is in stark contrast to that of the host nucleus with multiple, linear chromosomes and individual genes, each with its own regulatory elements and a single gene transcript. The organization of mitochondrial and plastid gene operons also reflects prokaryotic ancestry, with similar arrangements of genes. It is also clear that the organelles' genes are closely related to bacterial genes. Indeed, gene trees provide us with a clear picture of organelle evolution. Mitochondrial sequences are most closely related to those of alpha-proteobacteria, whereas plastid sequences are most closely related to those of cyanobacteria. The gene trees tell us several important things. Firstly, these organelles must have arisen from separate endosymbiotic events: one for the mitochondria and one for the plastids, because their genes obviously derive from different parts of the bacterial radiation. If both mitochondria and plastids emerged in one place from the bacterial tree, we might assume that they derived from a common endosymbiosis, but this is not so. Secondly, the alliances between mitochondria and alpha-proteobacteria, and between plastids and cyanobacteria, rationalizes the similarities we observe in their metabolisms. Cyanobacteria are photosynthetic. Alpha-proteobacteria possess a Krebs cycle and oxidative phosphorylation, although these respiratory processes occur widely in prokaryotes and are not restricted to alpha-proteobacteria. Further down the information chain we also see identity between mitochondria and plastids and prokaryotes. The transcription of RNA from DNA utilizes the same type of RNA polymerase components, although select mitochondria and plastids also appear to have recruited a viral RNA polymerase as well. The translation of protein from mRNA in mitochondria and plastids is also prokaryotic, using Shine–Dalgarno ribosome binding motifs, formyl methionine as the initiator amino acid, and ribosomes with a 70S sedimentation coefficient, which is the same as bacterial ribosomes and distinct from the 80S ribosomes of eukaryotes. Translation by mitochondrial and plastid ribosomes also exhibits similar sensitivities to pharmacological agents as prokaryotic ribosomes. This latter phenomenon reflects the similar mechanisms and components participating in this core process, and will be expanded upon later in the section exploring drug targets in mitochondria and plastids. In sum, the majority of what we know about mitochondria and plastids points to them being reduced bacteria now living inside eukaryotic cells.

MITOCHONDRIAL REDUCTION

The mitochondrial endosymbionts have undergone substantial modification. Massive reduction, such as the loss of the wall and large components of the biosynthetic capacity, probably stem from redundancy in the new role as endosymbiont. More interestingly there has also been massive depletion of the endosymbiont genome. Typical mitochondrial genomes only encode a small number of proteins, whereas a free-living alpha-proteobacterium probably encodes more than a thousand. Even accounting for the losses through redundancies, a typical mitochondrion utilizes an estimated 500 proteins. However, the human mitochondrion only encodes 13 proteins; yeast is little better with only eight proteins made by the mitochondrion. Where are the genes for the other mitochondrial proteins? In the nucleus. The host has confiscated most of the endosymbiont's genes. The reasons for this transfer of genetic responsibility are not certain, but are likely to do with the restricted capacity for organelle genomes to deal with mutation. Denied the opportunity for recombination with other members of their population, mitochondria have entered a genetic bottleneck. Locked within its host cell, the mitochondrion has apparently existed as a clonal line with little or no opportunity for genetic exchange. Although one can conceive of some scenario where mitochondria from two parents might combine in a newly formed zygote, this does not appear to occur. Indeed, in most organisms where mitochondrial descent has been studied we observe uniparental inheritance of the organelle. Accumulation of deleterious mutations may be inevitable, and one solution is to relocate these essential genes into the host nucleus, which presumably has occasional meiosis and the opportunity to remove deleterious mutations through recombination.

Relocation of the genes, which we refer to as intracellular gene transfer, can occur by one of several processes. A common mechanism seems to be the escape of fragments of the mitochondrial DNA from the organelle and their random incorporation into nuclear chromosomes. Another mechanism involves reverse transcription of mitochondrial mRNAs to produce cDNAs that become incorporated into nuclear chromosomes. These transfers are apparently ongoing, providing a steady trickle of genetic transfer between endosymbiont and host. Eventually some of these transfers result in active copies of the genes in the nucleus. Under select circumstances the endosymbiont gene becomes inactivated and the nuclear copy takes over.

Transferring genes to the host nucleus may have avoided some mutational problems, but it presents an obvious new problem. How can the gene product be returned to its site of operation? Messenger RNAs do not seem able to cross membranes. They exit the nucleus through nuclear pores, which in effect are very large openings in the double-membrane envelope

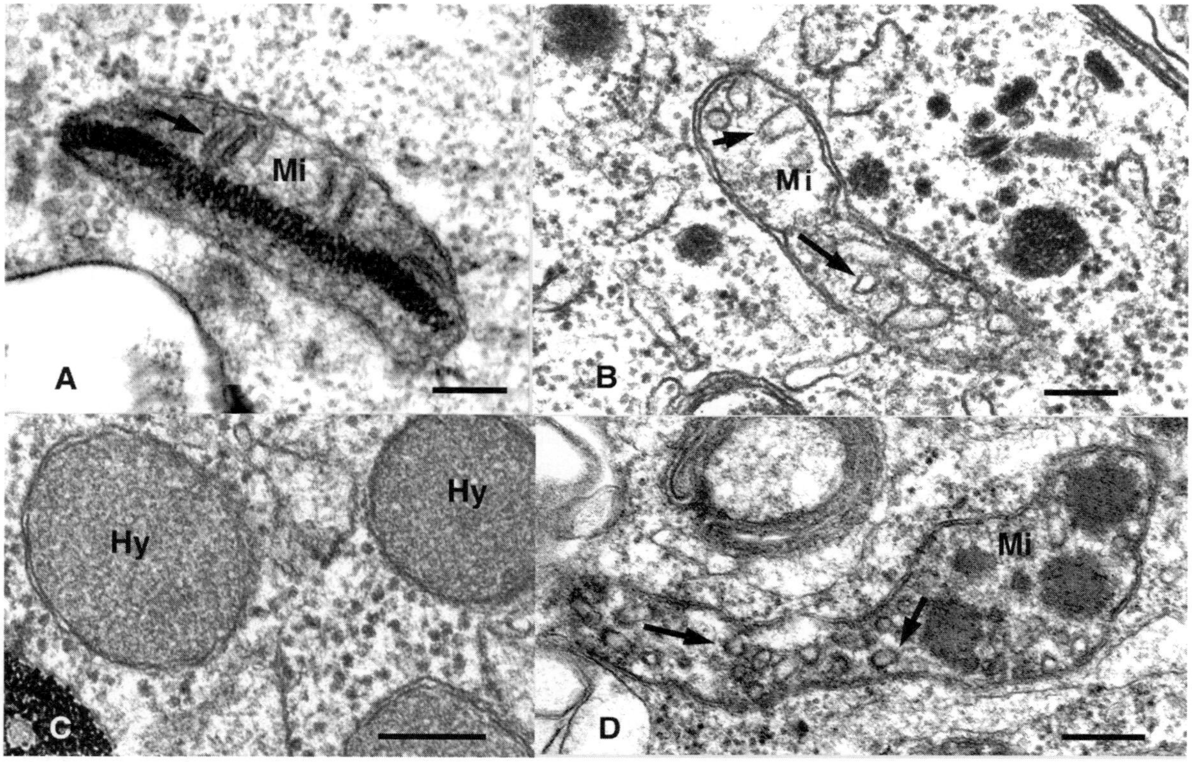

FIGURE 12.1 Various forms of the mitochondrion in parasites. (A) Mitochondrion (Mi) from *Leishmania mexicana* showing a conspicuous kinetoplast of concatenated DNA minicircles. (B) Mitochondrion (Mi) of *Toxoplasma gondii* showing double membrane and tubular cristae (arrows). (C) Hydrogenosome (Hy) of *Trichomonas vaginalis* showing double bounding membrane and amorphous contents. (D) Mitochondrion (Mi) of *Plasmodium falciparum* gametocyte showing tubular cristae (arrows) and unidentified dark bodies. Scale bars = 200 nm.

around the nucleus. Thus, the nucleus-encoded mRNAs for mitochondrial proteins cannot enter the mitochondrion but must be translated on ribosomes in the cytoplasm. The gene products are then specifically targeted into the organelle by an N-terminal extension that is recognized by complex machinery on the surface of the mitochondrion. Only proteins bearing these N-terminal extensions, known as transit peptides, are targeted into the organelle. Once established, this targeting mechanism facilitated the ongoing relocation of genes from organelle to host genome. Each transferred gene needs to acquire one of these transit peptides before the product can be returned to the organelle. Acquisition of a transit peptide is probably expedited by the fact that transit peptides have a fairly loose set of parameters to function as targeting motifs.

MITOCHONDRIAL METABOLISM

The canonical function of mitochondria is the oxidation of pyruvate to CO_2 and H_2O.

Cofactors of this reduction, such as NAD^+, temporarily hold the electrons from the oxidized carbon–carbon bonds prior to transfer into the electron-transport chain of oxidative phosphorylation. Further transfer of the electrons along the chain drives proton pumps that establish an electrochemical gradient across the inner and outer mitochondrial membranes. This gradient is used to rotate the ATPase, and the rotation energy is converted to chemical energy by phosphorylation of ADP to ATP. Ultimately the electrons are transferred to O_2 to produce H_2O. Do parasite mitochondria perform similar functions? Sometimes. Depending on the parasite in question, quite a range of mitochondrial functions is evident. In kinetoplastid parasites, for instance, the mitochondrion apparently performs similar functions to homologous organelles in other eukaryotes. In other parasites, though, the role of the mitochondrion is far from typical. Indeed, in many parasites the mitochondrion performs bizarre reactions not commonly seen elsewhere in eukaryotes. These reactions are probably adaptations to the extraordinary environmental conditions in which these parasites live. Additionally, these atypical metabolisms of parasite mitochondria are often relatively low in efficiency. Such low efficiency may reflect the fact that the success of the parasite may not require maximal efficiency in utilizing energy resources, which after all are being acquired from the host.

The atypical metabolisms of parasite mitochondria have proven to be useful targets for several drugs. For instance, mitochondria of malaria parasites are the target of the anti-malarial drug atovaquone. The exact role of malaria parasite mitochondria is unclear. The mitochondrial genome is the smallest known and encodes only five genes: two rRNAs and three proteins (CoxI, CoxII and Cytb). The rRNAs are highly fragmented, but careful reconstruction of the fragments suggests they have a viable secondary structure, which indicates that a translation system for expression is probably operative. All translation components other than rRNAs are presumably imported. Structurally the malaria parasite mitochondrion is also unusual. In erythrocyte stages virtually no cristae are present, and only in gametocytes does the organelle resemble other eukaryotic mitochondria, with so-called tubular cristae. Little is known about mitochondrial metabolism in *P. falciparum*. The parasite is considered to be a homolactic fermenter. Recently, it has been demonstrated that the *P. berghei* mitochondrion is capable of oxidative phosphorylation, but that the proton gradient is uncoupled. The malaria parasite mitochondria contain ubiquinones, the electron carrier molecules embedded in the inner mitochondrial membrane, and the anti-malarial atovaquone is a ubiquinone analog. Atovaquone apparently inhibits the cytochrome *bc*(1) complex of the cytochrome pathway. It is effective against malaria, toxoplasmosis and other parasites such as *Pneumocystis carinii* pneumonia. Utility of atovaquone as an anti-malarial is limited due to resistance arising via a mutation to the catalytic domain of the *bc*(1)complex (Chapters 16, 17).

Mitochondria of kinetoplastid parasites (e.g. trypanosomes and *Leishmania*) are also unusual. The name kinetoplastid derives from the unique mitochondria of these parasites. The mitochondrion, which stains heavily due to presence of extraordinary amounts of DNA, typically lies just posterior to the flagellar bases and was initially thought to have a special role in motility, hence kinetoplastid (motility body). The large quantity of DNA reflects a special type of genome organization in kinetoplastids. Kinetoplastid mitochondria depart from the so-called 'central dogma of molecular biology', which dictates that information is passed from

DNA to RNA to protein. However, in kinetoplastid mitochondria we observe a highly unusual detour from this linear information chain, in which editing of the RNA intermediate occurs. RNA editing involves the post-transcriptional alteration of the coding sequence in the mRNA for a protein. Hence, the information encoded within the DNA copy of a kinetoplastid mitochdondrial gene is not the same as the ultimate sequence of the protein. The final sequence of the mRNA is a combination of two disparate DNA templates encoded on two separate circular molecules in the mitochondrion. The first is a canonical, circular mitochondrial genome with the core gene. The second template is encoded by a number of so-called minicircles. The minicircles, which can be very numerous, encode what are known as guide RNAs. A complex known as the editosome then uses the guide RNAs as templates to edit the main mRNA to create the final version of the coding sequence by inserting and deleting numerous uridine (U) residues. The numerous minicircles that encode the guide RNAs are catenated into a substantial mass of DNA that is referred to as the kinetosome, a structure unique to kinetoplastids. Replication of the kinetosome DNA minicircles is beginning to be understood. Because kinetosomes are unique to these parasites, as is the mechanism of U insertion or deletion in mitochondrial transcripts, these processes are potentially good drug targets. As yet no drugs that specifically inhibit these processes have been identified.

'AMITOCHONDRIATE' ORGANISMS

Several major parasites were originally characterized as lacking mitochondria. *Entamoeba histolytica*, for instance, was considered amitochondrial. *E. histolytica* is the causative agent of amebic dysentry, which is a major parasitic disease globally. Very recently it was established that a vestigial mitochondrion, known as the cryptome or mitosome, is in fact present in these parasites. The function of the *E. histolytica* cryptome or mitosome is unknown. Indeed, it is not absolutely clear that the cryptome or mitosome is actually a mitochondrion. The classic hallmark of mitochondria is the mitochondrial genome, which ultimately derives from a reduced alpha-proteobacterial genome. The cryptome or mitosome has no known genome. Indeed, the only reason we know of its existence is that an alpha-proteobacterial gene product (Hsp60) is localized in the structure. Most mitochondria contain Hsp60 protein. However, the gene itself has apparently undergone intracellular gene relocation from the mitochondrion to the nucleus and the protein is typically targeted back to the organelle. In the case of *E. histolytica* it is believed that the mitochondrial genome has disappeared and only the organelle, essentially a double membrane bag in which some undefined process(es) occur, remains. The targeting of footprint proteins such as Hsp60 to the organelle is the give-away to the presence of the organelle. Only recently recognized, the cryptome or mitosome is not known to be the target of any drug treatments for amebic dysentery. Once its function is established, it may be a useful target.

Another major parasite thought to lack mitochondria is *Giardia lamblia*, the causative agent of so-called traveler's diarrhea or beaver fever. *Giardia* is a diplomonad that inhabits the alimentary canal of various animals including humans. The apparent lack of a mitochondrion in *Giardia* (and other parasites, see below) is part of an interesting chapter in the study of eukaryotic origins. The canonical theory of endosymbiotic origin of mitochondria, in which an alpha-proteobacterium was engulfed and retained by a nucleated host, invokes the existence of an early lineage of eukaryotes without mitochondria. Although, the mitochondriate derivatives of the endosymbiotic event have become more numerous and far more conspicuous, students of eukaryotic origins recognized that if any descendants of the amitochondrial stock are still alive, then these organisms would be valuable models in interpreting our origins. The quest was thus on to find the organisms that best represented the 'host', the premitochondrial eukaryote. Amitochondriate parasites were considered good candidates at the time.

Reconstruction of eukaryotic evolution underwent explosive growth with the advent of DNA sequencing. Most of this activity initially focused on one gene, rRNA, as a marker of relationships. Phylogenetic trees inferred from rRNA gene sequences have yielded valuable additions to our models of evolution, with clear depiction of relationships previously recognized on other grounds and also the revelation of other relationships not previously well understood. The initial trees incorporating parasites like *Giardia* were exciting for evolutionary biologists. *Giardia* emerged at the base of the eukaryotic tree, exactly where one would expect to find it if it were truly an early offshoot of eukaryotes prior to mitochondrial acquisition. Several other amitochondrial organisms, principally parasites, were also added to the phylogenetic trees, and three lineages (the diplomonads, including *Giardia*; the parabasalids, including *Trichomonas vaginalis*; and the microsporidia, including *Encephalitozoon*) consistently emerged at the base of these trees. This gene-tree topology was congruent with the endosymbiotic model of eukaryogenesis and provided us with several models to study. Different versions of the rRNA trees were not always in agreement as to which amitochondriate protist was the first to emerge from the eukaryotic lineage, but combined with the often simple ultrastructure of the organisms a cohesive story of an ancient relict lineage was beginning to emerge. These organisms became known as the kingdom Archezoa, a group defined primarily by the lack of mitochondria and their basal position in the eukaryotic tree. They were considered the first eukaryotes and an extant version of the kind of cells from which we ourselves are derived.

However, not all the data sat comfortably with the 'Archezoa' hypothesis. The original amitochondriate eukaryotes were envisaged as anaerobes. Indeed, mitochondrial acquisition is thought to have aided their coping with increasing O_2 levels. If the amitochondriate eukaryotes (Archezoa), which were primarily parasites inhabiting anaerobic niches within animals, were indeed descendants of the first eukaryotes, where had they survived prior to the advent of animal body cavities? These anaerobes would have had to inhabit anaerobic zones for many hundreds of millions of years and then to invade these new environmental niches as metazoa evolved. This scenario is plausible, but other arguments against the Archezoa hypothesis also appeared.

The ascendancy of rRNA as the key phylogenetic marker was not so much because it is the ideal marker but because of its practicability. rRNA genes are relatively easy to sequence (nowadays at least) and the enormous database provides a huge head start for interpreting relationships within the larger context. But interpreting evolution of organisms from a single gene sequence places a great deal of reliance on those sequence data reflecting the actual pattern of evolutionary branching. It is clear (with the benefit of hindsight) that factors other than simple fixation of mutations as a consequence of evolutionary divergence act on gene sequences. Thus, patterns in the gene sequences that do not necessarily reflect the divergence of the organism harboring the genes do occur. These non-phylogenetic patterns can confound tree inference. For these reasons it has always been desirable to obtain sequences of other genes to look for congruence in the phylogenies. Intriguingly, the initial extra genes, mostly translation-related proteins such as elongation factors and aminoacyl-tRNA synthetases, reaffirmed the rRNA trees. But other genes have provided a conflicting set of trees. Tubulin is a universal eukaryotic protein that forms microtubules, the cytoskeletal elements in spindles and flagella. Trees of tubulin sequences place some Archezoa in non-basal positions. Microsporidian tubulins, for instance, are strongly allied with those of fungi, a relatively late-emerging eukaryotic group. Similarly, *Giardia* tubulins do not seem to be basal. How do we rationalize these conflicting hypotheses? As mentioned above, patterns other than just phylogenetic patterns exist in gene sequences. For instance, some genes in some organisms appear to undergo accelerated evolution. In trees reflecting nucleotide (or amino acid) substitution events as branch lengths, genes with accelerated rates of evolution can show up as long branches. Long branches can group together in trees, even though the organisms from which the long-branch genes derive are not closely related. Thus the 'long-branches-attract' phenomenon can posit false relationships or positions in trees. Could such a 'long-branches-attract' phenomenon have confounded the early trees showing basal Archezoa? Perhaps. The rRNAs, elongation factors and aminoacyl-tRNA synthetases of these protists are certainly long branches, and their basal position might be an artefact, with their sequences being drawn down the tree to the similarly long-branch outgroup sequences (bacterial genes used to root the eukaryotic sequences). Tubulin genes, for instance, seem to have (on the whole) avoided such periods of acceleration and probably do not show artefactual basal positions.

Parasites of the parabasalid lineage deserve special mention in a discussion of mitochondrial origins and evolution. The best known parasite in this lineage is *Trichomonas vaginalis*, a widespread (though relatively benign) infection of the human genital tract. *T. vaginalis* is referred to as a parabasalid because of the conspicuous Golgi bodies adjacent the basal bodies of the flagella. Parabasalids lack classic mitochondria but they do possess a structure, the hydrogenosome, that appears to have the same origin as mitochondria. Hydrogenosomes are round organelles bounded by two membranes and, as the name suggests, they generate hydrogen. Like mitochondria, hydrogenosomes import pyruvate produced by glycolysis. Unlike mitochondria they do not oxidize this pyruvate to CO_2 and H_2O. Rather, they oxidize the pyruvate to acetate. Electrons from pyruvate oxidation are transferred not to NADH but to ferredoxin using a pyruvate ferredoxin oxidoreductase (PFO) (Chapters 7 and 17). Ferredoxin is then reoxidized by hydrogenase, which transfers the electrons onto protons, producing H_2. Similar metabolism occurs in anaerobic bacteria. PFO is the target of a major drug category. The 5-nitroimidazoles (Metronidazol™ and Flagyl™) have their nitrate group activated by PFO. This reactive nitrate then alkylates surrounding molecules, particularly DNA, which proves lethal for the parasite. The 5-nitroimidazoles are thus useful against *Trichomonas*. PFO also occurs in some of the previously discussed anaerobic parasites, such as *Entamoeba* and *Giardia,* and these infections also respond well to 5-nitroimidazole therapy. The PFO in *Giardia* and *Entamoeba* is not localized in a hydrogenosome structure and these organisms do not generate significant quantities of hydrogen (Chapter 7). Nevertheless, PFO is a useful drug target.

How does the hydrogenosome relate to mitochondria? Like mitochondria, parabasalid hydrogenosomes are bounded by two membranes and divide by binary fission independently of cell division. However, unlike mitochondria, parabasalid hydrogenosomes contain no DNA – all the proteins contained within are synthesized in the cytoplast from genes harbored in the parabasalid nucleus. The gene products are then targeted into the hydrogenosome post-translationally. This targeting mechanism closely resembles the targeting of proteins to mitochondria in that an N-terminal presequence mediates targeting and is removed within the organelle. Furthermore, the presequence shares similar properties to canonical mitochondrial targeting peptides from mitochondriate organisms. The question thus arises: is the hydrogenosome a modified mitochondrion with no DNA? A definitive answer is not yet available but some strong lines of evidence speak in the affirmative. Import of proteins into hydrogenosomes involves chaperones such as Hsp10, Hsp60 and Hsp70. Phylogenetic analysis of these proteins allies them unequivocally with those of eukaryotic mitochondria. Other components of the mitochondrial import machinery, namely Tim9, Tim10, Tim44 and Tom20, are also involved in hydrogenosome protein targeting. Finally, eukaryotic mitochondria possess a unique channel that exchanges ADP from the cytoplasm for ATP generated in the mitochondrion. Such a channel does not occur in bacteria (one wouldn't expect bacteria to export energy-rich molecules in exchange for ADP) so it appears to be an adaptation to the endosymbiotic relationship. A homologous ADP/ATP exchanger is present in hydrogenosomes, suggesting that they evolved from the same endosymbiotic progenitor as mitochondria.

If we accept that hydrogenosomes are a modified form of mitochondria, what happened to the genome? An attractive hypothesis centers around the gene content of mitochondrial genomes. A small number of proteins in the classical respiratory chain are always encoded by the mitochondrial DNA. Indeed, in the malarial mitochondrial genome, the most stripped down mtDNA known, the three remaining proteins are all members of the membrane complexes involved in respiration. Why these proteins are refractory to transfer of their genes to the nucleus is uncertain. It may be that the gene products, which are highly hydrophobic as membrane proteins, are not amenable to transport back into the organelle. Whatever the case, they are invariably encoded within the genome. However, hydrogenosomes lack this electron transport chain. The simple explanation is that the genome was able to disappear once these proteins were not required and the organelle was utilized solely for anaerobic respiration. Thus, parabasalids apparently have an alternate type of mitochondria, an anaerobic endosymbiont that has lost its genome. An interesting question yet to be resolved is whether the parabasalid mitochondrion represents a degenerate mitochondrion or an early offshoot of eukaryotes soon after mitochondrial acquisition. Various phylogenetic analyses identify parabasalids as the earliest diverging eukaryotic lineage. Hydrogenosomes may thus represent a primitive offshoot of the endosymbiotic origin of mitochondria.

A NEW HYPOTHESIS

The recent recognition of hydrogenosomes as a special type of mitochondrion provoked a rethink of the endosymbiosis theory. The classical endosymbiotic theory postulates mitochondrial acquisition being driven by efficiency of ATP generation. The endosymbiont with its oxidative phosphorylation pathway introduced a more efficient mechanism and perhaps a means to detoxify oxygen, thereby providing a selective advantage to the partnership. Could other drivers have been responsible for the association producing the mitochondrion? A new hypothesis, namely the hydrogen hypothesis, adopts this rationale. The hydrogen hypothesis suggests that the endosymbiont wasn't adopted for its ability to perform oxidative phosphorylation but for its ability (under anaerobic conditions) to generate hydrogen. Like the classical formulations of the endosymbiont hypothesis, the hydrogen hypothesis invokes a mutual benefit for the two partners and postulates an alpha-proteobacterial endosymbiont as the partner producing the mitochondrion. Where the hydrogen hypothesis departs from the paradigm is in its proposed host. Rather than a nucleated, amitochondriate phagotroph as a host (posited by the classical theory), the hydrogen hypothesis invokes a hydrogen-consuming, methane-evolving archaebacterium as the host. Referred to as a syntrophic (shared feeding) relationship, the hydrogen hypothesis has the endosymbiont feeding the archaebacterium with its wastes (hydrogen and CO_2). Initially the endosymbiont would have been outside the host (perhaps attached) and would have scavenged reduced carbon compounds from the environment. With the ongoing establishment of the interaction, the endosymbiont would have gradually become enclosed within the host, which would eventually have to assume responsibility for providing the endosymbiont with reduced carbon compounds for conversion into hydrogen and CO_2 that the host could use. The hydrogen hypothesis then goes on to postulate that the endosymbiont (now a protomitochondrion) was modified to activate its oxidative phosphorylation and perform aerobic respiration using the supplied carbon substrate (pyruvate). The consortium is thus envisaged to have started out in an anaerobic environment but to have switched to an aerobic metabolism, perhaps as the concentration of O_2 in the atmosphere increased.

Hydrogenosomes of parabasalids would thus represent one branch in this diversification where anaerobic metabolism was maintained. Conversely, most eukaryotes utilized the endosymbiont for aerobic respiration. The fact that parabasalids emerge early in the eukaryotic divergence is congruent with this scenario. A major objection to the hydrogen hypothesis (when it was first proposed) was that Archaea are not known to engulf other cells. However, it was recently demonstrated that some prokaryotes (beta-proteobacteria) harbor endosymbiotic gamma-proteobacteria, so bacteria-within-bacteria is not an impossibility. Moreover, the recent identification of cytoskeleton-type proteins equivalent to tubulin and actin in prokaryotes is making us rethink our models of prokaryotes as organisms unlikely to undergo cell-shape changes conducive to engulfment of other cells.

The hydrogen hypothesis thus places the origin of the mitochondrial endosymbiont before the origin of the nucleus, which would have occurred by an autogenous process similar to that invoked for nuclear origins in the classical formulation of the endosymbiotic hypothesis. This new version of the endosymbiont hypothesis, which challenges a long-standing paradigm in cell evolution, owes its conception to the study of hydrogenosomes in parabasalid parasites like *Trichomonas vaginalis*.

PLASTIDS

This is the first parasitology text to contain a chapter on plastids. Until recently plastids were the domain of botany and algology. No more! Plastids have been identified in apicomplexan parasites in recent years. Researchers have gathered a host of biological, biochemical and molecular information from malarial and toxoplasmodial parasites. However, by focusing on them as parasites, and the effects they have on humans as the host, we missed a key feature; these parasites started life as autotrophic, photosynthetic, alga-like organisms. The key to this revelation was the identification of a plastid, a small parasite organelle that shares the same evolutionary heritage as chloroplasts of plants and algae.

Just as the theory of endosymbiosis explains the origin of mitochondria from alpha-proteobacteria, it also explains the origin of a second organelle, the plastid, or chloroplast. Plastids clearly originated from engulfed cyanobacteria, which brought the power of photosynthesis into eukaryotes. Whereas the driver for mitochondrial endosymbiosis is posited to have been respiration or hydrogen syntrophy, in plastid origins it is argued to have been autotrophy that drove the partnership. Plastid acquisition followed mitochondrial acquisition and created the eukaryotic autotrophs. Reduction of the cyanobacteria-like endosymbiont followed a similar course to the reduction of the mitochondrial endosymbiont with transfer of many cyanobacterial genes to the host nucleus. Similar to nucleus-encoded mitochondrial gene products, the plastid proteins encoded by genes relocated to the nucleus must also be targeted to the plastid to function normally. These plastid proteins are also targeted using N-terminal extensions distinct from those of mitochondrial proteins. Once established this system again allowed wholesale intracellular gene transfer, and the great majority of genes for plastid proteins are now located in the nucleus. The residue, typically in the order of 100 to 200 genes in plastid DNA, is clearly cyanobacterial in origin with a circular architecture, a single origin of replication, genes in ancestral operons, −10, −35 promoters for the bacterial type RNA polymerase, and an absence of spliceosomal introns (though bacterial-like group I and group II introns occur in plastids). Thus, the typical plant or algal cell has three genomes, the mitochondrial, the plastid, and the nucleus, the latter also containing a large number of genes acquired from the organelles.

A mysterious genome

US researcher Araxie Kilejian started the trail of research clues that led to the discovery of a plastid in malaria. She used EM to examine a circular, extrachromosomal DNA molecule in *Plasmodium lophurae*, a malarial parasite of birds. Not expecting to see plastid DNA in a parasite, Kilejian naturally assumed these DNA circles were the parasite's mitochondrial genome. Similar circles were found in other malarial parasites and Iain Wilson's group in London commenced a study of the 35 kb circle from *Plasmodium falciparum*, which causes cerebral malaria in humans. The first sequence data from the circular genome indicated that the genes were prokaryotic in nature. Ironically, finding bacterial-type genes on the 35 kb circle only reinforced the misconception that it represented the mitochondrial genome, because mitochondrial DNAs are typically circular and harbor genes of bacterial origin. But when a third parasite genome was discovered, the penny dropped. The third genome, a linear 6–7 kb element existing as tandem repeats, carried mitochondrial-type cytochrome genes (see above), and in subcellular fractionations, only the linear element co-purifies with mitochondria. The 35 kb genome wasn't the mitochondrial DNA; alternative explanations had to be considered.

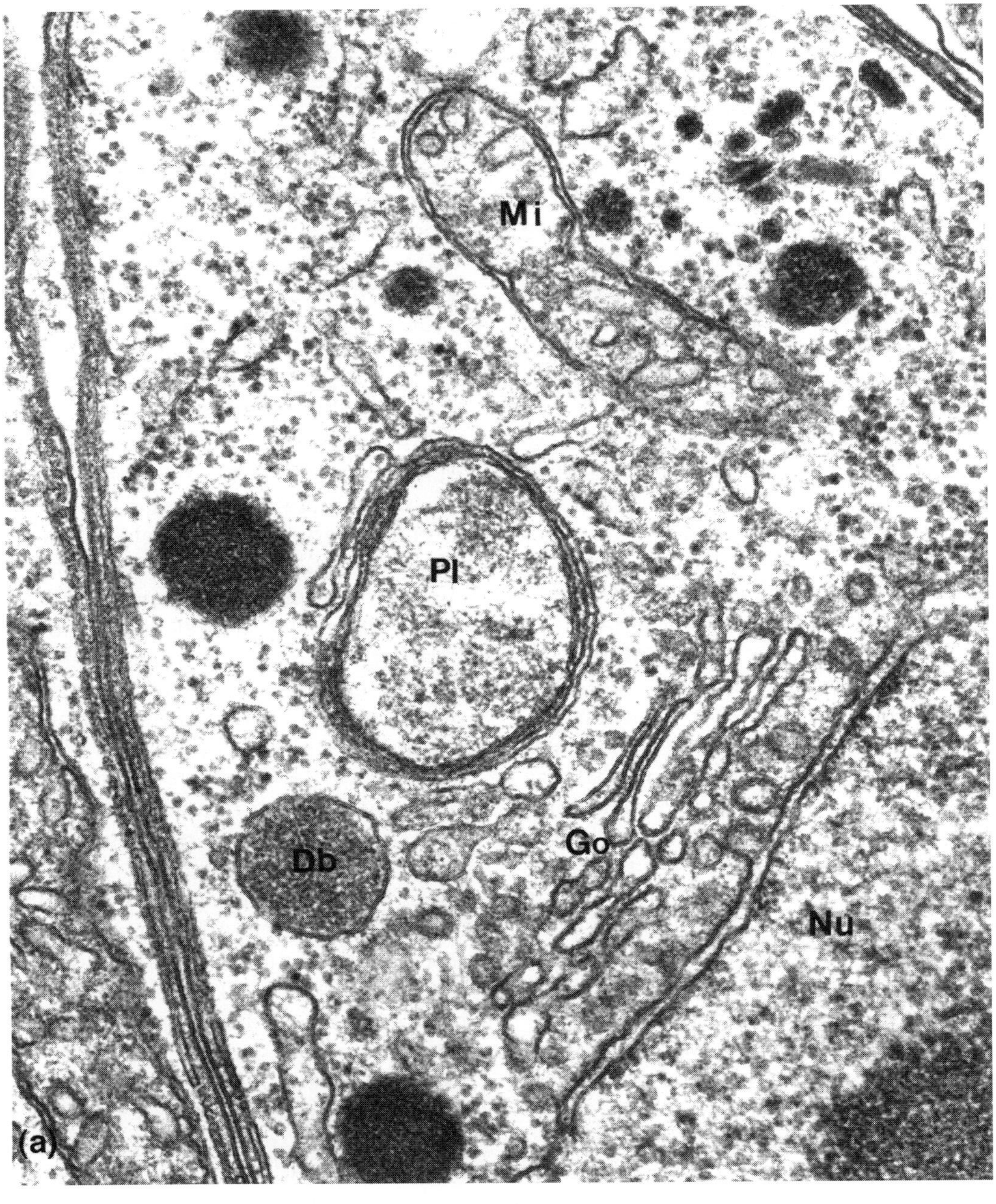

FIGURE 12.2 Transmission electron micrograph of the apicoplast from *Toxoplasma gondii*. The organelle is bounded by four membranes, contains a number of ribosomes, but lacks internal membranes. A smooth vesicle, likely ER, lies adjacent the apicoplast (Pl). Other structures such as the nucleus (Nu), Golgi (Go), mitochondria (Mi) and dense bodies (DB) are also visible. Scale bar = 200 nm.

Further sequence data from the 35 kb circle led to the conclusion that it is a relict plastid genome. First, the inverted repeats, intitally seen by Kilejian in the EM, were shown to contain the ribosomal RNA genes. Plastid genomes classically have rRNA genes in inverted repeats. Second, the circle encoded a bacterial-type RNA polymerase similar to the one in plastids. Since mitochondria do not normally use a polymerase like this, it was increasingly difficult to describe the 35 kb circle as a bizarre mitochondrial genome. Finally, the discovery of a plastid gene ycf24 (which probably functions in the assembly of iron–sulfur centers in electron carriers like ferredoxin) on the malaria 35 kb circle convinced many biologists that something plant-like was lurking in the malaria parasite.

In plants and algae, plastids are large and conspicuous, heavily pigmented, and packed with membranous folds known as thylakoids. While the molecular genetic data pointed to a vestigial plastid in malaria, researchers didn't find one. However, several reports of curious, membrane-bound organelles in malaria led parasitologists to speculate that these structures were relict plastids that could perhaps harbor the 35 kb genome. It was time to determine exactly where in the parasite cell this mysterious genome was located.

In situ localization of the 35 kb genome traced it to a single, roughly spherical, multi-membrane-bound organelle located adjacent to the nucleus. The labeling experiments clearly indicate that this organelle, which is bounded by several membranes (Figure 12.2), houses the 35 kb genome. The mitochondria were not labeled, suggesting they contained only the 6–7 kb linear genome. The contents of the 35 kb genome-containing organelle are largely homogeneous; the only discernible structures in the organelle are particles of size comparable to the 70S ribosomes of plastids, mitochondria and bacteria. As the *in situ* hybridization data show that the 35 kb genome is housed in a membrane-bound compartment distinct from the mitochondrion, and the phylogenetic analyses of genes and operon structure clearly ally this genome with plastids, it is clear that this organelle is a vestigial plastid, which is now referred to as the apicoplast.

How did we miss the plastid for so long? Well, we didn't. We just didn't recognize it as a plastid. Electron microscopists saw the plastid two decades earlier, but with no molecular genetic clues to expose their ancestry at the time, they could not recognize the mysterious structures as relict photosynthetic organelles.

Apicoplast origin

Molecular phylogenetics had clearly demonstrated that the malaria parasite, along with many other parasites such as *Toxoplasma* that together make up the phylum Apicomplexa, are the closest relatives of an algal group known as dinoflagellates. Given that Apicomplexa and dinoflagellates are close relatives, and both have plastids, we might assume that they acquired these plastids prior to their divergence. However, in protist evolution plastid acquisition has been a complex business. As discussed above, the origin of plastids through cyanobacterial endosymbiosis is well established, but this process is only the first chapter in a set of endosymbiotic events responsible for the acquisition of plastids in a range of eukaryotic lineages. This first chapter (cyanobacterium + eukaryotic host = photosynthetic eukaryote) is referred to as the primary endosymbiosis. A subsequent chapter in plastid acquisition is referred to as secondary endosymbiosis, and can be described by the equation (photosynthetic eukaryote + eukaryotic host = different photosynthetic eukaryote). In this chapter the product of the first

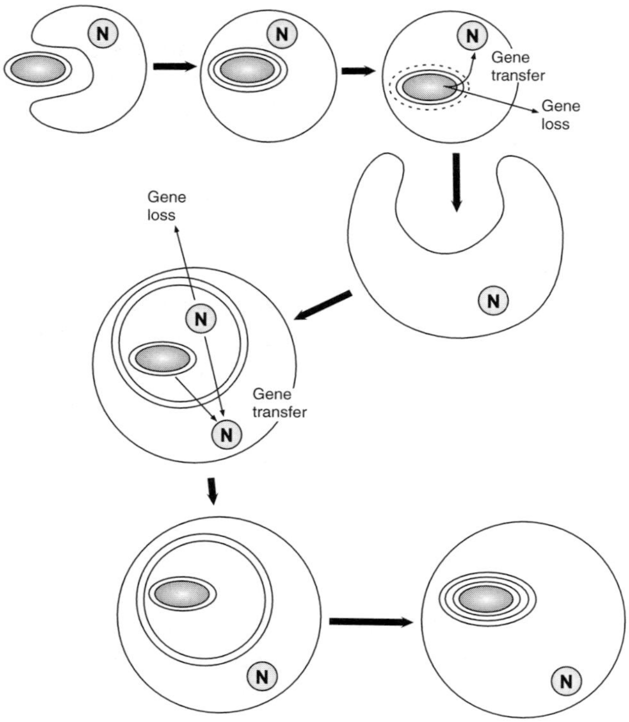

FIGURE 12.3 Sequential endosymbiotic events (primary and secondary) producing plastids. A primary plastid arises from engulfment and retention of a photosynthetic prokaryote. The phagosomal membrane ruptures releasing the endosymbiont, with its two membranes derived from the Gram-negative membranes, into the cytoplasm. Genes transferred from the prokaryote to the eukaryote host nucleus are targeted back to the endosymbiont using a transit peptide. Plastids acquired by secondary endosymbiosis typically have four bounding membranes. The inner pair originates from the primary plastid membranes (i.e. the two membranes from the bacterium). The outer membrane derives from the phagosome (food vacuole) and the membrane between the outer one and the innermost pair derives from the endosymbiont plasma membrane. Genes transfer to the secondary host nucleus from the endosymbiont nucleus, which eventually disappears leaving only the four membranes. Algal examples of each stage are known. In Apicomplexa the eventual loss of photosynthesis resulted in a four membrane-bounded, relict plastid.

equation (photosynthetic eukaryote) is the first component of the second equation (Figure 12.3). While primary endosymbiosis happened (to the best of our knowledge) only once, secondary endosymbiosis occurred at least twice and, some would argue, perhaps numerous times. Secondary endosymbiosis results in plastids with more than two membranes; three or four membranes is typical (Figure 12.3).

Although dinoflagellates and Apicomplexa are close relatives, they both have secondary plastids, which leaves open the possibility that they have two separate secondary acquisitions. Dinoflagellates almost certainly contain a red algal endosymbiont but an early analysis of apicomplexan plastid genes concluded that the apicomplexan plastid derived from a secondary endosymbiotic green alga. More recently trees incorporating plastid rRNA genes show dinoflagellate plastids to be most closely related to plastids of apicomplexan plastids. At face value this relationship suggests that the plastids were acquired by a common secondary endosymbiotic event, but the trees are not clear-cut. The branches of dinoflagellate and apicomplexan plastids are extraordinarily long and, as discussed above, their grouping must be regarded with abject caution. A potentially more telling piece of evidence for a common origin comes not from the difficult-to-tree plastid genes but from the nuclei of the hosts. Trees of glyceraldehyde-3-phosphate dehydrogenase (GAPDH) genes provide striking evidence that the plastids of dinoflagellates and Apicomplexa (as well as other algal groups) have acquired their plastids in one common secondary endosymbiosis of a red alga.

Understanding the origin of the apicoplast is more than just an academic exercise. Because the apicomplexan plastid is potentially an excellent target for anti-parasite drugs (see below), it is important that we understand its evolutionary history. If a common origin for dinoflagellate and apicomplexan plastids is confirmed, we will have deepened our understanding of the origins of one of the world's deadliest parasites. Dinoflagellates and Apicomplexa diverged at least 400 million years ago, but despite outward appearances they are not so fundamentally different. Both have the ability to associate closely with animals, dinoflagellates as endosymbionts of corals and other invertebrates and Apicomplexa as intracellular parasites. An attractive scenario is that this ability to associate with animals goes back to their common ancestor, and that one lineage (dinoflagellates) persisted with photosynthesis and commensal interactions while another (Apicomplexa) abandoned photosynthesis, instead converting to parasitism to exploit the host. This presumably happened quite early in animal evolution, but the parasites are still with us. Why Apicomplexa keep a vestige of their plastid is the next question.

Apicoplast function

Why do these parasites still have a vestigial plastid? As far as is known, apicomplexan parasites are not photosynthetic, but the plastid is indispensable. Clearly, evolution has failed to expunge the organelle despite several hundred million years of parasitic living. Moreover, lab-generated mutants in which the plastid is unable to divide (and therefore unable to be passed on to daughter parasites) are not viable. Clearly, the parasites depend on the plastid for some service or function. Interestingly, plants are dependent on their plastids too. Some plants, such as beechdrops or Indian pipe, have abandoned photosynthesis and are wholly parasitic. Importantly, these non-photosynthetic plant parasites always keep their plastids, although they are much reduced and lack chlorophyll. Why is this?

Although the main role of plastids in photosynthesis is well recognized, it is less well appreciated that a plant plastid is the sole site for several essential cellular processes such as heme, isoprenoid, essential amino acid and lipid synthesis. These functions probably make plastids indispensable for plants and many algae. Could parasites be dependent on their apicoplasts for one or more of these functions? It is a pervasive paradigm in parasitology that parasites don't make anything that they can procure from their hapless hosts. For instance, because malaria parasites can scavenge and modify lipids from the erythrocyte, it was believed that malarial parasites are unable to synthesise fatty acids *de novo*. This dogma has recently been overturned with the identification of a fatty acid synthesis pathway in the apicoplast. Similarly, pathways for isoprenoid and heme biosynthesis have also been partially characterized in malarial apicoplasts. At this stage no one has demonstrated that products of these plastid-like anabolic pathways are exported from the organelle for use by the parasite, but it seems likely that one or more of these products is essential to the parasite and the basis for apicoplast persistence.

Protein targeting to the apicoplast

Products of nuclear genes that accumulate in the plastid are translated on cytoplasmic ribosomes and then targeted into the organelle. Plastid targeting can be classified into two broad categories depending on evolutionary origin and ultrastructure, particularly the number of membranes surrounding the chloroplast. In plants, green algae, red algae and glaucophytes, whose chloroplasts are enclosed by two membranes, targeting is mediated by an N-terminal presequence known as the transit peptide, which is removed after import. Plant and green algal transit peptides are the best characterized, and while no consensus sequence or secondary structure is evident, they are typically 25–125 amino acids in length, basic, and enriched in particular amino acids.

Targeting presents a more complex problem in organisms with more than two membranes around the plastid. Details of protein trafficking to plastids containing multiple membranes is not well understood, but because the outermost membrane is part of the host's endomembrane system, targeting apparently commences via the secretory pathway into the endoplasmic reticulum (ER) using a classic signal peptide. Subsequent targeting across the inner pair of chloroplast membranes involves a downstream transit peptide.

Nuclear-encoded apicoplast proteins were recently identified, and all the genes are predicted to encode substantial N-terminal extensions when compared to the equivalent chloroplast and bacterial proteins. These N-terminal extensions are cleaved *in vivo*, and the extension is sufficient to target a reporter protein (green fluorescent protein, GFP) into the parasite apicoplast of *Plasmodium falciparum* or *Toxoplasma gondii*.

The extreme N-terminal regions (16–34 amino acids) of these extensions resemble classic signal peptides, containing a hydrophobic domain followed by a 'von Heijne' cleavage site. Downstream of the predicted signal peptides, the proteins exhibit the general features of transit peptides. The presequence thus has a bipartite structure similar to that proposed for algal models. Routing to the apicoplast thus occurs via the endomembrane system, and the transit peptide is somehow responsible for sorting within the endomembrane pathway for sequestration into the apicoplast.

It is not yet clear how many genes encode proteins for the apicoplasts. In plants, about 2000 nuclear genes code for plastid proteins. The number is undoubtedly less in apicomplexan parasites, and an estimate of 800 has been made. Completion of the parasite genomes should allow identification of these genes and the assembly of gene-based pathways for the organelle.

Medical and veterinary significance of apicoplasts

The Apicomplexa comprise at least 5000 species and, with the possible exception of *Cryptosporidium*, all appear to contain a plastid. This plastid-containing group of endoparasites is of enormous medical and veterinary significance. As well as malaria and toxoplasmosis, Apicomplexa are responsible for a variety of diseases including sarcocystosis in humans; babesiosis, theileriosis, sarcocystosis, and neosporosis in cattle; and coccidiosis in poultry. Treatments for most of these diseases are less than optimal (see Chapter 17).

What relevance does the plastid have for diseases caused by these parasites? Much of the current effort against malaria is based on vaccines directed against bloodstream stages of the parasite and typically directed at surface antigens. Malaria's photosynthetic ancestry probably has little or no bearing on these antigens or the ultimate success of the vaccines. However, more efficacious drug treatments for malaria and toxoplasmosis would be useful adjuncts to a vaccine, and it is in this area that the plastid could have enormous implications.

The apicoplast appears to contain ribosomes, and the genome encodes both protein and RNA components of ribosomes, making it likely that the parasite plastid contains machinery to express its information content. This machinery, which represents a third genetic compartment in these cells, is a potential target for therapeutics. A number of antibiotics inhibiting either transcription or translation in prokaryotic-like systems can block mitochondrial and/or plastid function. Interestingly, some also inhibit growth of apicomplexan parasites. For instance, bacterial translation blockers like doxycycline, thiostrepton, spiramycin, and clindamycin inhibit malaria growth. No direct evidence for action on apicoplast translation for these compounds is yet available, but it seems likely that the anti-malarial activity derives from plastid inhibition. Further up the information chain, rifampicin (a blocker of bacterial transcription) is also anti-malarial and ciprofloxacin (which blocks bacterial-type DNA replication) stops apicoplast genome replication and kills parasites. These pharmacological phenomena strongly suggest that the apicoplast is an excellent drug target.

The above-mentioned parasiticidal agents are thought to act on the apicoplast's genetic machinery, blocking genome expression at various levels. Further downstream there is also potential for pathway inhibition. Antibacterials with targets in the fatty acid biosynthetic machinery such as thiolactomycin (FabF and FabH), triclosan (FabI) and arylphenoxypropionate herbicides (ACCase) have anti-malarial or anti-toxoplasmodial activity. Isoprenoid biosynthesis has also been targeted by fosmidomycin, which blocks deoxyxylulose phosphate synthase, an enzyme unique to plastids and the bacteria from which they derive and absent from eukaryotic organisms without plastids. The apicoplast therefore presents itself as a promising, parasite-specific target with a wide selection of pathways that can be targeted with drugs. Significantly, a large body of detailed information about these pathways, the enzymes involved and the modes of action of a number of specific inhibitors is already available from studies of bacteria and plant chloroplasts. This information is a fantastic springboard for exploration of novel drug options for disease management. The tremendous success of antibiotics directed against bacteria stems from their specificity. Because they target prokaryote-specific functions, antibacterials generally have few contraindications. Protozoan, fungal and viral diseases, on the other hand, are more difficult to treat as they share many of the host's metabolic processes. The Apicomplexan plastid, therefore, presents a unique, and potentially exploitable, difference between host and parasite.

ACKNOWLEDGMENT

The author thanks David Roos for Figures 12.1B and 12.2, and Guy Brugeroke for Figure 12.1C.

FURTHER READING

Akhmanova, A., Voncken, F., Alen, T.V. *et al.* (1998). *Nature* **396**, 527–528.

Beeson, J., Winstanely, P., McFadden, G. and Brown, G. (2001). *Nature Med.* **7**, 149–150.

DeRocher, A., Hagen, C.B., Froehlich, J.E., Feagin, J.E. and Parsons, M. (2000). *Cell Sci.* **113**, 3969–3977.

Doolittle, W.F. (1998). *Nature* **392**, 15–16.

Dyall, S.D. and Johnson, P.J. (2000). *Curr. Opin. Microbiol.* **3**, 404–411.

Dyall, S.D., Koehler, C.M., Delgadillo-Correa, M.G. *et al.* (2000). *Mol. Cell Biol.* **20**, 2488–2497.

Embley, T.M. and Martin, W. (1998). *Nature* **396**, 517–519.

Feagin, J.E., Mericle, B., Werner, E. and Morris, M. (1997). *Nucl. Acids Res.* **25**, 438–446.

Gardner, M.J., Bates, P.A., Ling, I.T. *et al.* (1988). *Mol. Biochem. Parasitol.* **31**, 11–18.

Gardner, M.J., Williamson, D.H. and Wilson, R.J.M. (1991). *Mol. Biochem. Parasitol.* **44**, 115–123.

Gray, M., Burger, G. and Lang, B. (1999). *Science* **283**, 1476–1481.

Jomaa, H., Wiesner, J., Sanderbrand, S. *et al.* (1999). *Science* **285**, 1573–1576.

Keeling, P. and McFadden, G. (1998). *Trends Microbiol.* **6**, 19–23.

Keeling, P.J. (1997). *Bioessays* **20**, 87–95.

Martin, W. and Herrmann, R.G. (1998). *Plant Physiol.* **118**, 9–17.

Martin, W.F. and Müller, M. (1998). *Nature* **392**, 37–41.

McFadden, G.I., Reith, M., Munholland, J. and Lang-Unnasch, N. (1996). *Nature* **381**, 482.

McFadden, G.I. and Roos, D.S. (1999). *Trends Microbiol.* **6**, 328–333.

Muller, M. (1997). *Parasitol. Today* **13**, 166–167.

Palmer, J. (1997). *Science* **275**, 790–791.

Roger, A.J., Clarke, C.G. and Doolittle, W.F. (1996). *Proc. Natl. Acad. Sci. USA* **93**, 14618–14622.

Roger, A.J., Svard, S.G., Tovar, J. *et al.* (1998). *Proc. Natl. Acad. Sci. USA* **95**, 229–234.

Sato, S., Tews, I. and Wilson, R. (2000). *Int. J. Parasitol.* **30**, 427–439.

Sogin, M. (1993). *Nature* **362**, 795.

Surolia, N. and Surolia, A. (2001). *Nature Med.* **7**, 167–173.

Takeo, S., Kokaze, A., Ng, C.S. *et al.* (2000). *Mol. Biochem. Parasitol.* **107**, 191–205.

Tovar, J., Fischer, A. and Clark, C.G. (1999). *Mol. Microbiol.* **32**, 1013–1021.

Uyemura, S.A., Luo, S.H., Moreno, S.N.J. and Docampo, R. (2000). *J. Biol. Chem.* **275**, 9709–9715.

Vial, H.J. (2000). *Parasitol. Today* **16**, 140–141.

Waller, R.F., Cowman, A.F., Reed, M.B. and McFadden, G.I. (2000). *EMBO J.* **19**, 1794–1802.

Waller, R.F., Keeling, P.J., Donald, R.G.K. *et al.* (1998). *Proc. Natl. Acad. Sci. USA* **95**, 12352–12357.

Wilson, I. (1993). *Nature* **366**, 638.

Wilson, R.J., Williamson, D.H. and Preiser, P. (1994). *Infect. Agents Dis.* **3**, 29–37.

Wilson, R.J.M., Denny, P.W., Preiser, P.R. *et al.* (1996). *J. Mol. Biol.* **261**, 155–172.

Wilson, R.J.M. and Williamson, D.H. (1997). *Microbiol. Mol. Biol. Rev.* **61**, 1–16.

SECTION

III

BIOCHEMISTRY AND CELL BIOLOGY

HELMINTHS

C H A P T E R

13

Helminth surfaces: structural, molecular and functional properties

David P. Thompson and Timothy G. Geary
Animal Health Discovery Research, Pharmacia Corporation, Kalamazoo, MI, USA

INTRODUCTION

Helminth surfaces are where parasites interact with their environment. For infectious stages of animal parasitic species, the environment is the host. These surfaces include external and, for some helminths, internal structures. The external surface, termed the tegument in trematodes and cestodes and the cuticle in nematodes, is highly adapted in a stage- and species-dependent manner for the varied environments occupied by these organisms. Most obviously, the external surface provides a physical barrier that protects internal cells and organ systems from the external environment. The external environments of parasitic helminths living in a host pose many challenges, including various immune effectors, digestive enzymes and potentially toxic concentrations of protons and oxygen. Other essential roles of external surfaces include structural support and regulation of the transport of water, inorganic ions, nutrients and waste products. Cestodes lack an alimentary tract. For nematodes and trematodes, the intestinal surface also plays an important role in digestion, nutrient absorption, and transport of water and inorganic ions. Although incompletely characterized, the protonephridial system in trematodes, tubular system in nematodes, and lateral duct system in cestodes provide yet another surface implicated in osmotic/volume regulation and excretion. This chapter illustrates the structural and functional biology, biochemistry and molecular biology of these surfaces. Within each phylum of helminth, the best studied parasitic species (*Hymenolepis diminuta, Schistosoma mansoni, Ascaris suum*) is highlighted to establish general principles.

Molecular Medical Parasitology
ISBN 0–12–473346–8

Important exceptions are noted. For example, although *Caenorhabditis elegans* is a free-living species, insights it offers into the genetic bases for structural and functional properties of the cuticle, gut and tubular systems in nematodes justifies its inclusion as a model organism. Comments are primarily restricted to adult stages, which are usually the most pathogenic (with some notable exceptions, e.g. *Echinococcus* spp. and *Onchocerca volvulus*) and represent the principal targets for chemotherapy.

CESTODES

Although cestodes have been less thoroughly studied than trematodes or nematodes, we begin with these parasites because of their anatomical simplicity. The absence of both an alimentary tract and a distinct excretory system simplifies the interpretation of functional properties of the external surfaces.

External surface

Structure

Cestodes interact with their environment only across the tegument. In addition to the functional properties associated with the external surface in other helminths, the cestode tegument is also structurally adapted to perform the functions normally associated with an intestine. Indeed, the cestode body plan has been conceptualized as an 'inside-out intestine'. Consequently, research on the cestode surface has been somewhat biased toward features relevant to its role in nutrition.

The adult tegument forms a coherent boundary between the fluid compartment of the fibrous interstitium and the environment.

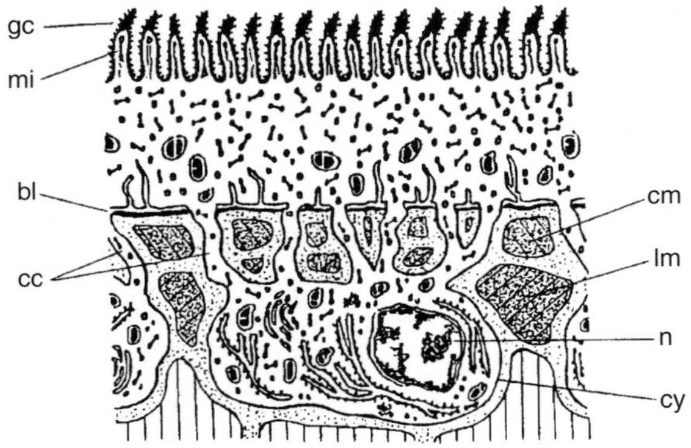

FIGURE 13.1 Schematic of the tegument and underlying tissues in *Hymenolepis diminuta*. Abbreviations: gc, glycocalyx; mi, microtriche; bl, basal lamella; cc, cytoplasmic channel; cm, circular muscle; lm, longitudinal muscle; n, nucleus; cy, cyton (drawing courtesy of J.W. Bowman).

Several structurally distinct components are evident (Figure 13.1). The outermost layer is the glycocalyx. This acellular, translucent mucopolysaccharide–glycoprotein coat, also referred to as the laminated layer, adheres to the tegument plasma membrane, which is the limiting boundary of the tegumental syncytium. The surface is evaginated into microtriches, structures that are reminiscent of the apical microvilli on vertebrate intestinal epithelial cells. Microtriches amplify the functional surface area of the parasite 2- to 12-fold, depending on the species and stage.

The inward-facing membrane of the tegumental syncytium is bounded by a fibrillar basal lamella. The cytoplasm and organelles within the syncytium are synthesized by cell bodies (subtegumental cells, cytons) that lie beneath the basal lamella. Somatic muscle cells are anchored to the fibrillar components of the basal lamella, so that contraction is coupled to movement of the body. Our analysis of the cestode surface focuses on the material bounded by the basal lamella and the glycocalyx.

The cestode tegument is not homogeneous; regions of specialized structures are found over most of the adult animal. These include adaptations that form the scolex, the region involved in attachment to the host intestinal tract. Putative excretory and secretory openings interrupt the continuity of the distal cytoplasm and plasma membrane. Sensory nerve processes terminating in external cilia also interrupt it, as does the genital atrium. Little is known about the biochemistry of these specialized structures. Our discussion of the functional properties of the cestode tegument will therefore focus on tegument typical of the proglottids.

Biochemistry and molecular biology of structural components

The cestode tegument contains many structural proteins and enzymes, some of which are clearly adsorbed from host fluids. *In vitro* studies measuring uptake and incorporation of [^{14}C]leucine by *H. diminuta* revealed at least 17 microtriche proteins synthesized by the parasite. Together, the glycocalyx and microtriches contain at least 30 distinct proteins and glycoproteins, with molecular masses of 12–237 kDa. These proteins impart a negative charge to the surface of the parasite. They are rapidly shed, exhibiting half-lives of only about 6 h, and are replaced by new proteins synthesized in the cytons. Actin, tubulin, collagen and keratin are present at high levels.

The cestode tegument contains a variety of glycolipids and phospholipids. In *H. diminuta*, lipids comprise 40% of the mass of the outer tegumental membrane, including the microtriches, with the major components being cholesterol, cardiolipids and phosphatidylethanolamine. The predominant fatty acids contain 16–18 carbons and are 50–60% unsaturated. Like other helminths, cestodes are unable to synthesize cholesterol or other long-chain fatty acids *de novo* from acetyl-CoA. Lipid droplets are formed in the cytons from fatty acids and sterols that accumulate following absorption from the host across the tegument.

Little is known of the physiological or biochemical characteristics that distinguish the tegument of the developing cestode from the adult. Endocytosis may play a more important role in the absorption of macromolecules in immature stages, though supporting data come only from studies on *Taenia crassiceps*. Marked structural changes in the tegument typically follow the molting process, some of which appear to be regulated by neuropeptides released beneath the basal laminella. In *Diphyllobothrium dendritium*, neuropeptides are released in response to the elevated body temperature of the definitive avian host. The possibility that the tegument actively responds to external signals is suggested by the presence of G proteins in the tegumental brush-border membrane of *H. diminuta*.

Proteins within the glycocalyx and the membrane of the microtriches are protected from digestion by host proteolytic enzymes. The underlying mechanism(s) is unknown. One possibility is that the glycocalyx helps maintain an unstirred water layer between the parasite and the contents of the host intestine. Another possibility is that the organic acid end-products of carbohydrate metabolism excreted by cestodes form a microenvironment in the immediate vicinity of the tegument that is too acidic for host intestinal proteases to function. Alternatively, the glycocalyx may contain factors, as yet undefined, that specifically inhibit host hydrolytic enzymes. For example, *H. diminuta* inactivates trypsin and chymotrypsin by releasing a protease inhibitor. All of these mechanisms, and others, may be important.

Functional biology

Digestion

Digestive enzymes, such as alkaline and acid phosphatases, disaccharidases, and esterases are associated with the outer tegumental membrane. However, it is not known if these enzymes are derived from the parasite or the host. These enzymes may participate in extracorporeal digestion of nutrient macromolecules, i.e. 'contact digestion', although this process has not been established conclusively in cestodes.

Nutrient absorption

Cestodes derive most of their energy from glucose, which is absorbed at high rates across the tegument, primarily by an active transport system. Glucose is absorbed against a steep concentration gradient by cysticerci and adults, and the process is inhibited by agents that block ATP synthesis, including *p*-chloromercuribenzoate, iodoacetate and 2,4-dinitrophenol. Glucose uptake is also blocked by phlorizin, a specific inhibitor of Na^+-coupled glucose transport in mammals and many other organisms. Galactose and several other monosaccharides competitively inhibit glucose absorption, indicating the presence of a common carrier system for hexoses. Though the mechanisms of carbohydrate absorption in cestodes and mammals are generally similar, substrate selectivity is greater in cestodes, and only glucose and galactose are accumulated. Based on immunohistochemical data, high levels of a protein recognized by antibodies to a vertebrate Na^+-dependent glucose transporter (SGLT) are found in the external cyst wall of *T. solium* neurocysticerci, and also in the apical membrane of the tegument of adult worms. The gene encoding this SGLT homolog has not been cloned, so it has not been possible to characterize its transport properties and pharmacology.

Cestodes also may absorb glucose by facilitated diffusion, though this process is less thoroughly delineated than active transport. However, genes for two facilitated glucose transporters, TGTP1 and TGTP2, related to the mammalian transporter GLUT1, have been identified in *T. solium*. Both proteins are localized in cytons and in the apical membrane of the tegument. When cRNAs for these transporters are functionally expressed in *Xenopus* oocytes, they confer functional glucose transport that is inhibited by natural stereoisomers of D-glucose and D-mannose, but not by L-glucose or ouabain. The significance of this facilitated glucose transport system in cestodes must be reconciled with pharmacological data showing that phloretin, a known inhibitor of facilitated glucose transport in other organisms, does not inhibit glucose transport in intact cestodes.

Glycerol is absorbed by *H. diminuta* by passive diffusion at high concentrations (>0.5 mM) and by a carrier-mediated process at lower concentrations. At low concentrations, absorption of glycerol is non-linear, dependent on temperature and pH, and competitively inhibited by α-glycerophosphate. The existence of two distinct glycerol carriers is suggested by studies which show that only about half the saturable component in *H. diminuta* is Na^+-sensitive and inhibitable by 1,2-propanediol. In *H. diminuta*, two distinct carrier-mediated transport systems for fatty acid absorption have been partially characterized, one specific for short-chain fatty acids and the other for longer-chain molecules. Acetate absorption by *H. diminuta* is mediated by the short-chain carrier at low concentrations, but passive diffusion occurs at high concentrations.

Cestodes have at least two choline uptake systems. In *H. diminuta* one predominates at pH $\geqslant$7, is highly sensitive to Na^+ and HCO_3^-, and is inhibited by hemicholinium-3 and amiloride, known Na^+ transport blockers. The other functions at pH <5, via a Na^+-independent mechanism, and is inhibited by benzamil, verapamil and 130 mM Na^+.

Amino acids are absorbed by adult cestodes through multiple carriers that are saturable and sensitive to temperature and a variety of inhibitors; accumulation typically occurs against steep concentration gradients. Cestode amino acid carrier systems are biochemically similar to those in mammals. There are separate carriers for acidic and basic amino acids and multiple carriers for the neutral amino acids, which overlap in their specificities. Unlike mammals, however, which show higher affinities for L-amino acids, cestode transporters are not stereoselective. Also, amino acid absorption in some cestodes is not dependent on cotransport of inorganic ions; replacement of Na^+ with choline or K^+ does not affect transport kinetics.

Cestodes synthesize pyrimidines *de novo*, whereas purines must be absorbed from host fluids across the tegument. Both types of nitrogenous base are absorbed *in vitro*, mediated in part in *H. diminuta* by at least three distinct carriers, two of which contain multiple substrate-binding sites. A passive diffusional component also contributes to both purine and pyrimidine absorption.

Endocytosis does not appear to play a major role in the absorption of proteins, though the capacity of cestodes to absorb macromolecules has not been systematically explored. Horseradish peroxidase and ruthenium red appear to be endocytosed by the pseudophyllidean cestodes *Lingula intestinalis* and *Schistocephalus solidus* and by *Hymenolepis nana*. The apparent lack of endocytosis detected in most studies may be attributable to the fact that only mature proglottids were examined.

Transport of inorganic ions

Homeostatic regulation of inorganic ions is a key function of all transporting epithelia. Close orthologs of proteins that serve this function in vertebrates are present in cestodes. Na^+/K^+-ATPase, a key regulator of intracellular levels of these cations in all metazoa, has been demonstrated in the tegument of *Echinococcus granulosus*, but has not yet been reported in other cestodes. However, ouabain, a specific inhibitor of this enzyme, blocks amino acid absorption by *H. diminuta*, suggesting that Na^+/K^+-ATPase is present in other species. The cestode tegument also contains a Ca^{2+}-ATPase activity which, by analogy with better characterized systems, maintains intracellular Ca^{2+} at low concentrations. Two forms of the enzyme occur in the outer tegumental membrane of *H. diminuta*, one of which is calmodulin-dependent. Though no direct electrophysiological measurements of an ion channel in a cestode tegument have been reported, at least two large conductance cation channels have been isolated and partially characterized from protoscoleces of *E. granulosus*. One is a 244 pS channel that resembles some Ba^{2+}-sensitive, Ca^{2+}-activated K^+ channels in vertebrates. The other is a 107 pS channel with a very high open probability under a broad range of recording conditions, resembling a cation channel that has also been characterized in the apical membrane of *S. mansoni* tegument. Neither the location nor physiological roles of these channels has been determined.

Osmotic/volume regulation

Control of hydrostatic pressure gradients across the body wall and across cell membranes in cestodes has been linked to bulk ion transport. Some cestodes are osmoconformers, adjusting internal osmotic pressure by varying fluid volume, thereby shrinking or swelling in direct relation to the osmolarity of their environment. A few species, including *H. diminuta*, regulate their volume (i.e. maintain their weight at a constant level) over a fairly wide range of osmotic pressures. However, even these worms rapidly swell or shrink in media that are below or above that range, respectively. The mechanisms by which *H. diminuta* regulates its volume are not known, but the ability to excrete organic acid end-products of glucose metabolism (succinate and acetate) appears to play a critical role. Osmoregulation is glucose-dependent, and both glucose absorption and metabolism increase linearly with osmotic pressure between 210 and 335 mosmol/liter, whereas rates of uptake of leucine, Na^+ and Cl^- are unaffected by osmotic pressure. As noted above, glucose transport proteins, including one that couples to Na^+ transport, are abundant in the apical membrane of the tegument. In mammals, Na^+/glucose cotransporters play a key role in water transport across membranes, and they probably serve a similar role in the cestode tegument. A duct-like structure in the terminal proglottid may also contribute to osmoregulation in cestodes (see below).

Immune evasion

Little is known about the role of the tegument in immune evasion by cestodes compared to trematodes. The glycocalyx is believed to provide some protection from host immune effectors and digestive enzymes. In cestodes in culture, disruption of this layer by drugs or solvents exposes the apical membrane of the tegument to damage from immune components. Another immune evasion strategy is referred to as molecular mimicry. This process is thought to occur when parasites evolve antigenic convergence with their host, and thereby camouflage proteins exposed on their surfaces. There are numerous examples of parasites that express host-related proteins and glycoproteins, including receptors, on the apical membrane of the tegument. Though definitive data for involvement of specific proteins are lacking, one possible example is the SGLT1-related protein associated with the apical membrane of the tegument. Antibodies raised against residues 564–575 of rabbit SGLT1 recognize a common epitope in the cestode homolog of this protein. This protein is strategically located for disguising the parasite from recognition by host immune effectors.

Putative osmoregulatory ducts: an internal surface in cestodes?

Adult cestodes possess a network of ducts that form within the terminal proglottid and appear to empty directly into the host intestine. In larval cysticerci, these ducts do not open to the outside, but rather empty into the central cyst, which enlarges with fluid over time. Recent immunohistochemical studies have localized high levels of a Na^+-dependent glucose cotransporter (a homolog of human SGLT1) to the inner lining of these ducts. Though no biochemical or physiological evidence has been reported for a role for these structures in osmoregulation, it has been hypothesized that Na^+/glucose cotransport may regulate water transport across the membanes that form these ducts. In humans, SGLT1 couples the transport of 210 water molecules per two Na^+ ions and one glucose molecule, and it is tempting to speculate that a similar process may occur within the putative osmoregulatory ducts in cestodes.

TREMATODES

In addition to a tegument, adult trematodes possess an internal surface, the gastrovascular cavity (gastrodermis), which has a single opening that serves as both mouth and anus. The gastrodermis appears to be an important organ for nutrient digestion and absorption, though definitive experimental support for the latter is lacking. Trematodes also possess a canal/flame-cell system that may play a role in osmoregulation. The belief that tegument-associated antigens in trematodes are good candidates for vaccine development has motivated considerable research on tegument proteins and the effects of immune effectors on the external surface, primarily in *Schistosoma* spp. Few surface proteins, however, have been characterized biochemically. Focus on the external surface of trematodes has also been spurred by the belief that the tegument is an important site of action for the anti-schistosomal drug praziquantel.

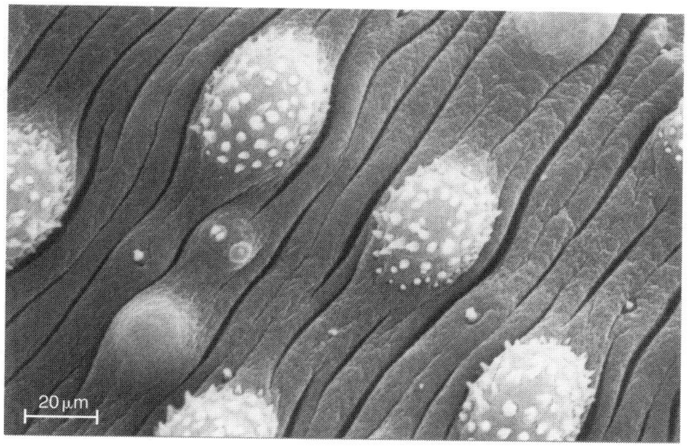

FIGURE 13.2 Scanning electron micrograph (×2000) of the dorsal tegument of adult male *Schistosoma mansoni*. Scale bar = 20 μm.

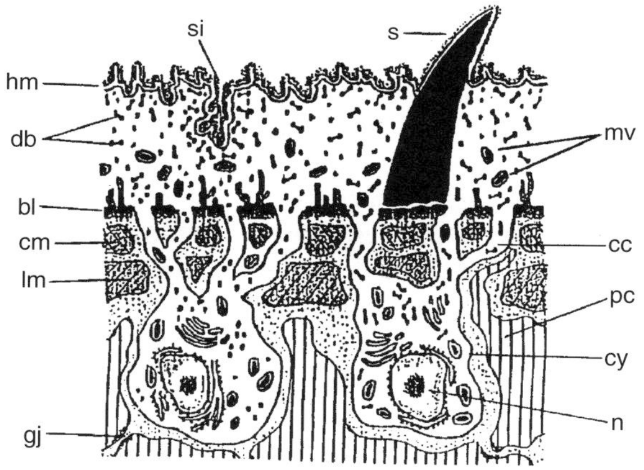

FIGURE 13.3 Schematic of the tegument in *Schistosoma mansoni*. Abbreviations: hm, heptalaminate membrane; db, discoid body; cm, circular muscle; mv, multilaminate vesicle; s, spine; si, surface invagination; bl, basal lamella; lm, longitudinal muscle; gj, gap junction; cc, cytoplasmic channel; pc, parenchymal cell; cy, cyton; n, nucleus (drawing courtesy of J.W. Bowman).

External surfaces

Structural considerations

Trematodes are bounded by a syncytial tegument that structurally resembles the cestode tegument (Figures 13.2 and 13.3). In schistosomes, the tegument is about 4 μm thick and is limited externally by a heptalaminate membrane consisting of two closely apposed lipid bilayers. The outer bilayer exhibits several properties that are atypical of plasma membranes. It contains few, if any, integral membrane proteins, exhibits no endocytosis, and is divided into two domains, one that sheds rapidly with a half-life of <11 h, and another that is shed very slowly. Based on the distribution of intramembranous particles observed in freeze-fracture studies, the inner bilayer is the true plasma membrane of the parasite. The double bilayer extends inward 1–2 μm along numerous pits that branch and interconnect to form a lattice which increases the surface area of the parasite at least tenfold. The tegument of *F. hepatica* is much thicker and less pitted than that of schistosomes, and is bounded externally by a standard lipid bilayer. In both schistosomes and fasciolids, a thin coat of glycoproteins with projecting side chains of gangliosides and oligosaccharides extends outward from the apical membrane of the tegument. The function of this structure, referred to as the glycocalyx, is unknown. Cytoplasm bounded by the tegumental membranes in schistosomes contains mitochondria and two major inclusions: discoid bodies and multilaminate vesicles. Discoid bodies are precursors of the dense ground substance of the cytoplasm as well as of the surface bilayer. They may also break down to form crystalline spines which are most prominent on the dorsal surface of male schistosomes (Figure 13.2). Multilaminate vesicles consist of concentric whorls of membrane which migrate to the surface and form the outer bilayer. Mitochondria in the syncytial tegument are sparse, small and contain few cristae.

In the teguments of schistosomes and fasciolids, nuclei, most mitochondria and ribosomes are contained in subtegumental cells (cytons) located beneath the layers of muscle that underlie the tegumental syncytium. The cytons are connected to the syncytium through cytoplasmic channels that contain numerous microtubules. Inclusion bodies synthesized in the subtegumental cells arise from Golgi bodies and are transported to the syncytium through these channels. This process is interrupted by protein synthesis inhibitors or microtubule poisons. The tegumental surface membrane is continually replaced by multilaminate vesicles that migrate to the apical region of the syncytium and fuse with the plasma membrane.

Biochemistry and molecular biology of structural components

The schistosome tegument contains several phospholipids, the most abundant being phosphatidylcholine, with relatively high levels of phosphatidylethanolamine, sphingomyelin, lysophosphatidylcholine, phosphatidylinositol and cerebroside glycolipids. Palmitic acid and oleic acid are the most abundant fatty acids. These molecules are derived from the host, as schistosomes, like cestodes, are unable to synthesize cholesterol or long-chain fatty acids *de novo*. Schistosomes cannot degrade long-chain fatty acids by β-oxidation. However, they can interconvert fatty acids and cleave the polar head from phospholipids. The apical membrane of the schistosome tegument is continuously replaced from multilaminate and discoid bodies. Estimates for the turnover rate of the apical membrane range from a few hours to several days. This wide range of estimates is probably due to the fact that some types of molecules (e.g. phosphatidylcholine) used as probes in these studies turn over much faster than others.

The schistosome surface membrane contains a range of enzymatic activities, including alkaline phosphatase, Ca^{2+}-ATPase and glycosyl transferase. Ouabain-binding sites on the surface of the tegument indicate the presence of Na^+/K^+-ATPase. A large number of associated proteins in schistosomes has been detected immunologically. Some are embedded in the tegumental membrane, whereas others are anchored by glycosylphosphatidylinositol (GPI). At least one is dependent on palmitoylation for membrane attachment. Unfortunately, little is known about the function of these surface-associated proteins. Paramyosin is an example of a protein that was identified and characterized after it was shown to be a partially protective antigen in the tegument. Biochemical studies on antigenic proteins have also revealed that the surface spines are composed of paracrystalline arrays of actin, and that glutathione S-transferase is present in the tegument.

Carbohydrates exposed on the surface of schistosome glycocalyx include mannose, glucose, galactose, *N*-acetylglucosamine, *N*-acetylgalactosamine and sialic acid. These sugars are linked to glycolipids and glycoproteins in the tegument, which also contains the enzymes required for their biosynthesis. Surface glycoproteins are anchored to the outer lipid bilayer by GPI through a process mediated by an endogenous phophatidylinositol-specific lipase.

Developmental biology

Schistosome cercariae undergo a profound structural and biochemical transformation during and after penetration into the definitive host. During penetration, cercariae lose their tails and begin to shed their fibrillar surface coat. Detachment of the glycocalyx follows secretion of proteases from the surface; these proteases may also play a role in tissue penetration. By one hour after penetration, multilaminate vesicles begin to migrate from the syncytium to the outer surface of the tegument. During this period, numerous microvilli form from the outer membrane of the tegument, then are rapidly shed in a process that leads to the formation of the heptalaminate outer membrane. This protein shedding process may play an important role in the development of concomitant immunity, since it may prime the host to mount immune responses against subsequent challenge by cercariae. Formation of the heptalaminate membrane is complete within three hours after penetration. At this point, the tegument of the schistosomulum and adult stages are morphologically similar, as are their sterol and phospholipid compositions. Much higher levels of several surface proteins are expressed by immature stages than by adult schistosomes, and many surface proteins are differentially expressed on the surface of adult males and females.

Functional biology

Biological functions of the structure

The trematode tegument is structurally adapted for immune evasion, nutrient absorption, ion transport and communication with the underlying neuromuscular system. Numerous pits at the surface of the tegument markedly increase the surface area of the parasite, which is consistent with a transport function (see below). In schistosomes freshly collected from host tissue, these pits contain erythrocytes, suggesting that they are open to the external environment. The tegument is less pitted in *F. hepatica*, but numerous invaginations of the surface effectively increase its surface area. The double layer of membrane at the surface of the tegument in schistosomes is common only among blood-dwelling trematodes; species that inhabit other environments, including *F. hepatica*, which resides in the bile duct, are limited by a standard lipid bilayer.

The tegument in schistosomes is electrically coupled to underlying muscle bundles, allowing changes in some ionic and electrical gradients at the surface of the parasite to be transmitted indirectly to the muscle fibers below. Gap junctions connecting these two tissue layers are believed to provide the morphological substrate for this low resistance pathway. This link between the surface membrane and underlying muscle may serve an important role in the ability of the parasite to respond rapidly to external stimuli, such as mechanical pressure, nutrient or ionic gradients. In addition, spines, which are predominantly on the dorsal surface of trematodes, may serve as holdfasts that maintain the parasites within blood vessels or, in the case of *F. hepatica*, the bile duct.

Immune evasion

Even a cursory summary of the tremendous body of evidence supporting a role for the schistosome tegument in immune evasion is beyond the scope of this chapter. Instead, we consider only a few examples that illustrate the complex array of evasion strategies employed by schistosomes to avoid destruction by natural or vaccine-induced immune effectors. Compelling evidence for the critical role of the tegument in immune evasion comes from studies showing that drugs that disrupt the outer bilayer of the tegument, such as praziquantel and oxamniquine, expose the parasite to extensive damage from the host immune response. Antisera that are toxic to schistosomes following exposure to low concentrations of these compounds have no intrinsic activity in the absence of tegumental disruption. Several distinct physical properties and proteins associated with the tegument contribute to immune evasion by schistosomes. However, definite evidence that any one of them is more critical than others is lacking. In this regard, several structural features of the tegument may contribute to immune evasion. The double membrane at the apical surface of the tegument in schistosomes may satisfy a structural requirement that allows continuous replacement of old or immunologically damaged membrane by new membrane derived from the multilaminate vesicles. The double membrane may also permit host antigens, largely blood group glycolipids and glycoproteins, to be incorporated into the surface of the parasite, thereby disguising it from immune attack. This form of camouflage is supplemented by another form, commonly referred to as molecular mimicry, wherein parasite-derived molecules that contain host-like epitopes are exposed at the surface. These proteins include, for example, homologs of the Na^+-dependent glucose transporter SLGT1, which are abundantly expressed on the surface of the apical tegumental membrane in schistosomes and fasciolids (see below). Some lipid components of this membrane, such as lysophosphatidyl choline, also protect the organism by lysing attached host leukocytes. Parasite-derived enzymes capable of degrading host proteins, including the cysteine protease SmCL1, are present at high levels on the surface. Some may be able to degrade host immunoglobulins. Paradoxically, the tegument of schistosomes also contains numerous proteins and glycoproteins that are potent immunogens. Several of these proteins, including the glycolytic enzymes glyceraldehyde-3P-dehydrogenase and triose-phosphate dehydrogenase, are released from the tegument and appear to play a role in concomitant immunity to reinfection (i.e. they stimulate production of antibodies that bind to schistosomula in subsequent infections).

Nutrient absorption

A role for the tegument in nutrient absorption can be inferred on the basis of structural, biochemical and ecological considerations. Numerous pits and channels provide an enormous surface area for absorption, and the tegument in adult stages is continuously exposed to host fluids, from which the parasite derives all of its nutrients. Several enzymes that function in amino acid absorption are located in the tegument, and in some cases (e.g. leucine aminopeptidase), these enzymes are absent from the intestine. Both *S. mansoni* and *F. hepatica* can survive extended *in vitro* incubations in the absence of detectable nutrient absorption across the gastrodermis. In addition, glucose absorption occurs in immature stages of trematodes, which lack an intestine.

Adult schistosomes are absolutely dependent on host serum glucose for energy. Glucose flux across the tegument occurs by an active, Na^+-dependent, carrier-mediated process. The inward-directed Na^+ gradient is maintained, in part, by at least one Na^+/K^+-ATPase in the tegument. This provides part of the electrochemical driving force for hexose transport.

Two schistosome glucose transporters have been cloned. SGTP4 is homologous to the human glucose transporter, GLUT1. It is only detectable in schistosomula and adults (i.e. stages where glucose is present in the environment), and is localized exclusively in the tegument. SGTP4 is integrated within membranous vesicles that are assembled within the cytons. The other glucose transporter, SGTP1, is localized exclusively in the intestine (see below). SGTP4 and SGTP1 have both been functionally expressed and partially characterized in the *Xenopus* oocyte system. This experimental approach should facilitate a thorough comparison of the biochemistry and pharmacology of parasite and host glucose transporters, perhaps setting the stage for the identification of parasite-selective inhibitors of glucose transport as leads for anthelmintic development.

Most glucose absorption occurs across the dorsal tegument in male schistosomes; there is no evidence for regional differences in females. An intriguing aspect of glucose uptake is that, for worm pairs *in copula* (the normal condition *in vivo*), all or most glucose obtained by the female is supplied by the male partner. In addition, mated pairs absorb more glucose than unmated schistosomes, even though the total worm surface area exposed to the external medium is much reduced in paired worms. This increase in glucose uptake in pairs may occur as a result of a greater number of transport sites on the dorsal surface of the male.

Amino acid uptake by trematodes occurs primarily across the tegument. Mechanical ligation of the intestine or drug treatments that induce regurgitation do not significantly affect amino acid uptake kinetics. Amino acid absorption by *F. hepatica* occurs by passive diffusion, as uptake is linear over a wide range of concentrations and there is no evidence for competitive inhibition among related amino acids. In schistosomes, it appears that methionine, glutamate, arginine and alanine are absorbed across the tegument by a combination of passive diffusion and a carrier-mediated system. The number of distinct amino acid uptake systems in schistosomes is unknown, though there is evidence for at least two and possibly three. These systems have a high affinity for amino acids only, as methionine uptake is not inhibited by other organic compounds, and moving the amino group to carbons adjacent to the α-carbon reduces the affinity of the transport system.

Trematodes do not synthesize cholesterol *de novo*, but acquire it from the host. Uptake occurs via the tegument, primarily at the dorsal surface of males, and cholesterol is then redistributed throughout the body. Cholesterol, like glucose, is transferred between partners *in copula*. Other lipids, including ceramide, are also transported rapidly across the tegument. Absorption may be restricted to distinct patches of membrane on the schistosome surface. Of particular interest are caveolae-like microdomains found mainly in the tubercles of male worms. In other eukaryotes, caveolae serve as pockets in which signal transduction occurs, via a linker protein called caveolin; this protein has not been identified in schistosomes.

Osmotic/volume regulation

During the process of transformation from cercaria to schistosomulum following penetration of the host skin, schistosomes become highly permeable to water. When placed in hyposmotic medium, schistosomula and adult schistosomes gain weight rapidly, then exhibit a slow recovery to control weight levels. This response indicates a capacity for osmoregulation, and has been interpreted as an initial osmotic uptake of water across the tegument followed by a slower efflux of ions from the tissue, with water following. In hyperosmotic medium, schistosomes slowly lose weight due to osmotic efflux of water. How trematodes re-establish osmotic equilibrium is not known, but their well-developed protonephridial system may play a role (see below). However, there is no direct evidence that the protonephridial system responds to ionic or osmotic stress. Osmoregulation could also be mediated by active transport of ions across the tegument. In *S. mansoni*, the Na^+ concentration in extracellular fluid is approximately equal to that in the bathing medium over a wide range. The Na^+ concentration in the tegumental syncytium, however, is maintained at a constant level that is much lower than that in the extracellular fluid or the external medium.

Transport of inorganic ions

The apical tegumental membrane of schistosomes and fasciolids has electrophysiological properties like those of other ion-transporting epithelia. The tegument is more permeable to K^+ than to Na^+ or Cl^-. This is a common feature in transporting epithelia of vertebrates and invertebrates. K^+ efflux contributes the hyperpolarized ground state of the tegumental electrical compartment. In both *S. mansoni* and *F. hepatica*, high K^+ concentrations are restored and low Na^+ concentrations (i.e. relative to the external environment) maintained in the tegument, in part, by Na^+/K^+-ATPase. Collectively, these biophysical properties contribute extensively to the maintenance of an electrochemical gradient which, in *S. mansoni*, results in a membrane potential of -60 mV. In addition, the schistosome tegument contains at least one voltage-dependent ion channel that is cation selective (conducts Na^+ and K^+, but not Cl^-) and exhibits an unusually large unitary conductance of 295 pS.

Excretion

Trematodes excrete metabolic end-products, such as lactate, amino acids, NH_4^+ and H^+. Both the tegument and protonephridial organs (see below) may serve important roles in the excretion of these molecules, though no direct physiological or biochemical evidence for this function has been reported. However, Na^+ levels within the tegumental syncytium become elevated after acidification of the external medium. Amiloride and low Na^+ medium interfere with recovery from an acid load, suggesting the existence of a Na^+–H^+ exchanger in the tegument that may serve an important role in the elimination of H^+.

Signal transduction

Evidence for a role for the schistosome tegument in sensory processing and signal transduction comes from anatomical and histochemical studies. Nerve fibers contact both ciliated and non-ciliated cells in the tegument that are believed to serve sensory functions, and two or three types of sensory papillae are localized within the teguments of schistosomes and fasciolids. Antibodies to inducible nitric oxide synthase (iNOS) localize at the surface of schistosomes, with NADPH-diaphorase staining being most prominent on the tubercles and papillae of male worms. Nitric oxide produced through iNOS in the tegument could serve a number of roles in sensation or locomotion, including the modulation of activity in underlying muscle fibers that form electrical connections with the tegument. G proteins have been isolated from the brush border of *F. hepatica*, and it is possible that they transduce signals involving chemicals in the environment. However, no evidence for G-protein-coupled receptors in the tegument has been reported.

More evidence that the tegument serves important roles in signal transduction come from behavioral studies on mating patterns. Male and female adult schistosomes are permanently paired within the vasculature of the host.

Recent evidence suggests that vitelline gland development in females, critical for egg production, may be controlled by chemicals released by the male. Vitellocytes in unpaired females are undifferentiated and few in number. Pairing with a male increases the number of vitellocytes and initiates terminal differentiation of a limited set of eggs. This process is mirrored by uptake of [^{3}H]thymidine in females, indicating that the factor released by the male stimulates both DNA synthesis and mitosis. Interruption of male/female pairing during this process leads to a profound reduction in transcription; re-mating restores the process. Terminal differentiation can also be stimulated in females by direct, surface-to-surface contact with only a small segment of a male, obtained from any part of the body. Acetone or ether extracts of male schistosomes also stimulate vitellocytes in females, suggesting that the key factor may be lipidic. Though there is no direct evidence for steroid hormone involvement in vitellocyte development, cholesterol is transferred from male to female schistosomes. Furthermore, a steroid hormone receptor that binds to promoter elements of the major eggshell-coding gene, p14, has recently been identified, and its transcription is stimulated by male extract.

The schistosome tegument also contains acetylcholine receptors and acetylcholinesterases. Acetylcholine is the major inhibitory neurotransmitter on schistosome somatic muscle. Acetylcholinesterase, the enzyme that inactivates acetylcholine, is covalently anchored to the apical membrane of the tegument through phosphatidylinositol; it is rapidly replaced *in vitro* following removal with PI-specific phospholipase C. The location of this enzyme at the surface allows it to destroy host acetylcholine in the circulation, protecting the parasite from its potential destabilizing effects on the body wall musculature. The presence of acetylcholine receptors on the surface of the tegument suggests a remarkably high level of communication and coordination between the tegument and the underlying somatic musculature, made possible by the low-resistance electrical pathway connecting these two tissue compartments.

Internal surfaces

Trematodes possess a gastrodermis and a protonephridial system, both of which are sites for the exchange of molecules between parasite and host.

Gastrodermis

The gastrodermis consists of a mouth, pharynx, and an esophagus that branches to form a pair of intestinal caeca that extend posteriorly and end blindly. The caeca are composed of a single layer of epithelial cells supported by a thin layer of longitudinal and circular muscle fibers. The epithelial layer is syncytial in schistosomes, but cellular in fasciolids. The surface area of the caeca in both schistosomes and fasciolids is amplified up to 100-fold by numerous digitiform microvilli and pleomorphic lamellae.

Developmental biology

Miracidia and sporocyts (stages that do not reside in the definitive host) lack a functional gastrodermis; this structure becomes patent 2–6 hours following penetration of the definitive host. *F. hepatica* begin to ingest host tissue immediately upon excystment of the metacercaria, with breakdown products of host tissue apparent in the caeca within a few hours. In newly excysted *F. hepatica*, the intestinal caeca appear to serve as reservoirs for enzymes used to penetrate the wall of the host intestine during migration to the liver. Upon reaching the liver, the immature flukes begin to feed on hepatic cells, and the caeca are transformed into digestive/absorptive organs.

Nutrient digestion and absorption

The importance of the trematode intestine for nutrient digestion and absorption is based on several observations. Schistosomes begin to feed on host blood within 4–6 hours after penetration. *In vitro*, immature stages can survive several days without erythrocytes, but their growth and development is severely retarded. Females ingest many more erythrocytes than males, presumably in response to the demand for amino acids and proteins used for egg formation. Lysis and digestion of erythrocytes begins in the esophagus, and appears to be initiated by the release of dense secretory bodies from gland cells in the posterior esophagus. Hemoglobin is released from the erythrocytes and passes into the intestine, where it is catabolized. Adult *F. hepatica* ingest mainly erythrocytes, but also other host tissues, including bile duct epithelium and mucus.

A wide range of enzymes involved in digestion and processing of nutrients are localized in gastrodermis epithelial cells, including several papain-like cysteine proteases, esterases, acid phosphatases and glucose-6-phosphatase. The most thoroughly studied hemoglobinases are the papain-like cysteine proteases SmCL1 and SmCL2, which are present in gut epithelial cells of *S. mansoni*. These enzymes are <50% identical at the amino acid level, and belong to separate lineages of papain-like enzymes. SmCL1 is expressed in the intestine and the tegument, and is related to human cathepsins F and W. SmCL2 is also present in the intestine, but may be expressed more abundantly in reproductive organs, and is more closely related to human cathepsins K, L and S. SmCL2 is also closely related to cathepsins L1 and L2, which are found in vesicles within epithelial cells of the intestine of *F. hepatica*. Other peptidases are present in schistosome and fasciolid intestines, but their specific roles in digestion have not been elucidated. Little is known about changes in the occurrence or distribution of these enzymes associated with development of the trematode gastrodermis.

The trematode gastrodermal epithelium is probably the principal site of absorption for large molecular weight nutrients that seem to be critical for growth and long-term survival and for egg production. Mechanisms that underlie the absorption of these molecules are uncharacterized. Endocytosis occurs to a very limited extent in *F. hepatica*, but there is no evidence for this process in schistosomes.

Ion transport and excretion

Little information is available on the role of the trematode gastrodermis in transport of inorganic or organic ions, or in volume regulation. Whether this tissue contributes to excretion of metabolic end-products is a matter of speculation. Given the high level of permeability of the tegument for water, inorganic ions and organic solutes, an essential role for the gastrodermis in excretion is doubtful. However, in schistosomes and *F. hepatica*, the by-products of erythrocyte digestion collect in the gastrodermis along with cellular organelles and macromolecules derived from the gastrodermis itself. This material is periodically regurgitated, a process that is probably more accurately equated to defecation in higher phyla. No careful analyses of the small molecules regurgitated by trematodes have been published.

Immune evasion

At least one schistosome gut-associated protein shed into the host's circulation is implicated in immune evasion. This large glycosylated protein, called circulating anodic antigen (CAA), binds host complement component C1q, but does not activate it or its precursor, C1. It may block the actions of C1q on schistosome surface proteins. Recent efforts to identify protective antigens in *F. hepatica* have focused on glycoproteins located on the surface of the intestine. Because these proteins are restricted to surfaces that uningested blood cells do not encounter, they are often referred to as concealed, hidden or protected antigens. Their potential role in vaccine development has been pursued further in nematodes (discussed in more detail below).

Protonephridial system

Trematodes possess a highly developed protonephridial system that is anatomically distinct from the tegument and the gastrodermis. In schistosomes, this system consists of flame cells that form at branch points along putative excretory canals. These structures have been anatomically localized using antibodies to calcineurin and a Ca^{2+}/calmodulin-dependent protein phosphatase, which is the molecular target of some cyclosporins. Based on histochemical studies using antibodies to these proteins, the number of flame cells in schistosomes increases with development from two in cercaria, to eight in schistosomula, to hundreds in adults. Direct physiological or biochemical evidence supporting a role for this system in excretion or osmotic regulation, however, has not been reported.

NEMATODES

Nematodes reside in diverse environments, including fresh and salt water, soil, plant roots and various animal tissues. At least one species of nematode is adapted for every environment on earth where organisms can be surrounded by at least a film of water. Aside from moisture, each habitat possesses particular challenges to nematode survival. These environmental challenges include various physicochemical stressors, such as high or low concentrations of osmolytes and inorganic ions, as well as pH and O_2 tensions that can exceed or fall below levels typically associated with metazoan existence. Nematodes that parasitize animals are also subjected to hostile reactions by the host, such as attack by a broad range of immune effectors and digestive enzymes. Considering the success of the phylum, it is clear that nematodes have evolved a host of strategies to overcome this wide variety of challenges. Principal among these strategies are those afforded by the barrier properties of the cuticle, a rigid structure at their surface that is quite distinct from the membranous teguments of cestodes and trematodes. Nematodes also possess a functional, one-way alimentary tract that ends in a muscularly-controlled rectum, and a tubule system that may provide another internal surface for exchange of water, ions and organic molecules between parasite and host.

Species- and stage-dependence

Nematodes that parasitize animals must adapt to two or more distinct and often harsh environments during their life cycle. For instance, embryonated eggs of *Ascaris suum* are shed in grass or mud where, upon hatching, they encounter salt concentrations of only a few mM, oxygen tensions of 100–200 mmHg, and near neutral pH. These eggs may be desiccated for long periods and even cooled to below freezing. Following ingestion, infective larvae enter the porcine stomach, a highly acidic environment that includes an array of powerful digestive enzymes, osmotic pressures from 200–500 mOsm and oxygen tension close to zero. From the stomach, they pass into the intestine, molt, then travel through the liver and the highly aerobic environment of the lungs. Larvae then migrate to the trachea, are swallowed and return to the stomach before eventually re-entering the anaerobic environment of the small intestine, where they mature and reproduce. At each site within the host, the parasite is exposed to immune effectors, many of which are designed specifically to destroy them.

Other species of nematode encounter remarkably different environmental challenges. Filarial nematodes parasitize internal tissue compartments and never experience a free-living stage; they are transmitted by insects that serve as intermediate hosts, moving from insect to vertebrate through the process of insect feeding. Some species evoke elaborate alterations of host cell morphology and function to create tailored feeding sites. This pattern is exemplified by the process of nurse cell creation by larval stages of *Trichinella spiralis* in a mammalian host muscle. Adult female *Onchocerca volvulus* elicit the construction of a nodule from the host response, in which worms can live and reproduce for up to a decade. Each of these animals is adapted to the peculiar conditions provided by its habitat. These include markedly different pH, ionic and osmotic conditions, and vary in the abundance, quality and type of food available.

Though nematodes have evolved a vast array of surface adaptations suited for specific environments, the general features of their surface biology are well conserved. That more is known about those surfaces in nematodes than in other helminths is attributable to several factors. Parasitic nematodes pose more prominent challenges to both veterinary and human medicine, and more research investment has been focused on them for medical and economic reasons. The large size of *Ascaris suum*, which is easily collected from the small intestines of swine, has greatly facilitated physiological and biochemical studies that would be very difficult using smaller species of nematode. Finally, much work has been performed on the free-living nematode, *Caenorhabditis elegans*, which in 1999 became the first metazoan for which the primary genome sequence was fully delineated. The power of *C. elegans* genetics illuminated many basic biological and biochemical phenomena in nematodes, including many related to external and internal surfaces. Fortunately, *A. suum* and *C. elegans* provide information that can generally be extrapolated to other species.

External surfaces

Structure

Gross and microscopic anatomy

Nematodes are bounded externally by a complex, multilayered cuticle that extends into and lines the pharynx, rectum, cloaca and other orifices. Molecules that form the cuticle are synthesized and secreted by the hypodermis, an anatomical syncytium that forms a continuous cellular layer immediately beneath the cuticle, and is specialized for transport and secretion (Figures 13.4 and 13.5). Though the cuticle and hypodermis are morphologically distinct, they represent a functional unit. Nematode cuticles are tremendously diverse in structure. Among species that parasitize vertebrates, the cuticle of the adult stage typically consists of six distinct layers: the epicuticle, outer and inner cortex, medial layer, fiber layer, and composite basal layer. Larval stages of some species also exhibit a loosely associated surface coat that is secreted through the excretory pore or from the esophagus; its functional properties have not been delineated.

The outermost layer of the adult cuticle, or epicuticle, is 6–30 nm thick and often appears trilaminate in electron micrographs. Considerable controversy exists about whether the epicuticle is the true limiting 'membrane' in nematodes. Several observations refute that concept. For example, the epicuticle is not dissociated by treatments that dissolve most membranes, and lipophilic markers do not exhibit the lateral mobility characteristic of membranes when inserted into the surface of nematodes. The epicuticle in most species examined is devoid of intramembranous particles that are normally associated with cell membranes. Finally, under *in vitro* conditions, charged solutes diffuse across the cuticle at relatively high rates, indicating the absence of a functional lipid barrier at the surface.

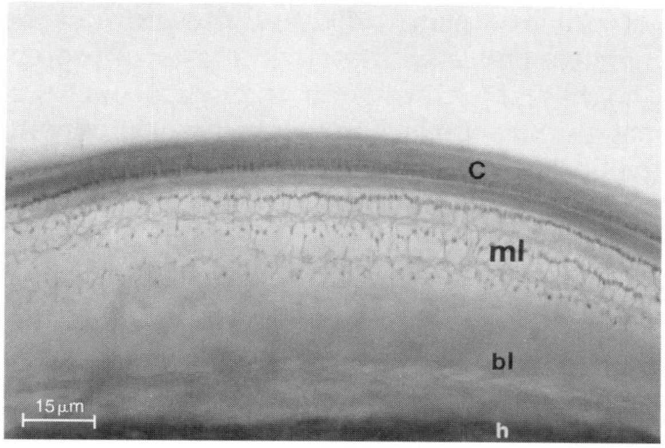

FIGURE 13.4 Light micrograph (×500) highlighting fine structure of the cortical and medial (aqueous) layers of the cuticle of *Ascaris suum*. Micrograph was stained with trichrome. Scale bar = 15 μm. Abbreviations: c, cortical layer; ml, medial layer; bl, composite basal layer; h, hypodermis.

However, the epicuticle contains at least some lipid components. In *Brugia malayi*, for example, a cuticular lipid fraction was isolated that contains high levels of lysophosphatidylethanolamine and a novel species that rapidly degrades to two or more products upon exposure to atmospheric oxygen. This lipid extract contains saturated and unsaturated fatty acids in the range 14–22 carbons, with the majority being mono- and di-unsaturated 18-carbon species. Furthermore, this parasite secretes a glutathione peroxidase across its

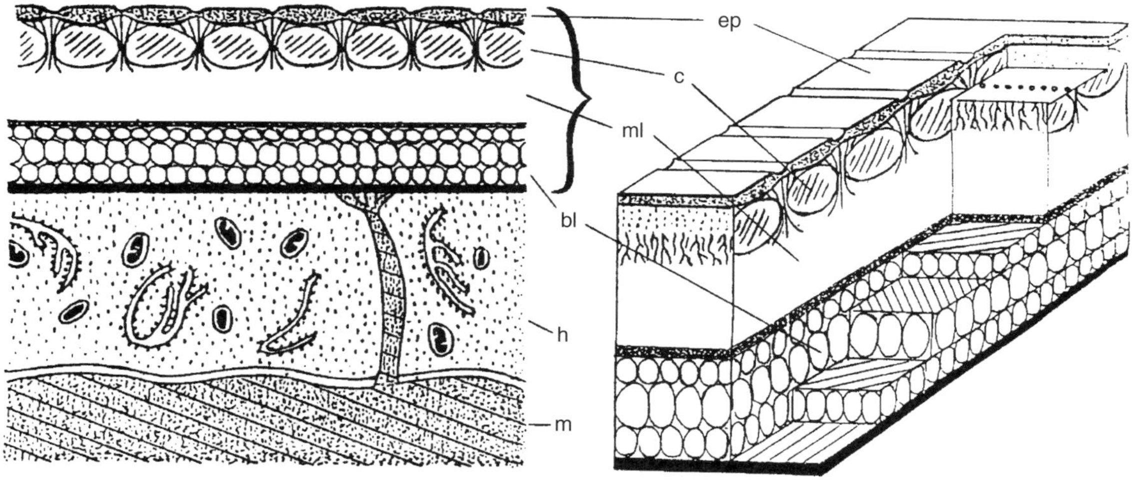

FIGURE 13.5 Schematic of adult *Ascaris suum* cuticle and hypodermis. Abbreviations: ep, epicuticle; c, cortical layer; ml, medial (aqueous) layer; bl, composite basal layer; h, hypodermis; m, muscle (drawing courtesy of J.W. Bowman).

cuticle, which detoxifies fatty acid and phospholipid hydroperoxides, but not hydrogen peroxide. These findings have been interpreted to suggest that the enzyme is secreted to protect lipids in the epicuticle from oxidative damage by reactive oxygen species released by the host. However, there is no direct evidence that the enzyme is secreted to protect lipid components of the epicuticle *per se* rather than those in some other site within the cuticle–hypodermis complex, or that this process even occurs *in vivo*. The surfaces of other species of nematodes contain different proportions of lipid components. In *T. spiralis*, for example, lysophosphatidic acid predominates, comprising 37% of the mass of surface coat lipids.

The cortex is an amorphous, electron-dense layer of the cuticle that is composed of at least two sublayers (outer and inner) which contain abundant keratin-like proteins and a highly insoluble protein termed cuticulin. The medial layer is an aqueous compartment that contains fine collagenous fibers that are less dense than those in the cortex. In some species, fluid within this layer contains hemoglobin. The composite basal layer contains crosslinked collagen fibers in two to three distinct sublayers that spiral around the nematode at an angle 75° to the longitudinal axis of the worm. Depending on the species, the cuticle may contain a wide range of gross structural elaborations, including annulae, lateral and transverse ridges, and spines. These structures are usually restricted to the cortex, though they may extend into the medial layer. In some filarial species, the cuticle is traversed by anatomical pores that extend to the hypodermis. There is no histological evidence for similar anatomically defined pores in the cuticles of parasitic intestinal nematodes, though the existence of functional pores is suggested by biophysical studies on the permeability characteristics of the cuticle (see below). In addition, infective larvae of most parasitic nematodes retain the cuticle of the second larval stage. The adherent former cuticle is referred to as a sheath, and is usually 400–600 nm thick. This structure retains the epicuticle and cortical layers, and provides the infective larvae with a second layer of protection that is highly resistant to physical or chemical damage, including insults associated with the host's immune system.

The hypodermis lies immediately beneath the composite basal layer. It is multicellular or syncytial, depending on the species. In most species, an outer hypodermal membrane is apparent, and it is likely that its absence in micrographs of other species is an artifact. The outward-facing membrane of the hypodermis probably represents the true limiting membrane of nematodes. Though typical cytoplasmic constituents are uniformly distributed in the hypodermis, nuclei are located only within lateral ridges that project into the pseudocelom at the mid-ventral, mid-dorsal and lateral lines. The ventral and dorsal ridges also contain the major nerve cords, whereas the lateral ridges in most species contain small canals (or tubules) that may serve excretory or secretory functions (see below). The hypodermis is also traversed by fibers that anchor the somatic muscles to the basal cuticle.

Biochemistry and molecular biology of structural components

The epicuticle is composed primarily of carbohydrate, including negatively charged residues and hydrophilic, sulfated proteins that are partially exposed to the environment. In some species, the epicuticle can be digested by elastase, suggesting that an elastin-like protein may be an important component. However, the precise chemical composition of the epicuticle is unknown.

Most tissue in the underlying layers of the cuticle consists of collagen or collagen-like proteins. Collagen is the most abundant protein in nematodes, as in other metazoa, comprising about 1% of the mass of adults. In *C. elegans*, two types of collagen have been identified. These correspond to two types of extracellular matrix, the cuticle and basement membranes, including a layer that surrounds the hypodermis. Collagens are characterized by repeats of the sequence X-Y-Gly, where X is usually L-proline and Y is 4-hydroxy-L-proline. Polypeptide chains containing these repeats are wound into tight triple helices. The hydroxylation of proline residues is catalyzed by prolyl 4-hydroxylase, a tetramer composed of 2α and 2β subunits. In *C. elegans*, selective deletion of *dpy-18*, which encodes the α-subunit, results in mutants with shorter (or 'dumpy') body shapes. In contrast, deletion of *phy-2*, which encodes the β-subunit, does not result in a phenotypic change. However, crossing *phy-2*$^-$ with *dpy-18*$^-$ worms yields progeny that are unable to synthesize cuticle collagens, including those that form the basement membrane of the hypodermis. In addition to expressing the dumpy phenotype, a high percentage of these worms explode following gradual structural breakdown or breaches of the cuticle. These mutations illustrate the crucial role of a resistive structure at the external surface of nematodes.

Nematode cuticular collagens contain higher levels of proline (11%) and hydroxyproline (12%) than vertebrate collagens. About 80% of the total cuticular protein in nematodes (*C. elegans, A. suum* and *Haemonchus contortus*) consists of 2-mercaptoethanol-soluble collagen. Most of the remaining 20% consists of highly insoluble proteins, such as a keratin-like protein and cuticulin, a protein that is highly crosslinked through tyrosine residues. Tyrosine crosslinkages are achieved by peroxidation, leading to dityrosine residues, but also include an unusual isotrityrosine component, the synthesis of which remains unresolved. These insoluble proteins are located primarily in the cortex. Structural analyses of ASCUT-1, a cuticulin found in *Ascaris lumbricoides*, demonstrate the critical nature of phenylalanine residues to intramolecular interactions between cuticulin molecules, which are essential for aggregation and cuticle formation.

Most cuticle protein assembly occurs in the hypodermis, and mRNAs for several genes involved in cuticle synthesis have been localized in the hypodermis of *C. elegans*. These include an amino acid hydroxylase that converts phenylalanine to tyrosine, which is critical to the process of protein crosslinking, and two cuticulins that localize to the external cortical layer. About 150 distinct collagen genes have been identified in *C. elegans*, and homologs are beginning to emerge from EST projects for parasitic species. Gene products found in cuticle are typically short (~300 amino acids) and fall into four families. Cuticle collagen genes in parasitic species, including *A. suum* and *H. contortus*, as well as the filariid, *B. pahangi*, also group into evolutionarily conserved families. However, most information about the genetics of nematode cuticle collagens has been derived from analyses of *C. elegans*. Mutations in collagen genes cause specific morphological (and sometimes behavioral) abnormalities that reflect the localization and function of the gene product. At least 50 genes in *C. elegans* affect body morphology, and eight of these encode collagens. For example, *sqt-1* encodes a cuticle collagen that is important in determining shape and locomotory behavior. Some mutations in *sqt-1* produce left-handed helical twisting of the nematode, which results in worms (left-rollers) that move only in circular paths. Careful observation of the structure of the cuticle in these mutants, using deep-etched replica analyses, show that the fibrous layer of collagen is composed of parallel fibers instead of the layers meeting at 60° angles seen in wild-type strains. At the biochemical level, mutations that effect this change result in removal of a conserved cysteine, which inhibits formation of tyrosine-derived non-reducible bonds between collagen proteins. Other mutations in *sqt-1* result in markedly different phenotypes. For example, *sqt-1* mutations that affect conserved arginines in a predicted cleavage site for a subtilisin-like protease cause right-handed helical twisting.

In addition to collagen and collagen-like proteins, nematode cuticles contain at least one glycoprotein, and small amounts of hyaluronic acid and lipid. Lipid components that are localized exclusively in the cortex have not been chemically characterized. Also present in the cuticle are chondroitin sulfate and several sulfated mucopolysaccharides, which convey a negative charge to the cuticle.

Developmental biology

The cuticle is a dynamic structure that is replaced during each of four molts. The biochemical events that control molting are not well understood, but the hypodermis appears to be the site of most cuticle protein formation. Selective laser ablation of *C. elegans* hypodermal cells before formation of the adult cuticle leads to abnormalities in cuticle regions directly above the lesions. Proteins recovered from the hypodermis of *A. suum* differ from those in the cuticle, indicating that considerable processing of proteins probably occurs in the cuticle layers prior to their incorporation. Protocollagen proline hydroxylase, an enzyme essential for cuticle protein synthesis, is present in the hypodermis of adult *A. suum*.

During synthesis of the collagen matrix, individual collagen genes are expressed in distinct temporal progression. Chemical signals that control the timing of cuticle molting are not well understood. In insects, molting is hormonally controlled, with the cholesterol derivative 20-hydroxyecdysone playing a critical role through its interaction with nuclear hormone receptor-class transcription factors, which are activated in a cascading manner. *C. elegans* contains about 270 genes that belong to the family of nuclear hormone receptor-like transcription factors. Most are uncharacterized, but at least one, *nhr-23,* is expressed in the hypodermis and has been implicated in molting. The megalin-related protein, LRP-1, is a putative sterol receptor that may play a critical role in this process. LRP-1 is expressed on the apical surface of the hypodermis. Mutations in *lrp-1* cause arrested growth, usually between the third and fourth molts, and an inability to shed and degrade the old cuticle (i.e. the sheath, in third stage larvae). The stage-specific switch from larval to adult cuticle appears to correlate with transcriptional activation of adult-specific collagen genes and simultaneous repression of larval-specific genes in hypodermal cells. Several other genes that participate in this process in *C. elegans* have been identified, including the heterochronic genes, *lin-29*, *lin-4*, *lin-14* and *lin-28*.

Most species of nematode grow extensively between molts, and some continue to grow even after the final molt. To accommodate that growth, the cuticle must expand concurrently. Generally, cuticle thickness increases in direct proportion to length of the parasite. Most of the increase in thickness occurs in the medial layer, which may become six to eight times thicker as the adult parasite grows from a length of 5 cm to 30 cm. The process by which cuticular proteins permeate the scaffolding of the growing cuticle to reach their destinations and become integrated into the structure is not known.

Functional biology

Biological functions of the structure

The cuticle–hypodermis complex provides a physical barrier that separates the cells and extracellular fluids (pseudocelomic fluid, PCF) of the worm from the aqueous environment in free-living stages, or from host body fluids in parasitic stages. Nematodes are pseudocelomates and possess a fluid-filled body cavity. The nematode pseudocelom differs from a true celom by the absence of both a mesentery (membranous lining of the body cavity) and a muscle layer surrounding the intestine. Fluid in the pseudocelom is in constant contact with almost all of the cells in nematodes, and only one cell layer separates PCF from the external environment. Externally, that cell layer consists of the hypodermis–cuticle complex; internally, it is formed by the intestinal cells, and in rostral regions by the tubular system.

The cuticle plays a critical role in the mechanics of nematode movement. Nematodes lack a rigid skeleton or circular muscles in the body wall to provide antagonistic systems against which longitudinal muscles can act to produce movement. They utilize instead a hydrostatic skeleton that is dependent on a high internal (or turgor) pressure within the pseudocelom, and limitations on expansion of the body imposed by the cuticle. Internal pressures recorded from the pseudocelom of *A. suum* average 70 mmHg above ambient, and values as high as 225 mmHg are recorded during locomotion. These values are much higher than transmural hydrostatic pressures recorded in other invertebrates. The biochemical mechanisms that underlie the high internal pressure in nematodes are unknown. Most species rapidly desiccate upon exposure to air, and transport studies using radiolabeled water indicate that the cuticle in some species, including *A. suum*, is highly permeable. Paradoxically, solute levels in pseudocelomic fluid are generally isosmotic or even hyposmotic to the fluid in the digestive tract of the host. No simple biophysical model can explain how nematodes maintain a high internal pressure in the absence of a measurable external driving force.

In addition to providing a scaffold against which the hydrostatic skeleton can be levered, the elasticity of the cuticle is essential for locomotion. Although the cuticle collagen fibers are inelastic *per se*, their unusual arrangement in the composite basal layer confers elasticity to the tissue. The fibers are organized into cylindrical helices which spiral around the body in opposite directions, such that fibers in the middle layer run at a 75° angle to the longitudinal axis of the parasite. The longitudinal muscles are attached to the cuticle at its inner surface. As these muscles contract, the helix is drawn apart. This changes the shape of the body, which becomes longer and thinner as localized volume reductions occur within regions of the pseudocelom beneath bundles of contracting muscle fibers. The importance of the organization of these helical layers to movement is illustrated by *C. elegans* mutants with helical coils that run in parallel, instead of at an angle to the top and bottom layers. One unusual phenotype exhibited by these worms is the inability to move in a straight line. Instead, they 'roll' to the left or right, and move only in small circles.

Most of the enzymes that catalyze crosslinking reactions, which in turn control the structural organization of proteins in the cuticle, are unknown. However, molecular mechanisms that underlie muscle attachment to the cuticle are partially delineated for *C. elegans*. The gene *mec-8* encodes a regulator of RNA splicing that controls processing of the gene *sym-1*. *Sym-1* encodes a protein that contains a signal sequence plus 15 contiguous leucine-rich repeats. The product of *sym-1* helps attach body wall muscle to the cuticle. Disruption of *mec-8* or *sym-1* leads to arrest during embryonic elongation, and to defects in attachment of somatic muscle to the basement membrane of the cuticle.

The high internal pressure within the pseudocelom of nematodes is also essential for ingestion. Nematodes lack circular muscles, and the gut normally collapses under the hydrostatic pressure of the pseudocelom. Ingested food is pumped caudally against this pressure by the pharynx, which is the only muscle in the digestive tract except for those in the rectum that regulate defecation. The internal pressure, which collapses empty regions of the gut and thereby prevents rapid passage of digesta, may contribute to the efficiency of nutrient absorption across the intestine.

Nutrient absorption

Transcuticular absorption of glucose, amino acids and other nutrients has been demonstrated in all filarial nematodes examined. In some cases, the cuticle is probably the principal site of nutrient absorption, with very little occurring across the intestine. This conclusion is supported by several observations. For example, microfilariae of most species possess a non-functional gut, yet absorb nutrients. In *O. volvulus*, significant atrophy of the gut occurs as the adult matures, coincident with thickening of, and increases in enzyme activity in, the hypodermis. The filarial hypodermis contains proteins associated with amino acid transport, whereas the intestine often does not. Also, direct measurement of nutrient transport across the cuticle/hypodermis complex in some filarial nematodes has demonstrated saturability, stereoselectivity and competitive as well as non-competitive inhibition, indicating the presence of specific carrier systems.

Evidence for transcuticular transport of physiologically relevant quantities of nutrients in intestinal parasites is less compelling than for filarial species. Most studies have used *A. suum*, which absorbs little glucose *in vitro*, and instead catabolizes stored glycogen, which is abundant in muscle cell bodies. The fact that some transcuticular absorption of glucose occurs *in vitro*, however, suggests that the process may be more important *in vivo*, where most glucose absorbed is converted quickly to glycogen. Other species, such as *H. contortus* and *Trichostrongylus colubriformis*, absorb and metabolize large quantities of exogenous glucose *in vitro*. The importance of the cuticle in glucose absorption is suggested by studies that demonstrate that metabolism of exogenous glucose by *H. contortus* is not altered when the pharynx is paralyzed with ivermectin.

How amino acid transport in nematodes is regulated is not yet understood. Significant transport of amino acids across the alimentary tract has not been demonstrated for any species, but this process has received little attention. Histochemical and enzymological data demonstrate that γ-glutamyl transpeptidase is present in the cuticle–hypodermis complex of *A. suum* in considerably greater abundance than in the intestinal epithelium. This enzyme is part of an amino acid transport system found in many organisms, including vertebrates. The cuticle–hypodermis complex may be more important than the intestine for uptake of some amino acids.

Transport of inorganic ions

Inorganic ions play many critical roles in cell function. Most important for homeostatic maintenance in cells are Na^+, K^+, Cl^-, Ca^{2+} and Mg^{2+}. Intracellular concentrations of these ions, particularly Ca^{2+}, are usually maintained within very narrow limits. Inorganic ion concentrations are regulated by an array of integral membrane proteins, including specific channels, energy-dependent pumps (e.g. Na^+/K^+-ATPase, Ca^{2+}-ATPase, Na^+/H^+-ATPase) and energy-independent ion transporters (e.g. Na^+/H^+ antiporter, Na^+/HCO_3^- symporter, HCO_3^-/Cl^- exchanger). Genes that encode most of these ion transport proteins have been identified in *C. elegans*. However, the functions of only a few have been studied *in vitro*, and their expression patterns in nematode tissues have not been reported. It is not yet known how many are expressed in the hypodermis or internal transport tissues.

The crucial importance of regulating inorganic ion concentrations across nematode cells is illustrated by the effects of several drugs that interfere with ion channels. The anthelmintic actions of ivermectin and other macrocyclic lactones, for example, are attributed to the ability of these compounds to open Cl^- channels associated with glutamate receptors in the pharynx (see below). By increasing the permeability of these cells to Cl^-, the pharyngeal membrane becomes hyperpolarized, usually by only a few mV, but this is sufficient to reduce its responsiveness to excitatory nerve impulses. This leads to flaccid paralysis of the pharynx and, eventually, starvation and loss of ability by the worm to remain at its site of predilection within the host. Two other classes of anthelmintics, the imidazothiazoles and tetrahydropyrimidines, exemplified by levamisole and pyrantel, respectively, target cholinergic receptors on somatic muscle membranes. These receptors are directly coupled to channels for Na^+ and Ca^{2+} which, when opened by these drugs, lead to depolarization and spastic paralysis of the somatic musculature. The actions of several other anthelmintics, such as piperazine and the paraherquamides, are attributable to their direct or indirect effects on ion channels associated with nerve and muscle membranes. Unfortunately, no data are available on the effects of anthelmintics on electrochemical gradients maintained across hypodermal or tubular cells in nematodes, and information on their effects on anterior regions of the alimentary tract is only just beginning to emerge. Thus, although pharmacological experiments have provided tremendous insights into ion transport mechanisms in nematode neuromuscular systems, we have a very limited grasp of related proteins that maintain ionic gradients within the pseudocel or across other internal membranes that regulate the environment around muscle and nerve cells.

The largest and most accessible structural barrier separating internal compartments in the nematode from the environment is the cuticle–hypodermis complex. Though the cuticle forms as an extracellular secretion from the hypodermis, it is possible to separate these structures by experimental procedures and study their individual contributions to the transmural transport of ions and other solutes. The cuticle is 50–100 μm thick in adult *A. suum*, but much thinner in other species. Aqueous pores that traverse the cuticle are negatively charged, and therefore present an electrostatic barrier to the transport of large organic anions. However, due to the large radius of these pores (15 Å in *A. suum*) compared to the size of relevant inorganic anions (Cl^-, radius = 1.2 Å), size has little influence on inorganic anion transport across the cuticle. This is best demonstrated by data from electrophysiological studies that measured electrical potential and current flux across isolated body wall segments of *A. suum*. An electrical potential of −30 mV (muscle-side negative) develops very rapidly when tissue segments containing living muscle and hypodermis are placed between the two half-cells of a diffusion cell system (Ussing chamber). When muscle and hypodermal tissues are mechanically scraped from this preparation, and residual lipid removed by extraction with chloroform and methanol, the transmural potential is abolished and there remains no resistive barrier to the passage of small inorganic ions, measured as electrical current, when an external voltage is imposed on the system.

The hypodermis is both an anatomical and electrical syncytium, with apical (cuticle-facing) and basal (muscle-facing) membranes separated by a cytoplasm-filled space of variable thickness. The transmural potential is maintained, in part, by the separate contributions of the cuticle-facing and pseudocelom-facing membranes. In *A. suum*, the electrical potential within the hypodermal compartment is 70 mV more negative than the external medium and 40 mV more negative than the pseudocelomic compartment, when tested in artificial media containing inorganic ions at concentrations approximating porcine intestinal fluid. Cations do not appear to play a major role in establishing the transmural potential. The concentrations of Na^+ and K^+ in PCF from freshly collected parasites are approximately equal to those found in swine intestinal fluid, consistent with the high permeability of the body wall to these cations. When parasites are placed in media containing elevated or reduced levels of these cations, concentrations in the PCF change fairly rapidly to approximate the outside concentrations. In contrast, the concentrations of the divalent cations Ca^{2+} and Mg^{2+} in PCF are maintained at constant levels in the face of large changes in their concentrations in the incubation medium. However, due to the low concentrations of these cations in cytoplasm or PCF, relative to Na^+, K^+, Cl^- and organic anions produced via carbohydrate metabolism, they do not contribute significantly to the transmural potential.

This situation contrasts markedly with that typical of transporting epithelia in other metazoa, where cation conductance is tightly regulated. Both passive and active transport of cations contribute extensively to the resting potential of most vertebrate and invertebrate cells, including somatic muscle cells and the pharynx in nematodes. However, electrogenic transport of ions contributes only about 36% of the transmural current in *A. suum*. Also, the concentrations of Cl^- in *A. suum* PCF and porcine intestinal fluid are nearly equivalent. Changing the external Cl^- concentration has little effect on the transmural potential in *A. suum*, suggesting that other anions play more important roles in this process. Given the high concentrations of organic anions in PCF (80 mM), relative to porcine intestinal contents (3–8 mM), it is likely that these species contribute extensively to the transmural electrical potential. A large conductance Cl^- channel that can also conduct organic acids has been described in *A. suum* muscle membranes, and a similar channel has been localized and partially characterized on both the inward- and outward-facing membranes of the hypodermis. This channel and other organic anion transporters identified in some nematode species probably play more important roles than inorganic ion channels in maintaining transmural electrical potential in nematodes.

Direct control of hypodermal or intestinal membrane permeability by the nervous system has not been demonstrated in nematodes. However, a receptor that recognizes bombesin-like neuropeptides, which serve an analogous function (i.e. regulation of Cl^- permeability across epithelial membranes) in some vertebrates, has recently been partially characterized. Bombesin-like peptides appear to be broadly distributed among nematodes. Immunohistochemical studies on *A. suum* and *Dirofilaria immitis* indicate that bombesin like immunoreactivity is localized primarily in the apical regions of the hypodermis, where the cuticle-facing membrane of the hypodermis appears to overlap with the cuticle. Competition binding assays using [^{125}I]GRP (a close structural analog of bombesin) and membranes prepared from *A. suum* and *C. elegans* reveal a saturable, high-affinity binding site with a K_d of 3 nM and B_{max} of 1 fmol/mg protein. Although the chemical structures of bombesin-like peptides in nematodes have not been determined, the GRP binding data suggest the peptides are similar to their putative vertebrate homologs. A molecule from *Panagrellus redivivus* extract that is recognized by antibodies to GRP has a molecular weight in the 1.7 kDa range, similar to that of vertebrate bombesin-like peptides. Demonstrating a role for this peptide and its receptor in the hypodermis of nematodes awaits determination of its sequence and identification of the gene that encodes its receptor.

Turgor pressure and osmotic/volume regulation

Processes that regulate turgor pressure and osmotic gradients in nematodes are distinct, but interdependent; both are critical aspects of the physiology of these organisms. Fluid in the pseudocelom, which is under pressure imposed by the limited ability of the cuticle to expand or contract, forms a hydrostatic skeleton that is essential to coordinated movement and to the digestive process. As noted above, hydrostatic pressure within the pseudocel of *A. suum* at rest is maintained at about 70 mmHg above ambient, and values may oscillate from 16 mm to 275 mmHg during contraction of the body wall muscle. The consequences of interfering with turgor pressure in nematodes can be deduced from *in vitro* experiments. For example, puncturing the cuticle leads to immediate evisceration of the intestine and ovaries, followed shortly by uncoordinated movement and death in *A. suum* and other nematodes. Loss of turgor pressure would thus likely be lethal to parasites *in vivo*. This effect can be partially mimicked by placing adult-stage nematodes in distilled water, which causes them to swell over several hours and eventually burst. Other clues to the important links between osmotic regulation and muscle function come from studies using *C. elegans* mutants that are resistant to levamisole and appear to lack pharmacological acetylcholine receptors. Among the phenotypes displayed by these mutants, referred to as '*uncs*' because they exhibit uncoordinated movement, is extreme sensitivity to hypotonic shock.

Osmotic and volume regulation are tightly linked processes in all metazoans. Marked fluctuations in environmental solute concentrations are encountered by parasitic species that reside in brackish water or on damp grass as larvae. Adults in the mammalian gastrointestinal tract may encounter osmotic pressures in excess of 500 mOsm. Adult stages of parasitic nematodes are less tolerant of hyposmotic conditions than are larvae. Adults swell and explode within a few hours when incubated in fresh water; however, they tolerate incubations for up to 4 days in dilutions of artificial sea water that range from 20–40%. During these incubations, osmotic pressure measured in PCF of *A. suum* is approximately equal to that of the external medium. The most spectacular examples of osmotic stress tolerance in nematodes come from studies that examine desiccation survival. Infective larvae of the gastrointestinal parasite *T. colubriformis* survive vacuum desiccation at 0% relative humidity for 9 hours. In most species, anhydrobiosis must be induced slowly for the worms to survive.

Though the mechanisms underlying anhydrobiotic survival in nematodes are unknown, several organic solutes, including trehalose, glycerol and inositol, may serve as replacement molecules for membrane-bound water. However, *Panagrellus silusae* and *A. suum* produce high levels of glycerol and trehalose, respectively, but also are highly susceptible to desiccation, suggesting that the presence of those solutes does not alone confer tolerance to anhydrobiosis.

The cuticle–hypodermis complex, intestine and tubular system all appear to participate in osmotic and volume regulation in nematodes. However, studies that define the relative contributions of each of these structures, or the underlying mechanisms, are lacking. Since osmotic regulation appears to occur over hours, while turgor pressure changes over seconds, it is unlikely that the two processes are directly linked. The ability of nematodes to withstand major osmotic changes in their natural environment, as opposed to those observed in the laboratory, probably depends more on the permeability of the cuticle to water than anything else. Marine nematodes are usually isosmotic with seawater, and their cuticles are highly permeable to water. Water permeability in the marine nematode *Europlus*, for example, is 4.4 μm^3 $H_2O/\mu m^2$ body surface/hour, a value over 100-fold greater than that recorded for *Aphelenchus avenue*, a soil-dwelling species, and over 200-fold greater than that for *C. elegans*, another soil-dweller, or larvae of the animal-parasitic species, *Nippostrongylus muris* (*brasiliensis*) and *Ancylostoma caninum*. The cuticle–hypodermis complex of adult *A. suum* is also highly permeable to water, as demonstrated by $[^3H]_2O$ flux studies. Water-filled pores that traverse the cuticle of *A. suum* and other species provide a highly accessible pathway for water in the environment to reach the outward-facing membrane of the hypodermis.

The rates of transcuticular water flux reported for most nematodes are hundreds-fold greater than achievable by simple diffusion across lipid membranes. In other organisms, water transport across membranes occurs via intrinsic membrane proteins called aquaporins. These proteins have been most thoroughly studied in human red blood cells and kidney proximal tubules, but also exist in other vertebrate tissues. Aquaporins are thought to form from two hemipores, each containing three transmembrane domains. Homologous proteins have been identified in other organisms, including bacteria, plants and insects. Each contains the highly conserved motif Asn-Pro-Ala in the putative pore-forming region of the protein. Recently, a cDNA encoding a *C. elegans* aquaporin was cloned and expressed in *Xenopus* oocytes. Expression of the channel endowed oocytes with greater permeability to water and, to a lesser extent, urea, but not glycerol. This gene was expressed only in early larvae of *C. elegans* and was completely suppressed before hatching.

Other genes predicted to encode aquaporins have been found in the *C. elegans* genome database and in cDNA libraries from various parasitic nematodes, including *Toxocara canis, Brugia pahangi* and *O. volvulus*. Definitive studies of the functional properties of the proteins encoded by these genes, as well as their pattern of expression in nematode tissues, are currently underway. Aquaporins probably play a critical role in the transport of water across the cuticle–hypodermis complex. Identification of the rate-limiting barrier to water diffusion across this tissue at either the hypodermis or the epicuticle (or both) awaits experiments on the localization and biophysical characterization of water channels in these organisms.

Though aquaporins may provide an important pathway for transport of water across the cuticle–hypodermis complex to compensate for changes in external osmotic conditions, it is important to note that when active regulation of water flux occurs, it is coupled in most metazoa to movement of inorganic or organic solutes across the same membranes, or to some metabolic activity that raises or lowers the concentration of organic solutes within the cells or tissues. This first process is usually controlled by ion channels that are specific for individual ionic species. Several ion channels have been suggested to contribute to osmotic and volume regulation in *C. elegans*. For example, 22 genes encoding Na^+ channels in the degenerin (e.g. *deg-1*, after 'degeneration') and mechanosensory (e.g. *mec-4* and *mec-10*) families have been identified. In other organisms, related channels form amiloride-sensitive Na^+ channels in Na^+-resorbing epithelial cells. In vertebrates, they are targets for aldosterone in apical membranes of kidney and intestinal epithelial cells. Evidence for their involvement in volume- or osmoregulation in *C. elegans* comes from studies showing that null mutations at these genes lead to enhanced Na^+ flux, swelling and eventual degeneration of mechanosensory cells in the amphids. If members of this channel family are expressed in the intestine, hypodermis or tubular system in nematodes, they would be positioned to affect ion and water movement across these surfaces.

Other channels in nematode surface membranes probably contribute to volume regulation, including those that regulate flux of other inorganic ions such as Cl^- and K^+. In vertebrates, volume regulation is partially regulated by CLC-2, a low conductance Cl^- channel that is activated by both swelling and acidosis. This channel is different from a Ca^{2+}-activated organic anion-conducting Cl^- channel that has been identified in hypodermal membranes (described below). Two other Cl^- channels present in vertebrates, with homologs in *C. elegans*, are activated by swelling: CLC-3, which conducts organic anions such as acetate and gluconate, and VRAC (volume regulated anion channel). Conductance of other, uncharged organic osmolytes by VRAC is greater than for Cl^- by a factor of at least 6. In addition, two genes encoding K–Cl cotransporters have been identified in *C. elegans*. In mammals, K–Cl cotransporters typically function to balance water flux across the opposite faces of epithelial cells to maintain cell volume. It is tempting to speculate that the nematode homologs could function in osmoregulation or in regulation of turgor pressure. Tissue localization and dsRNAi or gene deletion could be used to test that concept.

Nematodes may also adjust osmotic pressure by regulating the rate of transmural organic acid excretion (see below), or the intracellular concentrations of other organic osmolytes. During periods of dehydration and rehydration, for example, cuticle transport may function in concert with metabolism, including production of organic protectants, such as glycerol and trehalose, to maintain water balance. Trehalose is abundant in several parasitic species and in all life-cycle stages of *C. elegans*. Trehalose synthesis is catalyzed by trehalose 6-phosphate synthase and trehalose 6-phosphate phosphatase, and it is metabolized by trehalase. These enzymes have been cloned in *C. elegans*. Since vertebrates do not synthesize or utilize trehalose, nematode enzymes involved with its processing may be good anthelmintic targets.

Excretion

It is important to distinguish excretion from secretion. These two processes have often been considered together in the parasitology literature, under the combined heading 'excretory/secretory products'. We consider excretion to mean the removal of unusable or unnecessary material. Secretion is the active release of molecules that convey a survival advantage to the organism. Nematodes secrete an array of proteins (and possibly non-protein molecules) that assist in nutrient acquisition, defense against host responses and manipulation of the host environment. We focus discussion here on processes involved in excretion only. Animals typically excrete two kinds of products. Ingested material that is not digested is shed from the intestine as solid material called feces. In higher organisms, the soluble waste products of intermediary metabolism and catabolism are typically transported via a circulatory system from the tissue of origin to an excretory organ, where they are eliminated in a fluid (e.g. urine). Nematodes possess an intestine and exhibit defecation. However, nematodes lack a recognizable kidney (though see tubule system, below), and soluble waste products are apparently not concentrated prior to elimination. Late larval and adult stages of parasitic nematodes are primarily anaerobic, and oxidize only a small fraction of ingested carbohydrates to CO_2 prior to excretion. Instead, they produce and excrete low molecular weight organic anions, such as acetate, lactate and butyrate. The pseudocelom and PCF serve together as a circulatory system, transporting these molecules from muscle and reproductive tissue to the hypodermis, intestine and possibly the tubule system for excretion into the environment.

Nematodes derive most of their energy from the degradation of glucose (typically stored as glycogen or trehalose) or other sugars, including fructose. They excrete the end-products of energy metabolism, which would be toxic if allowed to accumulate. Little is known about the pathways through which CO_2 or alcohols are excreted from nematodes, even though these processes are found in most animal parasitic species. In vertebrates, CO_2 is exported from tissues via a Cl^-/HCO_3^- exchanger. No predicted *C. elegans* protein has been assigned this function. Higher alcohols, such as ethanol, are excreted from some nematodes, including *C. elegans*, presumably by simple diffusion. Glycerol, another major excretory product, is transported out of cells by aquaporin-type channels in other organisms. No data are available on whether nematode aquaporins can transport glycerol. In contrast to the absence of information on the mechanisms underlying CO_2 or alcohol elimination, many studies on organic acid excretion from adult stages of animal-parasitic nematodes have been published. The physiology of organic acid excretion in these species will be the focus of the remainder of this discussion.

In *A. suum*, the H^+ concentration in cells and extracellular fluids is regulated to a level that maintains the pH close to 7.0. In adult *A. suum* and other animal parasites, organic acid products of intermediary metabolism have pK_as that range from 2.0 (lactic acid) to 4.8 (branched-chain fatty acids, such as 2-methylvaleric acid). At pH 7.0, at least 99% of the organic acids produced by carbohydrate metabolism exist in the dissociated form, i.e. as H^+ and the conjugate organic anion. This point is important: non-dissociated organic acids are quite lipophilic, and can exit tissues via simple diffusion. However, organic anions are poorly soluble in lipids and almost certainly require a protein-mediated transport process for excretion.

Animal parasites lower the pH of incubation media to levels that are equivalent to the pK_a of acids they excrete, supporting the concept that they excrete organic anions and H^+ into their environment. The molecular mechanisms that mediate H^+ excretion through the hypodermis and intestine have not been defined, though it appears to occur through a different process than organic anion excretion. In vertebrates, protons are excreted by a Na^+/H^+ antiporter at pH <7.0, and by an HCO_3^-/Cl^- exchanger at pH >7.0. The electrical driving force for H^+ extrusion in the first case is provided by the steep, inward Na^+ gradient across vertebrate cell membranes. Orthologous H^+ antiporters have not yet been identified in nematodes. Less well characterized transport systems, such as an electrogenic H^+-translocating ATPase in vertebrate distal nephron, and H^+ conducting cation channels in vertebrate renal brush border cells and snail neurons, also play a role in H^+ excretion. Based on sequence homology, a vacuolar-type H^+-ATPase is present in the H-tubular cell of *C. elegans* (see below), but its functional properties have not been reported.

Direct evidence that H^+ and organic anions are excreted across the cuticle–hypodermis complex comes from *in vitro* studies using isolated *A. suum* cuticle segments in two-chamber diffusion cells and from patch-clamp recordings from cuticle- and muscle-facing membranes isolated from the hypodermis. Organic acid concentrations in the hypodermis and muscle compartments are maintained for extended periods in culture at levels that far exceed those in the incubation medium. The concentration and relative abundance of each acid are remarkably similar within the PCF, muscle and hypodermis compartments. Voltage-clamp studies on isolated hypodermal membranes reveal a large conductance, voltage-sensitive, Ca^{2+}-dependent Cl^- channel that resembles a channel in *A. suum* muscle membrane. The Ca^{2+}-activated Cl^- channel in *A. suum* muscle and hypodermal membranes is voltage-sensitive and is open at electrical potentials recorded across the membranes of these cells. It is an inward-rectifying channel that closes as the membrane becomes depolarized. Although the permeability of this channel is greatest for Cl^-, larger anions, including the organic anions excreted by *A. suum*, are also conducted. Permeability across the channel is inversely related to the Stokes' diameter of the anion, with 2-methyl butyrate (diameter = 5.62 Å) being only about 13% as permeable as Cl^- (diameter = 2.41 Å). Reducing external pH increases the probability of the channel opening at hyperpolarized membrane potentials. The negative electrical potentials (−60 to −80 mV) recorded across hypodermal membranes in *A. suum* could provide a driving force for the extrusion of organic anions through the Ca^{2+}-activated Cl^- channel. In fact, for each organic anion excreted by *A. suum*, there is a net outward-directed electrical driving force of 16–27 mV across the hypodermal membrane, which could supply potential energy for organic anion extrusion directly through the anion-selective channels, or indirectly via facilitated transport processes.

Other mechanisms, less studied in nematodes, may serve important roles in organic acid excretion. Several models have been proposed for volatile fatty acid (VFA) transport in vertebrates. At least some transmembrane movement of VFAs is mediated by a family of 12 transmembrane domain H^+-monocarboxylate cotransporters. Genes encoding these transporters have been cloned from several organisms, and at least four homologs have been identified in *C. elegans*. Parasitic nematodes decrease the medium pH suggesting that protons as well as organic acids are excreted, and implicating H^+-monocarboxylate cotransporters in this process. However, no information is yet available on the expression patterns or substrate specificity of the different *C. elegans* transporters, or if homologs are present in parasitic species.

Direct evidence for nitrogen excretion is available only for the intestine and cuticle–hypodermis complex of adult *A. suum*. However, nitrogen, in some form, is also likely excreted from other sites in nematodes, including the pharyngeal and reproductive glands, and the tubule system (see below). Depending on species and stage, 40–90% of nitrogen is excreted as NH_3 or NH_4^+ from the deamination of amino acids. The balance is eliminated as urea, amino acids and peptides. Since the pH in nematode cells and PCF is maintained at 6.5–7.0, ammonia in these compartments exists in a dynamic equilibrium that greatly favors the ionized (NH_4^+) over the non-ionized (NH_3) state. NH_4^+ transport across intestinal or hypodermal membranes should require a channel or carrier protein. While there is no functional evidence for an NH_4^+ transporter in nematodes, a homolog of the yeast NH_4^+ permease, MEP1, has been identified in the *C. elegans* genome. No information is yet available on the expression or function of this putative transporter. NH_3 diffuses freely across membranes, and most nitrogen is probably excreted from nematodes in this form, driving the conversion of additional NH_4^+ to NH_3. NH_3 excretion across the cuticle–hypodermis complex would be facilitated by the acidic microenvironment maintained within the aqueous pores of the cuticle, which would effectively ionize NH_3 as it enters the cuticle pores. This process would ensure low concentrations of NH_3 relative to NH_4^+ in the aqueous pores, providing a driving force favoring diffusion of the uncharged species out of the worm. Among helminths, this principle has been demonstrated experimentally only for *S. mansoni.* At least some members of the aquaporin family can also transport urea, and it is possible that an aquaporin contributes to nitrogen excretion from nematodes.

Some parasites, including *A. suum, T. spiralis* and *Nippostrongylus brasiliensis,* excrete low levels of short-chain aliphatic amines. Excreted amines were once thought to neutralize the acidic microenvironment of nematodes. The pK_as of these amines, however, fall between 10.05 and 10.65. Therefore, it is unlikely that they contribute substantially to the buffering capacity in the pH range of gastrointestinal worms, as they would be almost completely protonated at pH values below 7.5. An organic cation transporter cloned from *C. elegans* provides a potential route for the excretion of aliphatic amines, but no information is available on its expression. The classic substrate for this transporter is tetraethylammonium, but it also transports choline and a variety of basic drugs.

Role of the cuticle–hypodermis complex in drug absorption

The cuticle is an important site for the absorption of anthelmintics and other small organic molecules. For example, accumulation of levamisole by *A. suum* can be accounted for solely by transcuticular diffusion. For the filariae, *B. pahangi* and *Dipetalonema viteae,* absorption indices for a wide range of non-electrolytes show no obvious relationship with lipophilicity. When other physicochemical parameters, such as molecular weight, dipole moment and total energy are considered along with log K, quantitative predictions can be made about the absorption of non-electrolytes by filarial nematodes. This indicates that absorption is influenced by non-lipid components in the cuticle, and contrasts sharply with data from the trematode, *S. mansoni,* for which absorption kinetics can be predicted by the single variable log K.

A. suum and *H. contortus* have also been used to determine how physicochemical properties are related to absorption rates of drugs. As in filariae, log K alone does not accurately predict absorption rate, and collagen and lipid components of the cuticle each present a distinct barrier to diffusion of organic molecules. Diffusion across the collagen barrier is highly dependent on molecular size and charge of the permeant. The permeability of neutral solutes decreases with increasing size, and positively charged molecules penetrate faster, and anions slower, than neutral solutes of comparable size. These results indicate that the cuticle of *A. suum* contains functional, negatively charged aqueous pores that are about 15 Å in radius. The functional pores that mediate permeability through the cuticle of gastrointestinal parasites must consist of tortuous paths through the crosslinked fibers of the cuticle. Electron-dense solutes do not migrate through recognizable channels in *A. suum* cuticle.

The transcuticular excretion of organic acids has important implications for how charged molecules, including weakly acidic or basic drugs, are absorbed across the cuticle. Organic acid excretion drives the pH within the aqueous pores to values close to the aggregate pK_a of the excreted acids. The pH of weakly buffered medium containing adult *A. suum*, for example, which excretes predominantly branched chain fatty acids with pK_as in the range of 4.76–4.88, approaches pH ~5.0 during extended incubations. Species that excrete higher levels of lactic acid, such as *B. pahangi*, tend to acidify the medium even more because of the lower pK_a of lactic acid. This effect is most pronounced in the protected microenvironment of aqueous-filled pores that traverse the cuticle, as shown experimentally for *A. suum* and *H. contortus*. Rates of uptake of several model permeants, including weak acids and weak bases, do not vary as a function of pH in the medium. The rates of absorption and pattern of tissue distribution of model weak acids and bases are unaltered by mechanical or chemical ligation, or by marked changes in the pH or buffer capacity of the bulk medium, even though ligation blocks the uptake of large, hydrophilic molecules, such as inulin or dextrose, by >98%. These findings demonstrate that a large fraction of solutes is absorbed across the cuticle, and that absorption of weak acids and bases is dependent on pH within the aqueous pores, not that of the bulk medium (or, presumably, the pH of host body fluids).

Immune evasion

Adult stages of some parasitic nematodes can persist in immunocompetent hosts for many years. During the course of an infection, leukocytes, antibodies, complement and various oxidants are targeted to the worm surface, but generally have little effect on worm viability. This aspect of the nematode–host relationship is reason enough to conclude that parasites have evolved effective mechanisms for evading all components of the host's immune system. Consideration of the gross structural properties of the cuticle alone is instructive in this regard. The nematode cuticle, with its tough and relatively inert physical structure, has long been presumed to be a protective shell that excludes host immune effectors, most of which are far too large to penetrate the small aqueous pores that traverse it.

Cuticular glycolipids also contribute to immune evasion by protecting surface proteins from degradation by host proteases. *In vitro* studies on filarial nematodes show that surface proteins are degraded by host proteases only after the worm's surface is treated with agents that disrupt the glycolipid barrier. Other parasites, including *Leishmania* spp., employ glycolipids to mask their surface from host immune effectors, and it is possible that cuticular glycolipids serve a similar role. However, the surface of the nematode cuticle, by itself, is not highly immunogenic. For most species, the cuticle contains only a limited set of exposed proteins, glycoproteins and glycolipids. When purified and injected into animals, these molecules generally elicit weak immune responses and fail to significantly protect against infection. Indeed, molecules that adhere to or are released by the cuticle may suppress attack by the immune system or diminish damage caused by noxious agents released by leukocytes onto the parasite surface. For example, a selenium-independent glutathione peroxidase is expressed in the cuticle of late fourth stage larvae and adult *D. immitis*. This antioxidant counters the effects of oxidants released by host leukocytes. In addition to its putative role as an antioxidant, surface-associated glutathione peroxidase may also contribute directly to new cuticle synthesis and repair, as it may catalyze formation of tyrosine-based crosslinks between cuticle collagen-like proteins.

Several other proteins are thought to participate in immune evasion or suppression. For example, the surface coat of *Toxocara canis* cuticle is shed rapidly in response to binding of host antibodies. In this process, a glycoprotein (TES-120) is shed that contains an 86-amino acid residue domain that is conserved in all mucins. In vertebrates, mucins are immunosuppressants, and it is possible that mucin-like proteins shed from the cuticle serve analogous roles in nematodes.

These observations have led researchers to consider antigens not associated with the cuticle as candidates for vaccine development. Proteins that line the intestine of animal parasitic species have received considerable attention as candidate protective antigens. Because of their location within the organism, these are often referred to as 'hidden' or 'concealed antigens', and will be considered in discussion of the internal surfaces of nematodes (below).

Internal surfaces

Structural considerations

Gross and microscopic anatomy

The intestine and the tubule system provide surfaces that are important sites of molecular exchange between the nematode and host. The alimentary tract extends nearly the full length of the organism, opening at the mouth and anus. It consists of three regions: the stomodeum (mouth, buccal capsule and pharynx or esophagus), the intestine, and the proctodeum (rectum and anus, including the reproductive opening in males). The stomodeum and proctodeum are lined by cuticle. The buccal cavity of some intestinal nematodes contains teeth or cutting plates which are modifications of the cuticle. In some species, the walls of the buccal capsule are elaborated to form an extendable stylet used to inject digestive enzymes that aid in the extracorporeal digestion of food. These are most prevalent in species that feed on the mucosa of the alimentary or respiratory tracts of the host. Nematodes that have a small buccal cavity and rely on extracorporeal digestion of host tissues depend less heavily on the grinding action of teeth, and cuticle modifications in the buccal cavity are less extensive.

The intestine consists of a tube of epithelial cells, one cell layer thick, connecting the pharynx and anus. The number of cells that comprise the intestine varies extensively by species, ranging from as few as 20 in *C. elegans* to over a million in *A. suum*. The intestine has no muscular layer and so does not show peristalsis. It is normally collapsed due to the high internal pressure of the pseudocelom. Against this pressure, food particles are propelled backwards only by the pumping action of the pharynx. The intestine is bounded at its ends by muscular structures that control valves.

In most nematodes, microvilli line the intestine. In *A. suum*, microvilli increase the surface area for transport by almost 100-fold. They are coated by a unit membrane and a glycocalyx similar in appearance to that of vertebrate intestinal epithelium. The basal membrane of intestinal cells in nematodes may be smooth or folded. It forms a continuous layer which arises as a secretory product of epithelial cells; mitochondria are concentrated near the basal lamellae. Columnar cells, particularly in the anterior regions of the intestine, appear to serve glandular functions. Some parasitic species show markedly different structural patterns. For example, *Bradynema*, which parasitizes insects, lacks both a cuticle and a functional intestine. Whatever functions the gut serves in other nematodes must be adopted by other tissues. The hypodermis in *Bradynema* is elaborated by microvilli which greatly increase its surface area for nutrient absorption. A related phenomenon may occur in adult female *O. volvulus,* in which the hypodermis hypertrophies as the worm matures, coincident with atrophy of the intestine.

Biochemistry and molecular biology of structural components

Information on the biochemistry and molecular biology of structural components in the nematode gastrointestinal tract focuses mainly on the regulation of endodermal differentiation and on collagen synthesis. Proteins involved in these processes are reasonably well conserved among *C. elegans* and the few parasitic species examined, including *A. suum* and *H. contortus,* which should facilitate research in this area. The structural organization of collagen aggregates in the basement membrane of *A. suum* clearly distinguishes them from cuticular collagens, though biochemical explanations for the differences remain unresolved. The importance of tubulins in the nematode intestine is suggested by data showing that depolymerization of intestinal tubulin is the initial (and perhaps lethal) action associated with the benzimidazole anthelmintics.

Functional biology

The importance of the intestine for nutrient absorption, excretion, and ionic and osmoregulation varies among species and stages. In general, the intestine is more important in gastrointestinal than in tissue parasites. This generalization may be challenged as better culture systems for nematodes are developed or when transport studies can be extended to *in vivo* conditions. Since adult parasitic nematodes often do not survive extended *in vitro* incubations, it is likely that essential components are absent from the culture systems used, and these deficiencies may affect intestinal function. Little is known about biochemical changes in the intestine associated with development in parasites. However, considerable insight has been gained from *C. elegans.* The identification of genes that are only expressed in the gut (see examples below) permits a detailed analysis of the development of this specialized tissue.

Digestion

In intestinal parasites, digestion begins as food particles enter the pharynx, and even earlier in species that feed by extracorporeal digestion. *Nippostrongylus brasiliensis,* for instance, releases histiolytic enzymes from the pharyngeal and subventral glands that partially digest host mucosal cells, which are subsequently ingested by the pumping action of the pharynx. The release of digestive enzymes onto food particles is greatly reduced in saprophagous species, which feed on material already partially digested by the host. In most animal parasites, pharyngeal glands secrete digestive enzymes onto ingested food particles as they enter the pharynx. In *A. suum,* esterase, amylase, maltase, protease, peptidase and lipase are all secreted into the pharynx during feeding. The pharynx of species that feed on host mucosal cells may also contain hyaluronidase and additional enzymes specialized for digestion of host tissues.

A role for the nematode intestine in digestion can be inferred from the presence of numerous glands that are single cells that empty their contents directly into the lumen. Enzymes localized within the lumen of the nematode intestine include an array of hydrolases typical of digestive tissues; proteases are the best characterized. Carboxy and thiol proteases are found in the gut of many parasitic nematodes. A cysteine protease from *H. contortus* degrades fibrinogen and may help prevent clotting of ingested blood. The gene for a similar enzyme, along with an esterase and a vitellogenin, are expressed only in the intestine of *C. elegans*, and factors that regulate intestinal-specific expression have been identified. The alimentary canal in nematodes also contains proteins that inhibit digestive enzymes of the host, primarily proteases. These inhibitors may protect the intestine from damage. In *A. suum*, several disaccharidases are abundant on the microvilli, including sucrase, palatinase, maltase and trehalase, with maltase activity predominating. Lipases have been detected in intestinal cells of several species, and their activity depends extensively on the diet of the parasite. In *Strongylus edentatus*, which feeds on host mucosal tissue, lipase activity in the intestine is 12-fold greater than that for *A. suum*, a saprophagous feeder.

Nutrient absorption

The alimentary tract of intestinal parasites is an important surface for nutrient absorption. In *A. suum*, the intestine may be the only site where physiologically significant amounts of glucose and amino acids are absorbed. Unfortunately, almost everything we know about intestinal transport of nutrients in nematodes comes from studies using this species. The atypical size of this organism, and the fact that it ceases to absorb glucose following collection from the swine intestine, suggest that it may not be an ideal model for the study of nutrient absorption. Smaller species may be less reliant on the bulk transfer capacity of the intestine for the delivery of nutrients, other solutes and water to internal tissues. Nevertheless, among readily available nematodes, only *A. suum* is large enough to dissect and isolate segments of intestine to study by standard physiological techniques.

In isolated segments of *A. suum* intestine, glucose and fructose are absorbed much more rapidly than galactose. Glucose absorption occurs against a steep concentration gradient, is saturable and Na^+-sensitive. The non-metabolizable analog of glucose, 3-*O*-methylglucose, is absorbed unidirectionally (lumen to pseudocelom) across the intestine. Its absorption against a concentration gradient depends on the presence of authentic glucose in the incubation medium. Together, these findings suggest that glucose absorption across the intestine of *A. suum* is energy-dependent and involves a Na^+/glucose symporter.

The *A. suum* intestine also absorbs amino acids. Uptake of methionine, glycine, histidine and valine is stereospecific and non-linear with respect to concentration, indicative of a mediated transport process. Most nematodes excrete a wide range of amino acids. Some of these, such as alanine and proline, are true metabolic end-products, but others must be derived from ingested materials. Most evidence indicates that the absorption of fatty acids, mono- and triglycerides and cholesterol occurs mainly across the alimentary tract of intestinal nematodes. Absorption of albumin-complexed palmitic acid is a passive process that occurs in the absence of glucose or the presence of iodoacetate. Passage of this fatty acid from the intestine into the pseudocelom, however, requires glucose and is sensitive to metabolic inhibitors, indicative of mediated transport. Fatty acid absorption is stimulated by bile salts. Triglycerides are absorbed very slowly, whereas monoglycerides are absorbed rapidly by intestinal cells, then converted into free fatty acids before being actively transported to the PCF for dissemination to the body tissues.

Osmotic/volume regulation

The role of the intestine in osmoregulation appears to be stage- and species-specific. The intestine is fully collapsed in larval stages of some gastrointestinal and filarial species. Also, based mainly on morphological evidence, the intestine is non-functional in some animal-parasitic nematodes, including most tissue-dwelling species examined. In several filarial nematodes, water and solute transport appear to occur exclusively across the cuticle. There are undoubtedly exceptions to this generality. For example, in third stage larvae of *Pseudoterranova decipiens*, an anisakid found in the muscle of cod, the intestine appears to play a critical role in osmoregulation, as the oral opening is patent and water is consumed by the oral route. Intact worms accumulate water under hyposmotic conditions. Sacs composed of cylinders of body wall prepared without the intestine do not accumulate water under hyposmotic conditions, even though exchange of $[^3H]_2O$ between the medium and PCF occurs at a high rate. Similar results are obtained in worms mechanically ligated at the head and tail. Metabolic poisons, including cyanide and dinitrophenol, abolish the ability of the body wall segments or ligated worms to osmoregulate, again suggesting a possible role for active transport of ions in this process. These results indicate that, even though the cuticle–hypodermis complex provides a potential pathway for water flux, the process of regulating volume in this nematode is dependent on a patent and functional intestine.

Under hypotonic conditions, excess water can be removed through the intestine of *A. suum*. This process probably accounts for some of the capacity of *A. suum* and other species to survive hypotonic conditions, even though their cuticles are highly permeable to water. Under isosmotic conditions, water enters intestinal cells of *A. suum* from the lumen, then passes into intercellular spaces before entering the pseudocelom. It is not known if water enters or exits intestinal cells through aquaporins, through other channels, or via paracellular pathways. The rate of water transport from the gut into the pseudocelom is higher in the presence of oxygen. However, there are no reports that intestinal cells in *A. suum* or other parasitic nematodes are aerobic.

The molecular mechanisms of water diffusion across the intestinal epithelium from the pseudocelom are unknown. However, active Na^+ transport occurs across the intestine of *A. suum*, and it is likely that water follows the movement of Na^+ and possibly other inorganic ions in this system. In annelid and insect gut epithelia, as well as vertebrate intestine and kidney where these linked processes have been studied more thoroughly, epithelial Na^+ channels mediate the bulk flow of Na^+, with water following passively, probably through aquaporins. In vertebrates, epithelial Na^+ channels are regulated, in part, by vasopressin and aldosterone. The identification of a role, if any, for Na^+ channels in osmoregulation across the intestine of nematodes awaits the identification of candidate channels. A degenerin-like Na^+ channel, which is closely related to mammalian epithelial Na^+ channels, was recently cloned from *C. elegans*. This channel is expressed in the intestine in all stages of the life cycle. Mutations in this gene alter the defecation rhythm, but do not induce abnormalities in osmotic regulation. Further study of other *C. elegans* genes in this large family may identify one (or more) that is associated with water movement.

Other inorganic ion channels implicated in the regulation of water flux across the intestine include members of the family of ClC channels that conduct Cl^-. This family of voltage-gated Cl^- channels is found in organisms ranging from bacteria to humans. Some of these channels, including ClC-1 and ClC-2, play important roles in water transport (swelling) and ion transport. Six Cl^- channels related to the family of ClC channels are present in *C. elegans*. Four are close homologs, with CLH-1 being 40% identical to ClC-2. The *clh-1* gene in *C. elegans* is predominantly expressed in seam cells that run along the lateral ridges of the hypodermis.

Mutations in *clh-1* cause an increase in body width, which is reversed when the worms are placed in hypertonic medium. These findings suggest that CLH-1 may be required to maintain volume in seam cells by regulating ion balance. Two other ClC homologs in *C. elegans*, CeCLC-3 and CeCLC-4, are expressed in the H-cell of the putative excretory system (see below). CeCLC-3 is also expressed in the first four epithelial cells that form the intestine. CeCLC-3 and a splice variant (CLH-3b) have been functionally expressed and partially characterized in *Xenopus* oocytes and in insect (Sf9) cells, respectively, where they form inward-rectifying channels with complex gating that includes slow, anion-dependent activation by depolarization and very rapid inactivation at positive voltages. These characteristics are reminiscent of the prototype ClC channel, ClC-0 from *Torpedo*, which is a double-barreled channel with two identical pores, and serves an important role in cell repolarization after excitation. Additional studies, including the effects of null mutations on phenotype, will help define the roles served by CeCL-3 and CeCL-4.

Ionic regulation

It is important to distinguish ionic regulation across the pharynx, which controls the muscular activity of this part of the alimentary tract, from that which occurs across the intestine, which contributes to the maintenance of ion concentrations in the PCF. These processes are interrelated, however, to the extent that the pumping action of the pharynx controls the rate at which digesta pass through the intestine, thus influencing the extent to which ions and water are allowed to enter or exit the intestinal lumen. Though a considerable body of data has been published recently on mechanisms underlying ionic regulation across the nematode pharynx and the nerves that innervate it, direct evidence for transintestinal transport of inorganic ions is available only for *A. suum*. Concentration gradients for Na^+, K^+ and Cl^- across the intestine contribute to a 10–15 mV electrical potential, inside positive (i.e. the lumen is depolarized relative to the surrounding PCF), which is abolished below 20°C and inhibited by glycolytic poisons. These findings indicate the importance of active transport to the maintenance of this electrical potential. Flux measurements indicate that, among the major inorganic ions, only K^+ diffuses outward from the lumen more rapidly than it diffuses inward. Other ions must contribute to the electrical potential, although their relative contributions have not been determined. The electrical potential is also dependent, in part, on the presence of glucose on the luminal side, consistent with the concept that Na^+ flux from the intestine is coupled to the absorption of glucose into the epithelial cells of the intestine. The existence of a Na^+ gradient is essential for Na^+-coupled glucose transport in other systems. This gradient is usually maintained by a Na^+/K^+-ATPase that pumps more Na^+ ions out of the cell than K^+ ions in. Several genes encoding Na^+/K^+-ATPases are present in *C. elegans*; mutations in one of these, *eat-6*, impair pharyngeal pumping. However, Na^+ and glucose transport across the *A. suum* intestine is not affected by high levels of ouabain indicating that if Na^+/K^+-ATPase regulates the concentrations of these ions, in this species at least, the enzyme is pharmacologically distinct from that in vertebrates and most other invertebrates.

Other cation channels play a more important role in internal than external surfaces in nematodes. At least 75 genes encode K^+ channels in *C. elegans*. Among these are 20 voltage-sensitive and three inward-rectifying channels, which typically function to maintain the hyperpolarized state of cells. EXP-6, a voltage-gated K^+ channel expressed on pharyngeal muscle, underlies the rapid repolarization of the pharynx at the end of each action potential. This channel is biophysically similar to the HERG K^+ channel that repolarizes human heart muscle cells following contraction, though these two proteins are not highly related structurally. Several mutations in *exp-6* are lethal, suggesting a critical role. Pharyngeal function in *C. elegans* is also dependent on a K–Cl cotransporter, KO2, that is expressed in pharyngeal muscle membranes and is essential for survival under hyperosmotic conditions. Other K^+ channels have been implicated in more posterior regions of the intestine. The M-type K^+ channels, *kqt-1*, *kqt-2* and *kqt-3*, are expressed in the proximal and distal regions of the intestine. When expressed in *Xenopus* oocytes they exhibit slow activation kinetics resembling KCNQ-related M-type K^+ channels in vertebrate epithelial and cardiac cells. Suppression of *kqt-2* or *kqt-3* leads to prolongation of the defecation cycle. Though the mechanism underlying this response has not been determined, indirect evidence suggests that these channels regulate the timing of cytoplasmic Ca^{2+} oscillations in intestinal cells.

Many channels that regulate ion flux across the pharynx are under control of the nervous system. The rate of pharyngeal pumping in nematodes is controlled, in part, by cholinergic (MC in *C. elegans*), serotonergic (M3), glutamatergic (M1 and M4) and neuropeptidergic neurons, all of which innervate the pharynx. In some cases, receptors for these transmitters couple directly to ion channels, the biophysical and molecular characteristics of which are currently being delineated. In the case of the peptidergic inputs, the G-protein-coupled receptors and second messenger systems to which they couple have not been identified.

Excretion

That the intestine participates in excretion is supported by studies showing that each organic acid present in PCF is also present in the lumen of the intestine in adult *A. suum*. There is no evidence for a concentrating effect as digesta move into the caudal regions. However, indirect evidence that questions the importance of the intestine for organic acid excretion comes from studies on adult *A. suum*. Neither the rate of organic acid excretion into the medium nor worm viability, based on ATP and motility levels, is affected by ligation of the pharynx. Similarly, chemical ligation of adult *H. contortus* with ivermectin, which paralyzes the pharynx and inhibits flow of digesta along the intestine, does not affect the rates of excretion of propionic or lactic acids, or worm viability. These and related studies suggest that the cuticle–hypodermis complex may be a more important site than the intestine for excretion of some metabolic end-products.

Role for intestinal P-glycoproteins in drug elimination

The accumulation of some molecules by nematodes may be affected by drug transporting proteins, referred to as P-glycoproteins (pgps). Pgps actively ‘pump’ drugs out of some cells, thereby limiting exposure of the organism to their therapeutic actions. In vertebrates, overexpression of pgps has been linked to resistance to some anti-cancer drugs, though their relevance to this condition in the clinic remains controversial. Ivermectin is a substrate for pgp-mediated extrusion from murine kidney epithelial cells *in vitro*. Mice deficient in *mdr1a*, a pgp-encoding gene, are susceptible to the toxic actions of ivermectin, which accumulates in the brains of these mutants to levels over 100-fold greater than in wild-type mice. In *C. elegans*, pgps are expressed primarily on apical membranes of cells that form the intestine and tubular system, a pattern consistent with a role for these proteins (and internal surfaces) in protecting the organism from toxic xenobiotics. There are no reports describing pgp expression in the hypodermis. Fourteen pgp-encoding genes have been identified in *C. elegans*, some of which are probably pseudogenes. Deletion of two of these genes leads to increased susceptibility to chloroquine, colchicine and to heavy metals that are not toxic to wild-type *C. elegans*. To this point, however, no *C. elegans* anthelmintic-resistance trait has been mapped to a pgp locus.

The animal-parasitic nematode *H. contortus* expresses at least four *pgp* genes. In some strains, a pgp polymorphism is associated with macrocyclic lactone resistance. These observations, however, do not establish a causal link between pgp polymorphism and anthelmintic resistance. Resistant and sensitive strains of *H. contortus* do not absorb these drugs at different rates, or accumulate them in specific worm tissues in a different manner. Ascribing a role for pgps in anthelmintic resistance will require demonstrating that pgp-deficient strains accumulate drugs more rapidly, or that pgp over-expressing strains accumulate drugs more slowly, than wild types. In this context, a closantel-resistant strain of *H. contortus* accumulates closantel less rapidly than wild-type strains. Closantel, like ivermectin, is highly lipophilic, which is a hallmark of most substrates for pgps. However, no pgp polymorphism has been associated with the closantel-resistant phenotype. Because these experiments only tracked variation in a limited number of restriction sites, the findings cannot rule out pgp involvement in closantel resistance.

Whether or not alterations in pgp expression or sequence have been selected as a mechanism of anthelmintic resistance, there is ample evidence to conclude that these proteins contribute to the excretion behavior of nematodes. Their primary location in intestinal cells is consistent with a role in eliminating compounds that penetrate the cuticle, diffuse through worm tissue and accumulate in PCF. ATP-dependent transport of xenobiotics from PCF to the intestinal lumen would reduce internal drug levels and protect the organism from toxicity.

Defecation

Most adult nematodes defecate. Studies in *C. elegans* have illuminated the physiology and genetics of defecation, a rhythmic, patterned behavior that is controlled by the availability and quality of food. Transit of fecal material through the intestinal tract is rapid; the default defecation cycle in *C. elegans* is 45 seconds. A Ca^{2+} spike in posterior regions of the intestine precedes and probably initiates the first phase of defecation, referred to as the posterior body-wall contraction. Mutations in *aex-2*, which encodes a G-protein-coupled receptor expressed in the intestinal muscles NSM and AWB, result in defects in this response. Though the signaling mechanism that underlies this response has not been determined, available evidence suggests that it is peptidergic, as mutations in two enzymes, *aex-5* and *egl-21*, which process small neuropeptides, also produce the Aex phenotype.

Less is known about defecation in parasitic species. The contents of the gut in *A. suum* are turned over every three minutes, and defecation can propel fecal material up to 60 cm. The major muscle groups that regulate defecation (pharynx and rectum) are innervated, and it is likely that the nervous system plays a critical role in this process, as shown for *C. elegans*. However, not all nematodes possess a functional one-way gut, and defecation may be less important in these species. The extent to which the gut is functional in adult female *O. volvulus* is an open question. It appears that adults of this species, and perhaps other filariae, both acquire nutrients and excrete waste products across the cuticle–hypodermis complex. Though it would seem a simple matter to obtain samples for analysis, almost nothing is known about the composition of feces from any nematode. Until the composition and rate of production of this material is determined for several species, it is impossible to calculate the importance of defecation to the overall process of waste elimination by nematodes.

Considerations based on pharmacology

The importance of digestive tract function to nematodes, though difficult to quantitate by solute flux analyses, is readily deducible from data showing that the gut is the principal target for two of the three most widely used classes of anthelmintics. The macrocyclic lactones, such as ivermectin, target specific sets of motor neurons that innervate the pharynx. Pharyngeal pumping in *C. elegans* and various parasitic species is inhibited by concentrations of ivermectin that are 10–100-fold lower than those required to reduce motility. Inhibition of the ingestive process is probably sufficient for the therapeutic actions of avermectins, since host tissue levels of these drugs are much lower than those required to affect worm movement *in vitro*. The molecular target underlying pharyngeal paralysis has been localized in *C. elegans* to glutamate-gated Cl^- channels on two motor neurons, M1 and M4, that innervate the pharynx. These neurons, and their high level of sensitivity to macrocyclic lactones, are conserved in parasitic species. The importance of gut function for survival is also demonstrated pharmacologically by the benzimidazoles. The earliest signs of worm toxicity are found in the intestine, and include tubulin depolymerization and inhibited transport of digestive secretory vesicles in the anterior intestine. This is followed by disintegration of the anterior intestine, DNA fragmentation within anterior intestinal nuclei, and inhibition of host erythrocyte digestion.

Nematode intestinal proteins as targets for vaccine development

Infectious stages of most nematodes rely on intestinal functions for survival. Numerous proteins have been characterized from the microvillar surface of the nematode intestine. Some of these proteins are not generally exposed to the host (i.e. they are neither co-localized on the cuticle nor secreted from the worm) and elicited immune responses when used to vaccinate host animals. These gut proteins are often referred to as 'hidden' or 'concealed' antigens. Concealed antigens have been most thoroughly studied in the blood-feeding gastrointestinal species, *H. contortus*. Contortin, for example forms helical filaments that extend into the glycocalyx that lines the intestine, and contortin-enriched preparations used to vaccinate lambs confer up to 78% reduction in worm burdens following challenge with infective larvae. Because contortin is not usually exposed to the host, the response it elicits is described as artificial immunity.

Other hidden antigens have been characterized. A 110 kDa contaminant found in preparations of contortin confers an even higher level of protection (>90%) when administered to lambs in multiple immunizations. This glycoprotein, called H11, has the predicted structure of a type II integral membrane protein with an N-terminal cytoplasmic tail, a transmembrane domain and an extracellular region. It has aminopeptidase A and M activities in microsomal preparations, so probably functions as a digestive enzyme *in situ*. It is localized exclusively in the intestinal brush border, and close homologs have been identified in other gastrointestinal nematodes, including *Teledorsagia circumcincta* and *Ostertagia ostertagi*. Unfortunately, vaccine trials to determine the cross-species protection conferred by H11 and other hidden antigens have so far been disappointing. Also, H11 and other hidden antigens (including H-gal-GP and TSBP) are protective only in their native forms. When produced in recombinant systems, none has conferred protection, even though they retain enzymatic activity when expressed in yeast, suggesting that appropriate glycosylation may be required for the expression of protective immunity.

Tubular system

Tubular systems are present in most nematodes, and consist primarily of two lateral ducts that lie next to or within the paired lateral cords (extensions of the hypodermis that also enclose the nerve cords). In *A. suum*, canals that form part of this system are exposed to PCF along most of their length. These ducts or tubules, and the cells that form ampullae into which they empty before exiting the worm, were originally assigned an excretory function based on morphological evidence. The canals accumulate dyes injected into the pseudocelom, and a mechanism was proposed whereby the high internal (or turgor) pressure in the nematode could provide a driving force for filtering excretory products released from cells into the lateral canals. Given the relatively high molecular weights and charged nature of the dyes used in these studies, it is difficult to conceive of a filtering mechanism that would allow their passage while retaining water and other nutrients within the pseudocelom, and no morphological evidence exists for endocytotic or pinocytotic mechanisms in the canals that could underlie such a process.

The tubule system appears to play a critical role in osmoregulation of some nematodes, including *C. elegans*. The entire tubular system has been reconstructed from serial section electron micrographs, and consists of four cells with nuclei located on the ventral side of the pharynx. A pore cell encloses the terminus of an excretory duct cell, which leads to an excretory pore at the ventral midline. An H-shaped cell forms bilateral excretory canals that extend anteriorly and posteriorly along most of the length of the organism. These canals form numerous gap junctions with the hypodermis and are in direct contact with the PCF. In addition, an A-shaped excretory gland cell extends bilateral processes anteriorly to cell bodies located behind the pharynx. These processes fuse with the H-shaped cell at the origin of the excretory duct. Laser ablation studies on *C. elegans* have illuminated the role of these cells in homeostatic regulation. If the pore, duct or excretory cell is ablated, the nematode fills with water within 12–24 hours and dies within a few days. Ablation of the excretory gland cell results in no obvious developmental or behavioral defects. These results suggest a role for the tubular system in osmoregulation, and are consistent with microscopic examination of the excretory duct; its rate of pulsation under hyposmotic conditions is five- to six-fold higher than under isosmotic conditions.

C. elegans contains at least 12 ATPases, including several subunits of a vacuolar-type ATPase that transports H^+ out of cells in other organisms. The genes are expressed in the H-shaped excretory cell, the rectum, and in two cells posterior to the anus. One subunit has been used to complement a yeast mutant, conferring H^+ transport function to the organism. However, phenotypes associated with mutations in these genes have not been described. It should be straightforward to use dsRNAi techniques to determine if this ATPase contributes to pH regulation in nematodes, as it does in the vertebrale kidney.

Additional insights into the role of the tubular system in osmoregulation come from *C. elegans* mutants. Mutants defective in the posterior migration of canal-associated neurons (CANs) exhibit arrested development and excess fluid accumulation in the pseudocelom.

Also, a mutation in *let-653*, which encodes a mucin-like protein, results in lethal arrest concurrent with vacuole formation anterior to the lower pharyngeal bulb, in a position consistent with dysfunction of the tubular apparatus. The extent to which these physiological and genetic observations on *C. elegans* are relevant to parasitic species is largely undetermined. The existence of tubular structures in parasitic nematodes is well documented. Recent studies suggest that these structures may be more elaborate than originally thought. For example, anterior regions of the tubular system in *O. volvulus* include paired glomerulus-like structures in the lateral cords that appear to connect to the cuticle through canals formed by projections from the basal zone of the cuticle. Anatomically, these structures resemble organs with an osmoregulatory or excretory function. However, assigning them a function in homeostatic regulation will require additional evidence, such as localization within their membranes of ion channels or aquaporins.

There is no direct evidence that regulation of water or solute transport across the tubule system in nematodes is controlled by neuroeffectors or hormones as analogous processes are in higher eukaryotes. However, several observations suggest this possibility. As noted above for *C. elegans*, the excretory ampulla in larvae of some parasitic species, including *H. contortus*, pumps more rapidly under hyposmotic conditions. The rate of filling of the ampulla is controlled by the excretory valve, which is innervated and presumably under some form of neural control.

CONCLUSION

Helminth surfaces serve numerous critical roles. Indeed, it is possible to assign a role to surface structures in almost every aspect of helminth physiology. Knowledge of the biology of helminth surfaces has advanced more slowly than that of their nervous systems, probably because most modern anthelmintics target the latter specifically, and these compounds have provided an economic stimulus for many 'pharmacology-driven' investigations on helminth neurobiology. It is likely, however, that future exploitation of helminth surfaces in anthelmintic and vaccine discovery and development will be aided greatly by knowledge derived from *C. elegans* genomics. These approaches will facilitate the characterization of surface proteins in a variety of ways. Among these will be approaches referred to generally as 'reverse pharmacology', whereby genes for candidate target proteins, selected from electronic databases, will be cloned and then over- or underexpressed in *C. elegans*, which will allow their functions to be studied against the background of the full complement of 'normal' genes. This approach will also be extended to parasites as orthologous genes are identified. This genetic approach will be augmented by others, including selective laser ablation of cells that express proteins of interest. As illustrated by several examples in this chapter, these approaches are already being used with great success in *C. elegans*, and their extension to parasitic species requires only additional data on the genome sequences of those species.

FURTHER READING

Arme, C. (1988). Ontogenetic changes in helminth membrane function. *Parasitology* **96**, S83–S104.

Beames, C.G. and King, G.A. (1972). Factors influencing the movement of materials across the intestine of *Ascaris*. In: Van den Bossche, H. (ed.) *Comparative Biochemistry of Parasites*, New York: Academic Press, pp. 275–282.

Bird, A.F. (1991). *The Structure of Nematodes*, 2nd edn, New York: Academic Press.

Blaxter, M.L., Page, A.P., Rudin, W. and Maizels, R.M. (1992). Nematode surface coats: actively evading immunity. *Parasitol. Today* **8**, 243–247.

Court, J.P., Murgatroyd, R.C., Livingstone, D. and Rahr, E. (1988). Physicochemical characteristics of non-electrolytes and their uptake by *Brugia pahangi* and *Dipetalonema viteae*. *Mol. Biochem. Parasitol.* **27**, 101–108.

Cox, G.N. (1992). Molecular and biochemical aspects of nematode collagens. *J. Parasitol.* **78**, 1–15.

Fetterer, R.H. (1986). Transcuticular solute movement in parasitic nematodes: relationship between non-polar solute transport and partition coefficient. *Comp. Biochem. Physiol.* **84A**, 461–466.

Harris, J.E. and Crofton, H.D. (1957). Structure and function of the nematodes: internal pressure and the cuticular structure in *Ascaris*. *J. Exp. Biol.* **34**, 116–130.

Hobson, A.D., Stephenson, W. and Eden, A. (1952). Studies on the physiology of *Ascaris lumbricoides*. 11. The inorganic composition of the body fluid in relation to that of the environment. *J. Exp. Biol.* **29**, 22–29.

Howells, R.E. (1980). Filariae: dynamics of the surface. In: Van den Bossche, H. (ed.). *The Host Invader Interplay*, Amsterdam: Elsevier/North Holland, pp. 69–84.

Johnstone, I.L. (2000). Cuticle collagen genes: expression in *Caenorhabditis elegans*. *Trends Genet.* **16**, 21–27.

Jones, M. (1998). Structure and diversity of cestode epithelia. *Int. J. Parasitol.* **28**, 913–923.

Knox, D.P. and Smith, W.D. (2001). Vaccination against gastrointestinal nematode parasites of ruminants using gut-expressed antigens. *Vet. Parasitol.* **100**, 21–32.

Loukas, A., Hunt, P. and Maizels, R.M. (1999). Cloning and expression of an aquaporin-like gene from a parasitic nematode. *Mol. Biochem. Parasitol.* **99**, 287–293.

Pappas, P.W. (1988). The relative roles of the intestines and external surfaces in the nutrition of monogeneans, digeneans and nematodes. *Parasitology* **96**, S105–S121.

Pappas, P.W. and Read, C.P. (1975). Membrane transport in helminth parasites: a review. *Exp. Parasitol.* **37**, 469–530.

Pax, R.A. and Bennett, J.L. (1990). Studies on intrategumental pH and its regulation in adult male *Schistosoma mansoni*. *Parasitology* **101**, 219–226.

Pax, R.A., Geary, T.G., Bennett, J.L. and Thompson, D.P. (1995). *Ascaris suum*: characterization of transmural and hypodermal potentials. *Exp. Parasitol.* **80**, 85–97.

Sims, S.M., Magas, L.T., Barsuhn, C.L., Ho, N.F.H., Geary, T.G. and Thompson, D.P. (1992). Mechanisms of microenvironmental pH regulation in the cuticle of *Ascaris suum*. *Mol. Biochem. Parasitol.* **53**, 135–148.

Sims, S.M., Ho, N.F.H., Magas, L.T., Geary, T.G., Barsuhn, C.L. and Thompson, D.P. (1994). Biophysical model of the transcuticular excretion of organic acids, cuticle pH and buffer capacity in gastrointestinal nematodes. *J. Drug Target.* **2**, 1–8.

Skelly, P.J., Kim, I.W., Cunningham, I. and Shoemaker, C.B. (1994). Cloning, characterization, and functional expression of cDNAs encoding glucose transporter proteins from the human parasite *Schistosoma mansoni*. *J. Biol. Chem.* **269**, 4247–4253.

Smyth, J.D. and Halton, D.W. (1983). *The Physiology of Trematodes*, Cambridge: Cambridge University Press.

Smyth, J.D. and McManus, D.P. (1989). *The Physiology and Biochemistry of Cestodes*, Cambridge: Cambridge University Press.

Thompson, D.P. and Geary, T.G. (1995). The structure and function of helminth surfaces. In: Marr, J.J. and Muller, M. (eds). *Biochemistry and Molecular Biology of Parasites*, London: Academic Press, pp. 203–232.

Thompson, D.P., Ho, N.F.H., Sims, S.M. and Geary, T.G. (1993). Mechanistic approaches to quantitate anthelmintic absorption by gastrointestinal nematodes. *Parasitol. Today* **9**, 31–35.

Threadgold, L.T. (1984). Parasitic platyhelminths. In: Berieter-Hahn, J., Matolsky, A.G. and Richards, K.S. (eds). *Biology of the Integument*, vol. I, Berlin: Springer-Verlag, pp. 132–191.

Wright, D.J. and Newall, D.R. (1976). Osmotic and ionic regulation in nematodes. In: Zuckerman, B.M. (ed.). *Nematodes as Biological Models*, New York: Academic Press, pp. 143–164.

C H A P T E R

14

Carbohydrate and energy metabolism in parasitic helminths

Richard Komuniecki[1] *and Aloysius G.M. Tielens*[2]

[1] Department of Biological Sciences, University of Toledo, Toledo, OH, USA; and

[2] Faculty of Veterinary Medicine, Utrecht University, Utrecht, The Netherlands

INTRODUCTION

Parasitic helminths are a phylogenetically diverse group, encompassing both the acelomate flatworms and the pseudocelomate nematodes. Parasites from both groups inhabit a variety of habitats including the vasculature, lungs, tissue spaces, gut and bile duct. These habitats exhibit a wide range of physicochemical characteristics, including temperature, redox potential, pH, and oxygen concentration. The life cycles of most parasitic helminths are more complex than those of their free-living counterparts. In general, early larval stages of parasitic helminths are free living and use classic aerobic energy-generating pathways. In contrast, most adult parasitic helminths exhibit high rates of aerobic glycolysis when compared to their free-living counterparts. Not surprisingly, parasitic helminths from similar microenvironments exhibit similar metabolisms regardless of their phylogenetic origins. For example, blood- and tissue-dwelling helminths, such as *Schistosoma mansoni* and the filaria that are continuously bathed in glucose, excrete predominantly lactate, although mitochondrially based aerobic pathways have the potential to contribute significantly to energy generation. In contrast, lumen-dwelling helminths, such as adult *Hymenolepis diminuta, Haemonchus contortus* and *Ascaris suum,* which reside in environments where glucose availability is episodic (depending on host feeding) and oxygen tensions are low, rely on stored glycogen for energy and have modified their metabolic pathways to increase energy-generation in the absence of oxygen. Mitochondria from these organisms use unsaturated

Molecular Medical Parasitology
ISBN 0-12-473346-8

organic acids in addition to oxygen as terminal electron acceptors, and cyanide-insensitive electron-transport-associated ATP synthesis is coupled to the excretion of reduced organic acids as end-products of carbohydrate metabolism. The present chapter will consider these two metabolic strategies separately.

GLYCOGEN METABOLISM

Glycogen is present in all parasitic helminths, but its function can vary depending on species. Intestinal helminths, such as *Ascaris suum,* display a predominantly anaerobic metabolism and use stored glycogen as a key energy reserve. Not surprisingly, helminth glycogen levels fluctuate with host feeding, and glycogen stored during feeding is rapidly utilized when the host is in the post-absorptive state. In contrast, blood-dwelling parasites, like adult schistosomes, continuously reside in a glucose-rich environment, and the function of glycogen storage and metabolism in these organisms is still not completely clear.

Helminth glycogen synthesis is fuelled exclusively by host-derived carbohydrate (primarily glucose). Adult parasitic helminths are incapable of gluconeogenesis, except from intermediates at the level of triose phosphates. In mammals, gluconeogenesis operates when glucose is scarce, and then only in the presence of both a gluconeogenic substrate and second substrate other than carbohydrate that can be used for the production of the energy required for gluconeogenesis. These conditions are not found in adult parasitic helminths, as they rely almost exclusively on carbohydrate for energy generation. In contrast, gluconeogenesis may operate in aerobic larval stages that use substrates other than carbohydrates for energy generation. For example, during the development of *A. suum* larvae within the eggshell, glycogen is resynthesized from long-chain fatty acids through the glyoxylate cycle (see Section on the Aerobic/Anaerobic Transition during Helminth Development). This cycle, which is typically found in microorganisms and plants, has not been demonstrated in other parasitic helminths, although it is also present in the free-living nematode, *Caenorhabditis elegans.*

The simultaneous synthesis and degradation of glycogen would result in substrate cycling and a loss of energy. Therefore, not surprisingly, the activities of the key enzymes regulating glycogen metabolism, glycogen synthase and glycogen phosphorylase, are under tight regulatory control. These enzymes have been thoroughly studied only in adult *A. suum* muscle, where glycogen levels are up to 12-fold greater than in mammalian muscle and can account for up to 50% of the dry weight of the tissue. In adult *A. suum* glycogen is stored during periods of host feeding and is metabolized during host fasting. Both enzymes from adult *A. suum* muscle are remarkably similar in many respects to the well studied mammalian enzymes. For example, both glycogen phosphorylase and glycogen synthase activities are regulated by reversible phosphorylation/dephosphorylation, as are their mammalian counterparts. However, a novel form of glycogen synthase (GSII) has been identified in *A. suum,* and the regulation of the GSII complex, which appears to function as an intermediate between glycogenin and mature glycogen in muscle, may be important in maintaining the high glycogen levels in adult *A. suum* muscle. In contrast, the regulation of glycogen metabolism in other parasitic helminths is not always as tightly controlled. For example, in adult *Schistosoma mansoni,* glycogen degradation *in vitro* occurs even during periods of net synthesis, and vice versa. Similarly, glycogen reserves in schistosomes residing inside the veins of their hosts do not fluctuate between periods of glycogen degradation and synthesis, and the replenishment of glycogen reserves is not induced by a marked decrease in the glycogen levels, but instead occurs slowly and continuously. Therefore, the role of the glycogen metabolism in blood-dwelling parasites like adult schistosomes, that continuously reside in glucose-rich environments, is still not clear. Since the replenishment of the endogenous glycogen reserves in adult *S. mansoni* is not induced by a marked decrease in the glycogen levels, but occurs slowly *in vivo* and continuously in each and every worm pair, it is likely that glycogen is degraded intermittently, for instance for muscle contraction or tegumental membrane repair. This metabolism of glucose 'through' glycogen could help to maintain a low internal free glucose pool and so promote sufficient glucose import to deeper tissues through diffusion. In this respect it is noteworthy that adult schistosomes exhibit relatively high glucose 6-phosphatase activity, which would allow glucose derived from degraded glycogen stores in one cell to be exported to other cells within the adult. A balance between the rate of phosphorylation of imported glucose by hexokinase and the rate of glucose released from internal glycogen stores by glucose 6-phosphatase may help maintain an even distribution of glucose throughout the worm, and promote continuous, energy-independent movement of free glucose into the worm down a concentration gradient.

Trehalose is a soluble α-1-linked non-reducing disaccharide of glucose, and in many parasitic nematodes is more abundant than glucose. Trehalose is synthesized by the combined action of trehalose 6-phosphate synthase and trehalose 6-phosphate phosphatase, and is degraded to glucose by the action of trehalase. These enzymes have been detected in parasitic nematodes, but not extensively characterized. Trehalose levels are usually lower than those of glycogen, but trehalose still may function as a critical carbohydrate reserve in many nematodes and their eggs. However, since the storage of glycogen is energetically more favorable than trehalose, many authors have suggested that trehalose might also serve as an intermediate in glusose transport to the tissues, as has been well documented in insects. Trehalose is also clearly involved in stress responses in many nematodes, and may be important in resistance to desiccation and in cryoprotection.

GLYCOLYSIS

Glycolysis is the degradation of glucose to yield two molecules of pyruvate, and this almost universal pathway is also used by parasitic helminths for the catabolism of glucose, their main substrate for energy generation. Glycolysis occurs in the cytosol, and the compartmentalization of glycolytic enzymes within a membrane-bound organelle, like the glycosomes of some protozoan parasites, has not been observed in parasitic helminths. Based on the end-products of the glycolytic pathway and their further metabolism, parasitic helminths can be classified into three types:

1. Those parasitic helminths that use the complete glycolytic pathway and export the end-product, pyruvate, into the mitochondria for further degradation via the tricarboxylic acid cycle. This aerobic degradation occurs mainly in larval stages of cestodes, trematodes and nematodes (see section on the Aerobic/Anaerobic Transition).
2. Those that use the complete glycolytic pathway, but cytoplasmically convert its end-product, pyruvate, to more reduced end-products such as lactate and ethanol. This fermentation process, so-called

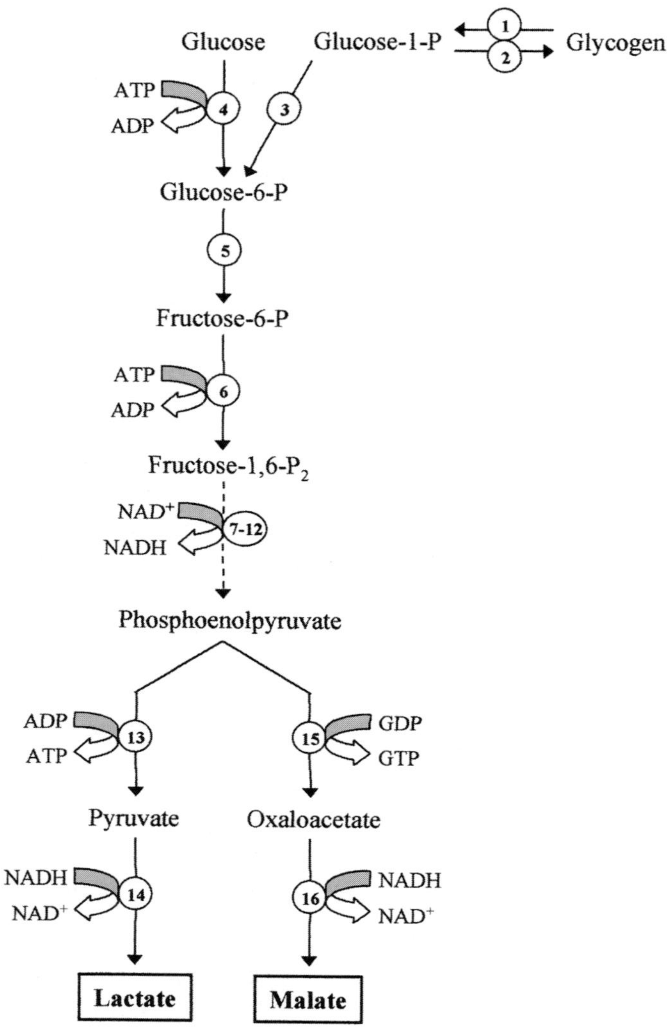

FIGURE 14.1 Glycogenolysis and glycolysis in parasitic helminths. (1) glycogen phosphorylase; (2) glycogen synthase; (3) phosphoglucomutase; (4) hexokinase; (5) glucosephosphate isomerase; (6) phosphofructokinase; (7) aldolase; (8) triosephosphate isomerase; (9) glyceraldehyde-3-P dehydrogenases; (10) phosphoglycerate kinase; (11) phosphoglyceromutase; (12) enolase; (13) pyruvate kinase; (14) lactate dehydrogenase; (15) phosphoenolpyruvate carboxykinase; (16) malate dehydrogenase.

'anaerobic' glycolysis, occurs in adult schistosomes and filarial nematodes.

3. Those that use the classical glycolytic pathway only up to phosphoenolpyruvate (PEP). Then instead of converting PEP to pyruvate, PEP is carboxylated by PEP carboxykinase (PEPCK) to oxaloacetate, which is subsequently reduced to malate (Figure 14.1). This part of the pathway occurs in the cytosol and is comparable to the formation of lactate or

ethanol; it is in redox balance and yields two mol of ATP per mol of glucose degraded. However, malate is not excreted like lactate or ethanol, but instead is transported into the mitochondria for further anaerobic metabolism, as discussed in greater detail below. This pathway is common in many parasitic helminths, especially the lumen-dwelling helminths like *A. suum*.

Several glycolytic enzymes have been purified from parasitic helminths and studied in great detail. These studies were stimulated by the early observation that the chemotherapeutic action of antimonials on schistosomes was associated with an inhibition of its phosphofructokinase (PFK), a rate-limiting glycolytic enzyme. Helminth PFKs appear to be much more sensitive to these anti-schistosomal drugs than the PFK of the host. Chemotherapeutic attack on the energy-generating systems of parasites is one of the more rational and promising approaches to combat parasitic diseases, since energy (ATP) is one of the few commodities that parasites cannot directly obtain from the host. Special attention has been, of course, directed towards enzymes that are absent in the host, or where differences are observed between enzymes of host and parasite.

The initial enzyme in the glycolytic sequence, hexokinase, controls the entry of glucose into the pathway, and is critical in the regulation of carbohydrate utilization. In mammals four hexokinase isoforms have been identified. Isoenzymes I–III are monomers with molecular weights of about 100 000, have a high affinity for glucose and are strongly inhibited by glucose 6-phosphate. Type IV, also called glucokinase, is a monomer of about 50 000 molecular weight, has a low affinity for glucose and is weakly inhibited by glucose 6-phosphate. Only a few hexokinases from parasitic helminths have been studied in detail. In adult *A. suum* only a single hexokinase isoform appears to be present, and its properties differ from the corresponding mammalian isoforms. The ascarid hexokinase is a monomer with a molecular weight of about 100 000, has a low affinity for glucose ($K_m = 4.7$ mM) and is only weakly inhibited by glucose 6-phosphate. The high K_m for glucose and the weak affinity for glucose 6-phosphate may facilitate glycogen synthesis in adult ascarid muscle. Hexokinase activity appears to be 'rate-limiting' in *S. mansoni* glucose metabolism. Cercariae and adults contain only a single form, which has a relatively high affinity for glucose and is moderately sensitive to inhibition by glucose 6-phosphate. The schistosomal enzyme is structurally related to other members of the hexokinase family. It is recognized by antisera against rat type I hexokinase, and the amino acid sequence shows a significant identity to the mammalian hexokinases. The relationship of this schistosomal enzyme with other hexokinases is interesting, as the schistosomal hexokinase, with its relatively high affinity for glucose and sensitivity for inhibition by glucose 6-phosphate, kinetically resembles the 100 kDa mammalian isoforms, although it is only about 50 kDa. The extensive sequence similarity of the schistosomal hexokinase with the mammalian isoforms is consistent with the view that the 100 kDa isoforms were formed via duplication of a gene encoding an ancestral hexokinase whose descendant is still present as a 50 kDa, glucose 6-phosphate-sensitive hexokinase in *S. mansoni*. Hexokinase appears to play a distinctive role in the rapid transition in energy metabolism that occurs in schistosomes when free-living cercariae enter the host (see section on the Aerobic/Anaerobic Transition).

Phosphofructokinase, another key glycolytic regulatory enzyme, catalyzes the ATP-dependent conversion of fructose 6-phosphate to fructose 1,6,-bisphosphate and is regulated by a variety of effectors. Both the mammalian and parasitic helminth PFKs are activated by fructose 2,6-bisphosphate and AMP and inhibited by ATP. In contrast to the mammalian PFK, phosphorylation of the helminth PFKs (at least in *Fasciola hepatica* and *A. suum*) results in activation of the enzyme. However, activation by phosphorylation is not unique to the helminth PFKs, as it also occurs in other invertebrates, such as earthworms and molluscs. Structurally, helminth and mammalian PFKs appear to be quite similar, although the sequence around the phosphorylation sites of the PFKs differs among helminth enzymes and is distinct from that of the mammalian-type PFK.

As described above, the final reactions in the cytosolic degradation of glucose can differ widely in various parasitic helminths (Figure 14.1). Pyruvate kinase (PK) may convert PEP to pyruvate, which can be then reduced in the cytosol to lactate or ethanol, or translocated into the mitochondrion for further oxidation by the tricarboxylic acid cycle. In contrast, PEPCK may carboxylate PEP to oxaloacetate which can then be reduced to malate by malate dehydrogenase. Not surpringly, the regulation of the PK/PEPCK branch point is potentially important in many parasitic helminths (Figure 14.1).

The PK-catalyzed conversion of PEP to pyruvate is coupled to ATP production. PKs from *F. hepatica* and *S. mansoni* closely resemble the L-type pyruvate kinase from mammalian liver. Both helminth enzymes show cooperative kinetics with PEP, but Michaelis–Menten kinetics in the presence of fructose 1,6-bisphosphate. Both helminth enzymes are inhibited by ATP, and this inhibition can be relieved by fructose 1,6-bisphosphate. Regulation of helminth PK via phosphorylation/dephosphorylation or by other effectors like fructose 2,6-bisphosphate, which stimulates trypanosomatid PKs, has not yet been reported.

In parasitic helminths, PEPCK physiologically fixes CO_2, and converts PEP to oxaloacetate (OAA) and, in common with PKs, is coupled to a substrate-level phosphorylation. This contrasts dramatically with the role of PEPCK in mammals, where PEPCK decarboxylates OAA as a prelude to gluconeogenesis. Therefore, not surprisingly, PEPCKs from parasitic helminths have been studied extensively in a search for molecular differences between host and parasites. However, the size and the kinetic properties of PEPCKs from parasitic helminths are similar to those of host enzymes with one exception: the K_m for HCO_3^- is significantly lower for the helminth PEPCK. This observation, coupled with the high pCO_2 in the habitat of many parasitic helminths, might explain why PEPCK acts in the carboxylating direction in helminth tissues. In contrast to PK activity, which is under tight allosteric control, PEPCK activity appears to be controlled primarily by the concentrations of enzyme, substrates and products.

MITOCHONDRIAL METABOLISM AND ENERGY GENERATION

One of the first observed metabolic differences between parasitic helminths and their hosts was the identification of novel anaerobic mitochondrial energy-generating pathways in the helminths by Saz and Bueding. However, the role played by oxygen in helminth metabolism has remained enigmatic. It is clear that almost all adult parasitic helminths generate a significant portion of their energy anaerobically, and that they also exhibit a wide variation in their ability to use oxygen as a terminal electron acceptor. In general, two basic metabolic schemes have emerged. Blood- and tissue-dwelling helminths, such as schistosomes and filariae, convert most of their abundant supplies of environmental glucose to lactate and survive well, at least in the short term, in the absence of oxygen. However, these organisms still contain cristate mitochondria capable of aerobic respiration and may preferentially use carbon derived from amino acids, such as glutamine, as substrates for mitochondrial metabolism. The contribution of aerobic metabolism to these lactate-forming helminths is difficult to quantify, although they do not appear to exhibit a marked Pasteur effect. However, the complete oxidation of glucose to CO_2 is energetically much more efficient than lactate formation (30 vs 2 mol ATP/mol glucose), so that even if only a small percentage of the glucose were oxidized completely, it still could contribute significantly to overall energy generation. In contrast to blood- and tissue-dwelling helminths, lumen-dwelling helminths, such as adult *A. suum*, contain unusual, less cristate mitochondria, which generate a significant portion of their energy anaerobically, by coupling phosphorylation to electron transport associated with the NADH-dependent reduction of fumarate and, in some cases, 2-methyl branched-chain enoyl CoAs.

Anaerobic mitochondrial metabolism in helminths

Helminths utilizing anaerobic malate dismutation produce a wide range of metabolic end-products of carbohydrate metabolism, but their mitochondrial pathways are surprisingly similar given their diverse phylogenetic origins. Typically, these organisms operate a portion of the tricarboxylic acid cycle (from oxaloacetate to succinate) and in some cases β-oxidation, in a direction opposite to that found in aerobic organisms. In fact, based on the midpoint potentials of the dehydrogenases involved, it is the helminth pathways that function in the energetically preferred direction. Succinate oxidation and β-oxidation are only energetically feasible if $NAD^+/NADH$ ratios are maintained in an oxidized state (>10) by a powerful oxidant, such as oxygen. In aerobic organisms, the removal of oxygen causes a rapid accumulation of NADH as the dehydrogenases approach equilibrium, and even ischemic vertebrate heart muscle accumulates succinate through a pathway similar to that operating in the parasitic helminths. In fact, every invertebrate phylum has members capable of anaerobic mitochondrial energy generation, but in most cases this ability is facultative and depends on the availability of oxygen. In the larger parasitic helminths, muscle mitochondria are committed almost exclusively to anaerobic energy generation and do little else. For example, mitochondria from adult *A. suum* body wall muscle lack functional tricarboxylic acid and urea cycles, are incapable of β-oxidation, are limited in their ability to metabolize amino acids and lack significant cytochrome oxidase activity.

All helminths capable of anaerobic metabolism use malate as the primary mitochondrial substrate, and the oxidative decarboxylations of first malate and then pyruvate generate intramitochondrial reducing power in the form of NADH (Figure 14.2). In contrast, the pathways used to reoxidize intramitochondrial NADH are quite diverse and depend on the stage or species of helminth under examination, but in all cases, redox balance is maintained and electron-transport-associated ATP is generated by the NADH-reduction of fumarate to succinate. In the cestode *H. diminuta*, succinate and acetate are the major end-products of anaerobic malate dismutation and are excreted in the predicted 2:1 ratio. In the trematode *F. hepatica*, succinate is then further decarboxylated to propionate with an additional substrate-level phosphorylation coupled to the decarboxylation of methylmalonyl CoA. *F. hepatica*

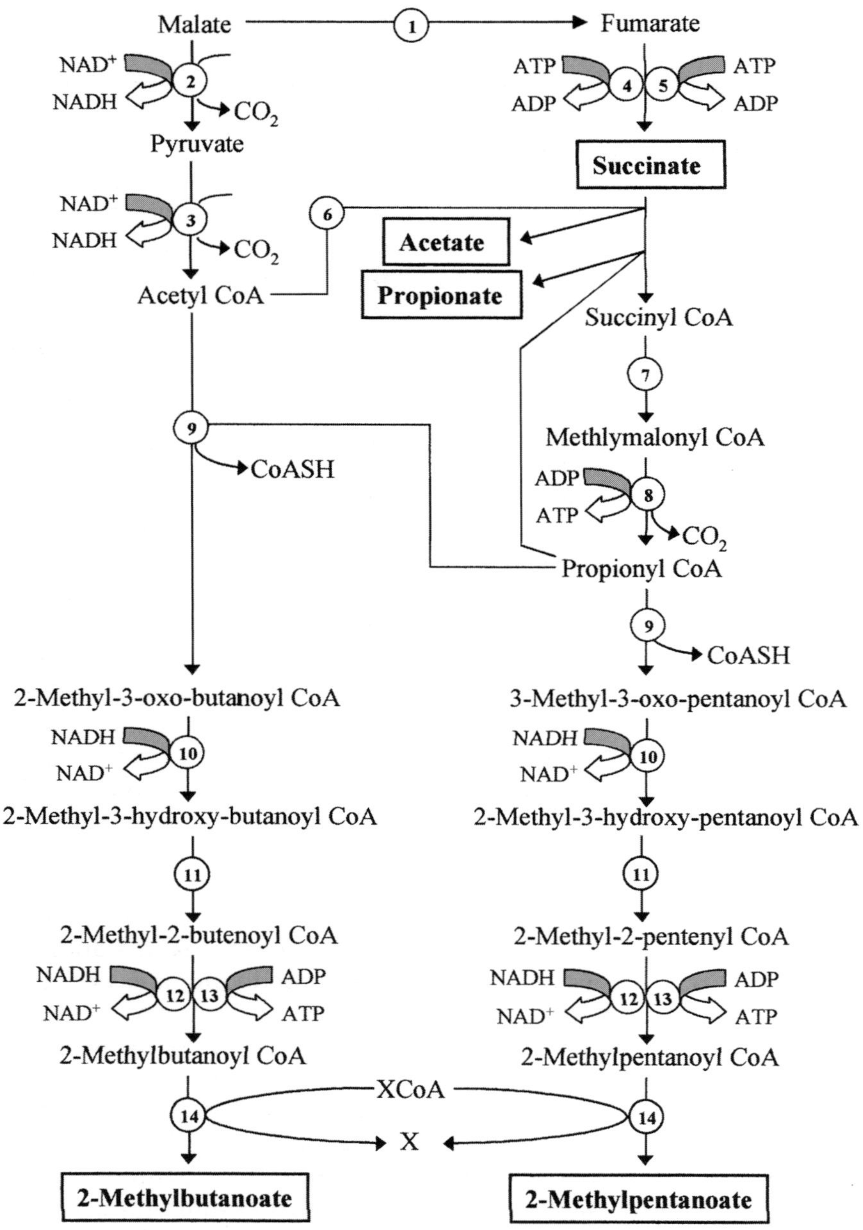

FIGURE 14.2 Malate metabolism in mitochondria from body wall muscle of adult *Ascaris suum*. (1) fumarase; (2) malic enzyme; (3) pyruvate dehydrogenase complex; (4) complex I; (5) succinate-coenzyme Q reductase (complex II, fumarate reductase); (6) acyl CoA transferase; (7) methylmalonyl CoA mutase; (8) methylmalonyl CoA decarboxylase; (9) condensing enzyme; (10) 2-methyl acetoacetyl CoA reductase; (11) 2-methyl-3-oxo-acyl CoA hydratase; (12) electron-transfer flavoprotein; (13) 2-methyl branched-chain enoyl CoA reductase; (14) acyl CoA transferase.

forms primarily propionate and acetate as end-products. In the nematode *A. suum*, acetate and propionate are further metabolized to the branched-chain fatty acids, 2-methylbutanoate and 2-methylpentanoate, through a pathway similar to reversal of β-oxidation, and a complex mixture of acetate, propionate, succinate, 2-methylbutanoate, and 2-methylpentanoate is excreted as end-products. Branched-chain fatty acid formation provides an additional avenue for the oxidation of excess reducing power, and is potentially an additional site of electron-transport-associated ATP synthesis coupled to the NADH-dependent reduction of 2-methyl branched-chain enoyl CoAs. Finally, some organisms, such as *H. contortus* and *Trichostrongylus colubriformis*, form neutral volatile compounds, such as propanol. The mechanism and site of propanol formation are not well understood, but since the reduction of propionate to propanol is unlikely, the pathway probably involves propionyl CoA as an intermediate. The factors dictating the types and ratios of anaerobic end-products are complex. However, habitat and worm size (as a function of surface to volume ratio), certainly play key roles in the selection of pathways used to generate energy and maintain redox balance, especially as they affect the availability of oxygen and glucose. In addition, the pK_as for many of these end-products, such as succinate or especially the volatile organic acids, are much lower than that of lactate, which may facilitate their excretion or minimize their effects on tissue acidification.

Ascaris suum is one of the few unusually large helminths from which substantial amounts of homogeneous tissue can be dissected for use in the isolation of mitochondria or protein purification, and for this reason its metabolism is probably the best studied of all of the parasitic helminths. However, it is important to note that because of its very large size *A. suum* is atypical, and generalizations from *A. suum* to its smaller cousins should be made with caution. Most of the enzymes involved in the dismutation of malate in anaerobic mitochondria from adult *A. suum* body wall muscle have been purified to homogeneity and at least partially characterized. As mentioned above, the tricarboxylic acid cycle is not significant in these novel organelles and the levels of citrate synthase, aconitase, isocitrate dehydrogenase, and α-ketoglutarate dehydrogenase are low or barely detectable. In fact, cytoplasmically generated malate, not pyruvate, is the primary mitochondrial substrate. Malate enters the mitochondrion through a phosphate-dependent porter system, and a portion is decarboxylated to pyruvate by the action of an NAD^+-linked 'malic' enzyme, generating reducing power in the form of NADH. This intramitochondrial NADH then reduces the remaining malate, via fumarate, to succinate. In *H. diminuta*, 'malic' enzyme is $NADP^+$-linked, and this helminth contains an active, energy-linked, NADPH:NAD^+ transhydrogenase activity that converts NADPH to NADH.

Mitochondria from adult *A. suum* body wall muscle catalyze the efficient, energy-linked, NADH-dependent reduction of fumarate to succinate (Figure 14.3). Complex I (NADH: rhodoquinone oxidoreductase), a novel quinone, rhodoquinone, and Complex II (succinate: rhodoquinone oxidoreductase) are all involved in NADH-dependent fumarate reduction. In fact, all mitochondria are capable of some succinate formation in the absence of oxygen, but anaerobic helminth mitochondria are modified to catalyze this reduction at a much faster rate and generate additional energy from the process. For example, although the polypeptide composition of both the *A. suum* Complex I and II are superficially similar to those of their aerobic counterparts, the ratio of fumarate reductase to succinate dehydrogenase

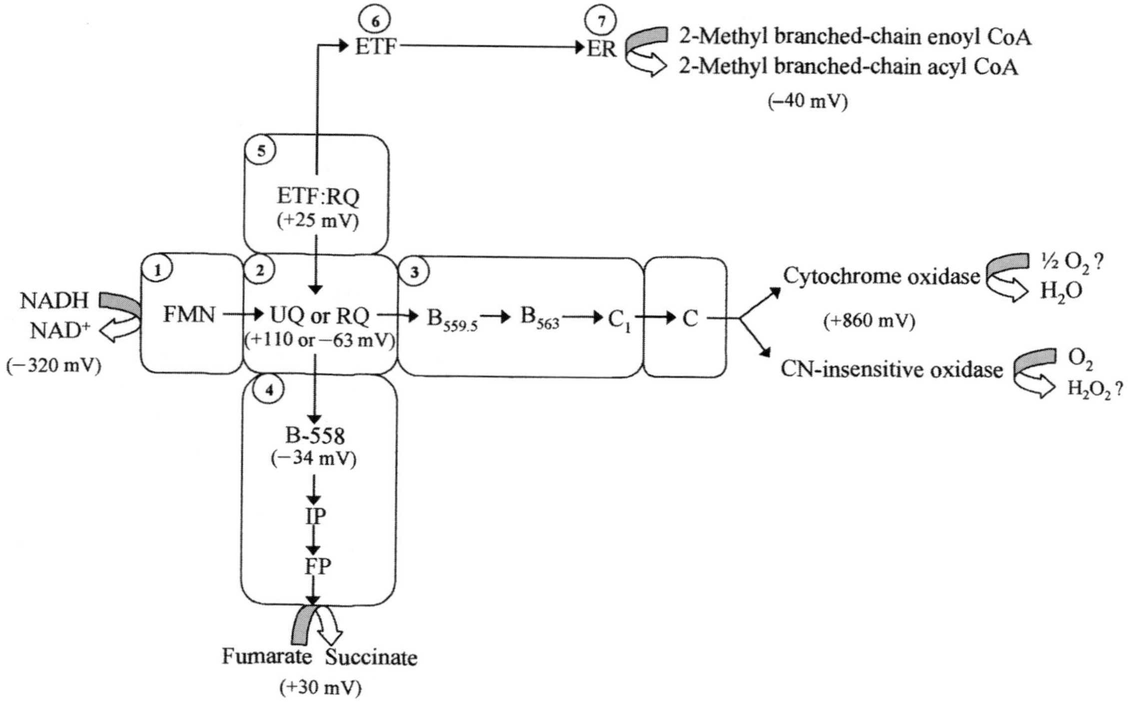

FIGURE 14.3 Electron transport in mitochondria from body wall muscle of adult *Ascaris suum*. (1) NADH: rhodoquinone reductase (complex I); (2) rhodoquinone or ubiquinone (in early aerobic larval stages); (3) reduced rhodoquinone cytochrome *c* reductase (complex III); (4) reduced rhodoquinone:fumarate reductase (complex II); (5) electron-transfer flavoprotein reductase; (6) electron-transfer flavoprotein; (7) 2-methyl branched-chain enoyl CoA reductase.

activity in adult *A. suum* mitochondrial membranes is about 400 times greater, and aerobic mitochondria catalyze the NADH-dependent reduction of fumarate much more slowly. A number of both subtle and more substantial differences have been identified to account for this ability, but the presence of two novel components, rhodoquinone and a cytochrome b_{558} appear to be essential for significant rates of fumarate reduction. Rhodoquinone is structurally similar to ubiquinone, the quinone commonly found in aerobic mitochondria (the substitution of an amino for a methoxy group), but its redox potential is much more negative ($E_o = -63$ mV vs $+110$ mV). In fact, the redox potential of rhodoquinone is much closer to that of the $NADH/NAD^+$ couple ($E_o = -320$ mV), and it appears that rhodoquinone is much more efficient at shuttling electrons from NADH to fumarate ($E_o = +30$ mV). As predicted, ubiquinone cannot replace rhodoquinone in catalyzing the NADH-dependent reduction of fumarate to succinate in reconstitution studies using pentane-extracted *A. suum* mitochondria or isolated respiratory complexes. Almost nothing is known about the synthesis of rhodoquinone and whether ubiquinone is an intermediate in the process. Certainly, the enzymes involved in rhodoquinone biosynthesis would be unique to helminths and obvious targets for chemotherapy. Similarly, Complex II, which comprises about 8% of the total membrane protein in adult *A. suum* mitochondria, contains a low potential cytochrome, b_{558}, whose properties differ significantly from those of cytochrome b_{560}, commonly found in Complex IIs isolated from aerobic mitochondria. Cytochrome b_{558} is reduced by NADH or succinate and reoxidized by fumarate in the presence of the appropriate *A. suum* respiratory complexes. These data suggest that cytochrome b_{558} also may play an important role in determining the direction of electron flow in these anaerobic organelles.

In addition to the oxidative decarboxylation of malate, intramitochondrial NADH also is generated from the oxidative decarboxylation of pyruvate to acetyl CoA, catalyzed by the pyruvate dehydrogenase complex (PDC). The PDC is a large multienzyme complex consisting of three catalytic components: pyruvate dehydrogenase (E1), dihydrolipoyl transacetylase (E2), and dihydrolipoyl dehydrogenase (E3) and, in eukaryotes, an E3-binding protein (E3BP) and an associated E1 kinase and E1 phosphatase. The reoxidation of NADH generated by pyruvate oxidation also may be coupled to fumarate reduction or used to drive the synthesis of branched-chain fatty acids. In *A. suum*, no other reactions have been identified which generate sufficient reducing power to account for the quantities of reduced products formed. During anaerobic metabolism, intramitochondrial NADH/$NAD^+$and acyl CoA/CoA ratios are elevated dramatically to levels that potently inhibit the activity of PDCs isolated from aerobic mitochondria. This inhibition results from both end-product inhibition and the stimulation of E1 kinase activity that catalyzes the reversible phosphorylation and inactivation of the complex. Given these observations, it was initially surprising to find a PDC with a functional kinase in adult *A. suum* muscle mitochondria, since under the elevated NADH levels present during microaerobiosis, maximal flux through the complex would be necessary to fuel the worms' relatively inefficient fermentative metabolism. In fact, PDC activity in many facultative anaerobes, such as *E. coli*, is downregulated during anaerobiosis, and other enzymes better suited to function in reduced environments, such as pyruvate:ferredoxin oxidoreductase and pyruvate:formate lyase are involved in pyruvate decarboxylation. Similar enzyme systems have been identified in anaerobic parasitic protozoa, such as *Giardia lamblia* and *Trichomonas vaginalis*, which also lack a PDC. In contrast, in some obligate anaerobes, such as *Enterococcus faecalis*, the PDC is dramatically overexpressed and more resistant to end-product inhibition.

Not surprisingly, PDC activity in adult *A. suum* body wall muscle appears to be regulated to maintain activity under the reducing conditions present in the host gut. For example, the PDC in adult body wall muscle is abundant; in fact the PDC is more abundant in *A. suum* body wall muscle than in any other eukaryotic tissue studied to date and approaches 2% of the total soluble protein. The PDC also is less sensitive to end-product inhibition by elevated NADH/NAD^+ and acetyl CoA/CoA ratios than PDCs from aerobic organisms. Interestingly, E3, the enzyme directly responsible for the NADH sensitivity of the complex, is identical in both aerobic second-stage larvae (L2) and anaerobic adult muscle. However, the PDC from adult muscle contains an unusual 'anaerobic' E3 binding protein (E3BP), which lacks the terminal lipoyl domain found in E3BPs from all other sources. The binding of E3 to this 'anaerobic' E3BP significantly reduces the sensitivity of the E3 to NADH inhibition, and helps to maintain PDC activity in the face of the elevated NADH/NAD^+ ratios associated with anaerobiosis. Similarly, the regulation of E1 kinase is modified in the adult muscle PDC. For example, E1 kinase activity is inhibited by pyruvate and propionate, metabolites that are elevated during anaerobic metabolism, and is less sensitive to stimulation by elevated NADH/NAD^+ and acetyl CoA/CoA ratios. The stoichiometry of phosphorylation and inactivation of the adult muscle PDC also differs substantially from its aerobic mammalian counterpart, and is designed to prevent the complete inactivation of the complex during anaerobiosis. In mammalian E1s, each E1α subunit of the $\alpha_2\beta_2$ tetramer contains three distinct phosphorylation sites and, given the specificity of the mammalian E1 kinase, inactivation *in vivo* is associated primarily with the phosphorylation of site 1. However, phosphorylation at any of the three phosphorylation sites is sufficient for inactivation. More importantly, inactivation is characterized by half-of-the-site reactivity, and phosphorylation in only one of the two E1α subunits results in the complete inactivation of the tetramer. In contrast, the E1α in the PDC purified from adult *A. suum* body wall muscle contains only two phosphorylation sites and inactivation is accompanied by substantially more phosphorylation than observed in the mammalian E1α, with both E1α subunits of the tetramer phosphorylated. These differences effectively prevent the complete inactivation of the complex *in vivo*, especially in the presence of E1 phosphatase activity. Interestingly, in contrast to mammalian E1 phosphatases, the *A. suum* enzyme is dramatically stimulated by malate, the major mitochondrial substrate in adult *A. suum* muscle. This represents yet another regulatory modification designed to maintain the PDC in an active dephosphorylated state. Little is known about the regulation of PDC activity in other parasitic helminths, although the structural modifications observed in the *A. suum* PDC do not appear to be conserved in the *F. hepatica* PDC, suggesting that while both organisms use identical pathways to form succinate, their enzyme systems have evolved independently.

The branched-chain fatty acids, 2-methylbutanoate and 2-methylpentanoate are the ultimate products of glucose degradation in *A. suum* muscle and accumulate to high levels (>50 mM total) in the perienteric fluid, where they rival Cl^- as the most abundant anions. Branched-chain fatty acids appear to exit the worm by diffusion through the cuticle, leaving open the possibility of their further metabolism as they pass through potentially aerobic mitochondria in the hypodermis. 2-methylbutanoate and 2-methylpentanoate are formed by the condensation of an acetyl CoA and a propionyl CoA or two propionyl CoAs, respectively, with the subsequent reduction of the condensation products (Figure 19.2). Enzymes in the pathway differ significantly from the corresponding enzymes of β-oxidation found in mammalian mitochondria. Differences might be anticipated since the ascarid enzymes function physiologically in the direction of acyl CoA synthesis, not oxidation. The final reaction in the pathway, the NADH-dependent reduction of 2-methyl branched-chain enoyl CoAs, is rotenone-sensitive and requires Complex I, rhodoquinone, and electron-transport flavoprotein (ETF) reductase and two soluble components, ETF and 2-methyl branched-chain enoyl CoA reductase (Figure 19.2).

Anaerobic energy generation

At least four potential sites for energy generation have been identified in anaerobic helminth mitochondria: substrate-level phosphorylations coupled to the decarboxylation of methylmalonyl CoA and the hydrolysis of acyl CoA; and site I, electron-transport-associated energy-generation coupled to the NADH-dependent reductions of fumarate and 2-methyl branched-chain enoyl CoAs. The substrate-level phosphorylation coupled to methylmalonyl CoA decarboxylation has been clearly demonstrated in both trematodes and nematodes. In contrast, energy generation from acyl CoA hydrolysis, although feasible, is less well documented in parasitic helminths, although thiokinase activities have been measured in helminth extracts. In fact, in adult *A. suum* muscle, the intramitochondrial levels of free CoASH are very low, and the transfer of the CoA moiety appears to be mediated by a number of distinct CoA transferases. Low free CoASH levels may be critical for the formation of branched-chain fatty acids, since the initial reaction in this sequence, catalyzed by propionyl CoA condensing enzyme, is potently inhibited by free CoASH. Low free CoASH levels are maintained by the PDC, whose apparent K_m for CoASH is much lower than values reported for complexes isolated from aerobic mitochondria. Whether similar constraints on CoA metabolism apply to organisms that do not form branched-chain fatty acids remains to be determined.

The NADH-dependent reduction of fumarate involves the generation of a proton gradient across the inner mitochondrial membrane and is coupled to a rotenone-sensitive, site 1 phosphorylation. The membrane potential is insensitive to cyanide and is dissipated with uncouplers, such as FCCP. More importantly, the membrane potential appears to be similar in magnitude to that reported for mammalian mitochondria, in spite of the fact that only a single site of proton translocation appears to be operative during anaerobic electron transport. Little is known about the movement of other ions across the mitochondrial membrane. An oligomycin-sensitive ATPase activity has been identified in isolated *A. suum* muscle mitochondria, which appears to be less sensitive to inhibition by uncouplers than the corresponding activity in mammalian mitochondria.

The NADH-dependent reduction of 2-methyl branched-chain enoyl CoAs also appears to be coupled to the generation of a mitochondrial proton gradient and associated phosphorylation, based on energetic considerations and the rotenone-sensitivity of the pathway. However, this proposed coupling has never been measured directly. For example, when isolated adult *A. suum* muscle mitochondria are incubated with malate *in vitro*, they form succinate and pyruvate, not the volatile acids characteristic of carbohydrate fermentation in intact muscle strips. Energy generation in these experiments, measured as the incorporation of ^{32}Pi into ATP, is associated exclusively with succinate formation, and gives no indication of the energy-generating potential of pathways leading to volatile acid formation. The inability of these isolated organelles to form volatile fatty acids is unclear; all of the enzymes necessary for their synthesis are present. In fact, complete carbon and energetic balances have never been measured experimentally for any physiologically functional helminth mitochondrial preparation. Based on theoretical calculations, the energy-generating efficiency of branched-chain fatty acid synthesis should actually be inferior to formation of acetate and propionate, since NADH that could be coupled to the energy generation associated with fumarate reduction is used unproductively to reduce 2-methylacetoacetyl CoAs. However, branched-chain fatty acid synthesis may provide increased flexibility in regulating intramitochondrial $NADH/NAD^+$ ratios by serving as an additional sink for excess reducing power in the anaerobic tissues of a large helminth like

A. suum. Alternatively, branched-chain fatty acids may be more fastidiously excreted from the organism, given their increased lipophilicity and elevated pK_as.

On a broader scale, little is known about the factors that regulate carbon flux and energy generation during muscle contraction in either flatworms or nematodes. Neither appears to possess the 'burst activity' characterized in other invertebrates, perhaps because predator avoidance or the active pursuit of prey is not a problem within the definitive host. However, recently it has become clear that the regulation of these processes may differ substantially from similar processes in the host. For example, during vertebrate muscle contraction, Ca^{2+} released from the sarcoplasmic reticulum is readily accumulated by muscle mitochondria through a high affinity transporter and activates the Ca^{2+}-sensitive E1-phosphatase and α-ketoglutarate dehydrogenase complex, thus effectively linking contraction and energy generation. In contrast, mitochondria from *A. suum* muscle are not uncoupled when incubated in Ca^{2+}, suggesting a limited capacity for high affinity Ca^{2+} uptake. In addition, the *A. suum* E1 phosphatase and *F. hepatica* α-ketoglutarate dehydrogenase are not stimulated by Ca^{2+}, suggesting that other factors are involved in the coordination of these two processes.

Oxygen as a terminal electron acceptor

Parasitic helminths have varying aerobic capacities, depending primarily on their size and the availability of oxygen in the surrounding environment. For example, some small lumen-dwelling nematodes, such as *Nippostrongylus brasiliensis*, which reside close to the mucosa where oxygen tensions are relatively high, have a functional tricarboxylic acid cycle and rely on oxygen as a terminal electron acceptor. In contrast, the aerobic capacity of the juvenile *F. hepatica* decreases with growth and is proportional to its surface area, since in the absence of a circulatory system oxygen must diffuse directly from the liver parenchyma. In adult *F. hepatica*, the outer aerobic layer of the parasite accounts for only 1% of the total volume and metabolism appears almost completely anaerobic. Muscle mitochondria from unusually large helminths, such as *A. suum*, which reside in especially microaerobic habitats, such as the vertebrate gut, form branched-chain fatty acids and represent the anaerobic extreme.

Branched electron-transport chains have been proposed for a number of helminth mitochondria, with one branch leading to fumarate as a terminal electron acceptor and the other leading to oxygen through either classical, CN-sensitive, cytochrome *c* oxidases or CN-insensitive, peroxide-forming, terminal oxidases. Certainly, the incubation of either intact helminths or isolated mitochondrial preparations in air alters the ratios of excreted fermentative products, with acetate formation usually increased and products of the reductive arm of the pathway, such as succinate, propionate and branched-chain fatty acids, reduced. The shift in end-product formation probably results from a decrease in intramitochondrial $NADH/NAD^+$ ratios. However, no terminal oxidase has ever been isolated or purified from any helminth mitochondrion, and at present there are no definitive data confirming the identity of any CN-insensitive terminal oxidase in any parasitic helminth. The existence of alternative oxidases has been inferred from studies using inhibitors of oxygen uptake, such as KCN or SHAM. These studies are limited by our understanding of the specificity of these inhibitors on helminth metabolism, and have often been conducted using either fumarate or oxygen as a terminal acceptor, without measuring relative rates of energy generation.

Perhaps more importantly, these studies have most often been conducted in air (about 20% oxygen), when oxygen tensions of 1–2% would probably be much more physiological.

In *A. suum* muscle mitochondria, perhaps the most anaerobic of all helminth mitochondria, reduced cytochrome *c* oxidase activity is barely detectable and oxygen utilization results exclusively in the formation of hydrogen peroxide. In fact, the ratio of Complex II to Complex III is about 100 times greater than in rat liver mitochondria, attesting to the importance of NADH-dependent fumarate reduction in the ascarid organelle. Substantial peroxide formation has also been observed in mitochondria isolated from other helminths, such as *H. diminuta* and *Ascaridia galli*. In isolated *A. suum* mitochondria, no additional energy generation appears to be associated with oxygen uptake, but succinate formation is slightly reduced, suggesting that oxygen uptake in these experiments may be unphysiological and simply the result of its direct reaction with the flavin moiety of the numerous flavoproteins present in the helminth electron-transport chain (i.e. the Fp subunit of complex II, ETF and ETF dehydrogenase). In mammalian mitochondria, these flavoproteins do not react readily with oxygen, but their helminth counterparts functioning in microaerobic environments may not be under the same selective pressures to maintain their oxygen insensitivity.

THE AEROBIC/ANAEROBIC TRANSITION DURING HELMINTH DEVELOPMENT

Two major generalizations may be made about the developmental aspects of energy generation in parasitic helminths. First, all adult helminths use fermentative pathways to some extent and excrete reduced organic acids as end-products of carbohydrate metabolism, even in the presence of oxygen. Second, at least some larval stages of most helminths are aerobic with active tricarboxylic acid cycles and CN-sensitive respiration. Therefore, a marked aerobic–anaerobic transition in energy metabolism often occurs during the development of most parasitic helminths. For example, schistosome miracidia and cercariae are free-living and aerobic. However, immediately after penetration of the definitive host during development from schistosomulum to adult, the schistosome increasingly relies on glycolysis for energy generation. Similarly, as mentioned above, while excysted juvenile *F. hepatica* develop to the adult, the ratio of tricarboxylic acid-cycle activity to acetate and propionate formation decreases, and there appears to be a direct correlation between tricarboxylic acid-cycle activity and the surface area of the fluke. A similar relationship has been suggested in nematodes. *N. brasiliensis*, *H. contortus* and *Ascaridia galli* each exhibit significant cytochrome oxidase staining in hypodermal mitochondria found immediately beneath the cuticle, but markedly decreased staining in mitochondria from internal tissues, especially in the larger helminths. Similar studies have not been conducted in *A. suum*, but an oxygen gradient in the worm has been suggested based on other data. *A. suum* muscle and perienteric fluid contain substantial amounts of hemoglobins with extraordinarily high affinities for oxygen (greater than 1000 times that of mammalian hemoglobin). It has been suggested that this hemoglobin functions as an O_2 buffer system, maintaining a low but constant pO_2 internally (i.e. preventing oxygen toxicity, but maintaining high enough oxygen tensions for oxygen-dependent biosynthetic reactions). For example, the synthesis of hydroxyproline from proline for collagen synthesis catalyzed by prolyl hydroxylase requires oxygen, and has been identified in all helminths examined to date. Whether oxygen can be used by tissues other than muscle for energy generation remains to be determined.

Helminth eggs leave the definitive host in widely varying stages of development. Many nematode eggs, such as those of *A. suum*, are undifferentiated and require oxygen for embryonation. Many cestode and trematode eggs leave fully embryonated, suggesting that embryonation occurs in the microaerobic habitat of the host. When unembryonated *A. suum* 'eggs' leave the host, cytochrome oxidase activity and ubiquinone are barely detectable and the 'eggs' appear to be transcriptionally inactive. After about 48–72 h in air, metabolism increases dramatically, fueled intitially by glycogen and then by stored triacylglycerols, as the worm begins a series of larval molts within the eggshell resulting ultimately in a quiescent infective larva. During this process cytochrome oxidase activity increases dramatically, the tricarboxylic acid cycle and β-oxidation are operative, and metabolism is aerobic. *A. suum* is one of the few metazoans capable of net glycogen synthesis from triacylglycerols, and possesses a functional glyoxylate cycle. At about day 10 of development, the activities of malate synthase and isocitrate lyase, two key enzymes in the glyoxylate cycle, begin to increase dramatically, triacylglycerol stores decrease, and glycogen, consumed earlier in development, is resynthesized. Once development is complete, the metabolic rate of the infective larva decreases significantly and the infective 'eggs' may remain dormant for long periods until ingestion by the definitive host. These quiescent larvae closely resemble the dauer larva of many free-living nematodes, such as *C. elegans*. Since dauer larva formation is linked exclusively with the transition to the third stage (L3) in all other nematodes, this quiescent ascarid larva may actually be an L3 and not an L2, as suggested by many of the early studies. Little is known about the factors regulating energy generation during this developmental process or in the arrested infective larvae. For example, the factors responsible for the initial burst of transcription and the synthesis of cytochrome oxidase, the regulation of glyoxylate cycle activity, or the decreased respiratory rate associated with dormancy are poorly understood. Similarly, the triacylglycerols of unembryonated eggs contain substantial amounts of 2-methylbutanoate and 2-methylpentanoate, major products of carbohydrate metabolism in adult muscle. However, nothing is known about the relationship between the enzymes catalyzing β-oxidation during early larval development and the reversal of β-oxidation in adult muscle. After ingestion by the porcine host, larvae hatch in the gut, then migrate through the liver to the lungs. Larval metabolism during this transition is aerobic and cyanide-sensitive. However, after the L3 migrate from the lungs back to the small intestine, they molt to the fourth stage (L4), respiration becomes cyanide-insensitive, and branched-chain fatty acids characteristic of the adult begin to be excreted. As predicted, the activities of enzymes associated with aerobic metabolism decrease dramatically, while those associated with anaerobic pathways increase in an equally dramatic fashion. Interestingly, many of the key enzymes regulating these metabolic pathways appear to exist as stage-specific isoforms.

Although we have a reasonably clear understanding of how mitochondria from adult helminths generate energy in the absence of oxygen, much less is known about how mitochondrial biogenesis is regulated during the various aerobic/anaerobic transitions that occur during helminth development. Clearly, a variety of different strategies have been identified, from the deletion of non-essential enzymes to the overexpression of key anaerobic-specific isoforms. For example, complex II functions as a succinate dehydrogenase in aerobic stages, but as a fumarate reductase in anaerobic stages, such as adult *F. hepatica* or *A. suum* body wall muscle. Not surprisingly, significant kinetic differences between complexes isolated from aerobic larval and anaerobic adult *A. suum* have been described. In fact, stage-specific isoforms of key Complex II subunits have been identified in both *A. suum* and *H. contortus*, whose differential expression parallels the aerobic/anaerobic transition. Similarly, stage-specific isoforms of the E1α subunit of the PDC (E1αI and E1αII) also have been identified in *A. suum*. These isoforms are over 90% identical at the amino acid sequence level, but are expressed in different stages and exhibit markedly different responses to phosphorylation. E1αII contains the three phosphorylation sites present in mammalian E1s, is most abundant in the aerobic L3 and exhibits a stoichiometry of phosphorylation/inactivation identical to that observed for E1s isolated from the host. In contrast, E1αI contains only two phosphorylation sites and is identical to the E1α isolated directly from adult muscle. More importantly, substantially more phosphate is incorporated into E1αI than E1αII as inactivation proceeds, dramatically decreasing the effectiveness of phosphorylation in inactivating of the complex (see section on Anaerobic Mitochondrial Metabolism). This difference helps to maintain PDC activity in the presence of the potentially inhibitory reducing conditions encountered in the host gut.

PERSPECTIVE

Of all the metabolic pathways operating in parasitic helminths, the reactions associated with carbohydrate metabolism and energy generation are, by far, the best understood. However, many key questions remain unresolved. For example, little is known about the physiological factors regulating glycogen metabolism, glycolytic flux, end-product formation or muscle contraction, and complete carbon balances have yet to be published for any physiologically functional anaerobic helminth mitochondria capable of volatile organic acid formation. In addition, the role of oxygen in helminth metabolism has remained enigmatic, and a CN-insensitive terminal oxidase has yet to be definitively identified. Little is known about individual ionic fluxes associated with electron transport or the pathway of rhodoquinone biosynthesis, and we are just beginning to understand tissue and stage-specific differences in energy generation at the molecular level. Clearly, energy generation in parasitic helminths differs substantially from that of the host, and remains an important target for chemotherapy. In fact, a potentially new anthelminthic, nafuredin, has recently been identified, that appears to specifically inhibit anaerobic helminth electron transport at the level of Complex I. Future studies will certainly continue to identify molecular differences between the metabolisms of host and parasite and hopefully will also identify additional new targets for chemotherapy.

FURTHER READING

Aggarwal, S.R., Lindros, K.O. and Palmer, T.N. (1995). Glucagon stimulates phosphorylation of different peptides in isolated periportal and perivenous hepatocytes. *FEBS Lett.* **377**, 439–443.

Amino, H., Wang, H., Hirawake, H. *et al.* (2000). Stage-specific isoforms of *Ascaris suum* complex II. II. The fumarate reductase of the parasitic adult and succinic dehydrogenase of free-living larvae share a common iron–sulfur subunit. *Mol. Biochem. Parasitol.* **106**, 63–76.

Barrett, J. (1981). *Biochemistry of Parasitic Helminths*. London: Macmillan Publishers.

Barrett, J., Mendis, A.H.W. and Butterworth, P.E. (1986). Carbohydrate metabolism in *Brugia pahangi*. *Int. J. Parasitol.* **16**, 465–469.

Barrett, J. (1976). Intermediary metabolism in *Ascaris* eggs. In: Van den Bossche, H. (ed.). *Biochemistry of Parasites and Host–Parasite Relationships*, Amsterdam: Elsevier, pp. 117–123.

Barrett, J., Ward, C.W. and Fairbairn, D. (1970). The glyoxylate cycle and the conversion of triglycerides to carbohydrates in developing eggs of *Ascaris lumbricoides*. *Comp. Biochem. Physiol.* **35**, 577–586.

Behm, C.A. (1997). The role of trehalose in the physiology of nematodes. *Int. J. Parasitol.* **27**, 215–229.

Behm, C.A., Bryant, C. and Jones, A.J. (1987). Studies of glucose metabolism in *Hymenolepis diminuta* using ^{13}C nuclear magnetic resonance. *Int. J. Parasitol.* **17**, 1333–1341.

Bryant, C. and Behm, C. (1989). *Biochemical Adaptation in Parasites*, London: Chapman and Hall.

Chojnicki, K., Dudzinska, M., Wolanska, P. and Michejda, J. (1987). Membrane potential in mitochondria of *Ascaris*. *Acta Parasit. Pol.* **32**, 67–78.

Davies, K.P. and Köhler, P. (1990). The role of amino acids in the energy generating pathways of *Litomosoides carinii*. *Mol. Biochem. Parasitol.* **41**, 115–124.

Ding, J., Su, J.-G.J. and Mansour, T.E. (1994). Cloning and characterization of a cDNA encoding phosphofructokinase from *Schistosoma mansoni*. *Mol. Biochem. Parasitol.* **66**, 105–110.

Duran, E., Kömuniecki, R., Komuniecki, P. *et al.* (1993). Characterization of cDNA clones for the 2-methyl branched chain enoyl CoA reductase: an enzyme involved in branched-chain fatty acid synthesis in anaerobic mitochondria of the parasitic nematode, *Ascaris suum*. *J. Biol. Chem.* **268**, 22391–22396.

Fairbairn, D. (1970). Biochemical adaptation and loss of genetic capacity in helminth parasites. *Biol. Rev.* **45**, 29–72.

Fioravanti, C.F. (1981). Coupling of mitochondrial NADPH:NAD transhydrogenase with electron-transport in adult *Hymenolepis diminuta*. *J. Parasitol.* **67**, 823–831.

Fry, M. and Beesley, J.E. (1985). Cytochemical localization of cytochrome oxidase in tissues of parasitic nematodes. *Parasitology* **90**, 145–156.

Geary, T.G., Winterrowd, C.A., Alexander-Bowman, S.J., Favreau, M.A., Nulf, S.C. and Klein, R.D. (1993). *Ascaris suum*: cloning of a cDNA encoding phosphoenolpyruvate carboxykinase. *Exp. Parasitol.* **77**, 155–161.

Geenen, P.L., Brescani, J., Boes, J. *et al.* (1999). The morphogenesis of *Ascaris suum* to the infective third-stage larvae within the egg. *J. Parasitol.* **85**, 616–622.

Ghosh, P., Heath, A.C., Donahue, M.J. and Masaracchia, R.A. (1989). Glycogen synthesis in the obliquely striated muscle of *Ascaris suum*. *Eur. J. Biochem.* **183**, 679–685.

Horemans, A.M.C., Tielens, A.G.M. and van den Bergh, S.G. (1991). The transition from an aerobic to an anaerobic metabolism in transforming *Schistosoma mansoni* cercariae occurs exclusively in the head. *Parasitology* **102**, 259–265.

Huang, X., Walker, D., Chen, W., Klingbeil, M. and Komuniecki, R. (1998). Expression of pyruvate dehydrogenase isoforms during the aerobic/anaerobic transition in the development of the parasitic nematode, *Ascaris suum*: altered stoichiometry of phosphorylation/inactivation. *Arch. Biochem. Biophys.* **352**, 263–270.

Kita, K., Hirawake, H. and Takamiya, S. (1997). Cytochromes in the respiratory chain of helminth mitochondria. *Int. J. Parasitol.* **27**, 617–630.

Kita, K., Takamiya, S., Furushima, R., Ma, Y. and Oya, H. (1988). Complex II is a major component of the respiratory chain in the muscle mitochondria of *Ascaris suum* with high fumarate reductase activity. *Comp. Biochem. Physiol.* **89B**, 31–34.

Kita, K., Takamiya, S., Furushima, R. *et al.* (1988). Electron-transfer complexes of *Ascaris suum* muscle mitochondria. III. Composition and fumarate reductase activity of Complex II. *Biochim. Biophys. Acta* **935**, 130–140.

Klein, R.D., Olson, E.R., Favreau, M.A. *et al.* (1991). Cloning of a cDNA encoding phosphofructokinase from *Haemonchus contortus*. *Mol. Biochem. Parasitol.* **48**, 17–26.

Klingbeil, M., Walker, D., Arnette, R. *et al.* (1996). Identification of a novel dihydrolipoyl dehydrogenase-binding protein in the pyruvate dehydrogenase complex of the anaerobic parasitic nematode, *Ascaris suum*. *J. Biol. Chem.* **271**, 5451–5457.

Köhler, P. (1991). The pathways of energy generation in filarial parasites. *Parasitol. Today* **7**, 21–24.

Komuniecki, R. and Komuniecki, P.R. (1995). In: Aerobic–anaerobic transitions in energy metabolism during the development of the parasitic nematode *Ascaris suum*. In: Boothroyd, J.C. and Komuniecki, R. (eds). *Molecular Approaches to Parasitology*, Wiley-Liss, pp. 109–121.

Komuniecki, R. and Komuniecki, P. (2001). Metabolic transitions and the role of the pyruvate dehydrogenase complex during development of *Ascaris suum*. In: Kennedy, M.W. and Harnett, W. (eds). *Parasitic Nematodes: Molecular Biology, Biochemistry and Immunology*, Wallingford: CABI Publishing, pp. 269–279.

Komuniecki, P.R. and Vanover, L. (1987). Biochemical changes during the aerobic–anaerobic transition in *Ascaris suum* larvae. *Mol. Biochem. Parasitol.* **22**, 241–248.

Komuniecki, R., Fekete, S. and Thissen, J. (1985). Purification and characterization of the 2-methyl-branched chain acyl CoA dehydrogenase, an enzyme involved in the enoyl CoA reduction in anaerobic mitochondria of the nematode: *Ascaris suum*. *J. Biol. Chem.* **260**, 4770–4777.

Lane, C.A., Pax, R.A. and Bennett, J.L. (1987). L-Glutamine: an amino acid required for maintenance of the tegumental membrane potential of *Schistosoma mansoni*. *Parasitology* **94**, 233–242.

Lloyd, G.M. (1986). Energy metabolism and its regulation in the adult liver fluke *Fasciola hepatica*. *Parasitology* **93**, 217–248.

Ma, Y., Funk, M., Dunham, W. and Komuniecki, R. (1993). Purification and characterization of electron-transfer flavoprotein: rhodoquinone oxidoreductase from anaerobic mitochondria of the adult parasitic nematode, *Ascaris suum*. *J. Biol. Chem.* **268**, 20360–20365.

MacKenzie, N.E., van de Waa, E.A., Gooley, P.R. *et al.* (1989). Comparison of glycolysis and glutaminolysis in *Onchocerca volvulus* and *Brugia pahangi* by ^{13}C nuclear magnetic resonance spectroscopy. *Parasitology* **99**, 427–435.

Mahrenholz, A.M., Hefta, S.A. and Mansour, T.E. (1991). Phosphofructokinase from *Fasciola hepatica*: sequence of the cAMP-dependent protein kinase phosphorylation site. *Arch. Biochem. Biophys.* **288**, 463–467.

Masaracchia, R.A., Rantala, M.R. and Donahue, M.J. (2000). Glycogenin-dependent organization of *Ascaris suum* muscle glycogen. *J. Parasitol.* **86**, 1206–1212.

Matthews, P.M., Foxall, D., Shen, L. and Mansour, T.E. (1986). Nuclear magnetic resonance studies of carbohydrate metabolism and substrate cycling in *Fasciola hepatica*. *Mol. Pharmacol.* **29**, 65–73.

Omura, S., Miyadera, H., Ui, H. *et al.* (2001). An anthelminthic compound, nafuredin, shows selective inhibition of complex I of helminth mitochondria. *Proc. Natl. Acad. Sci. USA* **98**, 60–62.

Rao, G.S.J., Cook, P.F. and Harris, B.G. (1991). Effector-induced conformational transitions in *Ascaris suum* phosphofructokinase; a fluorescence and circular dichroism study. *J. Biol. Chem.* **266**, 8884–8890.

Rao, G.S., Schnackerz, K.D., Harris, B.G. and Cook, P.F. (1995). pH-dependent allosteric transition in *Ascaris suum* phosphofructokinase distinct from that observed with fructose 2,6-bisphosphate. *Arch. Biochem. Biophys.* **322**, 410–416.

Saz, H.J. (1971). Anaerobic phosphorylation in *Ascaris* mitochondria and the effect of anthelmintics. *Comp. Biochem. Physiol.* **39B**, 627–637.

Saz, H.J. and deBruyn, B.S. (1987). Separation and function of two acyl CoA transferases from *Ascaris lumbricoides* mitochondria. *J. Exp. Zool.* **242**, 241–245.

Song, H.B., Thissen, J. and Komuniecki, R. (1991). Novel pyruvate dehydrogenase phosphatase activity from mitochondria of the parasitic nematode, *Ascaris suum*. *Mol. Biochem. Parasitol.* **48**, 101–104.

Takamiya, S., Kita, K., Wang, H. *et al.* (1993). Developmental changes in the respiratory chain of *Ascaris* mitochondria. *Biochim. Biophys. Acta* **1141**, 65–74

Thissen J. and Komuniecki, R. (1988). Phosphorylation and inactivation of the pyruvate dehydrogenase from the anaerobic parasitic nematode, *Ascaris suum*: stoichiometry, and amino acid sequence around the phosphorylation sites. *J. Biol. Chem.* **263**, 19092–19097.

Tielens, A.G.M. (1994). Energy generation in parasitic helminths. *Parasitol. Today* **10**, 346–352.

Tielens, A.G.M., van den Meer, P., van den Heuvel, J.M. and van den Bergh, S.G. (1991). The enigmatic presence of all gluconeogenic enzymes in *Schistosoma mansoni* adults. *Parasitology* **102**, 267–276.

Tielens, A.G.M., van den Heuvel, J.M. and van den Bergh, S.G. (1990). Continuous synthesis of glycogen by individual worm pairs of *Schistosoma mansoni* inside the veins of the final host. *Mol. Biochem. Parasitol.* **39**, 195–202.

Tielens, A.G.M., van den Heuvel, J.M., van Mazijk, H.J., Wilson, J.E. and Shoemaker, C.B. (1994). The 50-kDa, glucose-6-phosphate sensitive hexokinase of *Schistosoma mansoni. J. Biol. Chem.* **269**, 24736–24741.

Tielens, A.G.M., van den Heuvel, J.M. and van den Bergh, S.G. (1984). The energy metabolism of *Fasciola hepatica* during its development in the final host. *Mol. Biochem. Parasitol.* **13**, 301–307.

von Brand, T. (1973). *Biochemistry of Parasites*, New York: Academic Press.

CHAPTER

15

Neurotransmitters

Richard J. Martin[1], *Jennifer Purcell*[2], *Tim Day*[1] *and Alan P. Robertson*[1]

[1]Department of Biomedical Sciences, College of Veterinary Medicine, Iowa State University, Ames, IA 50011, USA; and

[2]Department of Preclinical Veterinary Sciences, R.(D.)S.V.S., Summerhall, University of Edinburgh, Edinburgh EH9 IQH, UK

NEUROTRANSMITTERS IN NEMATODES

Introduction

The study of neurotransmitters in parasitic nematodes has been encouraged by the need for the development of novel anti-nematodal drugs and a requirement for an understanding of the mode of action of these drugs. The study of neurotransmitters and the study of anthelmintics are therefore interconnected. An early illustration of this connection was the discovery of the action of the inhibitory neurotransmitter GABA and the effect of the anthelmintic piperazine in *Ascaris*.

We know now that the anthelmintics levamisole, pyrantel, piperazine and ivermectin (Figures 15.1, 15.2 and 15.3) mimic the effect of neurotransmitters that open membrane ion channels of nematodes (Table 15.1). Piperazine mimics the effect of the inhibitory transmitter

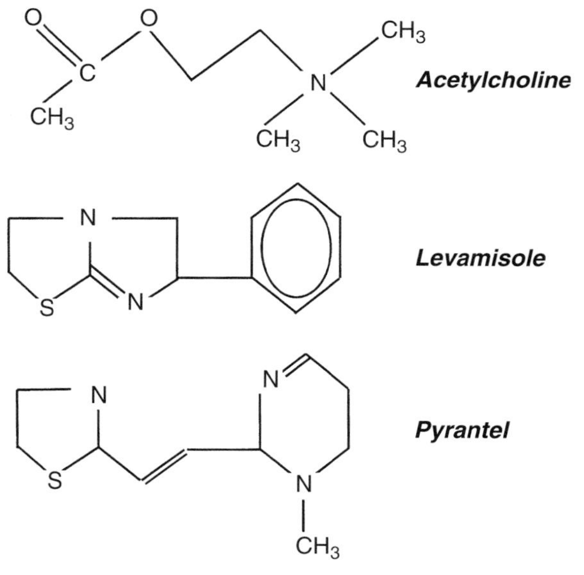

FIGURE 15.1 Chemical structures of acetylcholine and the nicotinic anthelmintics, levamisole and pyrantel. Note that levamisole is the L-isomer of the D-L mixture, tetramisole.

Molecular Medical Parasitology
ISBN 0–12–473346–8

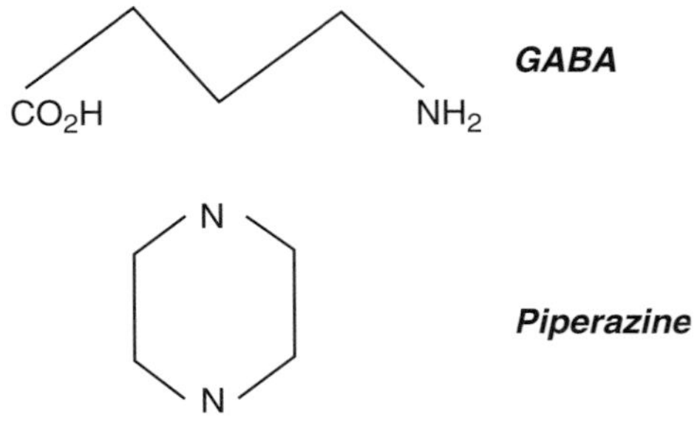

FIGURE 15.2 Chemical structure of γ-aminobutyric acid (GABA) and the anthelmintic piperazine. Note the absence of a carboxylic group in piperazine; the presence of CO_2 increases the potency of piperazine, presumably by combining with piperazine and mimicking the action of a carboxylic group.

GABA and produces muscle relaxation by opening chloride channels; levamisole mimics the effect of acetylcholine and produces muscle contraction by opening cation channels; ivermectin mimics the effect of L-glutamate and opens glutamate-gated chloride channels and inhibits nerves and pharyngeal muscle. The study of the neurotransmitters then, in nematodes, is encouraged by interest in the mode of action of anthelmintics, an interest in the mechanisms of anthelmintic resistance and also a desire to find new sites of action for novel drugs. Historically, the study of the mode of action of these neurotransmitters and anthelmintics in nematodes has been dominated by studies on the large intestinal nematode of the pig, *Ascaris suum*, and the model soil nematode, *Caenorhabditis elegans*. We examine the neurotransmitters of nematodes with an emphasis on their site of action (receptor sites) and effects of related anthelmintic drugs. The neurotransmitters we consider are: acetylcholine, GABA, L-glutamate, 5-HT, dopamine, NO and FMRFamide peptides (Figure 15.4). We start by introducing the anatomy and physiology of *Ascaris suum* (Figure 15.5) as a basis for explaining the actions of the putative neurotransmitters.

Ascaris suum

Jarman, in 1959, described this large, common porcine parasite for electrophysiological experiments. Humans are parasitized by a similar helminth, *Ascaris lumbricoides*, that is closely related to *A. suum*. *A. suum* is an excellent experimental model for several reasons: it is common in the domestic pig population so it can be readily obtained, post mortem, at abattoirs; its large size makes it easier to dissect than the smaller nematode species; and it is closely related to other parasitic nematodes, making it a representative model. Recent evidence suggests that *Ascaris* is also reasonably close, phylogenetically, to *Caenorhabditis elegans*, allowing genetic information obtained from *C. elegans* to be applied to *Ascaris*. For all of the above reasons, *A. suum* is a useful laboratory model for studies on parasitic roundworms.

The anatomy of Ascaris suum

Adult ascarids are 20–30 cm long and ~5 mm in diameter, which facilitates dissection. Like other intestinal parasites, *Ascaris* has a protective outer cuticlar layer. The cuticle prevents absorption of hydrophilic compounds, so to study the effect of hydrophilic compounds on *Ascaris* muscle cells, the cuticle is usually removed. This is done in electrophysiological studies of the somatic muscle cells, since access is required to the lumenal side of the body wall. Since nematodes are effectively a length of gut encased in a tube of muscle, dissecting the worm and making a flap preparation, which exposes the muscle layer, allows access to the muscle layer. A cross section of

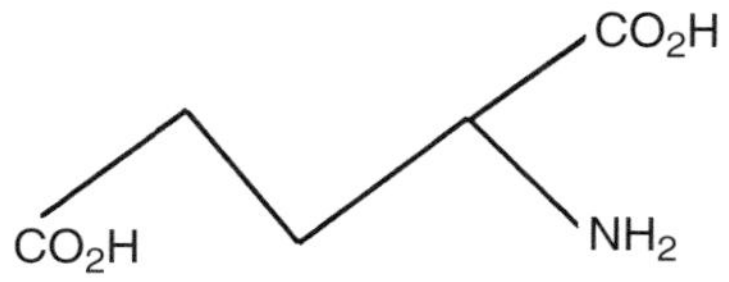

L-Glutamate

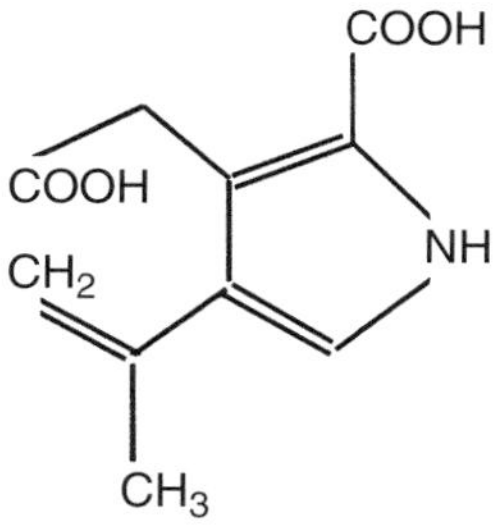

Kainate

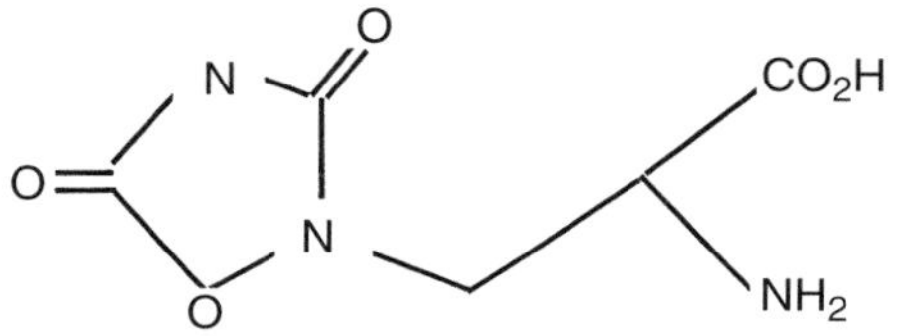

Quisqualate

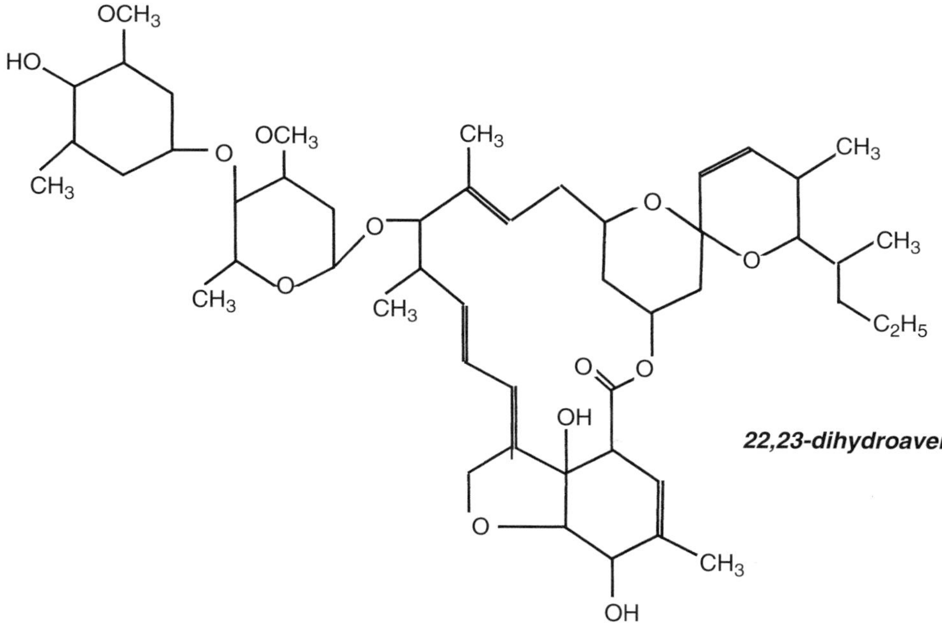

22,23-dihydroavermectin B_{IA}

FIGURE 15.3 Chemical structures of L-glutamate, the glutamate agonists kainate and quisqualate that have anthelmintic properties, and the macrocyclic lactone 22,23-dihydroavermectin B1a (ivermectin).

TABLE 15.1 Anti-nematodal drugs and their mode of action

Chemical group	Class	Mode of action
1 BZ	benzimidazoles, probenzimidazoles	Bind to the nematode structural protein, β-tubulin, adversely affecting nematode feeding and movement
2 LM	levamisole, morantel, pyrantel	Nicotinic receptor agonists, which activate nematode acetylcholine receptors to open sodium channels, causing muscular paralysis
3 AV	avermectins, milbemycin	Agonists at the glutamate-gated chloride channel in nematode nerve and muscle, causing inhibition
	organophosphates	Inhibit acetylcholinesterase, leading to increased stimulation of acetylcholine receptors and muscular paralysis
	piperazine	Agonist at GABA-gated chloride channels, resulting in muscular paralysis

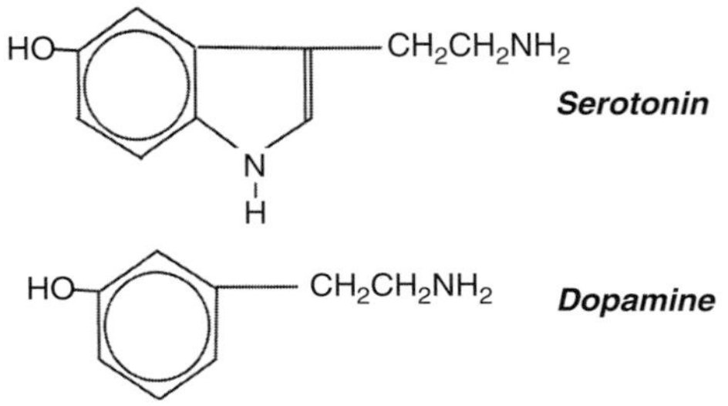

Phenylalanine-X-Arginine-Phenylalanine-NH_2
Original FMRFamide, FaRP

Now: Any Peptide with **RFamide** modif at the C terminal including FLFQPQRFamide

FIGURE 15.4 The structure of the neurotransmitters serotonin and dopamine, and general amino acid sequence of the FaRP peptides. X is any amino acid.

Ascaris and a diagram of an individual muscle cell are shown in Figure 15.6.

The muscle cell

Ascaris somatic muscle cells differ in structure from their vertebrate counterparts. As can be seen in Figure 15.6B, the muscle cells are composed of three distinct regions: the spindle or fiber; the muscle bag or belly; and the arm. The structure of the muscle cells is dictated by function, and was investigated initially by Rosenbluth in 1965.

Stretton described the innervation of the somatic muscle cells. They are arranged longitudinally and can be divided into two groups. The dorsal mass of somatic muscle is innervated by the motor axons in the dorsal nerve cord; the ventral mass receives innervation from the motor component of the ventral nerve cord. Due to this arrangement, waves of contraction are propagated along the nematode's body in the dorsoventral plane. By this means, the worm can move forwards or backwards.

The spindle

The spindle contains actin and myosin and is the contractile part of the cell. Actin and myosin are arranged in overlapping rows that produce striations, as in vertebrate skeletal muscle. However, in nematode body wall muscle, the striations are not aligned perpendicularly to the long axis of the muscle. Instead, the muscle is obliquely striated, which confers extensibility with high-speed contraction.

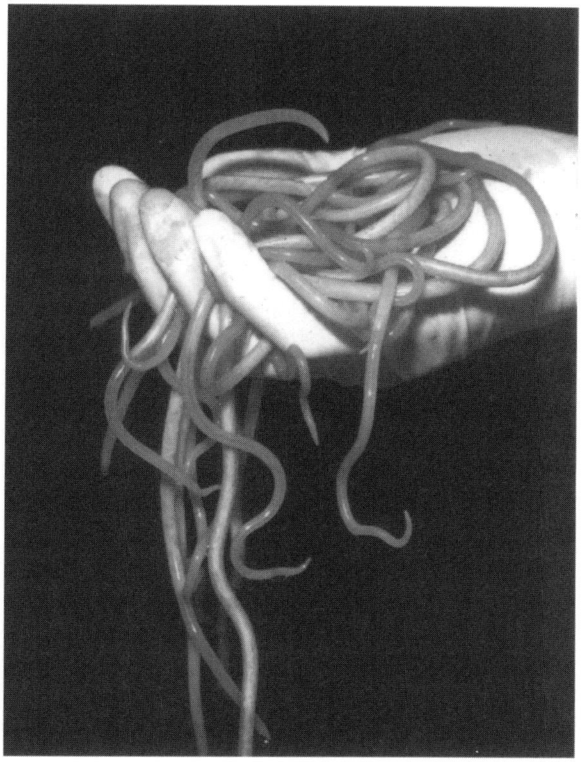

FIGURE 15.5 (See also Color Plate 10) Photograph of *Ascaris suum*. This photograph shows *Ascaris suum* within 24 hours of extraction from the host's gut. The pinkish color and the vitality of the nematodes indicate that they are in good body condition. The photograph also illustrates the large size of *A. suum*; adults are 20–30 cm long and about 5 mm in diameter.

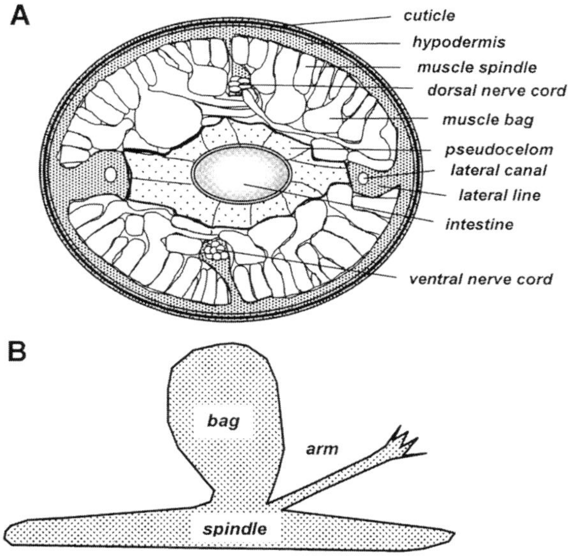

FIGURE 15.6 (A) Diagram of a cross-section through *Ascaris suum* in a region 4 cm from the head. Note the presence of the pseudocelom, dorsal and ventral nerve cords, the lateral lines, and the separation of the dorsal and ventral musculature. The hypodermis lies under the cuticle between the muscle layer and cuticle. (B) Diagram of the different regions of a muscle cell from *Ascaris*.

The muscle bag

The bag region of the cell contains the nucleus and is rich in glycogen and mitochondria. It therefore acts as the energy store for the muscle cell. There is also an argument for a support function for the muscle bag. In the intact parasite, the bag bulges inwards towards the gut; Harris and Crofton, in 1957, proposed that the turgor of the muscle bags supports the nematode, acting as an endoskeleton. Due to the restricted number of cells in nematodes, after the maximum number of cell divisions have taken place, growth continues by means of cellular expansion. As a result, all the nematode cells reach a relatively large size. The muscle bags of *Ascaris* are 100–200 μm in diameter in fully grown worms, which facilitates insertion of recording electrodes for electrophysiological study.

The arm

The muscle arm is the connection from the muscle cell to the nerve cord. The *Ascaris* neuromuscular system differs markedly from that found in vertebrates in that nematode muscle cells have projections to the neurons, rather than vice versa. Each muscle cell has one or more arms that extend from the muscle bag towards the nerve cord in the hypodermis. At their termination, each muscle arm branches into processes that join with those from other arms to form a syncytium, or functional network, around the nerve cord. The individual processes are separated from each other by tight junctions, which ensure the electrical coupling between muscle cells. Synapses are seen at the junction between the terminal divisions of the muscle arm and the nerve.

Pharyngeal muscle

The pharynx of feeding stages of nematode larvae and adults is an electrically coupled group of special muscle cells that form a muscular tube at the head region (Figure 15.7). The pharynx varies in precise shape with the species of nematode, and pumps food from the surrounding medium into the intestine. In *Ascaris* it is a large organ, nearly 1 cm in length. The electrophysiology of pharyngeal muscle was described initially by Del Castillo and Morales. It is different from that of most other excitable cells because it has a negative potential that repolarizes the cells. The triradiate lumen is closed when the muscle is relaxed, and contraction opens the lumen.

There are two types of synaptic potential recorded in the pharyngeal muscle. There is the depolarizing postsynaptic potential and a hyperpolarizing postsynaptic potential. We have used a two-microelectrode recording technique to record the effects of L-glutamate on pharyngeal muscle and found that it produces hyperpolarization associated with an increase in chloride conductance (Figure 15.7).

Ascaris *nervous system*

The neural anatomy of *A. suum*, like that of other nematodes, is consistent and relatively simple. According to Stretton, there are only

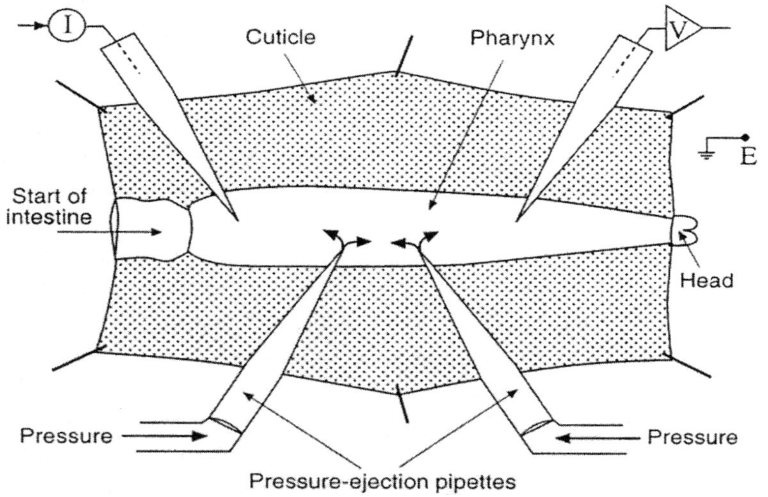

FIGURE 15.7 Diagram of an *Ascaris* pharynx preparation for two-micropipette current clamp recording. A current injection pipette (I) is placed in the pharyngeal muscle while a voltage recording pipette (V) records membrane potential and response to injected current for conductance measurement. Two additional pipettes are placed extracellularly for the application of drugs, including neurotransmitters.

approximately 300 neurons in *A. suum*. Sensory and motor neurons are concentrated at the head and the tail of the nematode. The rest, about 90 motor neurons, are arranged in repeated segments along the length of the body; they supply the innervation to the somatic muscle cells. Figure 15.8 illustrates a segment of the dorsal and ventral nerve cord with connecting commissures (three to the left and one to the right). Each segment has 11 motor neurons that are excitatory (cholinergic) or inhibitory (GABAergic). This layout is believed to be very similar in all species of nematode. Due to an elementary system of reciprocal innervation of the neural circuits, excitatory impulses to the ventral muscle are associated with inhibitory impulses to the dorsal muscle, and vice versa. This combination of signals results in contraction of one block of muscle whilst the opposing muscle mass relaxes, allowing the nematode to progress forwards or backwards with a wave-like motion.

Ascaris physiology

Membrane potential

The membrane potential of *Ascaris suum* muscle is near −30 mV, and is relatively insensitive to changes in the ionic composition of

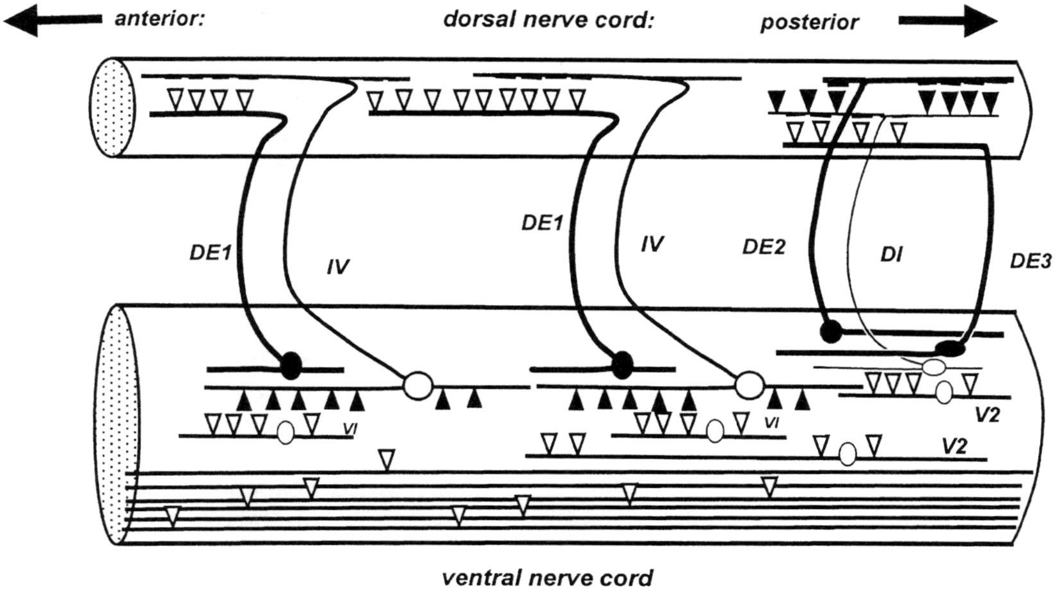

FIGURE 15.8 A 'segment' of the dorsal and ventral nerve cord of *Ascaris*, with 11 motor neurons and six longitudinal neurons. There are three right-hand commissures and one left-hand commissure in each segment. The diagram represents one of seven repetitive segments found in *Ascaris*. The filled, •, cells are excitatory motoneurons with acetylcholine as the fast transmitter and empty, O, cells the inhibitory motoneurons with GABA as the fast transmitter. FaRP peptides may act as co-transmitters. Each motoneuron is classified by the location of the cell body, and the shape and distribution of its processes. All cell bodies are found in the ventral cord. The excitatory, cholinergic outputs, ▽, are shown, as are the inhibitory GABAergic ouputs, ▼. The inputs, – are also shown. Dorsal excitatory motoneurons: DE1, DE2, DE3. Ventral excitatory neurons: V1, V2. Dorsal inhibitory neurons: DI. Ventral inhibitory neurons: IV. V1 and V2 do not have commissures. All commissures are right-handed except DE3.

the surrounding medium. A tenfold increase in the extracellular potassium concentration causes a 1.5 mV decrease in the *Ascaris* membrane potential. This small effect is in marked contrast to the equivalent situation in vertebrate muscle, where the membrane potential is dependent on potassium ions. Del Castillo and his co-workers considered that organic anions contributed to the membrane potential in *Ascaris* muscle, but ascribed the maintenance of the membrane potential to the intracellular chloride battery. Brading and Caldwell hypothesized that an electrogenic pump, operating across the muscle cell membrane, was responsible for maintaining the membrane potential. Our studies have led us to propose that there is a proton pump that allows the movement of carboxylic acids across the *Ascaris* muscle membrane and cuticle, in a manner closely linked to the membrane potential.

Neurotransmitters

Despite the topographical simplicity of the nematode nervous system, a large number of *A. suum* neurotransmitters have been identified. These include: acetylcholine; γ-aminobutyric acid (GABA); glutamate; serotonin (5HT); and FMRFamide-related peptides (FaRPs).

Acetylcholine

Initially, experiments with bath-applied acetylcholine showed that acetylcholine produces contraction of muscle strips of *Ascaris suum*, demonstrating the presence of acetylcholine receptors on muscle. Electrophysiological observations followed with bath-applied acetylcholine that was shown to produce depolarizations and changes in spike frequency and amplitude in muscle. The receptors responsible for these effects are located synaptically at the syncytial region, and extrasynaptically on the bag region of the muscle. The ionic basis of electrical responses to acetylcholine is due to the ability of acetylcholine to increase the nonselective cation conductance of the membrane: that is, acetylcholine opens ion channels permeable to both Na^+ and K^+.

Single-channel currents activated by acetylcholine

We have recorded single-channel currents activated by acetylcholine from cell-attached and isolated inside-out patches. The channels activated by acetylcholine have at least two conductances: the larger was 40–50 pS and the smaller 25–35 pS. The average open time was similar to that of the nicotinic channel at the frog neuromuscular junction and had a mean value of 1.3 ms. High concentrations of acetylcholine (25–100 μm) produced a reduction in open probability and caused single-channel currents to occur in clusters; this behavior has also been reported for the frog neuromuscular junction and described as desensitization.

Biochemistry of acetylcholine

Bueding was the first to demonstrate that cholinesterase, the enzyme inactivating acetylcholine, was present in *A. suum*. In the head region, activity is associated with the contractile region of the muscle in the extracellular matrix; enzyme activity was also observed in muscle arms near their endings on the nerve cords but not on the bag region or the nervous tissue. Johnson and Stretton showed that there are two types of cholinesterase, a 13 S and a 5 S form, which have different distributions in the body of *Ascaris*. These two forms of acetylcholinesterase are also found in *C. elegans*, where they are products of separate genes. Although cholinesterase is responsible for the breakdown of acetylcholine released from excitatory motoneurons, it is also secreted into the external environment by *A. suum* and may act on the host intestine, depressing the level of acetylcholine and its stimulation of host gut muscarinic receptors that would produce intestinal secretions and immune responses.

Cholinesterase antagonists as anthelmintics

Anti-cholinesterases potentiate the electrophysiological effects of acetylcholine in *Ascaris*. A number of organophosphorus anti-cholinesterases, originally introduced as insecticides, have been used for their anthelmintic properties. They include metriphonate (trichlorfon), dichlorvos, naphthalophos, crufomate and haloxon. The efficacy of these compounds against nematodes shows that the cholinesterases of nematodes are different from the mammalian hosts and that they are effective targets for anthelmintics.

Genetics of resistance to anti-cholinesterases produces modification of the synaptic vesicle cycle proteins or an acetylcholine transporter in *C. elegans*

The cycle by which the synaptic vesicle is formed and removed is divided into nine steps (Figure 15.9) consisting of: (1) Docking (contact between the vesicle and the active zone of the synaptic cleft); (2) Priming (a maturation process that makes vesicles competent for fast calcium-triggered membrane fusion); (3) Fusion/exocytosis produced by a calcium spike; (4) Endocytosis (empty synaptic vesicles are rapidly internalized by clathrin-coated pits); (5) Translocation (coated vesicles shed the clathrin, acidify and move to the interior); (6) Endosome fusion (recycling synaptic vesicles fuse with early endosomes, and in *Ascaris* vesicles can be seen attached in a circular arrangement to 'tubules'); (7) Budding (regeneration of synaptic vesicles); (8) Neurotransmitter uptake (transmitters are taken up into the vesicle); (9) Translocation (filled synaptic vesicles with transmitter move to the active zone). A number of proteins present in the synaptic vesicle have to be present for the synaptic vesicle cycle to take place. The proteins include synaptobrevins, synaptotagmins, SV2s, SCAMPs, synpatogyrins, synaptophysins, Rab3s, Rab5s, CSPs and synapsins. The synaptic plasma membrane

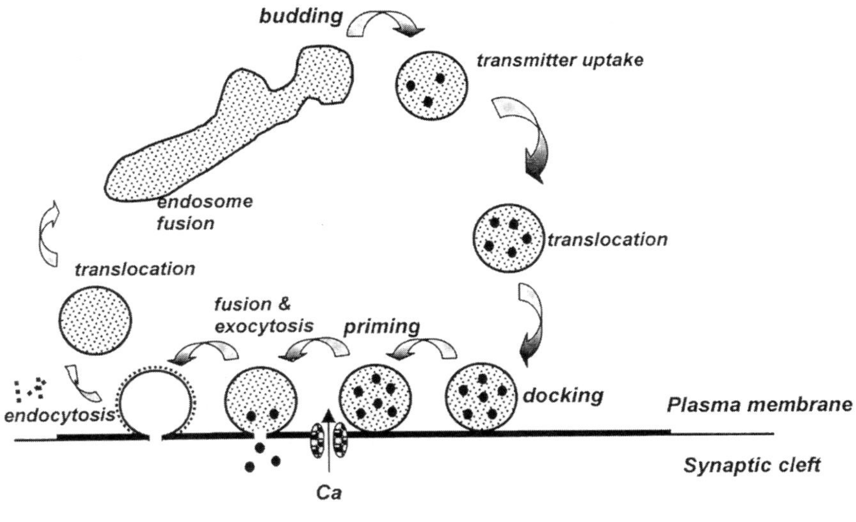

FIGURE 15.9 Diagram of the vesicle cycle with the nine stages illustrated.

proteins required for docking and priming include Munc13s, neurexins, SNAP-25 and syntaxins, and proteins that are associated with these proteins include complexins, Munc 18s, NSF, and $\alpha/\beta/\gamma$-SNAPs.

In *C. elegans*, *unc-17* codes for an acetylcholine transporter, and an *unc-17* null mutant is resistant to anti-cholinesterase. In a similar way, mutations in *unc-13* (which encodes MUNC-13), *unc-18* (which encodes MUNC-18S), and *syt-1* (which encodes synaptotagmin) in *C. elegans* also give rise to anti-cholinesterase resistance. The explanation for the resistance and the partial paralysis of the nematode relates to the impaired release of acetylcholine, which reduces motility but makes the worm less sensitive to the effects of cholinesterase antagonism.

Pharmacology of the acetylcholine receptor

Low concentrations of nicotine produce contractions of *Ascaris* body muscle strips like acetylcholine does. The acetylcholine-induced contractions are blocked by tubocurarine but not by atropine, so that the *A. suum* acetylcholine receptors have some of the pharmacological properties of vertebrate nicotinic receptors. For example, the potent ganglionic nicotinic agonist dimethylphenylpiperazinium is more potent than acetylcholine in *Ascaris*, and the potent ganglionic nicotinic antagonist mecamylamine is the most potent acetylcholine antagonist, more potent than tubocurarine. In contrast, hexamethonium, a potent ganglionic antagonist in vertebrates, had a low potency in *Ascaris*. Therefore, the *Ascaris* acetylcholine receptors cannot be classified as either ganglionic or neuromuscular, and can be regarded as a separate subtype of nicotinic receptor.

Electrophysiological techniques have also been used to examine effects of cholinergic agonists and antagonists on muscle. Levamisole and pyrantel are more potent agonists at the *Ascaris* muscle acetylcholine receptor than at vertebrate nicotinic receptors, where they have only weak nicotinic actions. The selective action of these drugs allows them to be used as effective anthelmintics, killing the nematode parasite but not the host. The degree of selectivity will obviously affect the safety and efficacy of any nicotinic drug selected for therapeutic purposes.

Pilocarpine and muscarine cause a slight hyperpolarization (up to 3 mV) rather than a depolarization. These observations raise the possibility that, in addition to the nicotinic-like receptors, muscle cells may possess muscarinic-like receptors for acetylcholine. Janet Richmond has demonstrated, with electrophysiological experiments on *C. elegans*, the presence of two types of acetylcholine-gated ion-channel receptor on muscle. Firstly, there is a non-desensitizing response to levamisole which requires the presence of UNC-28 and UNC-38 subunits; and secondly, a rapidly desensitizing response to nicotine that is blocked by dihydro-β-erythroidine. Thus there are at least two types of ionotropic acetylcholine receptor: a levamisole-preferring receptor, and a nicotine-preferring, dihydro-β-erythroidine-blockable receptor. Interestingly, the anthelmintic paraherquamide is a more potent selective antagonist and will also separate these two receptor types.

Cholinergic receptor anthelmintics

The anthelmintics that act as agonists to open acetylcholine ion-channel receptors of nematodes include the imidazothiazoles (levamisole and butamisole), the tetrahydropyrimidines (pyrantel, morantel and oxantel), the quaternary ammonium salts (bephenium and thenium) and the pyridines (methyridine).

Intracellular recording techniques on *A. suum* muscle were used by Aubry to observe effects of bath-applied tetramisole (a D, L racemic mixture of which levamisole is the laevoisomer). Tetramisole produced depolarization, an increase in spike frequency and contraction. He described the anti-nematodal drug action and pharmacological properties of pyrantel and some of its analogs in a variety of vertebrate and helminth preparations.

Single-channel currents activated by cholinergic anthelmintics

Figure 15.10 shows a kinetic diagram of the way in which levamisole-activated channel currents are thought to behave during the binding of the drug and opening of the channel. Also illustrated are some sample openings of levamisole-activated channels from *Ascaris suum*. The ion channels are cation-selective and have similar kinetic properties to acetylcholine-activated channels at low levamisole concentrations, but show evidence of channel block at higher concentrations. The receptor-operated ion channels have mean open times in the range 0.8–2.85 ms with conductances in the range 20–45 pS.

Likely structure of nematode nicotinic acetylcholine receptors (nAChR)

The detailed structural studies on vertebrate nicotinic receptor channels have been carried out on the nAChR derived from the *Torpedo*

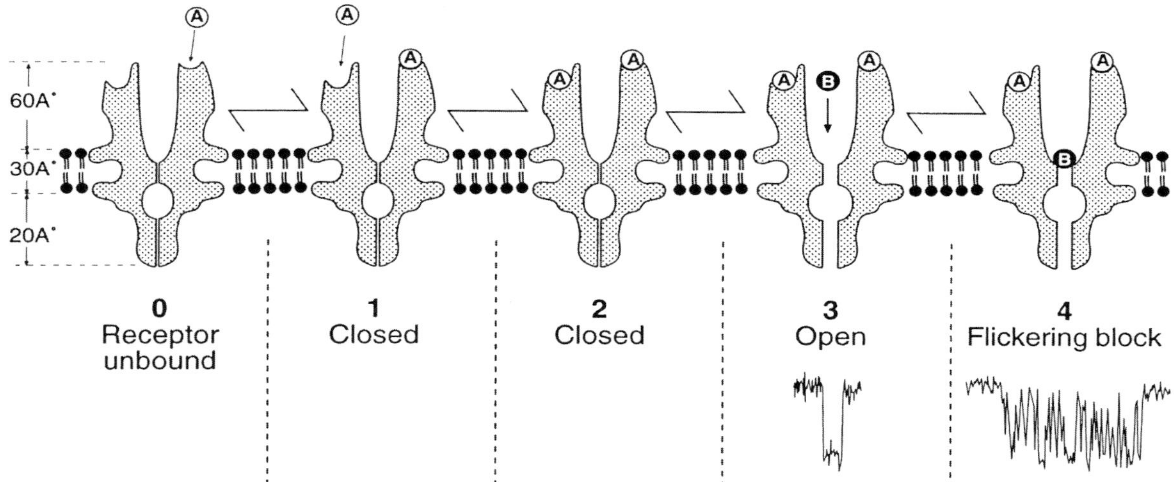

$$R \underset{k_{-1}}{\overset{k_{+1}\cdot x_a}{\rightleftharpoons}} AR \underset{k_{-2}}{\overset{k_{+2}\cdot x_a}{\rightleftharpoons}} A_2R \underset{\alpha}{\overset{\beta}{\rightleftharpoons}} A_2R^* \underset{k_{-B}}{\overset{k_{+B}\cdot x_B}{\rightleftharpoons}} A_2RB$$

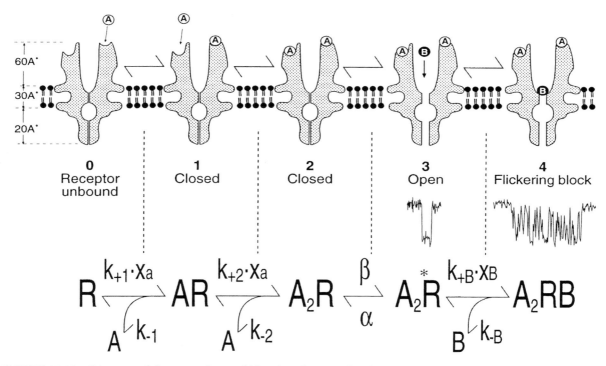

FIGURE 15.10 Diagram of the operation and kinetics of agonist binding, A, to a ligand-gated ion channel, and the opening and ion-channel blocking process. Two agonist molecules bind reversibly and in the sequences through the states, 0, 1 and 2, with the defined rate constants shown. The doubly bound ion channel then may open (3, rate constant β) briefly before closing (rate constant α), or the channel may become blocked with an agonist molecule entering the open channel pore to enter the state 4.

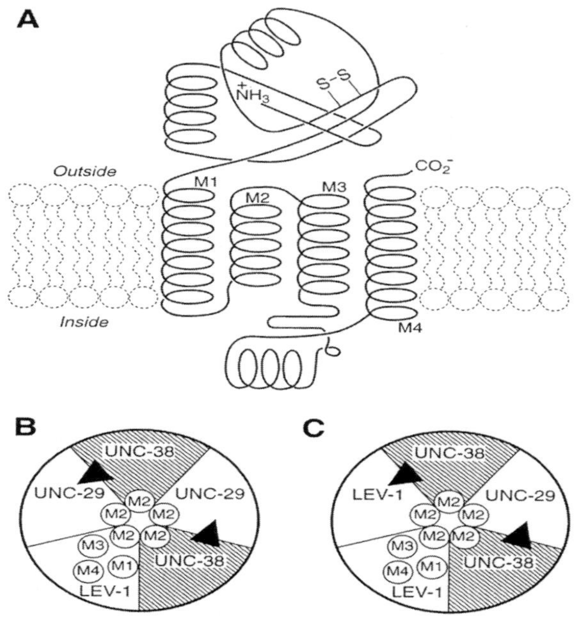

FIGURE 15.11 (A) Diagram of one of the subunits of the pentameric ion channel: α or UNC-38 subunit. There are about 500 amino acids starting at the N-terminal and four transmembrane α-helices, M1, M2, M3 and M4. The cytoplasmic loop between M3 and M4 may be phosphorylated and thus regulated. The M2 region forms the lining of the pore of the channel. (B) & (C) Five similar, subunits come together to form the ion channel. The subunits are identified by the products of the levamisole resistance genes, *unc-38, unc-29* and *lev-1* of *C. elegans*. The stoichiometry of the subunits of the ion channel is not known and may vary to produce different channels, (B) & (C). The agonist binding sites form at the junction of the UNC-38 and the adjacent subunit. Thus the two agonist binding sites on the ion channel may not be identical.

electric organ. The nAChR has a pentameric structure with five protein subunits (2α, 1β, 1γ and 1δ) arranged around the pore of the channel (Figure 15.11). Each subunit is about 500 amino acids and is glycosylated on the extracellular loop. The α subunit, but not the β (non-α) subunits, has two adjacent cysteine amino acids within the extracellular loop at locations referred to as 192 and 193. This region is believed to contain part of the agonist-binding site. Each subunit has four lipophilic regions (M1, M2, M3 and M4) that form α-helices that span the membrane. The M2 spanning unit lines the pore of the channels and has charged amino acids present, so that the channel selectively binds and carries either cation. A narrow region of the ion channel two-thirds of the way through the ion channel pore restricts the movement of cations greater than 6.5 Å through the channel. Large organic ions may try and enter the channel, becoming stuck and producing a voltage-sensitive channel block. Although the stoichiometry (arrangement and number of the subunits forming the channel) of vertebrate muscle nAChRs is fixed, the stoichiometry of vertebrate neuronal nicotinic receptors may not be fixed. The number and type of the different subunits in the neuronal nAChRs gives rise to changes in the sensitivity of the receptor to the agonist, changes in the amount of calcium that can pass through the pore, and changes in the rate of desensitization.

Phosphorylation of the different subunits affects the charge of the protein, and consequently affects opening, desensitization and synthesis of the new nAChRs. Regulatory phosphorylation sites for protein kinase C (PKC), protein kinase A (PKA) and tyrosine kinase (TK) are recognized in *C. elegans*. Phosphorylation of the levamisole receptor, then, is one of the most effective means available to the parasite to regulate and modulate the number and activity of the receptors, and so phosphorylation may play a role in the development of anthelmintic resistance. The level of phosphorylation is a balance between the opposing activities of protein kinases that phosphorylate the receptor and the phosphatases that dephosphorylate it. So if phosphorylation enhances channel opening of levamisole receptors, phosphatases should decrease activation and depress responses to nicotinic anthelmintics.

The presence of nAChR subunit genes in nematodes predict heterogeneity of receptors

Fleming recognized 11 genes associated with levamisole resistance in *C. elegans*. Three of these genes encode nAChR subunits in the nematode *C. elegans*, and are associated with strong levamisole resistance (Figure 15.11). These genes are *unc-38*, *unc-29* (both on chromosome I) and *lev-1* (on chromosome IV). *Unc-38* encodes an α subunit while *unc-29* and *lev-1* encode β subunits. It is also possible that there are additional genes (e.g. *acr-2* and *acr-3* on the X chromosome) encoding nAChR subunits associated with weaker levamisole resistance.

Other genes encoding subunits for the nAChR in *C. elegans* have been identified: the non-α subunits, *acr-2* and *acr-3* (on chromosome X); and the α-subunits *deg-3* and *Ce21* (on chromosome V). The varying combination of different subunits may produce significant heterogeneity of the ion-channel receptors activated by acetylcholine.

In *Oesophagostomum dentatum* the muscle nAChRs have conductances and mean open times that show variation between patches, and sometimes more than one subtype of nAChR is observed in one patch (Figure 15.12). We interpreted these data as indicating four different subtypes of levamisole receptor: G25, G35, G40 and G45 (Figure 15.13). nAChR channel currents from *Ascaris suum* with different conductances ranged between 19 and 50 pS with distinguishable peaks at 24 pS and 42 pS, but clearer separation of subtypes was more difficult, perhaps because of variations in the source of the *Ascaris*, which were derived from field infections. The functions of the different

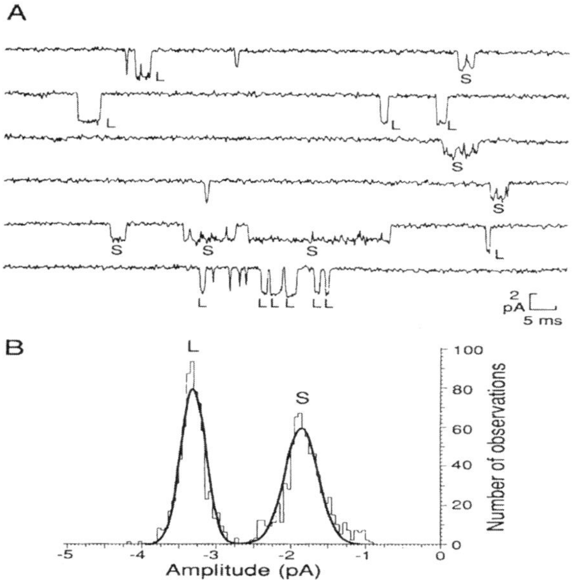

FIGURE 15.12 (A) Single-channel current recording from levamisole-activated channels from *Oesophagostomum dentatum*. Single-channel currents activated by 10 μM levamisole in a cell-attached patch at −75 mV. Note the presence of two types of channel: the larger openings (L) and the smaller openings (S). (B) The open-channel current histogram showing clear separation of the two channel openings (L & S).

subtypes of nAChR remain to be evaluated, but because the pharmacology of each subtype is predicted to be slightly different we might predict changes in the proportion of receptor subtypes associated with the development of resistance to nicotinic agonists.

Recently we have been able to compare the levamisole subtypes in levamisole-sensitive and levamisole-resistant isolates of *Oesophagostomum dentatum*. The G35 subtype was lost with the development of resistance, and resistance was associated with a reduction in the mean proportion of time the nAChR channel spent in the open state. Quantitatively the resistance is explained by the combination of the

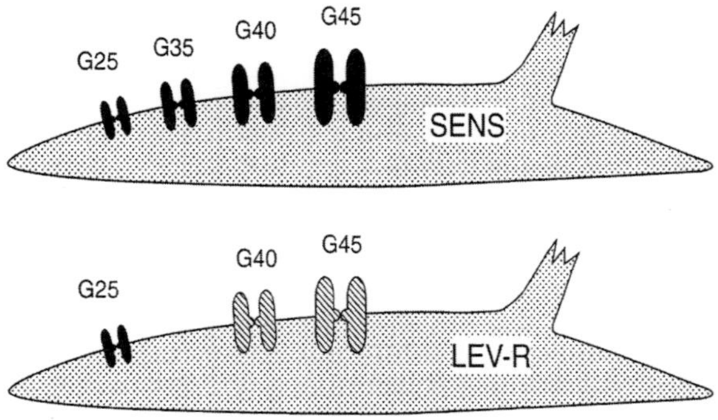

FIGURE 15.13 Diagram of the four subtypes of levamisole receptors seen in the sensitive isolate of *Oesophagostomum dentatum*. All four subtypes, G25, G35, G40 and G45, may not be present in all individual muscle cells but were found to be present in a sample of levamisole-sensitive female adults. In the levamisole-resistant isolate G35 was missing and G40 and G45 had different kinetics.

reduction in the number of active channels and the reduction in the average time the channel was open.

GABA

Effects of GABA on *Ascaris* muscle

We have observed GABA receptors that are located at the syncytium and extrasynaptically over the surface of the muscle cell including the bag region. Local application of GABA onto the bag, when examined under voltage-clamp, produces a current that is explained by the opening of ion channels permeable to Cl^-. The physiological function of the extrasynaptic GABA receptors is not known; however, they may be exploited in pharmacological experiments designed to look at receptor properties. They may also be activated by anthelmintics.

The agonist profile of the *Ascaris* GABA receptor is similar but not identical to that of the vertebrate $GABA_A$ receptor, but the antagonist profile is very different. The $GABA_A$ antagonists bicuculline, picrotoxin, securinine, pitrazepine and RU5135 are weak or inactive at the *Ascaris* muscle GABA receptor, but a series of arylaminopyridazine-GABA derivatives, which act as competitive $GABA_A$ antagonists, are competitive *Ascaris* muscle GABA-receptor antagonists with different potencies. A series of novel arylaminopyridazine derivatives has been synthesized and tested in *Ascaris*, where it was found that NCS 281-93 was the most potent competitive GABA antagonist in *Ascaris*. We may summarize the pharmacology of the GABA receptor derived from observations on *Ascaris* by concluding that the GABA receptor present in nematodes is unique.

Genetics of GABAergic transmission

In *C. elegans*, six genes, *lin-15, unc-25, unc-30, unc-43, unc-47* and *unc-49*, are required for GABAergic neurons to function. *Unc-49* functions postsynaptically and is necessary for an inhibitory effect on body muscles. A cDNA, HG1, has been recovered from *Haemonchus contortus* that has a high homology to the cDNA for vertebrate $GABA_A$ receptors.

GABA and piperazine single-channel currents

We have recorded GABA- and piperazine-activated channels in cell-attached and outside-out patches. The channels opened by both agonists had a main-state conductance of 22 pS. The average duration of the effective openings (bursts) produced by GABA was in the region of 32 ms, while the average duration of openings produced by the anthelmintic piperazine was 14 ms (Figure 15.14). Piperazine is ~100 times less potent than GABA in *A. suum*, and this difference in potency may be explained by the fact that higher concentrations of piperazine are required to achieve a similar opening rate to GABA, and the average duration of openings produced by piperazine is shorter.

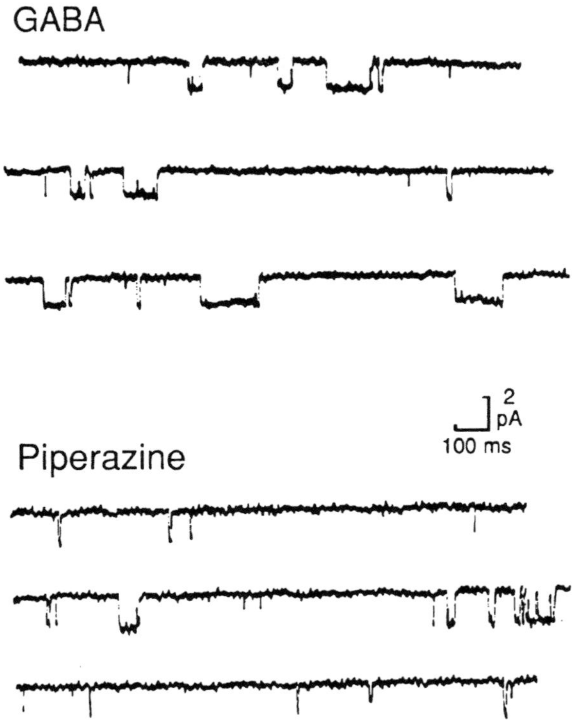

FIGURE 15.14 GABA- and piperazine-activated single-channel currents. The open times of the GABA-activated channels are longer on average (mean open time 32 ms) than the piperazine-activated channels (18 ms). Piperazine also requires a higher concentration to produce a similar opening rate to GABA, and so is less potent than GABA.

Glutamate

Excitatory effects

A number of anthelmintics, including kainate, have a chemical structure similar to the excitatory amino acid glutamate. Kainate, domoic acid and quisqualate have been used as anthelmintics in Asia for a long time, and are now usually used to study vertebrate glutamate receptors. DE2 motorneurons of *Ascaris* show depolarizing potentials and conductance changes in response to glutamate agonists (domoate > kainate >> glutamate > aspartate) that are similar to vertebrate kainate receptors. In the DI motoneurons, glutamate produces a hyperpolarization that may be mediated via an intervening inhibitory neuron. These observations suggest the presence of excitatory glutamate receptors in nematodes, and the cloning of the AMPA-like glutamate receptor gene *glr-1* from *C. elegans* confirms their presence. Mutants of *glr-1* are sluggish and defective in mechanosensory behavior. Electrogenic glutamate transporters are also present in the hypodermis. This hypodermal transporter may serve as a buffer, inactivating glutamate synapses throughout the nervous system in parallel to the hypodermal cholinesterase.

Inhibitory effects: the search for the ivermectin receptor

The success of ivermectin (Merck) as an anthelmintic, and its very potent effects against nematodes, led to a search for the mode of action of these compounds. One of the initial helpful observations was that ivermectin was found to bind specifically to membrane preparations of *C. elegans*. The lipophilic nature of the avermectins made detection of specific binding to receptors, present in low number, very difficult.

C. elegans was an excellent model for these mode-of-action studies because large numbers of worms could be grown in fermentation tanks to produce substantial amounts of RNA for extraction and expression. The nematode GluCl ion channels were first recognized by expression of a glutamate-activated chloride current, sensitive to avermectins, in *Xenopus* oocytes injected with *C. elegans* RNA. Expression cloning then led to the recognition of two *C. elegans* channel subunits, GluClα1 on chromosome V and GluClβ on chromosome I. The GluClα subunit is responsible for the sensitivity to ivermectin, and the GluClβ subunit is sensitive to glutamate. Subsequently, molecular genetic and cloning approaches have led to the recognition of additional GluCl subunits from *C. elegans* and parasitic nematodes (summarized in Table 15.2).

GluCl subunit structure and ion channels

The presence of N-terminal cysteines in the GluCl subunits and hydrophobicity analysis suggest that GluCl subunits have similar motifs common to all cys-loop ligand-gated channels. The GluCl subunits have a second pair of cysteine residues in the N-terminal extracellular domain that are characteristic of this family and the vertebrate glycine receptors. This sequence similarity has led to the suggestion that invertebrate GluCls are in fact orthologous (connected) to the vertebrate glycine receptor subunits. It is assumed that five GluCl subunits come together, as do the subunits of the nAChRs, to produce the GluCl ion channel, but the stoichiometric arrangement has not been determined for any native receptor. GluClα1 and GluClβ subunits may form homo-oligomers (five identical subunits) to form receptor ion channels as well as hetero-oligomers (five subunits not identical) to form channels when expressed in *Xenopus* oocytes.

GluCl subunits

In *C. elegans*, a family of genes encodes α-type subunits: *glc-1* encodes the subunit GluClα1, *avr-15* encodes GluClα2 subunits, *avr-14* encodes GluClα3A and GluClα3B, and *glc-3* encodes a fourth GluClα. The *avr-15* gene encodes two alternatively spliced channel subunits that are present on pharyngeal

TABLE 15.2 Cloned glutamate-gated chloride channel subunits from *Caenorhabditis elegans*

Gene	WormPep	EMBL/Genbank accession no.	Other name for subunit	*In vitro* expression data[1]	*In vivo* expression data[2]
glc-1	F11A5.10	U14524	GluClα	Forms ivermectin-gated channels	
glc-2	F25F8.2	U14525	GluClβ	Forms glutamate-gated channels	Pharyngeal muscle pm4 cells
glc-3	ZC317.3	AJ243914		Forms glutamate- and ivermectin-gated channels	
avr-14 (*gbr-2*)	B0207.12	U40573 U41113	GluClα3A GluClα3B	AVR-14B forms glutamate- and ivermectin-gated channels AVR-14A does not form functional channels	Multiple neurons in the nerve ring. Ventral cord motor neurons and mechanosensory neurons
avr-15	T10G3.7	AJ000537	GluClα2A GluClα2B	AVR-15L forms glutamate- and ivermectin-gated channels	Pharynx – pm4 & pm5 cells. RME & RMG ring motor neurons, DA9 & VA12 and other nerve cord neurons
	C27H5.5	U14635		Does not form functional channels	

[1] Expression in *Xenopus* oocytes.
[2] Reporter gene data.

muscle and in motoneurons of the nerve ring and nerve cords. The *avr-15*-encoded subunit (the long form) can form a homo-oligomeric channel that is ivermectin-sensitive and glutamate-gated when expressed in *Xenopus* oocytes. The *avr-14* subunits are expressed in a subset of 40 extrapharyngeal neurons in the ring ganglia of the head and motor neurons of the ventral cord and mechanosensory neurons. *Glc-1* appears to be represented in the extrapharyngeal neurons. The location of the expressed *glc-3* is not yet known. The GluClβ subunit was found on the pm4 pharyngeal muscle cells of *C. elegans*.

Electrophysiology of glutamate and ivermectin in nematode parasites

Avermectin-sensitive sites in *A. suum* have been identified on pharyngeal muscle using current-clamp; avermectins produce hyperpolarization and an increase in Cl^- conductance when applied onto pharyngeal preparations. Glutamate and macrocyclic lactones (MLs) increase the opening of expressed GluCl receptor channels in *Ascaris suum* pharyngeal muscle. They also increase opening of glutamate-gated chloride channels found in the pharynx of *Ascaris suum* (Figure 15.15).

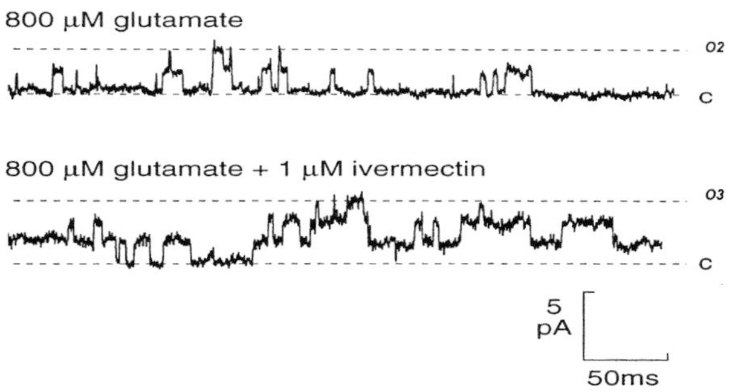

FIGURE 15.15 Glutamate-activated single-channel currents recorded from *Ascaris* pharynx using an inside-out patch. Addition of ivermectin to the inside-out patch increases the opening of the channels.

The split chamber technique allows the selective application of avermectin to the dorsal and ventral halves of *A. suum*. Under these conditions it was found that avermectin blocks the motor neuron DEI response to indirect stimulation but not direct stimulation. These observations suggest that avermectins block transmission between interneurons in the nerve and the excitatory neurons. Avermectins have more than one site of action, and these include ion channels in both muscle membranes and neuronal membranes. These observations are consistent with the immunological identification of GluCl subunits on the nerve cords and motoneuron commissures of *Haemonchus contortus* and *Ascaris suum* (Table 15.3). There are also inhibitory glutamate receptors on the muscle cells of the female reproductive tract of *Ascaris*. Macrocyclic lactones have inhibitory effects on egg laying and fertility. The effect may account for much of the efficacy of ivermectin treatment in controlling infections by *Onchocerca volvulus*.

Cloning of GluCl subunits from parasitic nematodes

Compared to *C. elegans*, we know less of the molecular biology of the GluCls from parasitic species. Our knowledge is summarized in Table 15.3. Three GluCl genes have been identified in *H. contortus*. Two of these are orthologous to *C. elegans glc-2* and *avr-14*, and the pattern of alternative splicing of *avr-14* is conserved between the two species. Not only that, but the expression of *avr-14* on ring and cord motoneurons appears similar in both *C. elegans* and *H. contortus*. The GluClα3B subunit from both species, but not the GluClα3A subunit, is able to bind ivermectin when expressed alone. *Avr-14* is the gene found most widely in parasitic nematodes, and is present in *A. suum* and the filariae, *O. volvulus* and *D. immitis*, though alternative splicing does not appear to

TABLE 15.3 Cloned GluCl subunits from parasitic nematodes

Species	Clone	Accession number	Other name	Closest *C. elegans* gene	Expression data[1]
Haemonchus contortus	Hc-gbr2A Hc-gbr2B	Y14233 Y14234	HG2 HG3	*avr-14* (*gbr-2*)	Hc-gbr2B binds ivermectin. Present in the nerve ring, on ventral and dorsal nerve cords, the anterior dorsal sublateral cord and motor neuron commissures[2]
H. contortus	HcGluClβ	Y09796	HG4	*glc-2*	Does not bind ivermectin. Expressed on motor neuron commissures
H. contortus	HcGluClα	AF076682 AJ131347	HG5	*avr-14* (*gbr-2*)[3]	Binds ivermectin. Expressed at higher levels in adults than larvae. Present on motor neuron commissures
Ascaris suum	As-gbr2	Y18347		*avr-14* (*gbr-2*)	Present on nerve cords
Onchocerca volvulus	GluClX	U59745		*avr-14* (*gbr-2*)	
Onchocerca ochengi	Putative α-subunit	AF054632		*glc-1*	
O. ochengi	Putative β-subunit	AF054631		*glc-2*	
Dirofilaria immitis	GluClX	U59744		*avr-14* (*gbr-2*)	

[1] Localization data obtained using immunofluorescence.
[2] The antibody used did not distinguish between Hc-gbr2A and 2B.
[3] The level of identity between this subunit and *avr-14* is lower than for the Hc-gbr2, As-gbr2 and the GluClX subunits.

be conserved in these species. The other *H. contortus* GluClα gene is interesting because it seems not to be orthologous with any of the four *C. elegans* GluClα genes, as it shares about 55–60% sequence identity with all of them. The *H. contortus* GluClα is expressed at higher levels in adult worms than in larvae, suggesting it may be most important in the parasitic stages of the life cycle. This subunit binds ivermectin, is found on motoneurons, and may be an important target for the drug *in vivo.*

The *H. contortus* GluClβ subunit, like its *C. elegans* counterpart, does not bind ivermectin. Though the pharmacology of the two subunits is similar, their distribution may be very different. The HcGluClβ subunit is found on motoneuron commissures in the anterior region of the adult parasite, but the *C. elegans* subunit is only found on pharyngeal muscle cells. If this difference in localization is substantiated by further studies, it may serve as an important caution against making generalizations among different species of nematodes.

Ivermectin resistance

We know from the studies of Dent that ivermectin resistance in *C. elegans* involves the genes *avr-15, avr-14, glc-1, unc-7, unc-9* and *dyf* (dye-filling defective). Simultaneous mutations of the GluClα genes *avr-15, avr-14* and *glc-1* were required for the establishment of high levels of immunity (>4000×). Little resistance is seen with mutations in individual genes. A combination of two of the genes (*avr-14* and

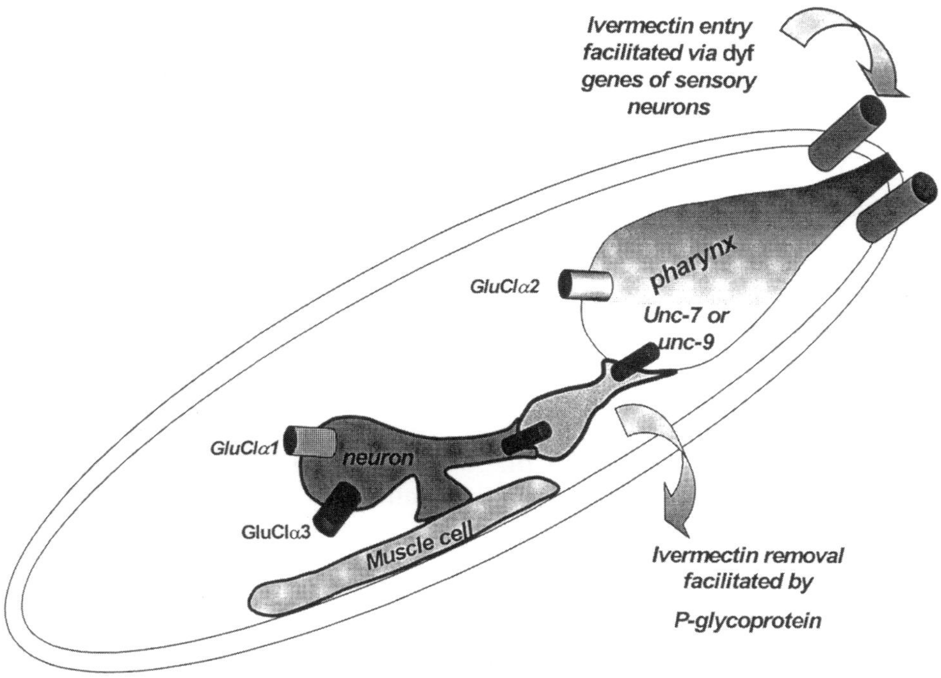

FIGURE 15.16 Diagram showing the interaction of 'ivermectin resistance' genes. Information on resistance to the macrocyclic lactones (MLs) in nematodes is derived mostly from *C. elegans*. The entry of ivermectin into the nematode is facilitated by sensory (amphidial) neurons on the head. Once the drug has gained entry across the cuticle, it is then able to interact with the GluClα receptors. There are at least three different genes encoding for at least three different GluClα subunits that form inhibitory ion channels on muscle of the pharynx, motor neurons and other neurons, and in addition perhaps on the female reproductive tract. The neurons that possess the GluCl channels connect via gap junctions that are made up of innexins, coded for by *unc-7* and *unc-8* genes. Thus the inhibitory effect of ivermectin on the pharynx may be direct via the GluClα2 subunit, or indirect, via the GluClα3 and GluClα1 subunits on extrapharyngeal neurons, and may require that effect to be mediated across gap junctions (*unc-7* and *unc-9*) to the pharynx. Removal of ivermectin and other MLs from the body of the nematodes appears to be mediated by p-glycoprotein excretion. The mode of action and genetics of resistance illustrates that the development of resistance requires the simultaneous mutation of several genes to develop a high level of resistance. Factors that increase the concentration of macrocyclic lactones in the nematode will increase susceptibility: genes (*unc-7* and *unc-9*) that increase the electrical effects of stimulation of the GluCl ion channel will increase susceptibility, and the presence of genes coding for GluClα-subunits will increase susceptibility. Note that the physiology and location of the target site determines how resistance genes may interact. For example the genes *unc-7* or *unc-9* form gap junctions that are able to pass on potential changes (hyperpolarization) associated with stimulation of GluClα1 and GluClα3 subunits in inhibitory channels.

avr-15) produced only a 13-fold increase in resistance.

Figure 15.16 summarizes information relating to the genetics of resistance to the macrocyclic lactones in nematodes, and is derived mostly from *C. elegans*. Entry of ivermectin into the nematode may be facilitated by sensory (amphidial) neurons on the head. Once the drug has crossed the cuticle, it is then able to interact with the GluClα receptors. At least three different GluClα subunits have been identified that form inhibitory ion channels on muscles of the pharynx, motor neurons and other neurons, and in addition perhaps on the female reproductive tract. Neurons that possess the GluCl channels connect via gap junctions that are made up of innexins, coded for by *unc-7* and *unc-9* genes. Thus the inhibitory effect of ivermectin on the pharynx may be direct via the GluClα2 subunit, or indirect, via the GluClα3 and GluClα1 subunits on extrapharyngeal neurons, an effect mediated across gap junctions (*unc-7* and *unc-9*) to the pharynx. Ivermectin removal may be influenced by p-glycoprotein excretion and is perhaps located on intestinal epithelial cells. The mode of action and genetics of resistance in *C. elegans* implies that the development of resistance in parasitic nematodes requires the simultaneous mutation of several genes to develop a high level of resistance. For example, factors that increase the concentration of macrocyclic lactones, or factors that increase the electrical effects of stimulation of the GluCl ion channel, such as *unc 7* or *unc 9*, will increase susceptibility. However, some studies suggest that ivermectin resistance in parasitic nematodes is simpler and may in fact be dominant.

5-HT

The majority of studies on nematode neuromuscular systems have focused on the body wall and nerve cord, but the pharyngeal muscle is also important for the survival of most nematodes. Pharyngeal pumping is necessary for feeding; in fact nematodes spend a large part of their time searching for food and feeding. Pumping is regulated by a range of neurotransmitters, including acetycholine, glutamate, GABA, 5-HT, dopamine and a range of neuropeptides (including KSAYMRFamide). In *C. elegans* 5-HT stimulates and dopamine depresses pharyngeal pumping. 5-HT also modulates pharyngeal pumping in *A. suum* and modulates activity of the vagina vera. The injection of 5-HT into *Ascaris* causes a rapid paralysis in animals that were moving. It causes depolarization of the male-specific transverse ventral muscle and curling in the male, mimicking mating behavior. It also abolishes slow potentials in VI motor neurons of *Ascaris*. 5-HT is localized in *Ascaris* in a pair of neurons in the pharynx of both sexes, and in five cells in the ventral cord of the male tail. The pharyngeal cells are probably neurosecretary cells and homologous to the *C. elegans* neurosecretary motor neurons (NSM).

A 5-HT receptor isoform that appears to be alternatively spliced has been identified in *A. suum*. AS1–3 are expressed in the pharynx, and AS1 and AS2 are expressed in the body wall muscle. Thus 5-HT has an inhibitory/ modulatory effect on most body muscle and stimulates pharyngeal pumping.

FMRFamide-related peptides

Short peptides referred to as FMRFamides may act in nematodes as co-transmitters along with the transmitters mentioned above.

A tetrapeptide FMRFamide was first isolated from the mollusc *Macrocallista nimbosa*. FMRFamide, short for Phe-Met-Arg-PheNH_2, is cardioexcitatory in *M. nimbosa* and is neuroactive in other invertebrates. FMRFamide-Related Peptides, or FaRPs, have been identified in species other than molluscs. Immunoreactivity to FMRFamide is present in rats, chickens, cattle, frogs and teleost fish, in addition to numerous invertebrates. Since the discovery of the original FaRP, FMRFamide, FaRPs have become recognized as an important group of neuroactive peptides, particularly amongst invertebrates. The widespread occurrence of FaRPs in nematodes underlines their important role in nematode biology (Table 15.4). Fewer FaRPs have been isolated from vertebrates, and they possess a structure different from the nematode FaRPs, suggesting that nematode FaRP receptors may be potential anthelmintic targets.

All of the FaRPs identified in nematodes act on the neuromuscular system. The majority

TABLE 15.4 Some FMRFamides from nematodes

FaRP[1]	Amino acid sequence[2]	Occurrence[3]
AF1	KNEFIRF.NH_2	A
AF2	KHEYLRF.NH_2	A, C, H, P
AF3	AVPGVLRF.NH_2	A
AF4	GDVPGVLRF.NH_2	A
AF5	SGKPTFIRF.NH_2	A
AF6	FIRF.NH_2	A
AF7	AGPRFIRF.NH_2	A
AF8/PF3	KSAYMRF.NH_2	A, P
AF9	GLGPRPLRF.NH_2	A
AF10	GFGDEMSMPGVLRF.NH_2	A
AF11	SDIGISEPNFLRF.NH_2	A
AF12	FGDEMSMPGVLRF.NH_2	A
CF1/PF1	SDPNFLRF.NH_2	C, P
CF2/PF2	SADPNFLRF.NH_2	C, P
CF3	SQPNFLRF.NH_2	C
CF4	ASGDPNFLRF.NH_2	C
CF5	AAADPNFLRF.NH_2	C
CF6	PNFLRF.NH_2	C
CF7	AGSDPNFLRF.NH_2	C
	(K)PNFMRY.NH_2	
	APEASPFIRF.NH_2	C
	KPSFVRF.NH_2	C
PF4	KPNFIRF.NH_2	C, P
AF13	SDMPGVLRF.NH_2	A
AF14	SMPGVLRF.NH_2	A
AF15	AQTFVRF.NH_2	A
AF16	ILMRF.NH_2	A
AF17	FDRDFMHF.NH_2	A
AF19	AEGLSSPLIRF.NH_2	A
AF20	GMPGVLRF.NH_2	A

[1] The compounds were originally named with a letter to denote the species from which the peptide was isolated, F for FaRP, and a sequential number; thus AF1 was the first FaRP found in *A. suum*. As more peptides were discovered, often due to the use of different solvents for extraction, some were found to exist in more than one nematode species. For example, AF8 is the same as PF3; PF1 is the same as CF1; PF2 is the same as CF2.
[2] The amino acid sequence is shown in single letter notation.
[3] A, *Ascaris suum*; C, *Caenorhabditis elegans*; H, *Haemonchus contortus*; P, *Panagrellus redivivus*.

have been tested on somatic muscle preparations, but some also act on ovarian smooth muscle. A range of responses can be observed including relaxation, contraction, and in some cases a biphasic relaxation–contraction response. The inhibitory and excitatory effects of FaRPs have been demonstrated at nanomolar concentrations, indicating the high potency of these compounds. Some act on the nervous system only, e.g. AF8 (PF3); others are independent of the nerve cord (e.g. PF1 and PF4).

FaRP nomenclature

No specific criteria have been agreed on for designating a peptide as a FaRP; initially only short peptides with a C-terminal motif of Phe-X-Arg-Phe-NH_2 qualified (Figure 15.4). X was most often methionine, as in the original FaRP, FMRFamide. The group was expanded to include peptides where the Phe at position four from the C terminal was replaced with Tyr, another aromatic amino acid. Longer peptides were accepted as FaRPs, and eventually any peptide with the RFamide motif at the C terminal, or YFamide plus either F or Y at position four from the C terminal, have been classified as FaRPs. Thus LPLRFamide from chickens, FLFQPQRFamide (known as NPFF) and AGEGLSSPFWSLAAPQRFamide (known as NPAF) from cattle have all been designated FaRPs.

The shortened FaRP names, such as PF4 or AF2, indicate when and from where each FaRP was isolated. For example, PF4 was the fourth FaRP found in *Panagrellus redivivus*: P for *P. redivivus*, F for FaRP and 4 because it was the fourth. The 'A' in AF2 stands for *Ascaris suum*; AF2 was the second FaRP to be isolated from *A. suum*.

Nematode FaRPs

The first nematode FaRP (AF1) was isolated from *Ascaris suum* in 1989 by Cowden. Since then, numerous FaRPs have been identified in the *C. elegans* genome, and isolated from the free-living nematodes *Panagrellus redivivus* and *C. elegans* as well as *A. suum*. FaRPs play an important signaling role in the nematode nervous system: FMRFamide-like immunoreactivity (FLI) has been found in sensory and motor neurons throughout the nervous system. The distribution of a neurotransmitter gives some indication of its contribution to neural signaling. According to this argument, nematode FaRPs make a sizeable contribution to nematode neurophysiology. The distribution of FLI in the *A. suum* nervous system has been variously estimated from 'more than half' the neurons showing FLI, to 75% of neurons showing FLI. In contrast, only 10% of neurons in *C. elegans* have FLI. FLI may not be an appropriate indicator of FaRP abundance in nematodes. Antibodies raised against FMRFamide may not be specific for FaRPs, although careful use of experimental controls maximizes specificity. Ultimately, the unequivocal test for a nematode FaRP is the direct isolation of the FaRP from the worm.

FaRPs have been isolated from nematode tissues using acetone or acid methanol as extraction solvents. The choice of solvent dictates which FaRPs will be obtained from the tissue, basic peptides (e.g. AF1 and AF2) being more soluble in an acidic solution. Initially, it was thought that parasitic and free-living nematodes had mutually exclusive complements of FaRPs. This hypothesis was disproved when AF2, originally isolated from *A. suum*, was extracted from the free-living *P. redivivus* and *C. elegans* and the parasitic nematode *H. contortus*. Similaraly, a *P. redivivus* FaRP, PF3, has been found in *A. suum* and *H. contortus*, and in *C. elegans*.

Given the strong evidence for conservation of FaRP structure among nematodes, it is not surprising that FaRPs isolated from one species are active in another. *A. suum* is commonly used as an experimental model for testing the effects of neuroactive peptides. Experiments investigating the action of nematode FaRPs have shown that these peptides have potent effects on *A. suum* somatic muscle and nerve tissue (Figure 15.17). The application of PF1, PF2 and PF4 produced muscle relaxation, while application of PF3, AF3 and AF4 caused muscle contraction. The FaRPs AF1 and AF2 elicit a biphasic response consisting of relaxation followed by contraction. *Ascaris* tissue responds to experimentally applied FaRPs derived from both *P. redivivus* and *A. suum*. The wide distribution of some FaRPs (e.g. AF2 and PF3), and the cross-species efficacy of others (e.g. PF4), suggests conservation of FaRP receptors among closely related species. This observation is consistent with the recent phylogenetic work that shows that A. *suum* is equally closely related to the free-living nematodes *P. redivivus* and *C. elegans*.

Nematode FaRP structure

There are some marked similarities between the structure of the nematode FaRPs. All the FaRPs that have been isolated from nematodes to date are between 7 and 14 amino acids long. All C terminals end with the RF-NH_2 motif; the amino acid at position 3 from the C terminal is methionine, leucine, isoleucine or valine; the position 4 amino acid is frequently aromatic.

However, despite their similar structure, the nematode FaRPs show variation in function that must reflect their constituent amino acids. To this end, some structure–activity relationships have been investigated. The structure–activity relationships so far elucidated illustrate that rules relating peptide function to structure vary between FaRPs. The between-FaRP variation underlines the complexity of the peptide interactions and suggests that *in vivo* FaRP–receptor reactions are highly specific.

Substitution of amino acids in AF1 and AF2 has revealed that both the N- and the

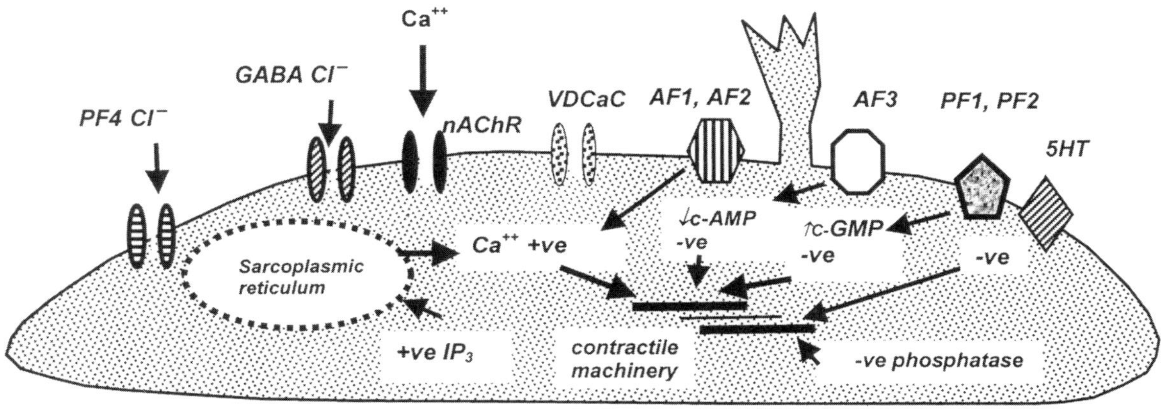

FIGURE 15.17 Diagram of the receptors that affect the contractility of *Ascaris* muscle. PF4 opens low (2–5 pS) conductance Cl^- channels. GABA opens higher conductance (20–40 pS) Cl^- channels. Acetylcholine opens 20–50 pS cation-selective channels. Depolarization opens voltage-dependent Ca^{2+} channels, VDCaC. AF1 and AF2 activate receptors that increase the excitability of muscle. AF3 activates receptors that decrease cAMP and also increases the excitability of muscle. PF1 and PF2 increase cGMP and decrease the excitability of muscle. 5-HT decreases the excitability of muscle.

C-terminals are essential for biological activity. The analog peptides had a similar or reduced level of potency to AF1, providing evidence for the existence of different receptors for AF1 and AF2. In PF1 and PF2 the N-terminal serine and aspartic acid residues are found to be unnecessary for biological activity. On the other hand, the phenylalanine and arginine residues (positions 1, 2 and 4 from the C-terminal) are essential. In PF4, leucine in place of isoleucine (position 3 from the C-terminal) did not affect the biological activity, and the proline (position 6 from the C-terminal) rendered the peptide resistant to enzymatic degradation. This proline also appeared to contribute to the biological activity of the molecule, probably due to its influence on tertiary structure.

For comparison, the structure–function relationship of the mammalian FaRP, NPFF, is presented. The two N-terminal amino acids are not essential for biological activity or receptor affinity. The remaining N-terminal amino acids were found to control molecular conformation and hence receptor affinity, while the C-terminal amino acids are necessary for biological activity. Hence, affinity for the receptor does not correlate with potency. In fact, substituted FMRFamides bound with lower affinity than FMRFamide, but had greater biological activity. These differences may be due to the enzyme-resistance of the FMRFamide analog, which could confer greater biological availability.

In a study of the molecular structure of related FaRPs from molluscs, it was observed that the number of turns in the molecule was inversely correlated with the receptor binding. Molecules that tended towards 'extended' (linear) conformations bind to receptors with higher affinity than FaRP molecules that tended towards a 'turn' (twisted) conformation. The turn conformation is most likely to be seen when an aspartic acid residue formed hydrogen bonds with the protonated amide groups of phenylalanine and leucine residues. Hence, the presence of aspartic acid conspicuously reduced the ability of the FaRP to bind to a receptor. This work demonstrates the important contribution of single amino acids to FaRP structure and, therefore, function.

FaRP mechanisms of action in nematodes

FaRPs can cause hyperpolarization or depolarization of *A. suum* muscle cells, and in some cases change the input conductance as well. Some of the nematode FaRPs have been investigated further to elucidate the ways in which signals from stimulated receptors are relayed to the effector mechanisms of the cell. The effects of PF1 and PF2 appear to be mediated by nitric oxide, and their muscle relaxant effect is blocked by compounds preventing nitric oxide synthesis. In contrast, AF1 and AF2 may act by increasing intracellular cyclic AMP, and AF3 by decreasing it. PF4 appears to open non-GABA chloride ion channels in the somatic muscle cell membrane.

Alternative splicing

Another feature of nematode genetics that is particularly relevant to the production of nematode FaRPs is alternative splicing. An example of alternative splicing in a FaRP-encoding gene has been documented in the nematode *C. elegans*. The gene *flp-1* can be transcribed in two different ways. Each transcript gives rise to eight FaRPs in *C. elegans*. Seven are the same in both sets; the transcripts differ in the eighth FaRP. One set has an extra copy of the FaRP SADPNFLRFamide (PF2), and the other has a copy of AGSDPNFLRFamide. It is postulated that the second FaRP may be age- or sex-specific.

Nematode FaRP genes

Twenty genes encoding 56 FaRPs have been identified in *C. elegans*. Of these, 13 FaRPs have been isolated. The *C. elegans* FaRP-encoding genes have been named *flp-1* to *flp-20*. *Flp-1*, *−6*, *−8*, *−9*, *−13*, *−14* and *−18* encode the FaRPs that have been extracted from *C. elegans*. Seven of the FaRP genes (in *C. elegans* and *A. suum*) encode multiple copies of their peptides, which is in marked contrast to the situation in mammals. The FaRP-encoding genes that have been found in vertebrates carry only one copy of each peptide.

So far, one FaRP-encoding gene (*afp-1*) has been identified in *A. suum*. It encodes six distinct FaRPs with the same C-terminal sequence, PGVLRFamide. This makes *afp-1* similar to, perhaps homologous with, *flp-18*. Of the FaRPs encoded by *afp-1*, three have been isolated from *A. suum*. The gene *afp-1* has single copies of each of the encoded peptides, like *flp-18*. Alternative splicing occurs in the transcription of *afp-1*, but unlike the process in *C. elegans*, the alternative transcripts of *afp-1* differ only in their untranslated regions (UTRs). The reason for alternative transcripts that vary in the UTRs only is not known; perhaps information contained in the UTRs may control gene expression by an unknown mechanism.

A directly-gated FaRP channel

FMRFamide, a molluscan cardioexcitatory peptide and the original FaRP, directly gates a sodium channel in the *Helix* C2 neuron. The FMRFamide-gated sodium channel has been cloned and expressed in *Xenopus laevis* oocytes. Its pharmacology and channel characteristics are similar to a superfamily of sodium channels, although sequence identity is low. The sodium channel superfamily is composed of epithelial sodium channels and degenerins, and has been named as the ENaCh/Deg superfamily.

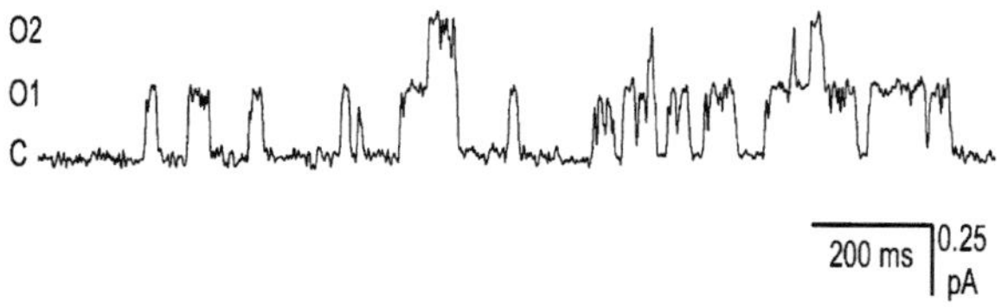

FIGURE 15.18 PF4-activated single-channel currents recorded from *Ascaris* somatic muscle. The channels have conductance of 3 pS and are small in amplitude.

The subunits of the FMRFamide-gated sodium channel show only 16% homology with the epithelial sodium channels and 13% with the degenerins. However, the protein structure of all three types of subunit is similar, having a large hydrophobic, membrane-spanning domain at each end of an extracellular loop, and one or two cysteine-rich regions. The FMRFamide-gated sodium channel subunits are arranged in groups of four, or tetramers, to form the functional channel. The other superfamilies of ligand-gated channels that have been characterized to date do not share these structural features. PF4 appears to directly gate a small conductance 2–5 pS Cl channel (Figure 15.18).

Diversity of FaRP receptors

It is clear that FaRPs are extensively involved in nematode muscle control. In view of the wide spectrum of functions mediated by FaRPs in nematodes, corresponding variation in nematode FaRP receptors would be predicted. The mechanisms of action of nematode FaRPs substantiate this prediction. Studies in other invertebrates demonstrate a variety of FaRP receptors linked to ion channels. It is likely that a similar range of FaRP receptors exists in nematodes, which would explain the range of mechanisms of action that are already known.

Davis and Stretton have proposed a tentative list of five major response types to guide the search for FMRF receptors in *Ascaris*. They examined the electrophysiological effects of the peptides on the dorsal excitatory motor neuron, DE2, and the dorsal inhibitory motor neuron, DI, and they examined the behavioral effects on injecting the peptides into the body. Their response types were:

1. Category 1: N-terminally extended – FIRF amide peptides (AF2, AF7 and AF1) produce strong behavioral effects
2. Category 2: Produce strong electrophysiological and behavioral effects (AF2, AF9 and AF8)
3. Category 3: Six peptides with a – PGVLR-Famide group which are encoded on the same gene, and AF11, which produce strong behavioral effects, but little effect on DE2 and DI
4. Category 4: Structurally unrelated AF peptides, with pronounced behavioral effects and similar electrophysiological effects on DE2 and DI. AF15 is excitatory but AF17 and AF19 are inhibitory on both the motor neurons
5. Category 5: Two peptides (AF6 and AF16) that are weak in their effect on behavior and electrophysiological effects.

Nitric oxide

As in vertebrates, the gas nitric oxide appears to be involved as a second messenger/transmitter in nematodes. The FMRFamides, PF1 and PF2, produce hyperpolarization of muscle membrane associated with flaccid paralysis. NO synthase is found in the hypodermis in twice the amount present in the muscle of *A. suum*, and NADH diaphorase, the NO-producing enzyme, has been demonstrated histochemically in neurons of *A. suum*. These observations support a role for NO in nematodes.

Summary

Despite the apparent anatomical simplicity of nematodes, with a cylindrical shape and a limited number of cells making up the nervous system and muscular system, there is considerable complexity revealed by the receptors for a range of classical and peptide transmitters. Except for glycine the classical transmitters of vertebrates appear to be present in all parasitic nematodes. The peptide transmitters, however, are not identical. The receptors for the classical and peptide transmitters in nematodes may have similarities to their vertebrate counterparts, but their pharmacology appears to be different. These differences can be exploited in the design of selective therapeutic agents.

NEUROTRANSMITTERS IN PLATYHELMINTHS

Introduction

In the previous section we have illustrated the connection between some important therapeutic agents and the receptors for neurotransmitters in nematodes. We described the presence of GABA receptors, nACh receptors, glutamate receptors and the effects of piperazine, levamisole, pyrantel and ivermectin. For the treatment of tapeworm and fluke parasite infestations, the economic pressure for the development of new therapeutic agents has not been as great, although major pathogenic parasites, including *Schistosoma mansoni*, cause chronic ill health in humans. Neither does the study of flatworms have model organisms like *C. elegans* to provide an added knowledge base. A further difficulty relating to the study of neurotransmitters in the study of flatworms is a greater diversity in the detailed neuronal arrangements of the different flatworms. For these reasons we know less about the structure and function of cestodes and trematodes and consequently less about the neurotransmitters of these parasites. Indeed, no single candidate that meets the classical criteria for inclusion as a neurotransmitter in flatworms has been identified. As with nematodes, we will consider first the organization of the nervous systems of the trematodes and cestodes.

The structure of the nervous system of trematodes and cestodes

The flatworm nervous system can be thought of as a melding of a complex central nervous system with a peripheral system of more primitive nerve nets, or plexuses. The primary features of the central nervous system are a bilobed anterior ganglion and two corresponding 'main' longitudinal nerve cords (Figure 15.19). Although a variable number of anterior-to-posterior nerve tracts can be present, there is most often a pair of predominant, or 'main' nerve cords. As the main nerve cords progress posteriorly, they are connected by a series of commissures, creating a ladder-like 'orthogonal' structure. The peripheral nervous system is composed of a number of nerve net-like plexuses, each a flat meshwork of nerve fibers confined to a limited area in the animal. Sub-epithelial, sub-muscular and intra-epithelial plexuses are present, as well as distinct plexuses associated with the suckers, the gut, the pharynx, and the reproductive structures. The cytology of the nervous system is consistent with this depiction of a complex brain joined to more primitive peripheral nerve nets, as the peripheral nervous system consists exclusively of multi- and bi-polar cells which are thought to be more ancient, while the central nervous system ganglia often feature more specialized unipolar cells.

The body structure of flatworms is without celomic cavities and does not contain a circulatory system, such as blood vessels or lymphatics. The distribution of regulatory hormones that stimulate growth and generalized tissue responses is presumed to occur via the nervous system. The nervous system therefore serves a neurosecretory function as well as the role of direct and more rapid control of tissue responses.

It is remarkable how little is actually known about the function of the nervous system in flatworms. The worms themselves are small and the individual cells are quite small and tightly packed, making rare any flatworm preparations where one can study the nervous system directly. Further, in the case of parasitic worms, the observation of *in situ* behavior is difficult at best, and the effects of isolation of the parasites from their hosts is largely undefined. Consequently, there are very few data elucidating how flatworm nervous systems function, or the role of the nervous system in various physiological processes or behaviors. Largely unknown are relationships between the central and peripheral nervous systems, and their relative roles in controlling various aspects of behavior. Free-living flatworms are capable of coordinated locomotion and feeding responses in the absence of CNS input, demonstrating that much of the neural control of basic flatworm behavior occurs in the peripheral plexuses. Similarly, in a number of parasitic flatworms, removal or damage to the central nervous system has little observable effect on the worms. Most of what is known about the roles of putative neurotransmitters in flatworms is focused on the somatic

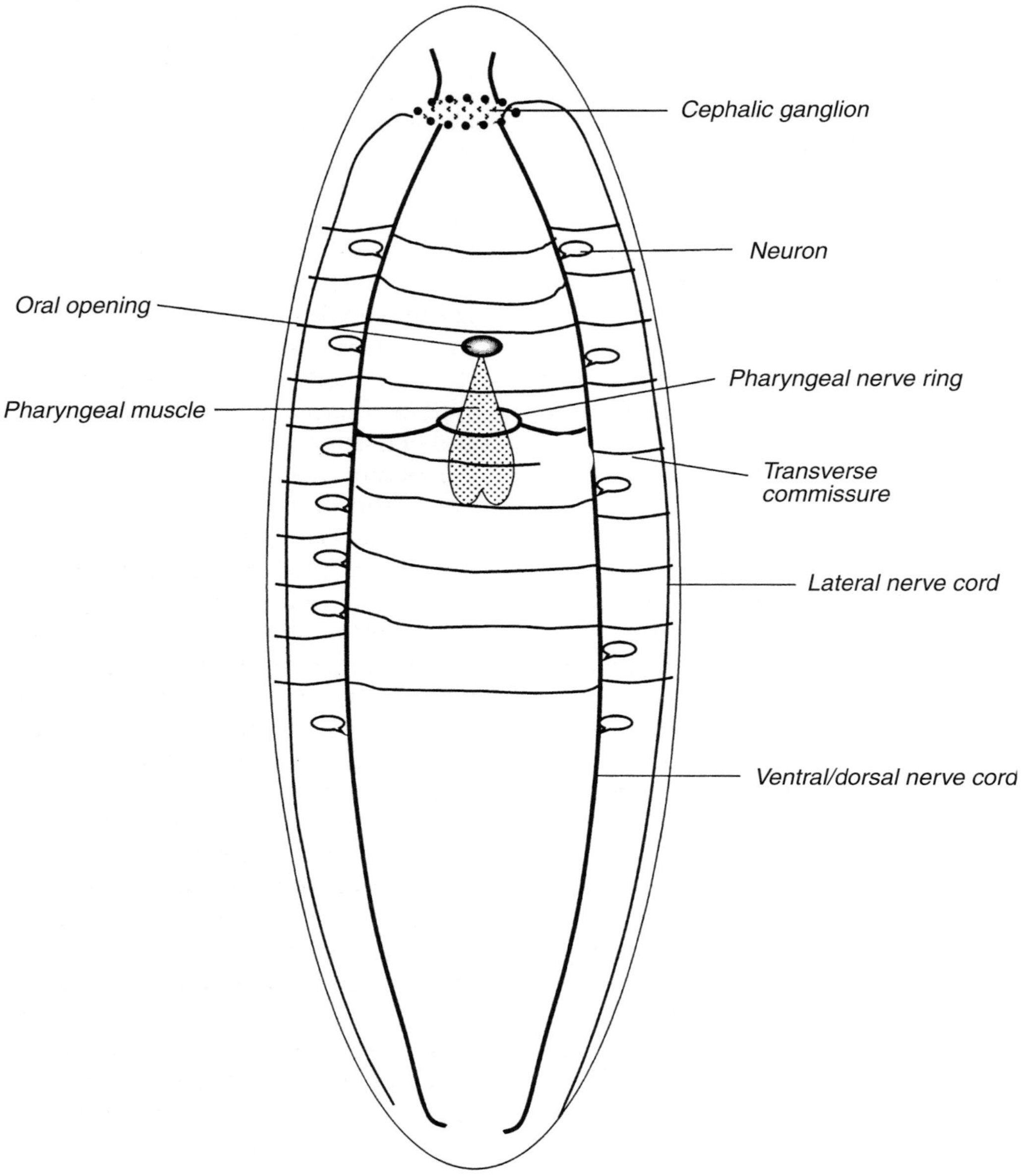

FIGURE 15.19 Diagram of the main features of the nervous system of trematodes and cestodes.

muscle system, simply because this has been the most amenable to bioassay.

Putative neurotransmitters in trematodes and cestodes

Acetylcholine

Acetylcholine has been a candidate inhibitory neurotransmitter in flatworms since the observation that cholinergic compounds induce flaccid paralysis in *Fasciola hepatica*. Following this early demonstration of acetylcholine's inhibitory action on the liver fluke, there have been numerous other studies showing the inhibitory action of acetylcholine on the musculature of trematodes and cestodes. Supporting these physiological observations, histochemical staining for acetylcholinesterase reveals a striking outline of the majority of flatworm nervous systems, including the entire central nervous system and most of the peripheral nerve plexuses. Antibodies targeting a choline conjugate support the widespread distribution suggested by acetylcholinesterase in the tapeworm *Hymenolepis diminuta* and the fluke *S. mansoni*.

The inhibitory actions of acetylcholine on flatworm musculature have been demonstrated on whole animal and muscle strip preparations derived from a number of different cestode and trematode species, including *H. diminuta*, *Dipylidium caninum*, *S. mansoni* and *F. hepatica*. Since these studies use either whole animal or muscle strips, they leave questions as to the location of the receptors mediating the inhibitory response. The response could be mediated by presynaptic neuronal receptors, receptors on the muscle itself, or receptors located in many other places in the preparation. Studies on muscle fibers isolated from schistosomes demonstrate that there are inhibitory acetylcholine receptors located on the muscle membranes; in these muscle fibers, without neural or paracrine input, acetylcholine potently inhibits the excitatory responses mediated by neuropeptides.

Another site of acetylcholine receptors in schistosomes appears to be the outer surface of the parasite, at the host–parasite interface. The role of these acetylcholine receptors in the biology of the parasite remains unclear, as do the molecular and biochemical nature of the tegumental receptors. A comparison of receptor-binding studies of the tegumental receptors and physiological contraction assays suggest that the tegumental receptors are pharmacologically different from the muscle receptors.

The pharmacology of flatworm acetylcholine receptors does not permit their classification into known subtypes. This is particularly puzzling, and some responses are associated with muscarinic (G-protein-coupled) acetylcholine receptors and some with nicotinic (intrinsic ion channel) acetylcholine receptors. Muscarinic and nicotinic cholinergic receptors are structurally not very closely related to each other, and the ambiguous pharmacology is unusual. To provide one example, the muscarinic agonists arecoline and carbachol are quite effective inhibitors of flatworm myoactivity (but muscarine itself is without effect), while the nicotinic antagonist α-bungarotoxin blocks the cholinergic action. Further, discrepancies often appear when trying to correlate results obtained using different preparations. For example, in schistosomes nicotine reduces muscle tone and myoactivity in whole animals, but has no inhibitory effect on isolated muscle fibers.

Another notable observation amplifying some of the uncertainty about the role of acetylcholine in flatworms is the excitatory action of acetylcholine on muscle fibers from the free-living flatworm *Bdelloura candida*. This myoexcitation is associated with an acetylcholine-induced inward current, indicating the presence of a nicotinic-like acetylcholine receptor on these muscles. As mentioned previously, acetylcholine effects on other flatworm musculature has been predominantly inhibitory.

The actions of acetylcholine on flatworm musculature are described most frequently as inhibitory, but the nature of the receptors mediating these responses is not clear. Cholinergic receptors on the tegument of schistosomes are also present, but the nature and function of these receptors also remain unknown.

Cholinergic compounds have been used therapeutically for the treatment of cestode infestations in the dog and horse. The selective muscarinic agonist, arecoline, is an old and still effective treatment for *Echinococcus granulosus* and *Dicrocoelium dendriticum* in the dog, but pilocarpine, another muscarinic agonist, is without effect on these cestodes. Although the effects of arecoline have been known for a long time, treatment is not without side-effects on the host. The muscarinic effects produce uncomfortable colic in the host as a result of the stimulation of peristalsis. The anti-nematodal drug, pyrantel, also has effects against tapeworms in horses. The effects of arecoline and pyrantel are further confirmation of the presence of cholinergic receptors in cestodes, and the anti-cestodal pharmacology confirms the unique nature of the receptors; atropine is not effective as an antagonist, and pilocarpine is not effective as an agonist like arecoline. The effect of cholinergic compounds as therapeutic agents against parasitic trematodes has not been demonstrated, and may be limited by the distribution of the drug following oral administration and the ability of the parasite to survive in a paralyzed state until the drug is cleared from the host. Parasites in the gastrointestinal tract may find it more difficult to survive such paralysis, because the continuous peristalsis would tend to drive the parasite out of the intestine. Such effects may not be observed in the bile ducts if cholinergic agonists were applied to *F. hepatica*. Below, three cholinergics with anti-cestodal activity are considered.

Pyrantel

Pyrantel, a cholinergic agonist more normally used in animals for the treatment of gastrointestinal nematode infections, is effective for the treatment of some cestode infections in the intestine. It is licensed for the treatment of *Anoplocephalia perfoliata* in horses. Effects against other cestodes and trematodes have not been described.

Arecoline

Arecoline (Figure 15.20) is an old treatment for cestode infections that is effective against *Echinococcus granulosus* and *Dipylidium caninum* in dogs. Arecoline stimulates the cholinergic receptors on the muscle of the scolex and body to produce paralysis. Arecoline also has a purgative effect on the host animal, due to its muscarinic action. The combination of purgation and paralysis of the tapeworm is effective. The purgation is severe in some animals, and so newer therapeutic agents are preferred.

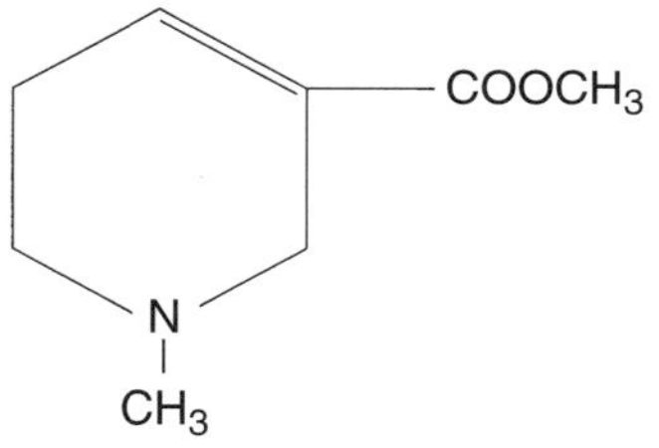

FIGURE 15.20 Chemical structure of arecoline.

Uredofos and metriphonate
The cholinesterases of cestode and trematode parasites may be antagonized selectively with less effect on the host. The partial selectivity of the organophosphorus drugs like uredofos and metriphonate has led to the use of these compounds for treatment. Uredofos was introduced in dogs and cats for the treatment of *Dipylidium caninum* but the compound has now been withdrawn because of toxicity. The design and use of more selective cholinesterases may be one approach for the development of novel anthelmintic agents.

Glutamate

A number of lines of evidence suggest a role for glutamate as a neurotransmitter in flatworms. Glutamate is one of the most abundant free amino acids in parasitic flatworms, and glutamate immunoreactivity has been described in the cephalic ganglia and longitudinal nerves, as well as in the peripheral nerve nets of the peripheral tissues, including somatic muscle.

In addition to the immunocytochemistry, there is convincing physiological evidence supporting a role for glutamate as a neurotransmitter. Tissue slices of the tapeworm *H. diminuta* take up glutamate through two distinct Na^+-dependent transport systems, and depolarization elicits a calcium-dependent release of glutamate. Exogenous glutamate elicits powerful rhythmic contractions of longitudinal muscle preparations, and it has biphasic effects on cAMP levels in *H. diminuta* homogenates. Glutamate or aspartate increase electrical activity in the nerve cords of the cestodarian *Gyrocotyle fimbriata.* The non-selective glutamate receptor antagonist 2-amino-4-phosphonobutyric acid antagonized the effect of the excitatory amino acids, suggesting an excitatory role for glutamate in these flatworms. In schistosomes, exogenous glutamate is also myoexcitatory. Surprisingly, at least some of the myoexcitation is mediated by electrogenic glutamate transporters located on the muscle membranes. The application of glutamate to isolated muscle fibers induces contraction. These contractions are not blocked by glutamate receptor antagonists, but they are blocked by glutamate transport inhibitors such as *trans*-pyrollidine-dicarboxcylic acid. Like some glutamate transporters, the contractions are dependent on extracellular Na^+.

GABA

GABA (and glycine) have inhibitory effects on the discharges of the lateral nerve cords of *Gyrocotyle fimbriata.* This effect is reversed by the GABA antagonists picrotoxin and bicuculline. However, these observations are only suggestive of a possible general role of GABA, since other authors have reported no effect of GABA in intact *S. mansoni.* Thus the role of GABA as a neurotransmitter in trematodes and cestodes remains to be established.

5-HT

Bennett and Bueding showed that 5-HT is present in the nervous system of both trematodes and cestodes, and in both classes the distribution is widespread throughout the central nervous system and the peripheral plexuses. One broad theme observed in flatworms is that 5-HT distribution is often distinct from the largely overlapping cholinergic and peptidergic staining patterns. 5-HT is the most studied of the flatworm neurotransmitter candidates, and a number of biochemical and physiological actions have been reported. 5-HT stimulates adenylate cyclase activity in homogenates of trematodes and cestodes, leading to the phosphorylation of a number of proteins. In trematodes, phosphofructokinase is phosphorylated, leading to a measurable increase in glucose utilization and lactate production.

5-HT produces increased motor activity in a wide range of flatworms, including important cestodes such as *D. caninum* and *H. diminuta*, as well as the trematodes *S. mansoni* and *F. hepatica*. When applied to intact animals and muscle strips, exogenous 5-HT has dramatic myoexcitatory effects on all flatworms that have been examined. Application of 5-HT to *S. mansoni* in high-Mg^{2+} solutions produces contraction, suggesting that the receptors responsible for the muscle contraction are present on the muscles themselves, rather than on the nerve cells. 5-HT has a direct effect on muscle fibers that have been dispersed from flatworms, confirming the muscle membranes as one location of 5-HT receptors in flatworms. These studies also reveal possible differences in the nature of 5-HT myoexcitation in different flatworms. 5-HT alone does not induce contraction of individual schistosome muscle fibers; instead it modulates the contractility of the fibers to other excitatory agents, such as depolarization or FMRFamide-related peptides. However, muscle fibers derived from free-living flatworms contract in response to 5-HT alone. Thus, although 5-HT myoexcitation is uniform throughout the flatworms, the exact nature of 5-HT-induced myoactivity remains unclear and may differ among the flatworms.

The molecular nature of the receptors mediating these biochemical and physiological effects remains obscure. The pharmacology does not allow for a tidy classification of the receptors into the subtypes derived from vertebrate pharmacology. In schistosomes, where there are the most data, the pharmacological profile associated with the biochemical effects is very similar to that associated with the physiological effects. Using either adenylate cyclase activity in homogenates, or the isolated muscle contraction assay, the order of 5-HT agonist potency is the same: methiotepin > metergoline > ketanserin. Further, the effect of 5-HT on the muscle fibers is mimicked by the adenylate cyclase activator forskolin and blocked by the protein kinase inhibitor H89. All these data point toward a common pathway for the activation of adenylate cyclase in homogenates and the myoexcitatory activity.

5-HT was the first viable candidate for a position as a myoexcitatory neurotransmitter candidate in flatworms, and it remains such. The wide distribution of 5-HT in flatworms suggests that it has physiological roles beyond muscle excitation. Simply, the availability of contraction assays and the lack of other meaningful biological assays have kept the focus on myoexcitation by 5-HT. A fuller understanding of the role of 5-HT in parasitic flatworms awaits more relevant bioassays, as well as the molecular identification and localization of flatworm 5-HT receptors.

Dopamine, noradrenaline and adrenaline

Dopamine receptors of vertebrates can be divided pharmacologically into D_1 and D_2 subtypes. Similar subtypes are suggested to be present in some trematodes, since D_1 agonists produce a turning–screwing hypermotility, while D2 agonists produce an arching–curling hypermotility. D1 antagonists block the turning–screwing hypermotility associated with the D1 agonist effects, and the D2 antagonists depress the effects of the D2 agonists that produce the arching hypermotility. The concentration of dopamine in neurons in planarians is reduced by haloperidol. In *Fasciola hepatica* noradrenaline is found in the cephalic region, and noradrenaline depresses motor activity in the fluke. Dopamine stimulates motor activity in *Fasciola hepatica*, but in *Schistosoma mansoni* dopamine produces an inhibitory effect on longitudinal and circular muscles.

Other putative neurotransmitters

In addition to the candidates considered thus far, there are a number of other compounds which may be neurotransmitters in flatworms, some examples being nitric oxide and histamine. In each of these cases, the information about the presence and role of these compounds is quite sketchy, so they will not be considered in great detail here.

A role of nitric oxide (NO) as a diffusible transmitter is suggested by NADPH diaphorase staining, which has been reported in the nervous systems of many flatworms, including the trematode *S. mansoni* and the cestodes *H. diminuta*, *D. dendriticum* and *Mesocestoides corti*. The staining is prominent in the nerve cords in each case, and sometimes includes nerves that appear to serve sensory structures in the tegument. Nitric oxide synthase (NOS) activity has been detected in homogenates of the tapeworm *H. diminuta*. In addition to the possibility that NO plays a role in worm physiology, there is the possibility that NO has a role in the host–parasite relationship. One response of schistosome-infected hosts might be to stimulate inducible NOS via IFN-γ, and that this might lead to the elimination of lung-stage somules. Conversely, it has been suggested (without significant supporting data) that schistosomes might use NO to alter the vasculature of the host.

The data regarding the presence of histamine in various flatworms do not create a clear pattern, with some worms apparently lacking histamine altogether and others with seemingly high concentrations. One puzzling example is the tapeworm *H. diminuta*, which allows exogenous histamine to diffuse across the outer surface, but does not itself synthesize histamine. Histamine has no (or very little) observable effect in most of the flatworm bioassays. One exception to this is *S. mansoni* cercariae, where H1 receptor antagonists immobilize the free-swimming stages, and high concentrations of exogenous histamine can block the immobilization. Therefore, the evidence that histamine serves as a neurotransmitter in platyhelminths is still rather scanty.

FMRFamides

Attention has recently focused on neuropeptides as neurotransmitter candidates in platyhelminths. This attention originated with studies in the 1980s demonstrating immunoreactivity to a plethora of antisera raised against neuropeptides from other animals. Amongst a number of potential neurotransmitters, the most data have accumulated supporting a role for short amidated peptides similar to the molluscan cardioexcitatory peptide Phe-Met-Arg-Phe-amide (using single letter annotation, FMRFamide). These peptides will be collectively referred to here as FaRPs, FMRFamide-related peptides.

Extensive FaRP immunoreactivity is present throughout the central and peripheral nervous system of every flatworm thus far examined (more than 40 species), including the plexuses associated with the somatic musculature, the reproductive structures, the holdfast organs and the alimentary system. The immunoreactivity suggests that FaRP distribution largely mirrors that of acetylcholine.

Despite the widespread distribution of FaRP immunoreactivity, the structural identification of FaRPs in parasitic flatworms has been difficult. The largest impediment to biochemical purification has been amassing sufficient quantities of tissue. Also, in comparison to many other animals, the levels of FaRPs in flatworms appears to be much lower, which may simply be a reflection of the lower nerve to body-mass ratio of platyhelminths. These limitations have dictated the species from which FaRPs have been structurally characterized. The only parasitic flatworm from which sufficient tissue has been gathered is the large sheep tapeworm *Monesia expansa*. This cestode yielded the first flatworm FaRP sequence, GNFFRFamide. All subsequent flatworm FaRPs have been identified from free-living flatworms, where tissue is more readily available: RYIRFamide from the predatory terrestrial turbellarian *Artioposhtia triangulata*, GYIRFamide from the freshwater planarian *Girardia tigrina*, and both GYIRFamide and YIRFamide from *Bdelloura candida*, an ectocommensal turbellarian of the horseshoe crab. To date, these four sequences are the only structurally characterized FaRPs from flatworms.

Flatworm-derived FaRPs are myoexcitatory in preparations from cestodes and trematodes, as well as in free-living flatworms. All three of the FaRPs derived from free-living flatworms contain the YIRFamide motif, and all are potently excitatory to muscle strips or isolated muscle fibers derived from free-living flatworms and trematodes. The tapeworm FaRP, GNFFRFamide, also has some activity in trematode muscle preparations, but it is much less potent on every preparation examined. One example is the dispersed muscle preparation from *S. mansoni*, where the half-maximal effects of YIRFamide, GYIRFamide and RYIRFamide are somewhere between 1 and 7 nanomolar, while GNFFRFamide is 500 nanomolar. The higher potency of peptides derived from free-living flatworms on trematode muscle is not surprising, since the trematodes are descended from and more closely related to the free-living flatworms than they are to the cestodes.

Conversely, the cestode-derived GNFFRFamide is markedly more potent than the YIRFamide peptides on cestode muscle. For example, GNFFRFamide is much more potent than YIRFamide on muscle strips from *M. expansa*, and it more potently stimulates motility in larval *Mesocestoides corti*.

It is important to stress that the focus of the existing data is on the myoactivity of FaRPs in flatworms, mostly because bioassays of somatic muscle are presently available. FaRP distribution throughout the reproductive structures also suggests that these neuropeptides might play a role in reproductive function in flatworms. In fact, FaRP immunoreactivity increases in the reproductive system innervation at times of reproductive activity in the monogenean *Polystoma nearcticum*. Nerves associated with the egg chamber express FaRPs during periods of egg production, but not during reproductively quiescent periods.

Given their widespread distribution and potent myoexcitatory activity in every flatworm examined, FaRPs are a serious candidate for the role of excitatory neuromuscular transmitter in the phylum, including both the parasitic trematodes and cestodes. It is also likely that FaRPs play other neurotransmitter roles in flatworms, based on their distribution in many of the specialized peripheral plexuses. The immediate challenges are to determine the precise structure of the FaRPs present in parasitic worms, and then to identify the receptors associated with these ligands.

CONCLUSION

In this chapter we have reviewed neurotransmitters of nematodes and platyhelminths, recognizing that one of the pressures for the development of new knowledge in this field is the requirement of new anti-parasitic drugs that is forced by the development of anthelmintic resistance. It is remarkable that the nematodes and platyhelminths seem at first glance to be so simple, but greater complexity is revealed as we study their neurotransmitters. The study of the parasite neurotransmitters and their receptor sites seems important for the development of new drugs, but is an interesting biological topic in it own right.

ACKNOWLEDGMENT

RJM and APR are supported by NIH: R01 A147194-01A1.

FURTHER READING

Aceves, J., Erliji, D. and Martinez-Marnon, R. (1970). The mechanism of the paralysing action of tetramisole on Ascaris somatic muscle. *Brit. J. Pharmacol.* **38**, 602–607.

Arena, J.P. (1994). Expression of *Caenorhabditis elegans* messenger-RNA in *Xenopus* oocytes: a model system to study the mechanism of action of avermectins. *Parasitol. Today* **10**, 35–37.

Byerly, L. and Masuda, M.O. (1979). Voltage-clamp analysis of the potassium current that produces a negative-going action potential in *Ascaris* muscle. *J. Physiol.* **288**, 263–284.

Day, T.A. and Maule, A.G. (1999). Parasitic peptides! The structure and function of neuropeptides in parasitic worms. *Peptides* **20**, 999–1019.

Dent, J.A., Davis, M.W. and Avery, L. (1997). Avr-15 encodes a chloride channel subunit that mediates inhibitory glutamatergic neurotransmission and ivermectin sensitivity in *Caenorhabditis elegans*. *EMBO J.* **16**, 5867–5879.

Dent, J.A., Smith, M.M., Vassilatis, D.K. and Avery, L. (2000). The genetics of ivermectin resistance in *Caenorhabditis elegans*. *Proc. Natl. Acad. Sci. USA* **97**, 2674–2679.

Geary, T.G., Marks, N.J., Maule, A.G. *et al.* (1999). Pharmacology of FMRFamide-related peptides in helminths. *Annals NY Acad. Sci.* **897**, 212–227.

Jarman, M. (1959). Electrical activity in the muscle cells of *Ascaris lumbricoides*. *Nature* **184**, 1244.

Mansour, T.E. (1979). Chemotherapy of parasitic worms: new biochemical strategies. *Science* **205**, 462–469.

Martin, R.J. (1985). Gamma-aminobutyric acid-activated and piperazine-activated single-channel currents from *Ascaris suum* body muscle. *Brit. J. Pharmacol.* **84**, 445–461.

Martin, R.J. (1995). An electrophysiological preparation of *Ascaris suum* pharyngeal muscle reveals a glutamate-gated chloride channel sensitive to the avermectin analogue milbemycin D. *Parasitology* **112**, 252.

Martin, R.J., Robertson, A.P., Bjorn, H. and Sangster, N.C. (1997). Heterogeneous levamisole receptors: a single-channels study of nictotinic acetylcholine receptors from *Oesophagostomum dentatum*. *Eur. J. Pharmacol.* **322**, 249–257.

Maule, A.G., Halton, D.W., Shaw, C. and Johnston, C.F. (1993). The cholinergic, serotoninergic and peptidergic components of the nervous system of *Moniezia expansa* (Cestoda, Cyclophyllidea). *Parasitology* **106**, 429–440.

Richmond, J.E. and Jorgensen, E. (1999). One GABA and two acetylcholine receptor function at the *C. elegans* neuromuscular junction. *Nature Neurosci.* **19**, 196–199.

Robertson, A.P., Bjorn, H. and Martin, R.J. (1999). Levamisole resistance resolved at the single-channel level. *FASEB J.* **13**, 749–760.

Rosoff, M.L., Burglin, T.R. and Li, C. (1992). Alternatively spliced transcripts of the flp-1 gene encode distinct FMRFamide-like peptides in *Caenorhabditis elegans*. *J. Neuroscience* **12**, 2356–2361.

Sangster, N.C. (1994). P-glycoproteins in nematodes. *Parasitol. Today* **10**, 319–322.

Stretton, A.O.W., Fishpool, R.M., Southgate, E. *et al.* (1978). Structure and physiological activity of the motoneurons of the nematode Ascaris. *Proc. Natl. Acad. Sci. USA* **75**, 3493–3497.

SECTION

IV

MEDICAL APPLICATIONS

C H A P T E R

16

Drug resistance in parasites

Marc Ouellette[1] *and Steve A. Ward*[2]

[1]Centre de Recherche en Infectiologie, Université Laval and Centre Hospitalier Universitaire de Québec, Québec, Canada; and

[2]Liverpool School of Tropical Medicine, Liverpool, UK

INTRODUCTION

Parasitic protozoa and helminths are responsible for some of the most devastating and prevalent diseases of humans and domestic animals. Protozoan parasites threaten the lives of nearly one-third of the worldwide human population, and are responsible for the loss of more than 50 million disability-adjusted life years (DALYs) and more than 2 million deaths a year (www.who.int). The parasites responsible for malaria (*Plasmodium* spp.) contribute most to DALYs and death, but the parasites responsible for the various forms of leishmaniasis (*Leishmania* spp.), African sleeping sickness (*Trypanosoma brucei gambiense, Trypanosoma brucei rhodesiense*), and Chagas disease (*Trypanosoma cruzi*) are also important contributors to global morbidity and mortality figures. Anaerobic parasitic diseases such as *Entamoeba histolytica, Trichomonas vaginalis* (vaginitis, urethritis) and *Giardia duodenalis* (diarrhea) as well as worms responsible for lymphatic filariasis, schistosomiasis and onchocerciasis also contribute significantly to the world's disease burden.

Parasitic protozoa and helminths result in considerable losses of life and productivity in humans and domesticated animals. Ideally, prevention would be the most efficient way to control infectious agents but despite considerable efforts, there are no effective vaccines against any of the clinically important parasites. For the moment, drugs are therefore the mainstay in our control of parasitic protozoa when simple prevention measures fail or prove impractical. Protozoa share with their hosts a number of cellular organelles and metabolic pathways, but they nonetheless have a number of unique and exotic biochemical pathways that should provide many exploitable targets for the development of parasite-specific drugs. Proof of this concept includes the effective use

Molecular Medical Parasitology
ISBN 0-12-473346-8

of some of the most successful drugs ever synthesized, such as chloroquine in the treatment of malaria, and ivermectin in the treatment of onchocerciasis. In stark contrast to this, many of the drugs directed against parasites have far from optimal pharmacological properties, with narrow therapeutic indices and limiting host toxicity. The arsenal of anti-protozoal drugs is thus limited, and this is exacerbated by the emergence of drug resistance. With the development of new drugs proceeding slowly, the emergence of drug resistance in parasitic protozoa and helminths is becoming a major public health problem, and several parasitic diseases were recently included in the World Health Organisation's list of diseases where antimicrobial resistance is a major issue (Chapter 17) (www.who.int/infectious-diease-report/2000/ch4.htm). In this chapter we will attempt to give an overview of the drug resistance mechanisms found to operate in parasites and helminths. We will attempt to focus on human parasites and emphasize resistance mechanisms as characterized in field isolates.

GENERAL MECHANISMS OF RESISTANCE

Parasites, like other living cells, can evade drug action by a number of diverse and elegant biochemical mechanisms. The general biochemical frameworks of resistance are often similar. Cells may evade drug action by hiding in sanctuary sites, by modifying drug uptake systems or alterating membrane composition, thereby thwarting drug uptake. Once inside the parasite, drugs may be inactivated, excreted, chemically modified to facilitate excretion, or routed into compartments away from the target site. Drug activation mechanisms may be suppressed or lost. The interaction of the drug with the target may be made less effective by increasing the level of competing substrates or by altering the target to make it less sensitive to the drug. Finally, the cell may learn to live with a blocked target by bypassing the block (Figure 16.1).

Resistance mechanisms have been studied mostly in parasites selected for resistance in the laboratory, or in a few cases, were deduced from work carried out in clinical isolates. Work on *in vitro* selected cell lines is justified. One can work with genetically defined cloned populations. Furthermore, by a detailed comparison of resistant mutants and the parental strain one can usually determine, or at least get important insights into, the resistance mechanisms. Resistance in clinical samples is less easily defined: parasite populations are often heterogeneous and the parent strain is not available for comparison. A further complication is that in the process of growing adequate amounts of material for analysis, the parasite resistance phenotype can change or a parasite subpopulation, which is not representative of the resistant population, expands to become the major genotype. It should nonetheless be possible to verify whether resistance mechanisms defined in the laboratory play a role in the field. This can be done by using sensitive tools, such as the polymerase chain reaction and monoclonal antibodies, to analyse potential resistance mechanisms in the few parasites that can be obtained from infected individuals.

In the following sections we will concentrate on the mechanisms of resistance against the main anti-parasitic drugs used against some of the most prevalent parasites. In particular, we will review the biochemical and molecular characterization of drug resistance in protozoa and helminths. We will also describe the possible implications of our increased understanding of drug resistance for the improved management and treatment of parasitic diseases.

DRUG RESISTANCE IN APICOMPLEXAN PARASITES

Plasmodium and chemotherapy

Of all parasitic diseases, malaria is at the forefront in terms of its ability to inflict devastating effects on the human population. In fact it has been eloquently argued that malaria represents one of the most profound influences in the shaping of the human genome, so devastating is its impact on human survival. Despite the optimistic ambitions of the 1960s to completely eradicate malaria, it is now clear that eradication is not realistically achievable. Furthermore, the gradual erosion in the effectiveness of all available control measures has seen a resurgence in the incidence of malaria worldwide.

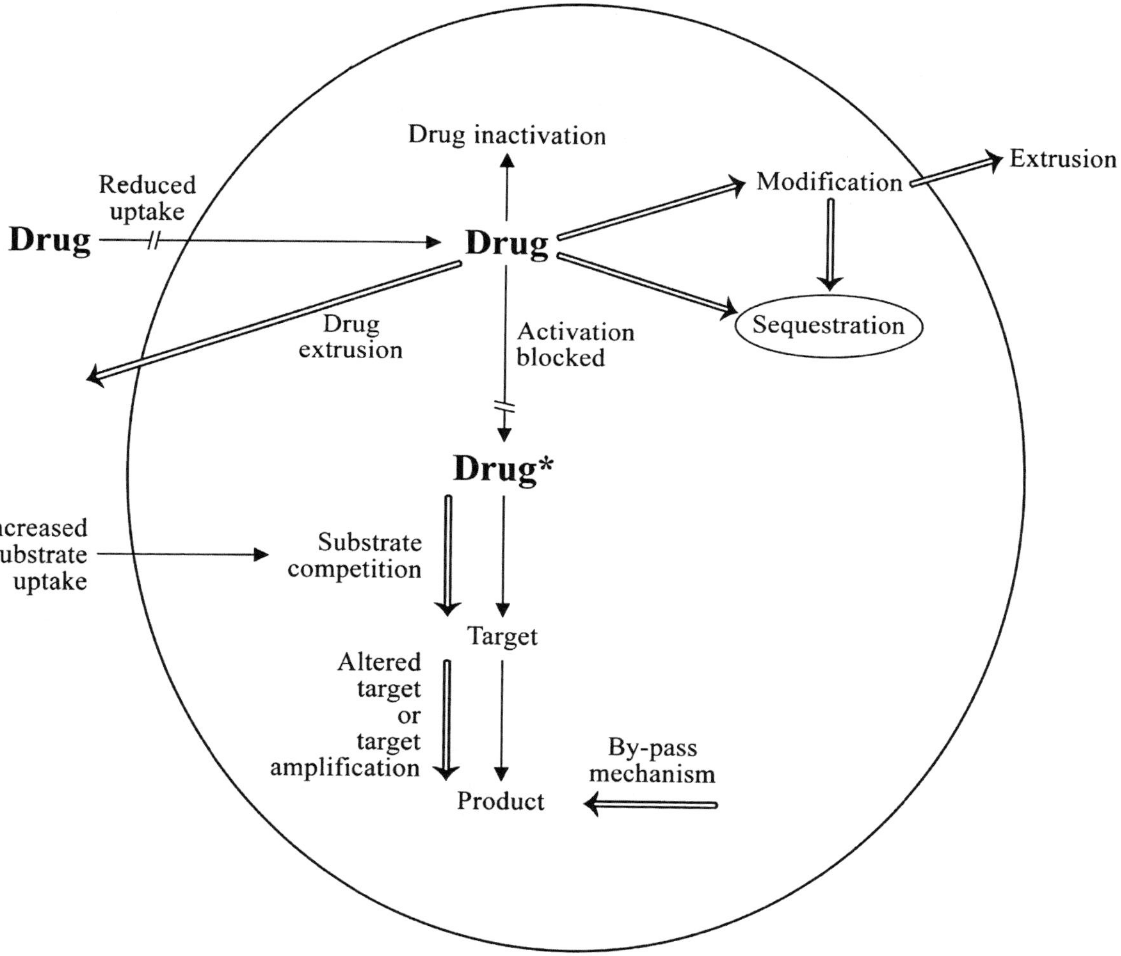

FIGURE 16.1 Main biochemical drug resistance mechanisms. A simplified scheme of drug-resistance mechanisms found in parasites is illustrated. Specific examples of each type of mechanism are discussed in the text.

It is now estimated that almost half of the world population is at risk from malaria. This risk translates into an annual incidence rate of some 400 million clinical cases and between 1.5 and 2.7 million deaths. Almost all of these deaths are due to infection with *Plasmodium falciparum*. Moreover the burden of this mortality is seen in sub-Saharan Africa with young children and pregnant mothers, deficient in an adequate immune response to infection, being by far the most at-risk groups.

Efforts to control malaria by targeting the mosquito vector continue, but represent a measure that at best may reduce malaria incidence. Malaria vaccines may offer a long-term solution to this disease but as yet a suitable vaccine remains many years away. Thus our principal tool in the control and treatment of malaria is chemotherapy. The practice of malaria chemotherapy has been in operation over many centuries. Despite this, as we enter the twenty-first century we find ourselves with only a handful of anti-malarial drugs with potential clinical utility. Moreover, parasite resistance to almost all of these drugs has been identified in wild parasite populations present in endemic areas of the world. As with all parasitic diseases targeting resource-poor areas of the world, the situation is exacerbated by issues of drug affordability and drug access.

Drug resistance in malaria is a major challenge. Failure to understand the factors contributing to resistance development in the field and failure to elucidate the basic biological mechanisms of resistance will have a devastating impact on our ability to deal with this life-threatening infection.

Chloroquine

The 4-aminoquinoline chloroquine (Figure 16.2) was introduced into clinical medicine in the 1940s and was the mainstay of malaria control efforts until the development and spread of resistance to the drug. Despite a general perception to the contrary there remain areas of the world where chloroquine is still effective. Furthermore, chloroquine remains the first choice anti-malarial drug in many parts of the world even where its clinical effectiveness has been seriously eroded. Against a parasite-susceptible backdrop, chloroquine's success can be explained in terms of its affordability, safety (when used as indicated), ease of use and effectiveness. Chloroquine's principal target is the intraerythrocytic feeding trophozoite stage.

The concept of chloroquine resistance was not generally appreciated until some 15–20 years after the drug's clinical introduction. It was not until the late 1950s that reports of clinical failure of chloroquine were reported independently from SE Asia and South America. Since these initial reports chloroquine resistance has gradually spread to encompass all malaria endemic regions of the world, with the late 1970s marking the period when resistance reached East Africa. In order to fully understand chloroquine resistance it is important to recognize that resistance acquisition took many years to develop. Even then it only occurred in a limited number of geographical foci with their own transmission characteristics and chloroquine treatment experiences. This contrasts with drugs such as pyrimethamine, where resistance has developed quickly and independently on a number of occasions. This difference must reflect the relative complexity of the chloroquine resistance mechanism compared to the simple point mutations in *dhfr* that are associated with reduced pyrimethamine sensitivity.

Parasites are generally referred to as chloroquine-resistant or chloroquine-sensitive. This definition may be acceptable from a clinical stance, but from a biological perspective it fails

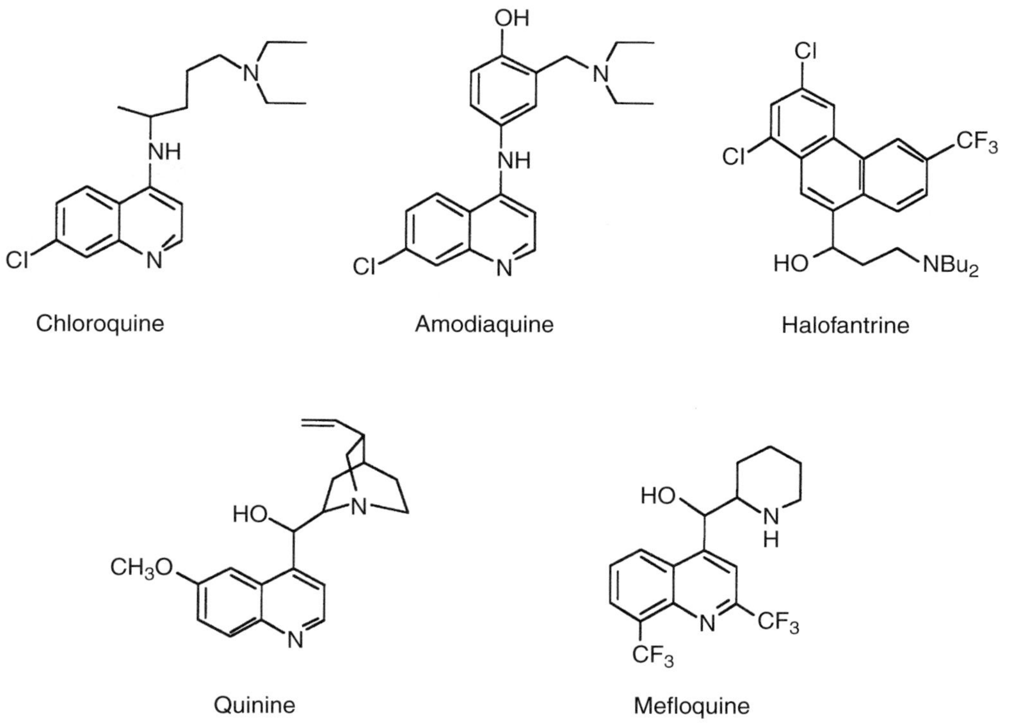

FIGURE 16.2 Structures of drugs used for the treatment of malaria.

to highlight the fact that drug susceptibility displays a graded distribution rather than two distinct but tightly distributed populations. In the case of *P. falciparum* an *in vitro* cut-off of 100 nM is often cited as the breakpoint between sensitivity and resistance. In fact it can be argued that within the drug susceptibility distribution it is possible to identify three general parasite populations based on phenotype (see below). In the discussions that follow it is important to appreciate that, in addition to the major gene effects contributing to chloroquine susceptibility status, additional genes must contribute to the overall drug susceptibility distribution.

Early attempts to define the chloroquine resistance phenotype in *P. falciparum in vitro* identified altered drug accumulation within the parasitized erythrocyte as a prominent feature. This is an observation that has been confirmed from many independent laboratories, and it appears to be a phenotypic characteristic independent of the geographical origins of the parasites tested. Chloroquine-susceptible parasites selectively accumulate drug several-thousand-fold compared to uninfected erythrocytes. As a diprotic weak base it was argued that all of this accumulation could be explained as a function of ion trapping within the terminal food vacuole of the malarial parasite, which is assumed to be acidic. Furthermore, experimental evidence has accrued from several sources indicating that the target for chloroquine action is the hemoglobin degradation pathway; this is again assumed to take place in the terminal food vacuole. Within this pathway host hemoglobin is enzymatically degraded, liberating potentially toxic ferriprotoporphyrin IX or heme. Under normal conditions the heme is converted into malaria pigment or hemozoin. Chloroquine (and related drugs) bind with high affinity to heme and the resulting complex, which cannot be incorporated into pigment, retains its cytotoxic potential.

Detailed investigation of this accumulation process has separated a high affinity, low capacity, saturable drug accumulation component from a low affinity, non-saturable process. The K_a of the high affinity process was shown to correlate with drug susceptibility in a range of parasite isolates. The kinetics of this uptake process are similar to those for heme–chloroquine binding. Furthermore, inhibition of heme generation using specific inhibitors of the aspartate protease involved in the initial hemoglobin cleavage results in a loss of chloroquine accumulation in parasitized cells. Thus the accumulation process can be largely explained in terms of drug binding to heme, and chloroquine resistance results from a reduced access to this binding site as measured by the apparent K_a of binding

It was rapidly recognized that this altered drug accumulation phenotype was similar to that observed in multidrug-resistant (MDR) cancer cells. In these cells it had been demonstrated that resistance was the result of over-expression of an ATP-dependent P-glycoprotein membrane transporter, capable of extruding drug from the cell. In addition a range of unrelated compounds, as exemplified by the calcium-channel blocker verapamil, were capable of reversing resistance *in vitro* through an undefined interaction with the MDR pump. Similar experiments in chloroquine-resistant parasites revealed a similar, although only partial, resistance reversal by verapamil. It was suggested that chloroquine resistance was, by analogy with MDR cancer cells, a drug efflux phenomenon, but detailed investigations have comprehensively failed to show enhanced drug efflux from resistant parasites.

The currently accepted chloroquine-resistance phenotype is one of reduced drug accumulation associated with reduced drug susceptibility ($IC_{50} > 100$ nM) that can be partially reversed by a range of 'reversers' such as verapamil. In fact a careful analysis of drug sensitivity data reported in the literature suggest that three phenotypes can be identified:

1. Fully chloroquine-sensitive parasites, $IC_{50} <$ 20 nM , no verapamil effect
2. Highly chloroquine-resistant parasites, $IC_{50} > 100$ nM, + verapamil effect
3. Partially chloroquine-resistant, IC_{50} 20 nM–100 nM, no verapamil effect or significantly reduced verapamil effect

Parasite isolates in the intermediate group are readily selected for high-level resistance with a greatly enhanced verapamil effect under drug pressure *in vitro*, whereas fully sensitive parasites cannot be selected for high-level resistance.

The initial studies aimed at elucidating the molecular basis of chloroquine resistance focused on the possible involvement of a multidrug resistance (MDR) P-glycoprotein-based mechanism. Two MDR homologs were identified in *P. falciparum*, *pfmdr1* and *pfmdr2*, with *pfmdr1* looking the most likely MDR gene. Pgh1, the protein product of *pfmdr1*, shares structural similarity with many members of the MDR family. The protein has two nucleotide binding sites, twelve transmembrane domains and has been localized to the parasite's digestive food vacuole membrane. It was originally argued that resistance resulted from an over-amplification of *pfmdr1*, by analogy with MDR cancer cells, but experimental

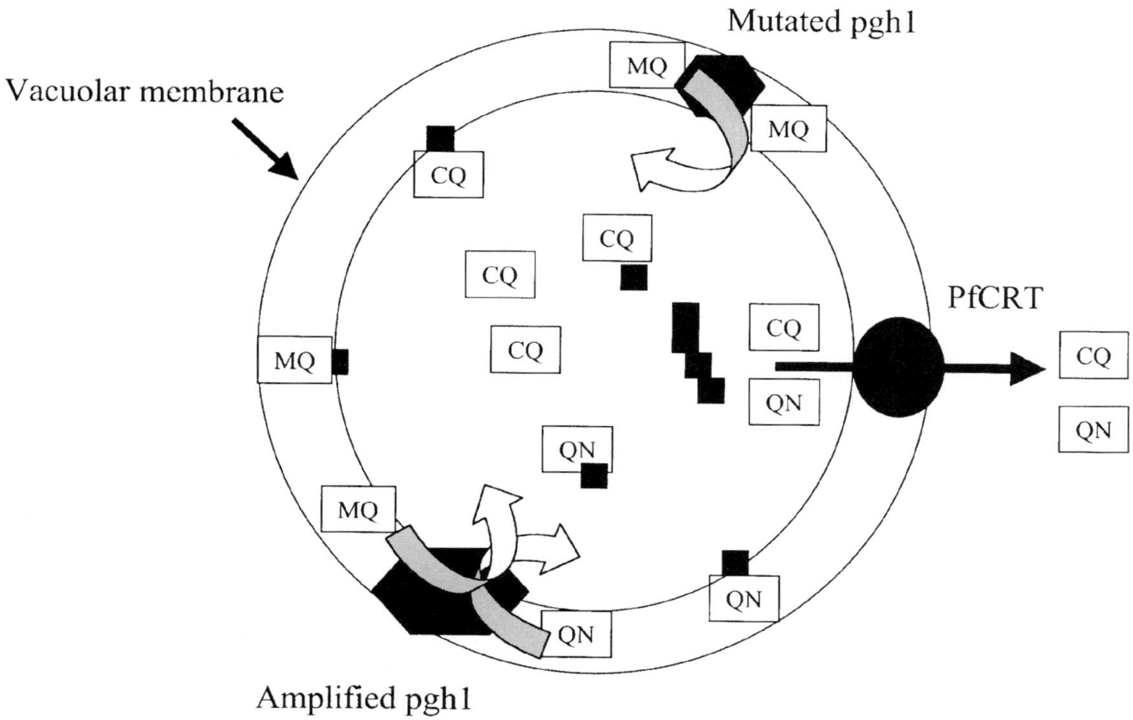

FIGURE 16.3 Proposed explanation for the transporter-dependent movement of anti-malarial drugs. Chloroquine (CQ), amodiaquine, and to a lesser extent quinine (QN) interact with free and membrane-bound heme (solid squares) at an aqueous interface and are substrates for PfCRT (solid circle), which removes drug from the intravacuolar aqueous space to the parasite cytosol. Mefloquine (MQ) (halofantrine and the artemisinins) and to a lesser extent quinine interact with membrane-bound heme at the lipid interface and are substrates for Pgh1 (large solid hexagon = amplification) which strips drug from the membrane into the vacuolar sap. Mutations in Pgh1 (small solid hexagon) preferentially display an increased recognition for mefloquine and halofantrine compared to quinine.

evidence failed to support this hypothesis. Subsequent studies have attempted to correlate chloroquine sensitivity with mutations of *pfmdr1*. Two *pfmdr1* haplotypes have been identified, the K1 allele carrying an Asn to Tyr codon change at position 86, and the 7G8 allele carrying codon changes at positions 184, 1034, 1042 and 1246. Extensive field studies have failed to clearly link these alleles alone with chloroquine susceptibility. However, allelic exchange experiments have demonstrated that the replacement of the mutations associated with the 7G8 allele with wild-type sequence results in an improvement in chloroquine sensitivity and a reduction in the magnitude of the verapamil effect, although the parasites remain phenotypically chloroquine-resistant. It is suggested that *pfmdr1* may act as a modulator of chloroquine resistance rather than as a primary determinant (Figure 16.3). This is supported by the observation that introduction of the 7G8 mutation into a chloroquine-sensitive genetic background failed to alter drug susceptibility.

An alternative laboratory strategy aimed at identifying the molecular basis of chloroquine resistance employed a genetic cross between a chloroquine-susceptible and a chloroquine-resistant parasite clone. The resulting drug sensitivity of the progeny was characteristic of one of the parent clones with no intermediates, suggesting simple monogenic inheritance of chloroquine resistance from these two parent lines. Linkage analysis mapped the chloroquine resistance locus to a 36 kb region of chromosome 7 (*pfmdr1* is located on chromosome 5). Detailed dissection of this chromosomal locus initially provided a list of ten candidate genes, and eventually uncovered the chloroquine-resistance gene *pfCRT* hiding in an assumed non-coding region between two candidate genes. The gene is contained within 13 exons, and difficulties in establishing the intron–exon boundaries are blamed for the delays experienced in tracking down the gene. It is proposed that the translated product of this gene, PfCRT, is a membrane transporter containing ten predicted transmembrane domains. In keeping with chloroquine's accepted mechanism of action, PfCRT has been immunolocalized to the boundaries of the digestive food vacuole (Figure 16.3).

Sequence analysis of PfCRT from the genetic cross have revealed eight point mutations (M741I, N75E, K76T, A220S, Q271E, N326S, I356T and R371I) capable of distinguishing chloroquine-susceptible from resistant parasites. Analysis of parasite isolates from different origins around the world has consistently identified the 76T and 220S mutations in all chloroquine-resistant isolates ($IC_{50} > 100$ nM). With the exception of the I356T mutation, the remaining six mutations were regularly found in association with resistant isolates from Africa and SE Asia. Based on the pattern of additional mutations that accompany the K76T mutation, it has been suggested that resistance must have developed from four independent foci. Interestingly, analysis of chloroquine-sensitive parasites, as defined by the 100 nM cut-off, indicated that with the exception of one isolate all isolates shared the wild type PfCRT sequence. The rogue isolate, which actually displays intermediate chloroquine sensitivity without the verapamil effect, contained all the PfCRT mutations associated with resistance with the exception of the critical K76T. Furthermore, under chloroquine pressure it was possible to select high-level chloroquine resistance with a verapamil effect from this parasite clone. The resultant parasites had acquired an additional mutation, K76I. This experiment highlights the importance of position 76 in PfCRT, and indicates that against the correct genetic backdrop selection of high-level chloroquine resistance is relatively easy. Episomal transfection studies claiming to provide further laboratory-based evidence for the role of PfCRT in resistance are compromised by the use of chloroquine as the selection agent and the acquisition of the K76I mutation in the resultant parasite line rather than the 76T encoded in the plasmid.

There is increasing field-based evidence from a number of distinct geographical settings including Mali, Cameroon, Sudan, Mozambique, Brazil, Laos, Thailand and Papua New Guinea, confirming that the PfCRT 76T mutation is universally present in patients failing standard chloroquine treatment. Equally of note are the observations of apparent treatment success against parasites containing the 76T mutation in semi-immune individuals, emphasizing the important contribution of host immunity to drug response. A number of studies have also suggested that the use of both *pfCRT* and *pfmdr1* as markers of chloroquine resistance improves predictive outcome. However, a recent multivariate analysis has found no improvement in predicting chloroquine failure rates by inclusion of *pfmdr1* status in association with *pfCRT*.

The biochemical function of PfCRT (and Pgh1) are unknown. Biochemical evidence indicating reduced access to heme as the basis of resistance is consistent with any mechanism that keeps heme and chloroquine apart within the parasite. It has been argued that mutations in PfCRT decrease the pH of the terminal vacuole. It was claimed that this reduction in pH, rather than increasing drug accumulation by ion trapping of the weak base, reduces heme solubility, thereby reducing the availability of the drug binding site. Unfortunately the experimental approach used to measure the pH in these experiments was fundamentally flawed. The pH probe used was localized exclusively in the cytosolic compartment rather than in the food vacuole and therefore could not be used to report the true vacuolar pH. Ongoing experiments are aimed at determining if PfCRT alters drug accumulation at the target site indirectly by influencing ion gradients within the parasite, or whether it can move the drug (or less likely heme) directly. This information is central to any strategies involving resistance reversal and rational drug design.

Chloroquine resistance is becoming an increasing problem in *Plasmodium vivax*. *P. vivax*, although not generally fatal, is responsible for considerable morbidity with some 75–90 million cases annually. The drug of choice against the erythrocytic stage is chloroquine. Chloroquine resistance in *P. vivax* was first reported in 1989 in Papua New Guinea. Despite comparable chloroquine use, clinical resistance in *P. vivax* has taken much longer to appear than resistance in *P. falciparum*. Little is known about the chloroquine-resistance phenotype in *P. vivax* due to difficulties in maintaining these parasites in continuous culture, although the drug is assumed to have the same mechanism of action. Analysis of the PfCRT homolog in *P. vivax* failed to identify any mutations in this gene that were associated with chloroquine resistance.

Amodiaquine

The 4-aminoquinoline amodiaquine is a structural analog of chloroquine (Figure 16.2). When used clinically amodiaquine undergoes efficient dealkylation to desethyl-amodiaquine. Both parent drug and metabolite contribute to antimalarial action *in vivo*. All of the available evidence supports the argument that amodiaquine and chloroquine share a common mechanism of action based on an interaction with heme. Amodiaquine is more active than chloroquine *in vitro*. This increased activity may be related to the drug's greater lipophilicity, but is independent of its weak base qualities (amodiaquine is a poorer weak base than chloroquine) or its ability to interact with heme (as determined from inhibition of heme crystalization). Amodiaquine and chloroquine share cross-resistance *in vitro*, and drug activity correlates with drug accumulation. Absolute parasite sensitivity to amodiaquine is always superior to chloroquine, and a verapamil effect has never been demonstrated. In comparison, the principal circulating metabolite desethyl-amodiaquine shares greater cross-resistance with chloroquine and there is a slight verapamil effect. Based on these observations it is probable that mutations in PfCRT also influence *in vitro* parasite susceptibility to amodiaquine and desethyl-amodiaquine. This argument has not yet been formally tested. However, the selection of the K76I mutation under chloroquine pressure in an isolate carrying the complete complement of additional PfCRT mutations resulted in a reduced susceptibility of the selected isolate to amodiaquine and quinacrine in addition to chloroquine. Despite these *in vitro* associations, there is compelling clinical evidence showing the superior efficacy of amodiaquine even in areas with a high incidence of chloroquine-resistant parasites. Taking all the available evidence it would appear that although amodiaquine and its desethyl metabolite are substrates for the chloroquine-resistance mechanism, the mechanism fails to limit drug access to heme to an extent sufficient to be therapeutically relevant. It will be interesting to see if additional mutations in PfCRT and/or changes in alternative resistance genes will compromise the clinical efficacy of amodiaquine as its use increases in chloroquine-resistant areas.

The quinoline and phenanthrene methanols

The quinoline methanols mefloquine and quinine and the phenanthrene methanol halofantrine share structural similarities (Figure 16.2) and all three drugs require the generation of intraparasitic heme in order to exert their full anti-malarial action. Although these three drugs are often grouped together, there are clear differences between quinine, mefloquine and halofantrine in terms of the speed of resistance development and cross-resistance pattern. Clinical resistance to quinine has been very slow to develop, although it is now recognized that many parasite isolates from SE Asia have reduced susceptibility to quinine which is of clinical relevance. In Africa most parasites appear to have retained sensitivity to quinine. In contrast to quinine, the development of clinical resistance to mefloquine was very rapid, occurring within a few years of the drug's introduction into SE Asia. In fact the loss of effectiveness was so profound that the recommended dosage regimen was doubled and still resistance was seen to develop. This is a situation that has only been arrested by the widespread use of mefloquine in combination with an artemisinin derivative. There are a number of reports showing clear cross-resistance between quinine, mefloquine and halofantrine. Drug sensitivity correlates with drug accumulation, and there is often an inverse relationship between sensitivity to these drugs and sensitivity to the 4-aminoquinolines. Interestingly, quinine-resistant parasites often show a verapamil effect. This has been reported once with mefloquine but never with halofantrine although there is some evidence that resistance to these drugs can be selectively reversed to a degree by a more limited number of agents, including penfuridol.

Support for the role of *pfmdr1* in resistance to the quinoline and phenanthrene methanols is certainly greater than that for the 4-aminoquinolines, but again the association is far from universal. Overexpression of *pfmdr1* has been associated with resistance (or reduced susceptibility) to these drugs in some field isolates of *P. falciparum*. Similarly, laboratory selection for mefloquine resistance under drug pressure, with a concomitant loss in susceptibility to quinine and halofantrine, has been associated with overexpression of *pfmdr1* (Figure 16.3). The link between sequence polymorphisms in *pfmdr1* and resistance is less clear. Extensive analysis of field isolates has failed to show any consistent relationship between mutations in *pfmdr1* and drug susceptibility. However genetic transformation studies in which the wild-type sequence of *pfmdr1* was replaced with three of the 7G8 mutations was associated with a reduced susceptibility to quinine and an enhanced susceptibility to mefloquine and halofantrine. Replacement of these mutations with wild-type sequence resulted in the reverse patterns of drug sensitivity, i.e. reduced susceptibility to mefloquine and halofantrine and increased sensitivity to quinine. One possible explanation for these observations would be that all three drugs are substrates for Pgh1, hence overexpression reduces drug susceptibility across the board, whereas mutations in pfmdr1 differentially alter the relative efficiency of the drug binding or transport process for each drug. The localization of an MDR-type drug pump on the food vacuole membrane has been seen as paradoxical, as its overexpression would have the effect of increasing vacuolar drug accumulation at the drug target site. However, this can be rationalized if we accept that mefloquine, halofantrine and to a lesser extent quinine are highly lipophilic and concentrate within membranes. Heme is even more highly lipophilic and thus will also concentrate in membranes. Thus by reducing the concentration of drug in the vacuolar membrane Pgh1 would limit drug access to the membrane-bound target heme.

Careful analysis of all the data suggests that the principal 4-aminoquinoline resistance mechanism controlled by PfCRT is distinct from the principal mefloquine/halofantrine resistance mechanism controlled by *pfmdr1*. Quinine in contrast appears to come under the control of both resistance mechanisms to some extent. This argument is further supported by the linkage between the development of the K76I mutation under chloroquine selection and a parallel reduction in quinine sensitivity (Figure 16.3).

Folate metabolism antagonists

The folate biosynthetic pathway in plasmodia has long been a target for chemotherapy. The unique susceptibility of this pathway was originally thought to arise from the parasites' basic need to produce folate cofactors *de novo*. Recent studies, however, indicate the presence of a folate salvage pathway in some parasites. This salvage pathway has obvious implications for antifolate drug action and synergy. Although a number of drugs capable of targeting the folate pathway have been employed as anti-malarial agents, it is the combination of pyrimethamine and sulfadoxine (SP) which has seen most clinical utility. SP is an affordable alternative to chloroquine in areas of chloroquine resistance. Several African countries have switched to SP as first-line treatment and it remains the second-line choice in many other African countries. Resistance renders this combination useless in SE Asia and South America. These drugs have limited anti-malarial activity when used alone, but when used in combination the drugs interact synergistically.

Pyrimethamine is a competitive inhibitor of dihydrofolate reductase (DHFR), a bifunctional enzyme with thymidylate synthetase. This is a key enzyme in the formation of tetrahydrofolate, which is essential in the biosynthesis of DNA and protein. Sulfadoxine, an analog *p*-aminobenzoic acid, is a competitive inhibitor of dyhydropteroate synthase (DHPS), another bifunctional protein with dihydrohydroxymethylpterin pyrophosphokinase, involved in the condensation of *p*-aminobenzoic acid with 2-amino-4-hydroxy-6-hydroxymethyldihydropteridine pyrophosphate to produce dihydropteroate (Figure 16.4).

The use of pyrimethamine for chemoprophylaxis in the middle of the last century provided the first indication that resistance to this drug was readily acquired. Furthermore, the observation that resistance occurred within a few months of the drug's introduction suggested a simple resistance mechanism. The molecular basis for pyrimethamine resistance has now been established. Unlike the situation in bacteria, where resistance is most commonly the result of plasmid-encoded *dhfr*, in malaria parasites it is point mutations in *dhfr* which are responsible. Laboratory studies have confirmed that the identified point mutations result in amino acid alterations around the enzyme's active site. This in turn reduces the affinity of the enzyme for the drug, thereby conferring

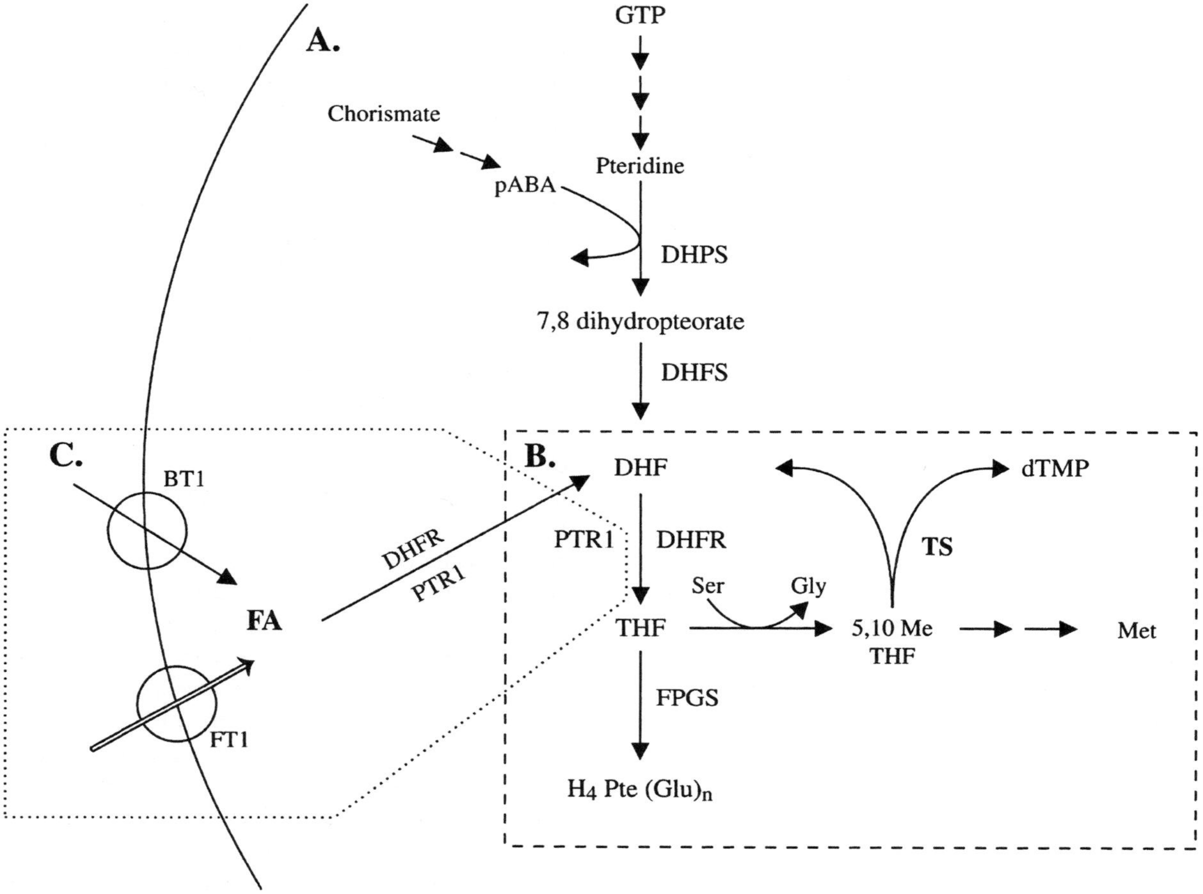

FIGURE 16.4 Folate and pterin metabolism. (A) *De novo* pterin biosynthesis. Pterins can be synthesized *de novo* from GTP in some parasites. Dihydropteroate is made by the conjugation of a pterin to *p*-aminobenzoic acid (pABA) which is catalysed by dihydropteroate synthase (DHPS), the target of sulfonamides. (B) Folate reduction and metabolism. The addition of glutamic acid by dihydrofolate synthase (DHFS) to dihydropteroate leads to dihydrofolate (DHF), which is reduced to tetrahydrofolate (THF) by dihydrofolate reductase (DHFR), the target of antifolates. Reduced folates are required for the biosynthesis of thymidine, and for the conversion of serine to glycine and in methionine biosynthesis. In most organisms folates are polyglutamylated by the enzyme folylpolyglutamate synthetase (FPGS). The reactions included in box B are likely to occur in most parasites. (C) Folate uptake. *Leishmania*, and possibly other Kinetoplastidae, cannot synthesize their folates *de novo* and rely on folate importers (FT1). Some folate can also be imported by a biopterin transporter (BT1). Folates are reduced mainly by DHFR but also by a pterin reductase (PTR1).

resistance to the actions of pyrimethamine. Analysis of field isolates confirms the view that these point mutations contribute significantly to clinical resistance. Importantly, under increasing pyrimethamine pressure, parasites develop stepwise reductions in susceptibility which are associated with the accumulation of additional mutations in *dhfr*. The formal confirmation of this relationship has been provided by molecular studies in which wild-type parasites were transfected with a range of mutant *dhfr* alleles.

Four key DHFR mutations have been identified to date in field isolates that are important in pyrimethamine susceptibility. In comparison with the wild-type enzyme an initial mutation at position 108 (S108N) appears to be a prerequisite for the acquisition of additional mutations. This single mutation alone confers a 100-fold loss of pyrimethamine activity. Additional mutations at positions 51 (N51I) and/or 59 (C59R) reduce sensitivity by another order of magnitude, and finally a mutation at position 164 (I164L) produces the highest level of pyrimethamine resistance observed to date. Triple and quadruple mutants harboring I164L rapidly evolved in SE Asia and South America, rendering SP useless within a few years. The 164 mutation has yet to be definitively identified in parasite populations collected from Africa. In addition to these mutations, an additional mutation at position 50 (C50R) and a five amino acid repeat (referred to as the Bolivia repeat) have recently been found in field isolates from South America. Transfection studies in yeast provide supportive evidence for a role of the C50R mutation in pyrimethamine resistance.

Sequence analysis of the *dhfr* domain from *Plasmodium vivax* (*pvdhfr*) parasites has identified point mutations that result in the replacement of key amino acids within the enzyme active site. Mutations in residues of the *P. vivax* enzyme at positions 15, 50, 58, 117 and 173 were predicted, by amino acid homology and predicted secondary structure, to correspond with positions 16, 51, 59, 108 and 164 in the *P. falciparum* enzyme. Notably, the mutation at codon 117 in *P. vivax* appears to correspond with the critical 108 mutation in *P. falciparum*, and is the initial step in the acquisition of pyrimethamine resistance. Analysis of field isolates from areas with differing exposure experience to SP indicated the predominance of double and triple mutations in *pvdhfr* (resulting in amino acid changes at positions 57, 58 and 117) in 99% of isolates taken from patients in Thailand, where SP has been extensively used in the past. In comparison, only 19% of parasites collected from India (moderate population exposure to SP) and 0% of parasites collected from Madagascar and the Comoros Islands (no population SP exposure) contained double or triple mutations. Importantly, the high incidence of the double and triple mutation in Thailand was associated with high levels of *P. vivax* SP resistance *in vivo*.

Cycloguanil, chlorcycloguanil and WR99210, the cyclic metabolites of the biguanides proguanil, chlorproguanil and PS15, are all potent inhibitors of plasmodial DHFR. The S108N mutation in DHFR has little influence on the anti-malarial activity of these drugs. Additional mutations at positions 51 and 59 produce a moderate loss in cycloguanil and chlorcycloguanil activity. In contrast, an A16V mutation in combination with S108N significantly compromises the activity of cycloguanil and chlorcycloguanil without affecting pyrimethamine sensitivity. Mutations at 108, 51 or 56, and 164 significantly compromise the activity of pyrimethamine, cycloguanil and chlorcycloguanil. All permutations of *dhfr* mutation appear to have little influence on the inhibitory potential of WR99210. This observation has important implications. The available data suggest that WR99210 interacts with critical features in the DHFR domain that cannot be modified readily without loss of parasite fitness. On a practical note, the ability of WR99210 to retain inhibitory activity against all recognised *dhfr* mutants supports the claims that this new drug will be a very effective anti-malarial even against a backdrop of high-level SP resistance.

Using similar strategies to those employed with *dhfr*, it has been confirmed that point mutations in the *dhps* domain confer resistance to the actions of sulfadoxine. There are five recognized mutations in *dhps*, all of which influence sulfadoxine activity. Although less clear than with *dhfr*, there does appear to be a hierarchy of mutations conferring increased resistance to sulfadoxine. A mutation at position 437 (A437G) confers very slight resistance to sulfadoxine and appears to precede other mutations in many instances. Additional mutations at positions 436 (S436A/F), 540 (K540E), 581 (A581G) and 613 (A613S/T) confer increasing levels of resistance. Unlike the situation with *dhfr*, mutations in *dhps* appear to have similar effects on susceptibility to related sulfonamides and sulfones.

It is assumed that a major selection pressure for these mutations of *dhfr* and *dhps* in parasite populations has been the widespread use of SP for malaria treatment, although the contribution of other antifolates (in widespread use for other conditions in these communities) cannot be ignored. The molecular basis for resistance presented above looks at each drug in isolation, whereas what the parasites actually see is a synergistic combination of the two drugs. Under these circumstances field data suggest that the selection pressure at each target site is different. Selection for mutations in *dhfr* occurs first, with mutations in *dhps* only being selected when most parasites have already acquired a double or triple *dhfr* mutation. One possible explanation for this difference in selection pressure may be related to the proposed ability of pyrimethamine to inhibit folate salvage in addition to its DHFR-inhibitory actions. If the pyrimethamine concentrations required to inhibit DHFR are much lower than those required to inhibit folate salvage, then at drug concentrations exerting selection pressure against *dhfr* the DHPS-dependent *de novo* pathway would not be essential to parasite survival. Reliance on the DHPS-dependent *de novo* synthesis would become important to survival when sensitivity of folate salvage and DHFR to pyrimethamine became equivalent. Only then is *dhps* under full selection pressure. As with chloroquine resistance, parasitological resistance to the antifolates does not necessarily predict clinical failure in countries in which malaria is endemic, where the influence of host immunity would contribute significantly to therapeutic outcome.

Atovaquone resistance

The anti-malarial potential of naphthoquinones has been known for more than half a century. Atovaquone represents the only naphthoquinone to be successfully developed for clinical use. Atovaquone exerts its anti-malarial effect at the level of the parasite mitochondria, selectively inhibiting electron transport via inhibition of the bc_1 complex. In addition to this inhibition of electron transport, atovaquone collapses the mitochondrial inner membrane potential, contributing to parasite death. Early studies proved that atovaquone was a highly effective anti-malarial capable of killing parasites in the nanomolar concentration range. However, unacceptable recrudescence rates were observed when the drug was introduced into clinical trials. Importantly, parasites from recrudescent infections were 1000-fold less sensitive to the drug than parasites from the same individual prior to drug administration. Interestingly, the frequency with which resistant parasites can be selected *in vitro* shows a geographical bias, with SE Asian parasites having more than a 100-fold higher frequency of resistance acquisition than parasites from South America or West Africa.

Resistance to atovaquone is associated with point mutations in the mitochondrial cytochrome *b* gene. These mutations encode amino acid alterations within the coenzyme Q binding site. Laboratory experiments in rodent malaria models have confirmed the importance of these point mutations in Q_0 to high-level atovaquone resistance.

Atovaquone is now used clinically in combination with proguanil. These drugs act synergistically, and the high frequency of recrudescence seen with atovaquone monotherapy has been eliminated. Importantly, it is proguanil and not the cyclic metabolite cycloguanil which acts synergistically. The mechanism of synergy is linked to the ability of proguanil to sensitize the parasite to the depolarizing effects of atovaquone. Experiments in rodent models suggest that the synergistic potential of proguanil is lost if parasites are first selected for high-level atovaquone resistance. The very rapid selection of atovaquone resistance after monotherapy is not readily explained, given the high degree of sequence conservation in the mitochondrial genome. It has been proposed that the drug may compromise its own future by promoting the emergence of resistance as a side-effect of the drug's mechanism of action. At therapeutic concentrations it is argued that atovaquone (as monotherapy) results in the formation of reactive oxygen species (ROS) within the mitochondria as a result of inhibition of electron transport and collapse of the membrane potential. These ROS act as local mutagens, increasing the potential for advantaged variants to arise. Proguanil, by reducing the atovaquone concentration required to collapse the membrane potential without affecting the drug's ability to inhibit electron transport, enables the atovaquone to kill parasites without generation of ROS.

Resistance to other anti-malarials

The artemisinins represent a novel class of anti-malarial drug containing a unique endoperoxide bridge which is central to their anti-malarial activity. All of the artemisinin derivatives currently available and in clinical use are based on the structure of artemisinin, the principal ingredient of a Chinese herbal remedy for fevers obtained from *Artemesia annua* or sweet wormwood (qunghao). The mechanism of action of these drugs is based on the iron-dependent cleavage of the peroxide bridge, liberating free radical species selectively within the parasitized erythrocyte. These drugs are fast acting, and can affect parasites throughout most of the erythrocytic cycle. There are no reports of clinical resistance to these drugs as yet, although there have been a few suggestions of slightly reduced, but clinically irrelevant, levels in absolute parasite sensitivity. Based on the accepted mechanism of action of these drugs it has been argued that resistance will be difficult to acquire, possibly requiring the intervention of a multigenic process. Attempts to induce resistance in laboratory isolates of *P. falciparum* have been unsuccessful, and resistance developed in rodent models is labile, requiring continued drug pressure for maintenance of the resistance phenotype. It is likely that the reduced susceptibility reported in these rodent strains is the result of an inductive process rather than a true stable resistance mechanism.

Primaquine is an 8-aminoquinoline, and is currently the only drug available for anti-relapse therapy that kills liver stage parasites. Although the drug also has activity against erythrocytic and gametocyte stages of the parasites' life cycle, the concentrations of drug required cannot be tolerated by the host. The mechanism of action is poorly understood but is thought to be mediated via the metabolic generation of highly reactive drug species within the liver. Although there are several reports of clinical resistance to primaquine, there are serious questions as to whether these represent true resistance and there is no information on potential mechanisms.

Resistance in other Apicomplexa

Resistance in Toxoplasma gondii

In addition to malaria, several Apicomplexa protozoan parasites are important pathogens of humans and animals. *Toxoplasma gondii* is widely distributed with a very high prevalence in many regions, and can cause serious infections in immunocompromised patients (particularly AIDS patients) and in the developing fetus. Treatment of *T. gondii* relies on the combination pyrimethamine–sulfonamide, with clindamycin and atovaquone as second-line drugs. Clinical resistance to these antimicrobial agents has not yet been described, and since there is no human-to-human transmission, the likelihood of resistant parasites emerging is low, although long-term therapy may select resistant organisms in infected individuals. Nonetheless, considerable work has been done on mechanisms of resistance induced *in vitro*. Single point mutations in the *T. gondii dhfr-ts* can produce pyrimethamine resistance in *T. gondii*. Atovaquone targets the cytochrome bc_1 complex of parasites, which functions as an acceptor of electrons. *T. gondii* mutants selected for atovaquone resistance had mutations in their cytochrome b gene in regions corresponding to regions likely to bind the drug, and in the same residues as found in atovaquone-resistant *Plasmodium*. Clindamycin is a macrolide inhibiting translation in bacteria. In *Toxoplasma*, prokaryotic-like translation takes place in the apicoplast, a plastid organelle similar to those found in plants and algae. One possible mechanism of resistance would be a mutation in the apicoplast rRNA.

Resistance in Eimeria

Eimeria is a genus closely related to *Toxoplasma* that causes coccidiosis, an intestinal tract disease of domestic fowl that infects several billion chickens reared annually. A large number of anticoccidials were developed in the 1960s and 1970s. The drugs mostly used are ionophores of the polyether type (e.g. monensin, maduramicin, salinomycin) that interfere with membrane integrity. Other drugs target the folic acid biosynthetic pathway, the purine biosynthesis or uptake pathways, or mitochondrial function, and for a number of effective drugs the mode of action is not understood. The usefulness of anticoccidial drugs is jeopardized by the development of resistance, and several of the strains in which resistance has been studied were multiply resistant, suggesting that in some of these cases broad-range substrate efflux systems such as P-glycoprotein could be involved in resistance. The biochemical or molecular basis of resistance for drug-resistant mutants either isolated *in vitro* or from field isolates has not yet been determined.

Resistance in Cryptosporidium

The protozoan parasite *Cryptosporidium parvum* can cause severe diarrhea in immunocompromised patients. Although many antimicrobial agents have been tested against this apicomplexan parasite none has been shown to be effective. The only drug with some limited activity is the aminoglycoside paromomycin, an inhibitor of the translational machinery. It thus appears that this parasite is intrinsically resistant to several different anti-protozoal agents. This has led to the suggestion that an ABC transporter efflux pump could contribute to intrinsic resistance, although concrete evidence for this is lacking. Other mechanisms are possible. For example antifolates are not useful against this parasite, but the sequence of the *C. parvum DHFR* active site contains novel residues at several positions analogous to those at which point mutations have been shown to produce antifolate resistance in other parasite DHFRs.

DRUG RESISTANCE IN KINETOPLASTID PARASITES

Leishmania and chemotherapy

Leishmania is a significant cause of morbidity and mortality, with a disease prevalence of 12 million cases and a worldwide annual incidence of 1.5–2 million cases. *Leishmania* manifestations range, depending on the species, from self-healing cutaneous lesions to life-threatening visceral infections. Although several experimental approaches (killed parasites ± BCG; subunit vaccines; DNA vaccines; attenuated organisms) have been used as vaccination strategies against *Leishmania,* chemotherapy is still the only effective way to control all forms of diseases. The first agents with a favorable therapeutic index, the pentavalent antimonials (SbV), were introduced in the 1940s, and despite the need for prolonged parenteral administration are still the mainstay of therapy for all forms of leishmaniasis. Formulations available include sodium stibogluconate (Pentostam) in which antimony is complexed with gluconic acid, and meglumine antimonate (Glucantime) where antimony is complexed with the sugar meglumine (Figure 16.5). Although antimonials are likely to remain the pillar of anti-leishmanial therapy for the foreseeable future, we remain uncertain of the mechanism of action or even the biologically active component of the most commonly utilized formulations. Over the years, in face of slowly developing clinical resistance in the 1970s and 1980s, both the daily dose of SbV and the duration of treatment were increased to maintain satisfactory levels of efficacy. Resistance to SbV is now endemic with more than 50% resistance rate in the state of Bihar, India. Undertreatment and non-compliance are likely to have contributed to SbV resistance, but in India anthroponotic transmission is possibly a determining factor in disseminating SbV resistance. Second-line drugs against *Leishmania* infections include pentamidine and lipid forms of amphotericin B (AmB), as well as a number of other drugs with more specific indications.

Pentavalent antimonials

Clinical unresponsiveness to SbV has been attributed to several factors, but resistant parasites have now been formally documented in the laboratory and contribute to treatment failure in several endemic regions. A strong correlation between clinical response and SbV sensitivity *in vitro* was observed only when strains were assessed as intracellular amastigotes, while no correlation with clinical response was found using extracellular promastigotes. This complicates the testing of resistance in the laboratory, as measuring and evaluating intracellular *Leishmania* is laborious. However, recent advances using *Leishmania* expressing reporter genes, such as firefly luciferase, may facilitate susceptibility testing within macrophages. It is also possible to grow amastigotes in axenic culture, and these could possibly be used for drug testing, although several species are refractory to axenization and it is possible that intracellular parasites will be required for drug testing.

The rate of appearance of resistance to SbV in the field has been slow, which suggests that resistance, as discussed for chloroquine resistance in *Plasmodium,* is the consequence of

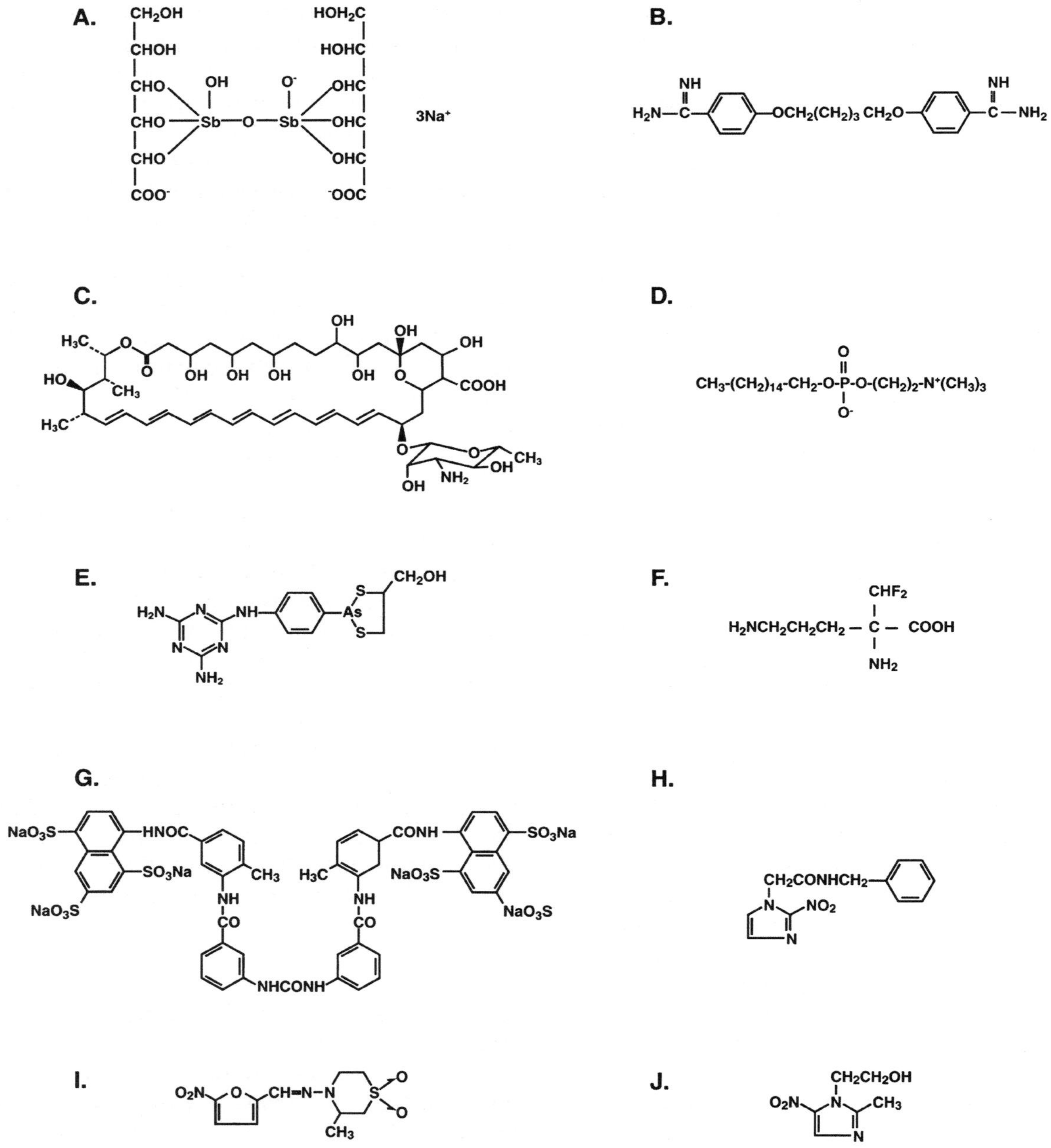

FIGURE 16.5 Structure of the main drugs used against kinetoplastid parasites. (A) Pentostam; (B) pentamidine; (C) amphotericin B; (D) miltefosine; (E) melarsoprol; (F) eflornithine; (G) suramin; (H) benznidazole; (I) nifurtimox; (J) metronidazole.

more than a single event. It is generally agreed that SbV is not toxic to *Leishmania* but that it must be reduced to the trivalent form SbIII. The site of reduction is unclear; some data are highly suggestive of the host macrophages while other data have highlighted a stage-specific amastigote intracellular SbV-reducing activity. If reduction of SbV to SbIII is rate-limiting, and provided that the enzyme does not have another essential role, inactivation of this enzyme would lead to SbV resistance. Interestingly, axenic *L. donovani* amastigotes selected for *in vitro* SbV resistance have lower SbV reductase activity than wild-type cells. The isolation of the putative reductase should help in determining whether reduction occurs predominantly within the parasite or the host, and measuring its exact role in drug resistance.

Work on metal resistance induced in *Leishmania* promastigotes has led to a coherent model of SbV resistance (Figure 16.6). Trypanothione (TSH), N^1, N^8-*bis*-glutathionyl spermidine, the major reduced thiol of Trypanosomatidae, is increased in several *Leishmania* promastigote species selected for metal resistance. This increase in TSH is due to the amplification or overexpression of genes coding for γ-glutamylcysteine synthase (γ-GCS) or ornithine decarboxylase (ODC), the respective rate limiting steps in glutathione and spermidine biosynthesis (Figure 16.6). Reducing the level of TSH by using the glutathione of spermidine biosynthesis inhibitors buthionine sulfoximine (BSO) and difluoromethylornithine (DFMO) was shown to revert resistance. Although high levels of TSH appear essential for metal resistance in *Leishmania* promastigotes, they are not sufficient, and the activity of membrane transporters that recognize SbIII conjugated to TSH is also modulated in drug-resistant mutants (Figure 16.6). One transporter is the intracellular ABC transporter PGPA that possibly confers resistance by sequestration of the metal conjugate inside vesicular elements; another transporter is an ATP-thiol-X efflux system located in the plasma membrane, that confers resistance by extruding the drug outside the cell (Figure 16.6). PGPA is part of a large gene family, and it is possible that one of these members corresponds to this plasma membrane efflux pump. In mammalian cells, resistance mediated by the MRP pump can depend on a high level of glutathione, high activity of the pump, and increased glutathione S-transferase activity. By analogy to the MRP pump, it is possible that a trypanothione S-transferase is required for achieving high-level metal resistance in *Leishmania*.

Additional work has allowed the isolation of novel antimony resistance genes that do not appear to be part of the resistance pathway shown in Figure 16.6. For example, a gene coding for a protein of 770 amino acids was found to be amplified in Pentostam-selected *Leishmania*, while overexpression of a leucine-rich repeat protein was shown to contribute to SbIII resistance in both the promastigote and amastigote stage of the parasite. Some antimony-resistant cells were also shown to have increased expression of the heat-shock protein HSP 70. In the coming years it should be possible to test whether any of the resistance mechanisms described above plays a role in clinical resistance to SbV in *Leishmania*.

Pentamidine

Pentamidine is a second-line drug against various *Leishmania* infections and was used successfully to treat infections refractory to SbV. With time, however, even by using higher drug concentrations, the efficacy of pentamidine rapidly declined in India, suggesting that parasites were becoming resistant. The cellular target of pentamidine is unknown, although it causes competitive inhibition of polyamine

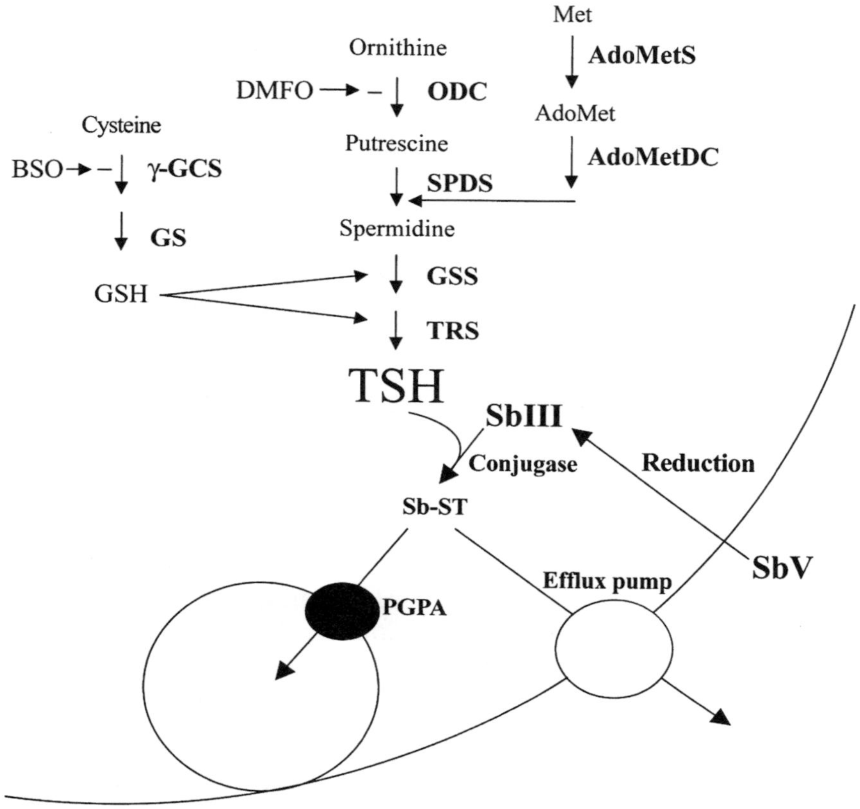

FIGURE 16.6 Trypanothione metabolism and antimony resistance in *Leishmania*. Trypanothione (TSH) is made of two molecules of glutathione (GSH) conjugated to the polyamine spermidine. The pentavalent antimony (SbV) used for treating *Leishmania* is reduced to the trivalent form (SbIII). The level of TSH is increased in *Leishmania* cells resistant to metals. TSH is conjugated to SbIII, possibly by a hypothetical conjugase. The metal–thiol conjugate can be extruded from the cell by a plasma membrane efflux pump or sequestered inside a vacuole by the ABC transporter PGPA. γ-GCS: γ-glutamylcysteine synthetase; GS, glutathione synthase; ODC, ornithine decarboxylase; SPDS, spermidine synthase; GSS, glutathionespermidine synthase; TRS, trypanothione synthase; AdoMetS, *S*-adenosylmethionine synthase; AdoMetDC, *S*-adenosylmethionine decarboxylase.

transport suggesting that it possibly enters into *Leishmania* through the parasite polyamine transport system. Resistance to pentamidine has been induced *in vitro* in several *Leishmania* species, both as promastigotes and as amastigotes. Resistance can be maintained stably throughout the life cycle of the parasite, and is associated with changes in the intracellular concentration of arginine, ornithine and putrescine. This may be due to alterations in polyamine carriers, but perhaps also to modulation of the activity of key enzymes in the polyamine biosynthesis pathway, such as ornithine decarboxylase and spermidine synthase (Figure 16.6). Analysis of pentamidine-resistant mutants indicated additional defects in the uptake and efflux of several substrates, which may be due to a perturbation of lipid constituents, thereby increasing membrane permeability. The exact contribution of this membrane defect to pentamidine resistance needs to be assessed, however.

Resistance to amphotericin B

The polyene antibiotic amphotericin B (AmB), a standard antifungal drug used in systemic fungal infections, is a second-line drug for the treatment of visceral leishmaniasis. The toxicity of AmB has limited its use, but liposomal formulations of AmB are highly effective against *Leishmania,* although high cost may be an issue for treating visceral leishmaniasis in several areas where it is endemic. However, short course treatment of lipid AmB formulations may help in the treatment of patients not responding to antimony therapy. AmB interacts with fungal membrane sterols (notably ergosterol), and the mode of action against *Leishmania,* which also have ergosterol, is apparently the same.

Resistance to AmB in field isolates is not widespread, although successive relapses in HIV-infected patients could enhance the emergence of AmB resistant *Leishmania* isolates. Resistance to AmB has been induced *in vitro* in *Leishmania donovani* promastigotes. The mutants showed increased membrane fluidity with changes in lipid composition including an ergosterol precursor. Changes in membrane fluidity may also be important in AmB resistance in yeast. Resistance to AmB has also been induced in *L. tarentolae* and two gene amplification events have been observed and characterized, although the link between these amplicons and resistance is yet to be confirmed. Some of the AmB-resistant mutants were cross-resistant to azole drugs such as ketoconazole, which inhibits steps in ergosterol biosynthesis. It is thus possible that *Leishmania* responds to alterations in ergosterol biosynthesis by modulating the expression of several genes of the ergosterol biosynthetic pathway. In yeast the expression of several genes was modulated in response to ergosterol depletion.

Miltefosine

Miltefosine, an alkylphosphocholine, possesses potent *in vitro* and *in vivo* anti-kinetoplastid activity, and has proved highly effective in visceral leishmaniasis clinical trials. *Leishmania* have high levels of ether-lipids present on the surface of the parasite. The glycosyl phosphatidyl inositol-anchored glycolipids and glycoproteins are perturbed in *Leishmania* promastigotes put in contact with miltefosine. Studies on the mode of action suggest that miltefosine may inhibit the glycosomal enzyme alkyl-specific-acyl-CoA acyltransferase involved in ether-lipid remodeling. Alkyl-lysophospholipids were initially developed as anti-cancer agents, but resistance to these agents developed readily against them. Possible resistance mechanisms include reduced drug uptake, differential plasma membrane permeability, faster drug metabolism and efflux mediated by the P-glycoprotein *MDR1*. *Leishmania* cells selected for resistance to the anti-cancer agent daunomycin overexpress their *MDR1* gene. These cells are cross-resistant to miltefosine, and it is highly possible that this ABC transporter efflux pump is involved in resistance to this new promising anti-leishmanial agent. *In vitro* work shows that it is relatively easy to select miltefosine-resistant mutants, which may suggest that miltefosine should be used in combination with other anti-leishmanial agents to reduce the emergence of resistant strains.

Resistance to other anti-leishmanial drugs

Leishmania cells are sensitive to a number of aminoglycosides. It is this susceptibility which has permitted efficient gene transfection studies using the bacterial neomycin phosphotransferase gene (*NEO*) as a selectable marker and the drug G418 as the selective drug. The aminoglycoside paromomycin (aminosidine) is a potentially useful anti-leishmanial topical agent against cutaneous leishmaniasis. The *NEO* gene confers cross-resistance to aminosidine. The mode of action is not known but resistance to paromomycin has been induced in *Leishmania* promastigotes. In several organisms resistance is due to mutations in ribosomal RNA, although this was not the case for *Leishmania*. In one study, resistance appears to be linked to decreased drug uptake.

Leishmania, like other parasitic protozoa, are incapable of synthesizing purines *de novo*, whereas the mammalian cells that they infect can. This has led to the clinical testing of some purine analogs such as allopurinol that can be effective in the treatment of certain forms of *Leishmania* infections. Biochemical and genetic studies in *Leishmania* have identified several purine salvage enzymes, some of which are considered as interesting drug targets against parasitic diseases. Resistance to nucleoside analogs has been induced in *Leishmania*, and point mutations in one high affinity adenosine-pyrimidine nucleoside transporter were shown to be responsible. *Leishmania* can also respond to purine analogs by amplifying a specific portion of its genome. A gene known as *TOR* with some similarities to transcription factors was shown to confer resistance to several purine derivatives including allopurinol. The TOR gene product appears to modulate the activity of transporters by a mechanism that remains to be determined.

Work on resistance to the model antifolate drug methotrexate (MTX) has contributed to our understanding of folate and pterin metabolism in *Leishmania*. It has also led to the discovery of novel but elegant resistance mechanisms and has pinpointed novel intracellular drug targets. *Leishmania* is auxotrophic both for folates and pterins. Mutations in the folate transporter (FT1) lead to MTX resistance. Alternative routes of entry must meet the folate requirements of these mutants. One of these corresponds to the biopterin transporter BT1, which is often overexpressed in MTX resistant mutants, and which is structurally related to the folate transporter (Figure 16.4). Amplification of the target gene *DHFR-TS*, and more rarely point mutations in positions known to correlate with MTX resistance in mammalian DHFRs, can also contribute to MTX resistance in *Leishmania*. The amplification of the gene coding for the pterin reductase PTR1 is a frequent mechanism of antifolate resistance. The main function of PTR1 is to salvage oxidized pterins; a secondary function is reduction of folates. PTR1 therefore serves as an alternative route for the synthesis of reduced folates and its amplification will lead to MTX resistance (Figure 16.4). In certain *Leishmania* species, MTX is polyglutamylated and a reduction in the polyglutamate chain length of MTX, which may modulate cellular retention, is correlated to MTX resistance. These experimental studies clearly indicate that several distinct mechanisms can lead to resistance to the same drug, and often these mechanisms coexist in the same cell.

African trypanosomes and chemotherapy

Two *T. brucei* subspecies, *T. b. gambiense* and *T. b. rhodesiense*, cause sleeping sickness (human African trypanosomiasis, HAT). Trypanosomes remain a major scourge of sub-Saharan Africa with a prevalence of 300 000–500 000, and infection is invariably fatal if untreated. These parasitic infections can be treated with only a handful of drugs, including suramin and pentamidine for early-stage HAT. Because of the highly ionic nature of suramin and pentamidine these drugs do not cross the blood–brain barrier, and the only effective drugs against late-stage HAT are the trivalent arsenical melarsoprol, and eflornithine, the only new drug developed in the past 50 years against HAT. Eflornithine is highly effective against *T. b. gambiense* but many clinical isolates of *T. b. rhodesiense* are naturally resistant to this drug. As a last resort, in cases of melarsoprol failure, a combination of drugs is used. A number of antiquated drugs are also available against veterinary African trypanosomes (*T. congolense, T. vivax, T. b. brucei*) and the phenomenon of drug resistance is well established in these parasites, as is the existence of multi-drug resistant isolates which have been validated by measuring drug sensitivity in *in vitro* assays. Resistance to the few anti-trypanosome drugs appears also to have made its way into human medicine. Indeed, the number of patients relapsing after treatment with melarsoprol has increased significantly in the last few years. Relapse can be due to a combination of factors, including host metabolism and immunological factors, but there is increasing evidence that it can be due to drug-resistant trypanosomes.

Arsenicals

The development of organic arsenicals by Ehrlich and colleagues to treat HAT and other infections has led to several of the current modern concepts in chemotherapy of infectious diseases and its associated resistance. One of the first aromatic arsenicals used was atoxyl, followed in 1910 by Salvarsan (the 'magic bullet'). Resistance to these first-generation arsenical compounds occurred rapidly, which led in 1949 to the synthesis of the melaminophenyl arsenical melarsoprol (MelB), the drug still in use today in the treatment of late stage HAT (Figure 16.5). Despite more than 50 years of use, the mode of action of the arsenicals is uncertain. A number of primary targets have been considered, including several glycolytic kinases, and trypanothione (Figure 16.6). The melarsoprol–trypanothione adduct is a potent inhibitor of trypanothione reductase (TRYR) but neither trypanothione nor TRYR seemed to be involved directly in resistance. Resistant cells were shown to contain much less lipoic acid, which led to the suggestion that this cofactor of the pyruvate dehydrogenase complex might also be involved in resistance to arsenicals, although direct evidence for this is lacking.

Resistance to melaminophenyl arsenicals has been induced and studied *in vitro*. Wild-type trypanosomes were shown to have two adenosine/adenine transporters, P1 and P2. Through competitive inhibition studies, P2 was found to transport melaminophenyl arsenicals, including MelB. Trypanosome cells resistant to arsenicals displayed a loss of P2 transporter function and a decreased uptake of the drug (Figure 16.7). The gene coding for the P2 transporter was recently cloned and yeast over-expressing this gene became more sensitive to arsenicals. The P2 transporter shows substantial homology to a number of eukaryotic nucleoside transporters. Recent studies have revealed several mutations in the gene that codes for the P2 transporter, TbAT1, in a laboratory-derived melarsoprol-resistant stock of *T. brucei*. The mutated transporter is incapable of adenosine transport. Analysis of *T. b. gambiense* isolates not responding to melarsoprol, which were collected from a focus in north-western Uganda, indicated that the majority of parasites contained mutant TbAT1 genes. Nonetheless a considerable proportion of wild-type TbAT1 were also found among relapse patients, strongly indicating that TbAT1 may contribute

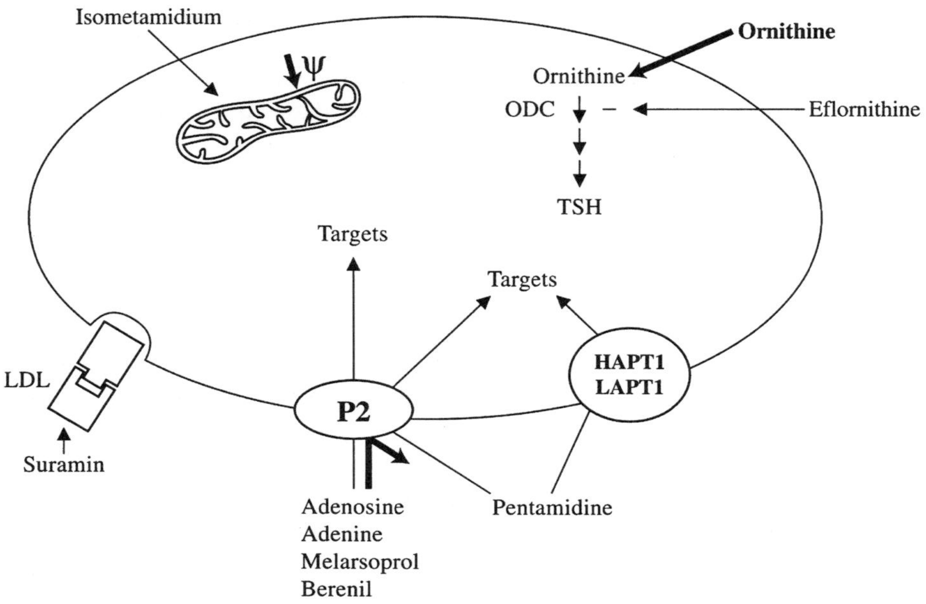

FIGURE 16.7 Drug transport and targets in *Trypanosoma brucei*. Melarsoprol is taken up by the P2 adenosine transporter. Mutations in P2 lead to reduced adenosine uptake and reduced melarsoprol uptake. The diamidines pentamidine and berenil are also taken up by P2, and mutations in P2 lead to reduced accumulation of pentamidine, although pentamidine is also taken up by high affinity (HAPT1) and low affinity (LAPT1) pentamidine transporters. Eflornithine inhibits the enzyme ornithine decarboxylase (ODC). In eflornithine-resistant mutants the rate of ornithine uptake was augmented. Suramin is proposed to enter the cell bound to low density lipoprotein (LDL), through receptor-mediated endocytosis. The veterinary drug isometamidium is accumulated into the mitochondria, and resistance to this class of drug has been associated with changes in the mitochondrial electrical potential (ψ).

to melarsoprol refractoriness, but is not the only gene product involved. Other markers of resistance may correspond to ABC proteins. Indeed, several ABC transporters have been described in *T. brucei*, and one gene highly similar to the *Leishmania* PGPA was recently cloned and transfected in *T. brucei* and was shown to confer melarsoprol resistance. The mechanism by which it confers resistance (efflux, drug sequestration) is unknown, but it will be highly interesting to test whether mutations in this gene are correlated to melarsoprol resistance in clinical isolates. The availability of drug resistance markers could be used in diagnostic tests for evaluating resistance in the field.

Eflornithine

Eflornithine, or D, L-α-difluoromethylornithine (DFMO) is used against MelB refractory late-stage *T. b. gambiense* infections. Its price and its availability limit its use in countries where infection is endemic. The drug, first developed as an anti-cancer agent, is a suicide inhibitor of the cellular enzyme ornithine decarboxylase (ODC), the limiting step in polyamine biosynthesis. Polyamines are essential for eukaryotic cell division and play a role in differentiation. Polyamines are also a component of trypanothione (Figure 16.6). Although DFMO is active against both the parasite and the host ODCs, its selective toxicity is the result of the longer half-life of the parasite enzyme. The higher specific ODC activity and faster enzyme turnover in *T. b. rhodesiense* compared to *T. b. gambiense* has been suggested to contribute to the intrinsic resistance of the East African species to DFMO. Other differences in polyamine biosynthesis were also suggested to contribute to the higher intrinsic resistance of *T. b. rhodesiense*. Field isolates of *T. b. gambiense* resistant to eflornithine have not yet been reported, but resistance was induced *in vitro*. In mutants selected for DFMO resistance, the ODC activity was shown to be similar between wild-type and resistant trypanosomes, in contrast to *Leishmania* in which the *ODC* gene was overexpressed in DFMO-resistant mutants. Instead, an important increase in the uptake of ornithine was found in DFMO-resistant *T. brucei*. The high level of the substrate ornithine was proposed to compete with DFMO for ODC, leading to sufficient polyamine biosynthesis (Figure 16.7).

Pentamidine

The isethionate salt of pentamidine is the drug of choice in early-stage Gambian HAT. The cellular target of pentamidine in trypanosomes is unknown, although its uptake is carrier-mediated. As pentamidine accumulates in the millimolar range in trypanosomes, it is likely that it will interact with numerous targets. Although MelB resistance seems to be on the rise, very few cases of pentamidine-refractory *T. b. gambiense* have been reported. Most of our understanding of pentamidine resistance mechanisms stems from work carried out *in vitro*. Trypanosomes resistant to melaminophenyl arsenicals are often cross-resistant to pentamidine. Mutations in the P2 nucleoside transporter lead to decreased uptake of both melarsen oxide and pentamidine. The common structural feature for recognition is likely to reside in the melamine and benzamidine moieties of the two classes of drugs. P2 (also known as ASPT1, for adenosine-sensitive pentamidine transporter) is not the only route by which pentamidine can enter *T. brucei*. Indeed, evidence for both a high-affinity and a low-affinity pentamidine transporter, biochemically distinct from P2, has been provided. This apparent multiplicity in routes of uptake may provide an explanation for the low rate of resistance to pentamidine. The analysis of some pentamidine-resistant *T. brucei* mutants showed that they accumulated pentamidine to the same levels as the drug-sensitive parental strain, and that the properties of the P2 transporter were unaltered. Mechanisms other than reduced uptake are therefore likely to contribute to pentamidine resistance.

Suramin

Suramin is a dye-related drug, clinically effective in the treatment of early *stage T. b. rhodesiense* HAT but with an unknown mechanism of action. It contains six negative charges at physiological pH, which renders its diffusion across membranes difficult. It appears that suramin enters the cell while bound to low density lipoprotein (LDL) through receptor-mediated endocytosis (Figure 16.7). Resistance to suramin is rare and usually not a problem in clinical practice. Suramin resistance can be induced in mice by administration of subcurative doses of the drug, and the resistance phenotype appeared stable. The resistance mechanisms have not been investigated, but suramin-resistant cells are not cross-resistant to arsenicals and diamidines.

Resistance to other classes of drugs

Only four drugs are available against HAT, and since there are no trypanocidal drugs in phases I–III of clinical development, it is unlikely that new chemotherapeutic agents will arise in the near future. The challenge of MelB-resistant HAT has led to the empirical use of drug combinations, and some case reports show successful treatment of refractory parasites. For example, a number of nitroheterocycle-containing drugs like nifurtimox (see below) and metronidazole (see below) have been used in combination with a number of classical anti-trypanosome drugs. Megazol is an experimental 5-nitroimidazole drug, which was shown to partially cross the blood–brain barrier, and was effective in curing chronic experimental infections in primates. However, megazol is mutagenic in the Ames test, so other non-mutagenic analogs need to be identified. Megazol appears to enter cells by simple diffusion. Once inside the parasite, megazol and related molecules need to be activated by reduction of the nitro group. Change in the enzymatic reduction of megazol leading to a loss in its activation would potentially lead to drug resistance.

Drug resistance is also a major problem in the treatment of veterinary trypanosomiasis. Treatment relies mostly on isometamidium chloride (Samorin) and diminazene aceturate (Berenil). Isometamidium, an ethidium-containing compound, crosses the plasma membrane, possibly by facilitated diffusion, and is accumulated in the mitochondria using the mitochondrial electrical potential as the driving force. Changes in the mitochondrial electrical potential have been demonstrated in isometamidium-resistant trypanosomes. It is thus possible that isometamidium resistance is linked to a reduction in mitochondrial accumulation, where possibly its target is located (Figure 16.7). Berenil is a diamidine, structurally related to pentamidine. Its mode of action is unknown and the P2 transporter seems to be the main route of entry. A loss of P2 could correspond to one mechanism by which parasites become resistant. This has not been tested, however, in parasites directly isolated from animals.

T. cruzi and chemotherapy

The protozoan kinetoplastid parasite *Trypanosoma cruzi* is the cause of Chagas disease. It is endemic from Northern Mexico to Argentina with an estimated 16 to 18 million people in Latin America infected. Considerable progress has been made in vector control (vectorial transmission is now eradicated in some countries) but several million people are still infected. The parasite invades a variety of cells including muscle and nerve cells of the heart and the gastrointestinal tract, leading to potentially fatal cardiomyopathy and gastrointestinal tract lesions. There is no prospect for a vaccine in the near future and chemotherapy, as with several other parasitic diseases, is unsatisfactory and relies on nifurtimox and benznidazole. These molecules are effective in acute cases, but are not effective against the chronic form of the disease. Both drugs are toxic and their efficacy varies, probably as a consequence of variation among parasite strains. Indeed, the efficacy of nitrofurans and nitroimidazoles varies greatly across Latin America, and this appears to be due to biological differences between the *T. cruzi* strains, some of which are naturally refractory to the existing drugs. More recently, azole antimicrobial agents were shown to have excellent activity in infected mice with both the acute and the chronic diseases. Work *in vitro* suggests, however, that resistance against azoles can develop rapidly in *T. cruzi*.

Benznidazole

Benznidazole is a nitroimidazole, and its mode of action is unclear. It has been suggested that it interferes with the synthesis of *T. cruzi* macromolecules, including RNA and protein. How the latter is achieved, however, has not yet been elucidated. Benznidazole also has numerous effects on host cells, and it is possible that some of its anti-*T. cruzi* activity is partially mediated through the immunomodulatory properties of the drug. Clinical resistance to benznidazole is frequent, but this may be because there are geographically diverse strains that are intrinsically resistant to the drug. A number of strains were analysed using a range of epidemiological polymorphic markers, which allowed the parasites to be separated into three groups. Resistant isolates were clustered in two of the three groups, but did not segregate with any of the markers used, including a number of ABC transporter genes. Resistance to benznidazole has been induced *in vitro* in epimastigotes, and resistance was retained in parasites differentiated to the amastigote stage. *T. cruzi* benznidazole-resistant strains were also selected in animals and again the clones retained their resistance even in the absence of drug pressure. The exact mechanism(s) of resistance in these laboratory-induced cases of resistance is(are) currently unknown.

Nifurtimox

Nifurtimox is a nitrofuran whose anti-*T. cruzi* activity may be due to the generation of oxygen-reduction products, which will ultimately lead to the generation of highly reactive hydroxyl radicals that will be poorly detoxified by the parasites. As for benznidazole, a number of *T. cruzi* isolates are naturally resistant to nifurtimox, although no specific markers have yet been linked to the resistance phenotype. Nifurtimox resistance has been induced in both epimastigotes and in tissue-derived trypomastigotes and a variety of karyotype changes, some accompanied by a wide spectrum of DNA changes, occurred in the nifurtimox-resistant strains. It remains to be seen, however, whether these genetic changes can lead to drug resistance.

Resistance to other drugs

The existence of *T. cruzi* strains naturally resistant to benznidazole and nifurtimox has led to the search for novel inhibitors. *T. cruzi* is extremely sensitive to ergosterol biosynthesis inhibitors. The antifungal agents ketoconazole and itraconazole have good activity against *T. cruzi* although they failed to cure experimental Chagas disease and chronic-phase human clinical disease. The search for new ergosterol biosynthesis inhibitors led to the discovery of the bis-triazole derivatives D0870 and SCH56592 (posaconazole), which are cytochrome P-450-dependent $C_{14}\alpha$ sterol demethylase inhibitors. These compounds could cure both acute and chronic Chagas disease. This new class of inhibitors is active against *T. cruzi* strains naturally resistant to nitrofurans and nitroimidazoles. A number of *T. cruzi* strains were shown, however, to be naturally resistant to these new azole molecules. Resistance to azole drugs has been induced *in vitro*. Resistance to fluconazole led to cross-resistance against other azole drugs but not against benznidazole. The resistance induced *in vitro* was stable even in absence of the drug, and was maintained in animal models. As therapy against Chagas disease requires a long treatment period, the potential for the emergence of drug resistance is real. It would thus be reasonable to think of using azole drugs in combination with benznidazole or nifurtimox. In fact, such combinations have already shown their usefulness in experimental animal models.

DRUG RESISTANCE IN ANAEROBIC PARASITES

Trichomonas vaginalis

Trichomonas vaginalis is a sexually transmitted protozoan parasite with an estimated 180 million new infections annually. The consequences of infections can range from asymptomatic to an acute inflammatory disease. The parasite is an anaerobic protozoan flagellate which lacks mitochondria and peroxisomes but has a specialized double-membrane-bounded organelle called the hydrogenosome. The hydrogenososme is critically involved in metabolic processes that extend glycolysis. An important metabolic function of the hydrogenosome is in oxidative decarboxylation of pyruvate to acetyl CoA via a ferredoxin-mediated electron transport system. The key enzyme involved is the iron–sulfur-containing pyruvate ferredoxin oxidoreductase (PFOR), and the transport of electrons generated by PFOR requires [2Fe–2S] ferredoxin and [Fe] hydrogenase. The drug of choice against *Trichomonas vaginalis* is the 5-nitroimidazole metronidazole (Figure 16.5J).

Metronidazole resistance

Metronidazole enters the parasite and the hydrogenosome by diffusion. Within this key organelle metronidazole competes with hydrogenase for the electrons produced from the PFOR system. This results in the one-electron reduction of the nitro group within the drug. The consequent production of highly reactive nitroso species is considered to be central to the mechanism of action of this drug.

Metronidazole resistance has been identified in field parasite isolates and in parasite lines selected under drug pressure both *in vitro* and *in vivo*. There is evidence for altered hydrogenosome morphology in resistant parasites, the relevance of which is not clear. Two types of resistance have been characterized biochemically, depending on the conditions required to demonstrate altered susceptibility. These are an anaerobic mechanism and an aerobic mechanism. The anaerobic mechanism of metronidazole resistance is associated with a loss or elimination of the PFOR activating pathway, notably a loss of PFOR activity. As metronidazole resistance is increased, ferredoxin, hydrogenosomal malic enzyme and NAD: ferredoxin reductase levels decrease. Loss of this key pathway is associated with a progressive increase in 2-oxoacid oxidoreductases as a compensatory mechanism incapable of activating metronidazole. There remains controversy as to whether this resistance mechanism, readily induced *in vitro*, operates in field isolates.

Aerobic resistance is only seen in the presence of oxygen and is associated with impaired oxygen scavenging. It is argued that the increased intracellular oxygen concentrations are involved in the re-oxidation of nitro free radicals or the competitive removal of electrons, thereby interfering with drug activation. Importantly, most, but not all, strains responsible for clinical resistance display aerobic resistance. These organisms remain susceptible to metronidazole under anaerobic conditions. There is evidence that, during the acquisition of anaerobic resistance, parasite isolates first develop the aerobic resistance mechanism which was detectable in the presence of oxygen but not under anaerobic conditions. PFOR activity in these isolates was significantly lower than that in the parent strain. In addition to these mechanisms, there is evidence to support a role for reduced levels or altered redox characteristics of ferredoxin and altered hydrogenase activity in some drug-resistant trichomonads. Several clinical isolates have been characterized as having reduced ferredoxin levels due to a reduction in transcription associated with alterations in upstream regulatory regions of the gene. An alternative resistance mechanism based on a *Trichomonas* P-glycoprotein transporter gene (*tvpgp*) has so far failed to shed any further light on the basis for metronidazole resistance.

Giardia duodenalis

The protozoan flagellate *Giardia duodenalis* is a common intestinal protozoan parasite and is the causal agent of giardiasis. The WHO have estimated that up to 1000 million people are infected with *Giardia* at any time, and the infection contributes to 2.5 million deaths annually from diarrheal disease. The consequences of infection can range from asymptomatic through to extreme weight loss and death. Infection rates are higher in children than in adults. The drug of choice for the treatment of giardiasis is metronidazole, although the acridine quinacrine, the nitrofuran furazolidine and the benzimidazole albendazole have all been used therapeutically. *In vitro* resistance has been documented with all four classes of drug.

Metronidazole

Clinical resistance to metronidazole has been reported to be as high as 20% and resistant organisms have been isolated from patients. Cross-resistance between metronidazole and the structural analog tinidazole has also been demonstrated. The mechanism of action of metronidazole against *Giardia* is essentially the same as that against *Trichomonas*, although drug activation occurs in the cytosol as these parasites do not have a hydrogenosome or hydrogenase activity. Resistance is again associated with downregulation of PFOR and ferredoxin. The downregulation of PFOR in *Giardia* is again compensated for by other oxoreductases which do not activate the drug. In addition to this reduced activation model of metronidazole resistance in *Giardia* there is evidence for an efflux-based drug resistance mechanism which is yet to be defined in molecular detail.

*Resistance to other anti-*Giardia *drugs*

Quinacrine resistance in *Giardia* is characterized by enhanced drug efflux and is more rapidly achieved *in vitro* from furazolidone-resistant parasite lines. Furazolidone is a nitrofuran that requires activation via NADH oxidase to produce free radical species. Resistance to this drug is readily achieved *in vitro* under drug selection, and resistant isolates have been recovered from patients refractory to the actions of the drug. Resistance to the benzimidazole albendazole is readily achieved *in vitro* and again selection is easier to establish in parasites already resistant to furazolidone. The mechanism of albendazole resistance is not known, but it is not related to mutations in tubulin genes as reported in other resistant organisms.

Entamoeba histolytica

Entamoeba histolytica is a protozoan parasite of the large intestine. Symptoms of infection range from mild acute amebic colitis with bloody stools and fever to severe life-threatening disease involving parasite invasion of tissues and organs. Many of the statistics relating to infection rates and disease burden from *E. histolytica* have been compromised by the failure to recognize and separate non-pathogenic *Entamoeba dispar* from invasive *E. histolytica*. Using this definition it is estimated that *E. histolytica* causes severe disease in some 50 million people each year, of whom 70 to 100 000 die. As with *Giardia* and *Trichomonas* the drug of choice for the treatment of amebiasis is metronidazole.

Metronidazole

The development of high-level metronidazole resistance in *E. histolytica* is much more difficult than has been the experience with *Trichomonas* and *Giardia*. Also in contrast to *Trichomonas* and *Giardia, E. histolytica* selected for metronidazole resistance in the laboratory is found to have levels of PFOR equivalent or slightly greater than those in the susceptible parent isolate. Furthermore there is no evidence for the involvement of an active drug efflux transporter in metronidazole-resistant *E. histolytica*. Resistance in these parasites is related to the overexpression of iron superoxide dismutase and peroxiredoxin, and a decrease in the expression of ferredoxin and flavin reductase. The critical involvement of iron superoxide dismutase and peroxiredoxin was confirmed by episomal transfection of these antioxidant enzymes into metronidazole-susceptible isolates, which was associated with reduced drug susceptibility.

Emetine resistance

Laboratory-induced resistance to emetine in *E. histolytica* displays a classic MDR phenotype associated with reduced drug accumulation which can be reversed by agents such as verapamil. Molecular analysis has revealed the involvement of two MDR-type drug efflux pumps: *Ehpgl*, which is upregulated in resistant parasites by a C/EBP-like factor and a multiprotein complex; and *Ehpgp5*, which is upregulated by the drug through AP-1 and HOX factors. Transfection studies suggest that *Ehpgp1* is responsible for low-level or basal emetine resistance, with *Ehpgp5* contributing to high-level resistance.

HELMINTHS AND CHEMOTHERAPY

Helminth infections, which are responsible for debilitating diseases such as schistosomiasis, filariasis, intestinal helminth infection and onchocerciasis affect more than half a billion people, contributing significantly to the world disease burden. Helminths are also important parasites of man's livestock and cause serious economic losses. In the absence of effective vaccines, worm infections in man and animals are controlled most efficiently by treatment with anthelmintics. The availability of broad-spectrum anthelmintics as single dose therapy and their use in mass treatment control programs has helped in the control of worm infections. Bona fide drug resistance in common worm infections is not yet widespread in human medicine. However, drug resistance is widespread in veterinary medicine, and thus the potential for resistance is important in worms infecting humans.

Resistance to praziquantel and other anti-schistosome drugs

Schistosomes are multicellular parasites, whose life cycle includes both human and snail hosts. It is estimated that 200 million people are infected, of whom more than 50% are symptomatic. Chemotherapy is presently the only efficient way to control schistosome infections. Praziquantel (Figure 16.8) is the drug of choice for treatment since it is active against all the *Schistosoma* species, particularly against *S. mansoni, S. haematobium* and *S. japonicum*. Praziquantel also has good activity against cestode infections. The mechanism of action

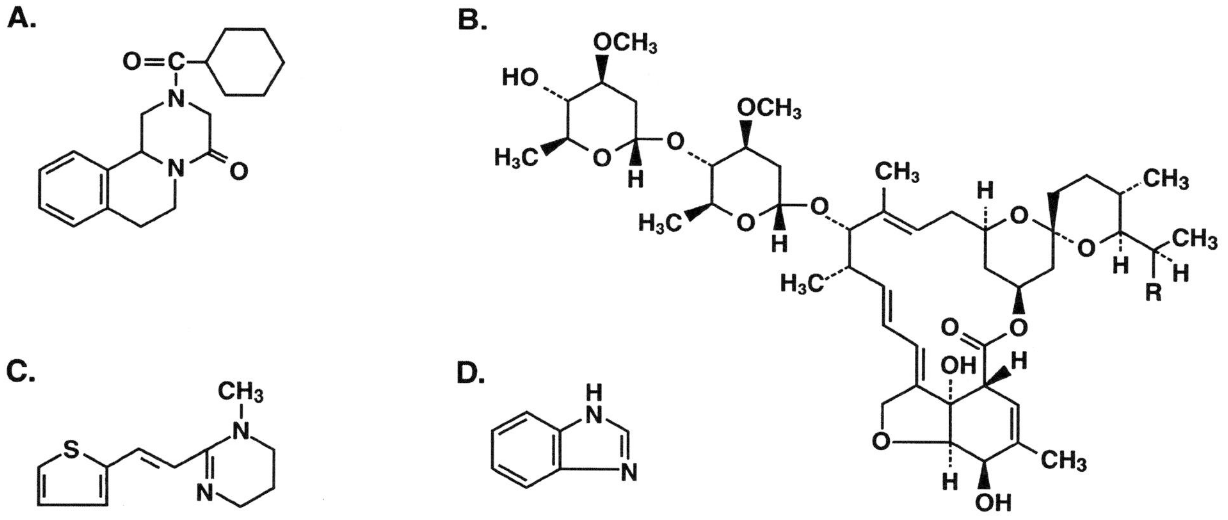

FIGURE 16.8 Structure of the main drugs used against worm infections. (A) Praziquantel; (B) ivermectin; (C) pyrantel; (D) benzimidazole.

of praziquantel is not known precisely. It is known, however, to cause muscle contraction that is linked to calcium flux, although no targets have yet been pinpointed. The description of resistant isolates from the field has not yet been unambiguously described, although there is evidence that suggests resistance is indeed a possibility. Notably, a lower praziquantel cure rate was noted in Senegal and Egypt. Several factors, such as the infection intensity or immunological factors, could contribute to this reduced efficacy but arguments in favor of resistant parasites are available. For example, both non-responding clinical isolates and praziquantel-resistant worms selected under laboratory conditions had a lower percent reduction in worm burdens in mice than controls. The mechanisms by which resistance occurs is unknown, but randomly amplified polymorphic markers were circumstantially linked to sensitive and resistant worms isolated from Egypt. A difference between the morphology of sensitive and resistant worms has been observed in response to praziquantel, and this change has been suggested as an *in vitro* test to monitor praziquantel resistance. Hycanthone and oxamniquine have been used in the treatment of schistosomes but the use of the former has been discontinued. The drugs are probably converted in the schistosome to an active molecule capable of alkylating DNA. Resistance seems to be linked to the absence of the enzymatic activity, that seems to be a sulfotransferase capable of activating the precursors of the two drugs.

Resistance to macrocyclic lactones

The macrocyclic lactones, avermectins, are broad spectrum anthelmintics with a wide margin of safety in vertebrates. Ivermectin is the drug of choice in the treatment of *Onchocerca volvolus*, the organism responsible for onchocerciasis. It is also widely used in veterinary medicine and agriculture. Ivermectin opens the worms' chloride channels and paralyses the pharynx, the body wall and the uterine muscles of nematodes, leading to starvation and paralysis. The evidence suggests that several gene products are involved in the mechanism of action of ivermectin. Intensive usage of ivermectin has selected for resistant parasites in the field, but fortunately for human beings, resistance has so far only been detected in nematodes of goats and sheep. Ivermectin-resistant strains are sometimes cross-resistant to structurally unrelated drugs, and multidrug-resistance reversing agents can partially reverse ivermectin resistance in *Haemonchus contortus*. Genetic analyses have linked a number of P-glycoprotein homologs to ivermectin resistance in a number of worm genera. This makes sense since ivermectin is an excellent substrate of the mouse MDR1a P-glycoprotein. In the nematode *Caenorhabditis elegans* simultaneous mutations in three genes encoding glutamate-gated chloride channel α-type subunits confers high-level resistance to ivermectin. A number of similar channels have been found in pathogenic worms, and polymorphisms in one glutamate-gated chloride channel was linked to resistance in *H. contortus*. Both target mutations and transport alteration can therefore lead to ivermectin resistance in worms, and other studies suggest that increased glutamate uptake can also be linked to ivermectin resistance. It is possible that different worm species will have different mechanisms of resistance towards ivermectin, and at present resistance has not been reported in *O. volvulus*, the etiological agent of onchocerciasis.

Resistance to levamisole and pyrantel

Levamisole and pyrantel are nicotinic anthelmintics that act as cholinergic agonists in nematodes, most likely by inhibiting the nicotinic acetylcholine receptors of the worms. This results in depolarization of nematode muscle cell membranes. In *C. elegans* a number of mutations in some of the subunits (lev 1, unc-29 and unc-38) of the nicotinic acetylcholine receptor have been correlated with levamisole resistance, although this remains to be shown for pathogenic worms. Resistance appears to be multigenic, although a reduction in the number or affinity of receptors has nonetheless been proposed as one possible mechanism for resistance. Pyrantel also acts on the nicotinic acetylcholine receptors, and patch-clamp experiments suggested that pyrantel resistance, like levamisole resistance, is associated with a modification of the target.

Resistance to benzimidazoles

Benzimidazoles, the most widely used anthelmintics, are used mainly in the treatment of veterinary intestinal nematode infections. They selectively inhibit tubulin polymerization, possibly by binding specifically to the parasite β-tubulin. Resistance to benzimidazole is now a problem, particularly in the ruminant parasite *H. contortus* and in small strongyles of horses. Point mutations in the β-tubulin gene, particularly at position 200, are associated with resistance to this class of drug. The role of these mutations in resistance was confirmed by gene transfection in *C. elegans* and by biochemical means. Other alterations in the tubulin genes can occur in highly resistant worms. The understanding of the main resistance mechanism has led to several PCR-based diagnostic tests for the detection of resistance to benzimidazole in a number of species of gastrointestinal worms.

CLINICAL IMPLICATIONS AND OUTLOOK

Drug resistance complicates the treatment of parasite infections. Studies on drug resistance

TABLE 16.1 Genes implicated in drug resistance in parasites for which DNA-based tests have been developed

Parasite	Genes	Drugs
Plasmodium falciparum	*PfCRT, Pfmdr1*	Chloroquine
Plasmodium falciparum	*DHFR, DHPS*	Antifolates and sulfonamides
Trypanosoma brucei	*TbTA1* (*P2*)	Melarsoprol
Haemonchus contortus	*TUB*	Benzimidazole

can help in finding strategies to increase the efficacy or the life span of the few drugs available. From reading the various sections of this chapter, it is clear that considerable progress has been made in recent years in our understanding of drug resistance in parasites. The availability of gene transfection for most parasites has facilitated the testing of putative resistance genes in sensitive parasites allowing an unambiguous reconstruction of resistance mechanisms. With genomes of parasites rapidly finding their place in databanks, the identification of genes involved in drug resistance will be simplified, and will become a driving force in the improvement of the use of existing drug combinations and in the development of new drugs. We anticipate that transcriptomic (DNA microarrays) and proteomic approaches will accelerate the pace by which we will understand the mode of action of drugs and their resistance mechanisms.

The first obvious applied outcome of understanding resistance mechanisms is to develop tools and assays for rapidly detecting resistance in clinical isolates. Clinical studies will first be required to test whether a resistance genotype correlates with the clinical response of the infected patient under drug treatment. In the case of a positive correlation, a rapid DNA-based diagnostic test can have tremendous impact for treatment. Indeed, in the case of a resistance genotype, useless and toxic drug formulations would not be used, and alternative treatments would be sought early during infection. Where resistance is mediated by numerous genes or mutations it could be possible to use low-density DNA microarrays, whereas if that resistance is mediated by a change in gene expression, real-time PCR assays could prove practical.

The monitoring of drug resistance by the use of DNA-based diagnostic procedures is at present limited by our lack of knowledge about the molecular and biochemical mechanisms of action and resistance of several of the antiparasitic agents. There are nonetheless notable exceptions for which molecular markers are now available to detect some forms of resistance occurring in field isolates (Table 16.1). The list is still short, and it is clear that for chloroquine and melarsoprol, resistance genes other than those listed in Table 16.1 are implicated in resistance. Thus, to be useful for therapeutic interventions, all the possible resistance mechanisms should be targeted in diagnostic assays. Once these tests are available and validated, and if they can be used in areas where the diseases are endemic, they would have the potential to decrease dramatically treatment failure and human suffering.

These DNA markers and diagnostic assays could also be used to monitor the pool of resistant parasites in diverse geographical regions, and to carry out large epidemiological studies. The high prevalence of antibiotic resistance in bacteria, and our relatively good understanding of several of the genes implicated in antibiotic resistance, have mobilized international efforts for global surveillance programs. It is hoped that these programs will contribute to reduce the dissemination of resistant isolates. We are far from this ideal for parasitic diseases. Indeed, in addition to our incomplete understanding of the genetic basis of resistance, networks of surveillance for resistant parasites are not yet organized and in place to affect national policies and treatment guidelines in defined geographical regions.

The understanding of resistance mechanisms and mechanism of action of drugs may also point the way to more rational use of drugs and drug combinations to minimize development of resistance and to achieve more effective chemotherapy. Indeed, there is now considerable evidence demonstrating that drug combinations, with different mechanisms of action and resistance, will on the one hand increase the efficiency of treatment and on the other hand reduce the probability of selecting for resistant parasites. Thus the understanding of the mode of action of drugs, of their pharmacology, and of resistance mechanisms should facilitate the choice of drugs to be used in combination in clinical trials. The concept of combination chemotherapy is already well established for treating HIV, cancer, tuberculosis and leprosy, and it is expected to grow in popularity in the coming years for other microorganisms. Few drugs are available in the treatment of parasitic diseases, and there is increasing agreement that the use of simultaneous or sequential combinations of agents must be implemented in order to protect the limited precious resource of therapeutically effective drugs. Several studies using drug combinations against malaria have highlighted their potential, and combination chemotherapy is rapidly being seen as the norm; this is at least in the planning stage for all the parasites causing disease in man and animals. Drug combinations have already been used empirically to treat (successfully) refractory isolates. Obviously, controlled clinical trials will be required to assess the efficacy and safety of such drug combinations.

Finally, studies of drug resistance will permit the identification of key intracellular targets and parasite defense mechanisms, which can be exploited for the rational development of drug analogs not affected by the most common defenses. There is a large body of evidence demonstrating efflux-mediated resistance in a number of parasites. Strategies and inhibitors to modulate the activity of efflux pumps are being developed, and analogs of these, in combination with our currently available drugs, could be useful in the treatment of parasitic diseases when a transport-related mechanism is the main resistance mechanism. *In vitro* work on antimony resistance in *Leishmania* has demonstrated that an increase in trypanothione and in transport systems are implicated in resistance. The use of trypanothione biosynthesis inhibitors such as buthionine sulfoxime (BSO) and DFMO (Figure 16.2) (two drugs already approved for use in humans) in combination with antimony was shown to reverse antimony resistance *in vitro*, and if warranted this combination could be used *in vivo*. Similarly, since arsenicals and trypanothione metabolism appear to be linked in *T. brucei*, MelB may turn out to be more effective when used in combination with BSO or DFMO. Our understanding of pyrimethamine resistance in *Plasmodium* and of its associated mutations in DHFR has led to the synthesis and testing of new antifolates such as WR99210 that have less potential for the selection of resistance. Using pyrimethamine and WR99210 in combination may serve to slow down the emergence of resistance, as was demonstrated for the combination AZT–3TC against HIV.

The structure of protein targets at the atomic level will permit a detailed structure–function analysis, which can lead to the rational design of selective inhibitors. In this 'post-genomic' era there are already major research efforts for several of the main parasites to carry out high-throughput protein-structure determination. These structures should prove to be a fruitful source for the development of novel anti-parasitic agents. Development of these drugs is unlikely to be led by the pharmaceutical industry, which usually does not invest in neglected diseases of poor countries. Nonetheless, large philanthropic organizations and public and parapublic national and international organisations could make investments to develop new anti-parasitic agents, and initiatives in that direction, although still modest in scale, are tangible. With the development of drug resistance and the limited numbers of drugs available, it is indeed necessary that such efforts be encouraged in our fight against parasitic diseases.

ACKNOWLEDGMENTS

MO is a CIHR investigator and a Burroughs Wellcome Fund Scholar in molecular parasitology. Work on drug resistance in the MO lab is supported by the CIHR and the Wellcome Trust.

FURTHER READING

Barrett, M.P. and Fairlamb, A.H. (1999). The biochemical basis of arsenical–diamidine crossresistance in African trypanosomes. *Parasitol. Today* **15**, 136–140.

Borst, P. and Ouellette, M. (1995). New mechanisms of drug resistance in parasitic protozoa. *Annu. Rev. Microbiol.* **49**, 427–460.

Bray, P.G., Mungthin, M., Ridley, R.G. and Ward, S.A. (1998). Access to hematin: the basis of chloroquine resistance. *Mol. Pharmacol.* **54**, 170–179.

Carter, N.S. and Fairlamb, A.H. (1993). Arsenical-resistant trypanosomes lack an unusual adenosine transporter. *Nature* **361**, 173–176. [erratum in *Nature* (1993) **361**, 374.]

Cotrim, P.C., Garrity, L.K. and Beverley, S.M. (1999). Isolation of genes mediating resistance to inhibitors of nucleoside and ergosterol metabolism in *Leishmania* by overexpression/selection. *J. Biol. Chem.* **274**, 37723–37730.

de Koning, H.P. (2001). Transporters in African trypanosomes: role in drug action and resistance. *Int. J. Parasitol.* **31**, 512–522.

Fidock, D., Nomura, T., Talley, A. *et al.* (2000). Mutations in the *P. falciparum* digestive vacuole transmembrane protein PfCRT and evidence for their role in chloroquine resistance. *Mol. Cell* **6**, 861–871.

Geerts, S. and Gryseels, B. (2000). Drug resistance in human helminths: current situation and lessons from livestock. *Clin. Microbiol. Rev.* **13**, 207–222.

Kohler, P. (2001). The biochemical basis of anthelmintic action and resistance. *Int. J. Parasitol.* **31**, 336–345.

Légaré, D., Richard, D., Mukhopadhyay, R. *et al.* (2001). The *Leishmania* ABC protein PGPA is an intracellular metal–thiol transporter ATPase. *J. Biol. Chem.* **276**, 26301–26307.

Maser, P., Sutterlin, C., Kralli, A. and Kaminsky, R. (1999). A nucleoside transporter from *Trypanosoma brucei* involved in drug resistance. *Science* **285**, 242–244.

Matovu, E., Seebeck, T., Enyaru, J.C. and Kaminsky, R. (2001). Drug resistance in *Trypanosoma brucei* spp., the causative agents of sleeping sickness in man and nagana in cattle. *Microbes Infect.* **3**, 763–770.

McFadden, D.C., Camps, M. and Boothroyd, J.C. (2001). Resistance as a tool in the study of old and new drug targets in *Toxoplasma*. *Drug Resist. Updat.* **4**, 79–84.

Murta, S.M., Gazzinelli, R.T., Brener, Z. and Romanha, A.J. (1998). Molecular characterization of susceptible and naturally resistant strains of *Trypanosoma cruzi* to benznidazole and nifurtimox. *Mol. Biochem. Parasitol.* **93**, 203–214.

Ouellette, M. (2001). Biochemical and molecular mechanisms of drug resistance in parasites. *Trop. Med. Int. Health* **6**, 874–882.

Prichard, R. (2001). Genetic variability following selection of *Haemonchus contortus* with anthelmintics. *Trends Parasitol.* **17**, 445–453.

Reed, M.B., Saliba, K.J., Caruana, S.R., Kirk, K. and Cowman, A.F. (2000). Pgh1 modulates sensitivity and resistance to multiple antimalarials in *Plasmodium falciparum*. *Nature* **403**, 906–909.

Shahi, S.K., Krauth-Siegel, R.L. and Clayton, C.E. (2002). Overexpression of the putative thiol conjugate transporter TbMRPA causes melarsoprol resistance in *Trypanosoma brucei*. *Mol. Microbiol.* **43**, 1129–1138.

Shaked-Mishan, P., Ulrich, N., Ephros, M. and Zilberstein, D. (2001). Novel intracellular Sb(V) reducing activity correlates with antimony susceptibility in *Leishmania donovani*. *J. Biol. Chem.* **276**, 3971–3976.

Sibley, C.H., Hyde, J.E., Sims, P.F. *et al.* (2001). Pyrimethamine–sulfadoxine resistance in *Plasmodium falciparum*: what next? *Trends Parasitol.* **17**, 582–588.

Upcroft, P. and Upcroft, J.A. (2001). Drug targets and mechanisms of resistance in the anaerobic protozoa. *Clin. Microbiol. Rev.* **14**, 150–164.

Vaidya, A.B. and Mather, M.W. (2000). Atovaquone resistance in malaria parasites. *Drug Resist. Updat.* **3**, 283–287.

Wellems, T.E., Walker-Jonah, A. and Panton, L.J. (1991). Genetic mapping of the chloroquine-resistance locus on *Plasmodium falciparum* chromosome 7. *Proc. Natl. Acad. Sci. USA* **88**, 3382–3386.

C H A P T E R

17

Medical implications of molecular parasitology

Richard D. Pearson[1], *Erik L. Hewlett*[2] *and William A. Petri, Jr.*[3]
[1]Department of Medicine and Department of Pathology, University of Virginia School of Medicine, Charlottesville, VA, USA;
[2]Department of Medicine and Department of Pharmacology, University of Virginia School of Medicine, Charlottesville, VA, USA; and
[3]Department of Medicine and Department of Microbiology, University of Virginia School of Medicine, Charlottesville, VA, USA

BACKGROUND

The history of anti-parasitic therapy dates to antiquity. Healers have for millennia known of the therapeutic efficacy of various leaves, roots and bark. Effective therapy for malaria, for example, was identified long before the etiology and life cycle of *Plasmodium* spp. were defined. While the Greeks were attributing intermittent fevers to bad air, native South Americans were successfully administering extracts of the bark of the cinchona tree to treat malaria.

As the science of chemistry advanced in the nineteenth and twentieth centuries, biologically active molecules such as quinine from the cinchona tree were isolated and characterized. In other instances organic compounds were found empirically to have therapeutic activity. Congeners were subsequently synthesized and screened. Those with the greatest efficacy and least toxicity were adopted for therapy, usually without an understanding of the mechanism of action. Governments seeking to protect military or colonial personnel often supported these efforts. Many of the anti-parasitic drugs that are currently in use were identified in this manner (Table 17.1).

One of the most interesting examples began with the observation by Paul Ehrlich, the father of modern medicinal chemistry, and his colleagues in 1891 that methylene blue dye was effective in the treatment of malaria. Two decades later, difficulty in acquiring quinine during WWI led German scientists to synthesize

Molecular Medical Parasitology
ISBN 0-12-473346-8

TABLE 17.1 Anti-parasitic drugs discussed in this chapter

Drug	Use
Treatment of malaria and other Apicomplexa	
Quinine	Treatment of chloroquine-resistant *P. falciparum* and other *Plasmodium* spp.; effective against asexual erythrocytic phase Used with clindamycin for treatment of *Babesia* spp.
Quinidine	Treatment of chloroquine-resistant *P. falciparum* when parenteral therapy is required; effective against asexual erythrocytic phase
Chloroquine	Suppressive prophylaxis and treatment of the asexual erythrocytic phase of *P. vivax, P. ovale* and *P. malariae,* and susceptible *P. falciparum*
Mefloquine	Prophylaxis and occasionally treatment of chloroquine-resistant *P. falciparum*; effective against asexual erythrocytic phase of other *Plasmodium* spp.
Doxycycline	Used for suppressive prophylaxis in areas where there is endemic chloroquine-resistant *P. falciparum* and with quinine for the treatment of chloroquine-resistant *P. falciparum*
Atovaquone/Proguanil (Malarone)	Used for suppressive prophylaxis and treatment of chloroquine-resistant *P. falciparum*
Primaquine	Radical cure of the pre-erythrocytic hypnozoites of *P. vivax* and *P. ovale*
Artemisinin derivatives (Qinghaosu) Artesunate Artemether	Treatment of *Plasmodium* spp. in the asexual erythrocytic phase, including chloroquine-resistant *P. falciparum*
Pyrimethamine/short-acting sulfonamides	Used with quinine to treat asexual erythrocytic phase of chloroquine-resistant *P. falciparum* acquired in areas where resistance is not common Also used for the treatment of *Toxoplasma gondii*
Pyrimethamine-sulfadoxine (Fansidar)	Presumptive treatment of chloroquine-resistant *P. falciparum* in areas where isolates are sensitive
Clindamycin	Used with steroids for treatment of ocular *Toxoplasma gondii* in immunocompetent hosts; used with pyrimethamine for *T. gondii* encephalitis in persons with AIDS who cannot tolerate sulfonamides Used with quinine for treatment of *Babesia* spp.
Spiramycin	*Toxoplasma gondii* during pregnancy and in the neonate
Treatment of Kinetoplastida: trypanosomes and *Leishmania*	
Nifurtimox	*Trypanosoma cruzi*
Benznidazole	*Trypanosoma cruzi*
Eflornithine	*Trypanosoma brucei gambiense* (hemolymphatic or central nervous system involvement)
Suramin	*Trypanosoma brucei gambiense* and *Trypanosoma brucei rhodesiense* (hemolymphatic stage)
Pentamidine isethionate	*Trypanosoma brucei gambiense* and *Trypanosoma brucei rhodesiense* (hemolymphatic stage) *Leishmania* spp. (alternative) *Pneumocystis carinii* (alternative)
Melarsoprol B	*Trypanosoma brucei gambiense* and *Trypanosoma brucei rhodesiense* (late disease with central nervous system involvement)

(*Continued*)

TABLE 17.1 (*Continued*)

Drug	Use
Stibogluconate sodium	*Leishmania* spp.
Meglumine antimoniate	*Leishmania* spp.
Amphotericin B deoxycholate	*Leishmania* spp. (alternative)
Liposomal or lipid-associated Amphotericin B	Visceral leishmaniasis (only FDA approved drug for this indication)
Treatment of luminal protozoa	
Metronidazole	*Entamoeba histolytica* (invasive disease)
	Entamoeba polecki
	Trichomonas vaginalis
	Blastocystis hominis
	Giardia lamblia
	Balantidium coli (alternative)
Paromomycin	*Entamoeba histolytica* (asymptomatic luminal infection)
	Dientamoeba fragilis
Iodoquinol	*Entamoeba histolytica* (luminal infection)
	Dientamoeba fragilis
	Blastocystis hominis
	Balantidium coli (alternative)
Diloxanide furoate	*Entamoeba histolytica* (asymptomatic luminal infection)
Furazolidone	*Giardia lamblia* (alternative)
Treatment of intestinal nematodes	
Albendazole	*Ascaris lumbricoides*
	Ancylostoma duodenale
	Necator americanus
	Enterobius vermicularis
	Trichuris trichiura
	Strongyloides stercoralis (variably effective)
	Cutaneous larva migrans
	Trichostrongylus spp.
	Capillaria philippinensis (alternative)
Mebendazole	*Ascaris lumbricoides*
	Ancylostoma duodenale
	Necator americanus
	Trichuris trichiura
	Enterobius vermicularis
	Capillaria philippinensis
	Angiostrongylus cantonensis
	Trichostrongylus spp. (alternative)
	Trichinella spiralis (recommended by some; used with steroids)
Pyrantel pamoate	*Enterobius vermicularis*
	Ascaris lumbricoides
	Ancylostoma duodenale
	Necator americanus
	Trichostrongylus spp.

(*Continued*)

TABLE 17.1 (*Continued*)

Drug	Use
Treatment of filariae	
Ivermectin	*Onchocerca volvulus*
	Strongyloides stercoralis
	Mansonella ozzardi and *M. streptocerca*
Diethylcarbamazine	*Wuchereria bancrofti*
	Brugia malayi
	Brugia timori
	Mansonella streptocerca
	Loa loa
	Tropical pulmonary eosinophilia
Treatment of trematodes and cestodes	
Praziquantel	*Schistosoma* spp.
	Clonorchis sinensis
	Opisthorchis viverrini
	Paragonimus westermani
	Fasciolopsis buski
	Heterophyes heterophyes
	Metagonimus yokogawai
	Metorchis conjunctus
	Nanophyetus salmincola
	Taenia solium (adult worm and cysticercosis)
	Taenia saginata
	Diphyllobothrium latum
	Hymenolepis nana
	Dipylidium caninum
Bithionol	*Fasciola hepatica*
Albendazole	Neurocysticercosis
	Echinococcosis

and evaluate a series of related organic compounds. The result led to the introduction of quinacrine following the war and eventually chloroquine, which became one of the most successful anti-parasitic drugs.

The modern era of parasitology began with the introduction of the light microscope. Anthonius von Leeuwenhoek is credited with identifying the first protozoal pathogen, *Giardia lamblia*, in his stool during a bout of acute diarrhea in 1681. By the second decade of the twentieth century the major parasitic pathogens had been identified and their epidemiologies and life cycles described. While these advances had little immediate impact on the development of chemotherapy, they provided the foundation for recent developments in molecular parasitology that hold great promise for the future.

For basic scientists and clinical investigators, the rational design of new drugs has always engendered great enthusiasm. The strategy is to identify an essential parasite molecule or pathway (Table 17.2). Then, once a potential target is characterized, the medicinal chemist is called upon to synthesize specific inhibitor(s). Ideally the resulting compound is: (a) selectively active against the target in the parasite and not against

TABLE 17.2 Ideal properties of anti-parasitic drugs

Selectively active against a defined parasite target(s) and not against related molecules in the host
Free of undesirable side-effects and interactions with other drugs
Stable under a broad range of environmental conditions
Easy to synthesize and inexpensive
Favorable pharmacodynamics
Effective in one or few doses

related molecules in the host; (b) free of undesirable side effects and interactions with other drugs; (c) stable under a broad range of environmental conditions; (d) easy to synthesize and inexpensive; and (e) characterized by favorable pharmacodynamics.

While many anti-parasitic drugs have been identified through mass screening programs, a number were developed through a rational approach. The recent advances in the molecular and cell biology of parasites discussed earlier in this book offer great hope for the future. It is reasonable to predict that these, coupled with advances in medicinal chemistry such as catalytic asymmetric synthesis and microarray screening technologies, will result in dramatic improvements in anti-parasitic chemotherapy in the decades to come.

PRINCIPLES OF ANTI-PARASITIC CHEMOTHERAPY

Given the biological diversity of parasites, it is not surprising that a broad armamentarium of drugs has arisen to treat them. When discussing chemotherapy, it is helpful to group parasites with similar sensitivities. The taxonomic classification is a useful starting point, with further subdivisions based on genetic differences and/or common adaptations to survival in humans.

Protozoa are single-celled organisms that multiple by simple binary division. Many have complex life cycles with sexual and asexual stages. Arthropod vectors transmit some, while others survive in the environment as cysts and reach their host through contaminated food or water. Theoretically, infection with one protozoan can result in overwhelming infection, and as a corollary, eradication of all protozoa by the combined effects of drugs and the immune system may be necessary for cure.

As noted in Chapter 7, protozoa can be divided into the Apicomplexa, which include *Plasmodium* spp., *Babesia* spp. and *Toxoplasma gondii*; protozoa that reside under anaerobic conditions in the lumen of the intestinal tract or vagina, such as *Entamoeba histolytica, Giardia lamblia* and *Trichomonas vaginalis*; and those that live in the bloodstream or body tissues under aerobic conditions, such as *Leishmania* spp., *Trypanosoma cruzi, Trypanosoma brucei rhodesiense* and *Trypanosoma brucei gambiense*. With a few exceptions, eosinophilia is not a characteristic of protozoal diseases.

In contrast to the protozoa, helminths, or worms, are highly complex, multicellular organisms with discrete organ systems. Many have complex life cycles. Although a few helminths such as *Strongyloides stercoralis, Hymenolepis nana* and *Capillaria philippinensis* are capable of autoinfection, most require development outside their mammalian host either in the environment or in an invertebrate vector before they become infective.

Adult helminths have finite life spans. The clinical severity of the infection is usually related to worm burden. Reduction of the parasite load is often sufficient to ameliorate the manifestations of disease. It is not necessary to eradicate all helminths in an individual patient, but with fewer parasites, there is decreased likelihood of transmission to others. Eradication of parasites in an entire population remains the long-term goal. Eosinophilia, which is not typically present in protozoal infections, is often observed when worms migrate through tissues.

The helminths are divided into nematodes (roundworms) and platyhelminths (flat worms). Roundworms can be further subdivided into those that reside as adults in the human intestinal tract and those that reside in tissue. The platyhelminths are further subdivided into the trematodes or flukes, and cestodes or tapeworms.

The outcome of any parasitic infection depends on the genetically determined virulence of the infecting parasite and the genetically determined innate and acquired immune responses of its host. Much needs to be learned about the virulence factors of parasites, but advances have been made. For example, the mechanisms by which hookworms anticoagulate host blood so that it will flow through their gastrointestinal tracts have been worked out. As virulence traits are understood, potential points of intervention will likely emerge. The sequencing of genomes of major parasitic pathogens should expedite these efforts. The genetic approach to analysis of virulence is given in Chapter 6.

With virtually all parasitic infections, there is a spectrum of disease. Some humans mount effective host defenses and either eradicate the invading pathogen or substantially limit its numbers. For example, under conditions of apparently comparable transmission, intestinal helminths are not evenly distributed in the population. Some persons have heavy burdens while others have a limited number of worms. Recent epidemiological data from family studies of intestinal helminthic infections suggest that genetically determined variations in human susceptibility account for much of the variation. A review of the innate and acquired human host defenses effective against various parasites is beyond the scope of this chapter, but it is important to remember that the effects of chemotherapy must be viewed in the broad context of the host–parasite interaction.

Some anti-parasitic drugs act on molecules that are unique to the parasite. Others act preferentially against parasite enzymes. For example, pyrimethamine inhibits both plasmodial and human dihydrofolate reductase, but it is greater than 1000-fold more active against the parasite enzyme. Other drugs are effective by virtue of differences in their distribution between parasites and humans. In the case of chloroquine, the drug is concentrated in infected erythrocytes, resulting in high concentrations in parasitophorous vacuoles. Differences in metabolism render some drugs toxic to parasites, but not to humans. For example, electron transport proteins with low redox potential in anaerobic organisms like *Entamoeba histolytica* reduce metronidazole. It thus acts as an electron sink, depriving the parasite, but not the host, of the required reducing equivalents.

An overview of the major parasitic diseases and the drugs that are currently used to treat them reveals substantial differences in efficacy and toxicity. Protozoal pathogens currently provide the greatest challenge for chemotherapy. As discussed in Chapter 16, evolving drug resistance has been a major problem among *Plasmodium* species that cause malaria. The drugs now used to treat Chagas disease, African trypanosomiasis, and leishmaniasis are variably effective and have substantial toxicity. In contrast, while the helminths are among the least studied of the parasitic pathogens, highly active, broad spectrum, relatively non-toxic, recently recognized drugs are currently available to treat most of them.

The historical background, current status of chemotherapy and potential future contributions of molecular parasitology are discussed below in reference to each group of parasites. The mechanisms of action and major untoward effects of selected drugs are reviewed. Specific therapeutic regimens, less common side-effects, and the nuances of treatment are summarized in *The Medical Letter on Drugs and Therapeutics* and major textbooks on infectious diseases and travel and tropical medicine.

TREATMENT OF PARASITIC DISEASES

Chemotherapy of malaria

History

Malaria is considered by many to be the most important human parasitic disease. It is a major cause of morbidity and mortality, particularly among children in the tropics. It is responsible for more than a million deaths each year in Africa alone. Malaria also poses a substantial threat to non-immune travelers, workers, immigrants, military personnel and diplomats in endemic areas.

The first written reference to the treatment of malaria in the Western literature referred to the bark of the cinchona tree, which is indigenous to South America. The Augustinian monk, Calancha of Lima, Peru, wrote in 1633: 'A tree grows which they call "the fever tree" in the country of Loxa, whose bark, the color of cinnamon, is made into powder and given as a beverage, cures the fevers and tertians; it has produced miraculous results in Lima'. For the next two centuries, an extract of cinchona bark was administered in Europe as well as Latin America for the treatment of presumed malaria. In 1820 Pelletier and Caventou isolated the active agent, quinine.

Similarly, the Chinese for centuries used the herbal medication, qinghaosu, which is an extract of the wormwood plant, *Artemisia annua,* for the treatment of febrile illnesses, many of which were likely due to malaria. Only in the past decades have artemisinin, a sesquiterpene, and its derivatives been systematically studied. Artemisinin derivatives are now extensively used in Thailand for the treatment of multi-drug resistant *Plasmodium falciparum* and are being increasingly used elsewhere although the US Food and Drug Administration has not yet approved them.

Quinine remained the treatment of choice for malaria for approximately 300 years. While the description of the life cycle, biology and epidemiology of *Plasmodium* species by Ross and others constituted a major scientific advance and resulted in improved strategies for disease prevention through vector control, they did not dramatically alter therapy. The next major therapeutic advance came from an extensive research program in Germany that started when supplies of quinine were interrupted during World War I. Guttman and Ehrlich had reported in 1891 that methylene blue dye (Figure 17.1) had anti-malarial effects. The subsequent evaluation of a series of related organic compounds led to the identification of pamaquine and then quinacrine following the war.

Chloroquine, one of a large series of 4-aminoquinolines, was first produced in 1934, but its value was not immediately recognized. The toxicity associated with quinacrine eventually fueled the quest for less toxic alternatives. During WWII, chloroquine was re-evaluated in the United States. It quickly proved to be highly effective with minimal toxicity, and became the drug of choice for the treatment and prophylaxis of malaria by the end of the war.

Unfortunately, chloroquine resistance emerged, first among isolates of *P. falciparum* around the world and more recently in some

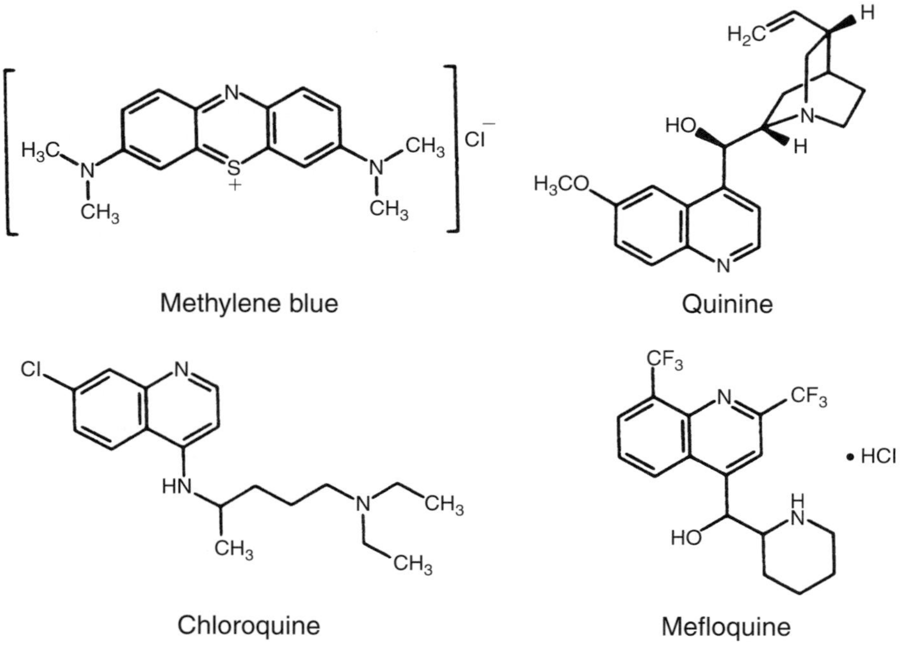

FIGURE 17.1 Drugs used for the treatment and/or prophylaxis of malaria. The discovery of the anti-malarial activity of methylene blue dye by Guttman and Ehrlich (1891) eventually resulted in the synthesis and development of chloroquine. Modifications of quinine led to mefloquine.

Plasmodium vivax. This led to further exploration and the screening of compounds structurally related to quinine. Mefloquine was found to be active against chloroquine-resistant isolates. It has been the mainstay of chemoprophylaxis for chloroquine-resistant *Plasmodium* species for more than a decade, although increasing concerns about its neuropsychiatric effects and the development of mefloquine resistance have led to the search for new drugs.

Attempts at the rational design of anti-malaria drugs have led to some success. Dihydrofolate reductase inhibitors such as pyrimethamine and proguanil, that act preferentially against the plasmodial enzyme, used alone or in combination with sulfonamides, have been available for a number of years for prophylaxis and treatment of malaria. They have been used in many geographic areas, but the development of resistance and allergic reactions to sulfonamides have limited their use in recent years.

The most recent anti-malarial to be licensed by the US Food and Drug Administration is the fixed-dose preparation, proguanil and atovaquone (Malarone). The rationale for combinations of drugs is that compounds with different mechanisms of anti-malarial action might act synergistically or additively and might delay the emergence of resistance. Together, proguanil and atovaquone have synergistic activities against *P. falciparum*, but as discussed below, this effect appears to be due to their combined ability to alter electron transport in the mitochondrion, rather than inhibition of plasmodial dihydrofolate reductase by proguanil.

Current status of malaria prophylaxis and treatment

Malaria remains a major health problem for residents of endemic regions and for travelers to them. *P. falciparum* is the principal cause of mortality. Intensive efforts have also been directed toward the development of a malaria vaccine. While the possibility of transient immunity has been substantiated, the chance of finding a vaccine that stimulates the development of permanent or long lasting immunity is still more a hope than a reality.

Plasmodium species are transmitted by the bite of female anopheline mosquitoes. Sporozoites are inoculated and spend approximately half-an-hour in the bloodstream before entering hepatocytes. The parasite then undergoes development in the liver, the pre-erythrocytic stage, which takes a week or longer, and is asymptomatic. Eventually, merozoites are released into the bloodstream and infect circulating red blood cells, initiating the erythrocytic stage of infection. The typical symptoms of malaria, fever, myalgia, headache and malaise are associated with the destruction of erythrocytes and release of malarial antigens. In the case of *P. vivax* and *Plasmodium ovale,* hypnozoites, a latent form that causes no symptoms, can persist in the liver for months to years before maturing to release merozoites into the bloodstream. This scenario can result in the development of symptomatic malaria months after a person departs from an endemic area.

Vector control should not be overlooked as a means of protection against malaria. Bed nets impregnated with permethrin or other insecticides, and personal protective measures, such as long-sleeved clothing and the topical application of DEET, have been shown to decrease the transmission of malaria, but they are only partially protective.

Recommendations for the prophylaxis and treatment of malaria are complicated and continuously changing as drug resistance evolves in different parts of the world. The optimal regimen depends on the causative *Plasmodium* species, its likelihood of being susceptible to specific drugs (which varies among geographic regions), the past medical history of the patient, and the likelihood and severity of side-effects.

Prophylactic and treatment strategies for malaria vary (Table 17.3). Chloroquine, mefloquine and many of the other anti-malarial drugs act only on the erythrocytic stage. *Suppressive treatment* refers to the continuous administration of a drug to kill parasites in erythrocytes after completion of the pre-erythrocytic stage, thus preventing the development of symptomatic malaria. *Suppressive cure* is achieved when treatment with a drug active against

TABLE 17.3 Prophylactic and treatment strategies for malaria

Suppressive treatment refers to the continuous administration of a drug to kill parasites in erythrocytes after completion of the pre-erythrocytic stage, thus preventing the development of symptomatic malaria.
Suppressive cure is achieved when treatment with a drug active against parasites in erythrocytes is continued beyond the lifespan of the pre-erythrocytic stage.
Clinical cure refers to the elimination of intra-erythrocytic malaria with drugs such as quinine, chloroquine or mefloquine resulting in resolution of symptoms, but depending upon the circumstances, leaving the subject susceptible to recrudescence.
Radical cure refers to the elimination of both pre-erythrocytic and erythrocytic forms and requires the addition of a drug such as primaquine that is active against the pre-erythrocytic stage in the liver.

parasites in erythrocytes is continued beyond the lifespan of the pre-erythrocytic stage. In the case of *P. falciparum*, the exo-erythrocytic stage is completed in less than 4 weeks, so continuation of drugs such as chloroquine, mefloquine or doxycycline for 4 weeks after departure from an endemic area results in suppressive cure. In the case of symptomatic malaria, *clinical cure* refers to the elimination of intra-erythrocytic malaria with drugs such as chloroquine or mefloquine resulting in resolution of symptoms but, depending upon the circumstances, leaving the subject susceptible to recrudescence. In contrast, *radical cure* refers to the elimination of both pre-erythrocytic and erythrocytic forms and requires the addition of a drug such as primaquine that is active against the pre-erythrocytic stage in the liver.

Drugs that are currently used for the prophylaxis and treatment of malaria are discussed below. Prophylactic drugs are usually begun one week prior to departure to evaluate for untoward effects and to allow therapeutic levels to be reached. Drugs are continued during exposure and for 4 weeks after the last possible bite of an infected mosquito. That allows for suppressive cure of *P. falciparum*. The combination of atovaquone and proguanil is somewhat of an exception since it has activity against pre-erythrocytic organisms as well as those in the bloodstream. It is continued for 1 week after exposure. In the case of *P. vivax* and *P. ovale*, terminal prophylaxis with primaquine, an 8-aminoquinoline, is used to kill intrahepatic hypnozoites.

A number of drugs that were once widely used for prophylaxis or treatment will not be discussed in detail. In the case of pyrimethamine, a dihydrofolate reductase inhibitor, resistance is now widespread. Although there is still a role for empiric therapy with the fixed drug combination, pyrimethamine and sulfadoxine, in selected settings, resistance and serious allergic reactions to the long-acting sulfonamide have markedly limited the use of this combination. Specific regimens are discussed in *Health Information for International Travel, 2001–2002* published by the Centers for Disease Control and Prevention (CDC), at the CDC web site and in major textbooks of medicine, infectious diseases and tropical medicine.

Quinine

As described above, quinine, a cinchona alkaloid (Figure 17.1), has been used for the treatment of acute malaria for centuries. It is active against erythrocytic stages of all four *Plasmodium* species. Its short half-life, 5–15 h, and toxicity keep it from being employed prophylactically. Quinine is currently used for the treatment of acute malaria caused by chloroquine-resistant *P. falciparum*. It is usually administered orally; a parenteral preparation is available in some developing areas. Its stereoisomer, quinidine, is of comparable efficacy and can be used intravenously in persons unable to take oral medications due to alterations in consciousness or vomiting. The mechanism(s) of action of quinine and quinidine remain uncertain. They and other quinolinemethanols bind to high-density lipoproteins. They are thought to act on the hemoglobin-containing digestive vesicles of intraerythrocytic parasites.

Quinine and quinidine have the poorest therapeutic-to-toxic ratios of the anti-malarial drugs. Quinine administered orally produces a syndrome known as cinchonism, which includes nausea, vomiting, temporary hearing loss, headache, dysphoria, tinnitus and mild visual disturbances. Vomiting can be severe and interfere with therapy. A very important adverse effect in patients with severe *P. falciparum* infection is hypoglycemia, due in part to the parasite's consumption of glucose and also to the release of insulin from pancreatic β-cells by quinine.

Intravenous quinidine or quinine can produce severe myocardial depression, hypotension, prolonged QT interval and/or ventricular arrhythmia if infused too rapidly. They can result in respiratory depression in persons with myasthenia gravis. In addition, quinine stimulates uterine contraction during pregnancy and can precipitate abortion.

Chloroquine

Chloroquine (Figure 17.1) remains the drug of choice for the prophylaxis and treatment of malaria in the few remaining areas of the world where resistance has not yet emerged. It is active only against the erythrocytic stages of the parasite. Chloroquine accumulates in the acidic vesicles of *Plasmodium* species. It inhibits plasmodial heme polymerase, an enzyme that polymerizes and thereby detoxifies ferriprotoporphyrin IX, a potentially toxic breakdown product of hemoglobin. The result is an increase in the concentration of this toxic molecule and death of the parasite. Resistance occurs by parasite-mediated transport of chloroquine from malaria-infected erythrocytes. In *P. falciparum*, chloroquine resistance is linked to multiple mutations in PfCRT, a protein that appears to function as a transporter in the parasite's digestive vacuolar membrane. Rapid diagnostic assays have been employed as surveillance tools to detect chloroquine resistance in the field. The transport of chloroquine out of parasite-infected erythrocytes can be blocked by calcium channel inhibitors, but not at concentrations that can be safely attained in humans. Chloroquine-resistant *P. vivax* isolates appear to differ in their mechanism of resistance.

Chloroquine's long half-life, approximately 7 days, permits once weekly administration. It is generally very well tolerated, but the drug has a bitter taste and can cause gastrointestinal symptoms. It has been associated with exacerbation of psoriasis and other dermatoses. Retinal degeneration has been reported with high doses administered for the treatment of collagen vascular diseases, but it rarely if ever occurs with doses recommended for malaria prophylaxis or treatment.

Unfortunately, resistance to chloroquine has become widespread among *P. falciparum* and is now well documented in *P. vivax* in some areas. Only in Latin America west and north of the Panama Canal Zone, Haiti and the Dominican Republic and in some areas of the Middle East can *P. falciparum* be presumed to be sensitive. Terminal treatment with primaquine is needed in persons infected with *P. vivax* or *P. ovale* who are treated with chloroquine to prevent relapses.

Mefloquine

Mefloquine (Figure 17.1), like chloroquine, is only active against the erythrocytic stages of *Plasmodium* species. Despite being a quinine derivative, the precise mechanism of mefloquine's action is not known. It cannot be assumed to be the same as that of chloroquine. Mefloquine is concentrated in membranes and may interfere with the parasite's food vacuole. It has a long half-life, ranging from 6 to 23 days, and is administered weekly for prophylaxis. Higher doses are used for the treatment of acute malaria. Mefloquine resistance is now prevalent in the rural rim of Thailand and in adjacent areas of Southeast Asia. It has been reported in the Amazon River basin and elsewhere.

The biggest problem with mefloquine is its side-effects. The drug has neurostimulatory activity and frequently produces vivid dreams. It can result in seizures in persons with a history of epilepsy. Although not frequent, serious neuropsychiatric responses, including severe depression and frank psychosis, are well documented in previously healthy recipients. Many travelers now refuse the drug, and some physicians have become increasingly wary of prescribing it. As in the case of chloroquine, primaquine must be administered in those infected with *P. vivax* and *P. ovale* to eradicate hypnozoites in the liver and thus to prevent late relapses.

Doxycycline

This drug acts at the ribosomal level to inhibit protein synthesis, and is an important alternative to mefloquine for prophylaxis among travelers to areas of chloroquine resistance. Doxycycline and other tetracyclines act too slowly to be used alone in the treatment of acute malaria. The short half-life of doxycycline necessitates daily administration. Like chloroquine and mefloquine, doxycycline is continued for 4 weeks after exposure to malaria in order to ensure suppressive cure of *P. falciparum.* It is not active against hypnozoites of *P. vivax* or *P. ovale*. The major side-effects include gastrointestinal symptoms, which are lessened if the drug is taken with meals. Photosensitivity dermatitis occurs in approximately 3% of recipients, candidal vaginitis is a potential problem in women, and pseudomembranous colitis can occur in anyone taking broad-spectrum antibiotics, but the risk is greatest in the elderly.

Atovaquone/proguanil

The most recent FDA-approved anti-malarial is the fixed combination of atovaquone/proguanil (Malarone). Although proguanil is a dihydrofolate reductase inhibitor, it appears to act synergistically to enhance the effects of atovaquone on electron transport in the parasite's mitochondrion. Atovaquone/proguanil has activity against erythrocytic and pre-erythrocytic stages of *P. falciparum*, and only a week of post-exposure prophylaxis is recommended. Data on the efficacy of the combination for the prophylaxis and treatment of *P. vivax* and *P. ovale* are limited, but relapses are well documented after treatment of acute *P. vivax* infections. Therefore, terminal prophylaxis with primaquine is needed.

Both atovaquone and proguanil have short half-lives, so the combination must be administered daily. Although gastrointestinal side-effects can occur, they are usually mild. These drugs have not been associated with the neuropsychiatric side-effects of mefloquine or the photosensitivity dermatitis or candidal vaginitis of doxycycline. The major concerns with atovaquone/proguanil are cost and the potential for the emergence of resistance, which can occur to either atovaquone or proguanil with single point mutations. The widespread use of the combination may be followed by the rapid emergence of resistance.

Primaquine

Primaquine, an 8-aminoquinoline, is the only drug licensed in the United States with activity against the hypnozoites of *P. vivax* and *P. ovale*. The mechanism of action is uncertain, but primaquine is active only after metabolism by the host. It is administered after treatment of *P. vivax* or *P. ovale* malaria to prevent late relapse and in travelers who have had significant exposure to these *Plasmodium* species. Relapses occur in a variable percentage of persons after primaquine and are more common in some geographic areas, such as Southeast Asia, than others. The recommended dosage varies with geographic location. Daily primaquine has also been used successfully in a limited number of persons as prophylaxis against

P. falciparum. Primaquine has gametocytocidal activity against all four *Plasmodium* species that infect humans, but this does not have therapeutic relevance for the infected individual.

Primaquine is well absorbed orally. The major toxicity is hemolysis in persons with glucose-6-phosphate dehydrogenase (G-6-PD) deficiency. It is recommended that recipients be tested for G-6-PD deficiency before use unless they have previously taken the drug without adverse effect. Gastrointestinal side-effects occur, but usually do not necessitate the discontinuation of therapy. Rare complications include granulocytopenia or agranulocytosis.

Artemisinin and its derivatives

The emergence of multi-drug resistant *P. falciparum* has fueled the search for new anti-malarial compounds. Extensive study of the Chinese herbal anti-malarial, qinghaosu, led to the characterization of artemisinin and a number of derivatives including artesunate and artemether. They act rapidly against multi-drug resistant *P. falciparum*, are relatively inexpensive to produce, and are generally well tolerated. They are usually administered with mefloquine or another anti-malarial to prevent relapse and the development of resistance. Their short half-lives and association with neurotoxicity after chronic administration in a dog model mean that they are unlikely to be used for malaria prophylaxis. Although not yet licensed in the United States, artemisinin derivatives are now used for *P. falciparum* infection in Thailand, other areas of Southeast Asia, and elsewhere in the world.

Artemisinin and its derivatives are endoperoxide-containing drugs. Depending on the compound, they are administered enterally, intramuscularly or intravenously. Their mechanism(s) of action are not entirely clear. In the presence of intra-parasitic iron, they are thought to be converted to free radicals and other intermediates that alkylate malaria proteins or damage parasite membranes. The artemisinin derivatives arrest the development of parasites in erythrocytes and prevent cytoadherence and rosetting of infected erythrocytes.

Artemisinin derivatives are generally well tolerated. Side-effects are infrequent and mild. They include transient liver enzyme elevations, gastrointestinal complaints, drug fever and leukopenia. Central pontine demyelination has been reported in dogs treated chronically at high doses.

Future considerations

The propensity of *Plasmodium* species to develop drug resistance suggests that it is only a matter of time until they become resistant to currently available medications. Resistance has already developed to chloroquine, mefloquine, quinine, pyrimethamine and sulfonamides among *P. falciparum* and to a lesser degree *P. vivax* in various regions of the world. In fact, cross-resistance between pyrimethamine and trimethoprim appears to be occurring as a result of the extensive use of trimethoprim–sulfamethoxazole prophylaxis in patients with AIDS in Africa. In response to the known development of resistance, combinations of anti-malarial drugs are now being considered in order to treat organisms potentially resistant to one or the other and theoretically to reduce the development of resistance. Artemisinin derivatives have been administered with mefloquine or other anti-malarial drugs for this reason.

Efforts are urgently needed to identify new anti-malarial drugs, even among known anti-parasitic drugs. For example, diamidine compounds including pentamidine have recently been shown to have anti-malarial activity, apparently by binding to and preventing breakdown of toxic ferriprotoporphyrin IX. In addition, empirical mass screening has worked in the past, and with advances in medical chemistry and microarray technology, it may work in the future. Alternatively, one directed approach would be to determine the mechanism of resistance to specific drugs, and then to develop a strategy to circumvent it. For example, chloroquine resistance in *P. falciparum* is due to efflux of chloroquine from erythrocytes mediated by multiple mutations in the PfCRT protein. Chloroquine efflux can be inhibited by calcium channel blockers, but unfortunately not at concentrations that can safely be achieved in humans. It is possible that other parasite-specific chemosensitizing agents will be designed in the future.

A different approach is to target the pre-erythrocytic stage of malaria. In limited studies primaquine, which is active against *Plasmodium* species in the liver, has appeared promising when used daily for prophylaxis. The possibility of hemolysis in persons with G-6-PD deficiency prevents primaquine from being used on a mass basis, but a new compound with activity against pre-erythrocytic stages of malaria, but free of hemolytic side-effects, is attractive. Further exploration of the cell biology and molecular biology of *Plasmodium* species may identify currently unappreciated targets for chemotherapeutic attack.

Treatment of other diseases caused by Apicomplexa: babesiosis and toxoplasmosis

Two other apicomplexans are important human pathogens, *Babesia* spp., which are transmitted by ticks, and *Toxoplasma gondii*, which is acquired though the ingestion of oocysts from cat feces or inadequately cooked contaminated meat. *Babesia* spp. infect erythrocytes while *T. gondii* can infect virtually all nucleated cell types. The combination of quinine (see Chemotherapy of Malaria) and clindamycin is used to treat babesiosis. In the case of toxoplasmosis pyrimethamine is usually administered with sulfadiazine (see Chemotherapy of Malaria). Clindamycin is an alternative. Prednisone or another corticosteroid is added during the treatment of toxoplasmic retinochoroiditis to decrease the inflammatory response in the eye. Spiramycin, a macrolide, is used to treat women who acquire toxoplasmosis during pregnancy in order to prevent congenital infection. It is also used for the treatment of congenital toxoplasmosis.

Clindamycin and spiramycin act at the 50S ribosomal binding site in bacteria and probably have the same mechanism of action in susceptible protozoa. They are well absorbed orally and generally well tolerated. The major side-effect with clindamycin is pseudomembranous colitis, which increases in incidence with age. Spiramycin is not licensed in the United States.

Treatment of the Kinetoplastida: trypanosomes and *Leishmania*

The trypanosomes and *Leishmania* are closely related. They have in common a specialized mitochondrial structure, the kinetoplast, that contains a substantial amount of extranuclear DNA in catenated mini- and maxi-circles. They are adapted to survival at ambient temperature in their arthropod vectors and at body temperature in humans and other mammalian hosts. Among parasites, the Kinetoplastida pose the greatest challenge to therapy. The drugs currently used to treat them are limited by variable efficacy and substantial toxicity.

Trypanosoma cruzi *(Chagas' disease)*

Trypanosoma cruzi, the cause of Chagas' disease, is transmitted by reduviid bugs to persons sleeping in adobe dwellings in rural, endemic areas in Latin America. The incidence of new infections has decreased in Brazil and other countries due to urbanization and improved living conditions, but millions of persons remain infected. *Trypanosoma cruzi* can also be transmitted via blood transfusion, which is a major concern in endemic countries in Latin America as well as in North America, where a substantial number of chronically infected immigrants now live. Less commonly, infection occurs across the placenta or following a laboratory accident.

Acute Chagas' disease can be life threatening, but it is usually a non-specific febrile illness in children and others living in endemic areas that resolves spontaneously without diagnosis or therapy. After the resolution of acute symptoms, infected persons enter a prolonged, asymptomatic, *indeterminate* phase. Some of those chronically infected have no further evidence of disease, while others develop chronic Chagasic cardiomyopathy or mega-disease of the esophagus or intestines decades later. The cardiomyopathy is associated with histopathological changes of chronic inflammation of the heart. Mega-disease is associated with destruction of the myenteric plexus in the gastrointestinal tract. These late manifestations of infection are thought to be due at least in part to toxic effects of the host immune response.

In the treatment of acute Chagas' disease, anti-parasitic chemotherapy is effective in shortening the duration of symptoms and decreasing the likelihood of death. Until recently, chemotherapy was not thought to have a role in the treatment of persons with indeterminate or chronic Chagas' disease. Data from Latin America now suggest that treatment of young persons in the indeterminate phase of infection cures more than half and may reduce the likelihood of chronic Chagas' disease developing later in life. Once symptoms of cardiomyopathy or mega-disease develop, anti-parasitic chemotherapy is thought to have minimal value in reversing the chronic pathological changes. Unfortunately, the only drugs now available for Chagas' disease, nifurtimox and benznidazole, are relatively toxic and require prolonged administration.

Nifurtimox

Nifurtimox has been the mainstay of therapy for Chagas' disease in the United States. Although it is currently out of production, remaining supplies are available through the Centers for Disease Control and Prevention. Treatment with nifurtimox reduces the duration of symptoms and the likelihood of death from acute Chagas' disease. Treatment failures occur, and there is geographic variation in responsiveness. Therapy for 90 days is recommended, but treatment must often be terminated prematurely because of drug toxicity.

Nifurtimox is well absorbed orally. The mechanism of action is uncertain, but it may be related to free radical formation and oxidative damage to the parasite. Its effects are not specific to the parasite, and as a result, nifurtimox is associated with multiple side-effects including nausea, vomiting, weight loss, severe neuritis, weakness, paresthesias and sleep disturbances.

Benznidazole

Benznidazole is used for the treatment of Chagas' disease in Latin America. It is well absorbed orally and is administered for periods of 30 to 90 days. The mechanism of action is not known, but it, too, is a nitroimidazole and may act by free radical formation. Benznidazole causes dose-dependent polyneuropathy, gastrointestinal side-effects, psychological disturbances and allergic skin rashes.

Future considerations

Chagas' disease currently poses one of the greatest therapeutic challenges in tropical medicine. If treatment during the indeterminate stage can prevent the manifestations of chronic disease, as suggested by recent experience in South America, then it would be advisable for millions of *T. cruzi*-infected residents of Latin America as well as immigrants in the United States and elsewhere to be treated. Unfortunately, the prolonged treatment course and side-effects with benznidazole or nifurtimox make the prospects of such a mass treatment program daunting. New, effective, non-toxic therapy is desperately needed.

The kinetoplast, a defining structure of the trypanosomes and *Leishmania*, has drawn attention as a potential target for chemotherapy. It is a specialized mitochondrion with catenated mini- and maxi-circles containing a substantial amount of extranuclear DNA. Topoisomerase II has been shown to play an important role in kinetoplast (kDNA) and nuclear DNA function and replication. Some anti-parasitic drugs appear to act as topoisomerase II activators or inhibitors, shifting the equilibrium from circular to linearized kDNA. It is possible that specific inhibitors of trypanosomal topoisomerase II will be developed, as discussed below in reference to leishmanial topoisomerase II inhibitors. It is also possible that new *T. cruzi* targets will be identified and strategies developed to inhibit them.

Trypanosoma brucei rhodesiense *and* Trypanosoma brucei gambiense *(African sleeping sickness)*

The past decade has seen a dramatic increase in the incidence of sleeping sickness in endemic areas of equatorial Africa. *Trypanosoma brucei gambiense* is responsible for disease across West and Central Africa while *Trypanosoma brucei rhodesiense* is endemic in East Africa. Both occur in Uganda. The African trypanosomes are transmitted by tsetse flies. In *T. b. gambiense* infection there is a prolonged hemolymphatic stage before the organism crosses the blood–brain barrier to infect the central nervous system. In the case of *T. b. rhodesiense,* infection progresses much more rapidly and there is early involvement of the central nervous system.

Eflornithine is the drug of choice for all stages of *T. b. gambiense* infection although its high cost has dramatically limited its use and production has been sporadic. Eflornithine is not effective against *T. b. rhodesiense.* Suramin and pentamidine are useful in the treatment of the hemolymphatic stage of both pathogenic species. These drugs are less expensive than eflornithine, but more toxic. Melarsoprol is used in patients with central nervous system infection by *T. b. rhodesiense* or *T. b. gambiense* when eflornithine is not available. Decreased sensitivity of *T. b. gambiense* to melarsoprol has been reported from Central Africa. The major problem with melarsoprol is its association with potentially life-threatening neurological toxicity.

Eflornithine

A major advance in the development of anti-parasitic chemotherapy was the development of eflornithine. It has been called the 'resurrection drug' because of its efficacy in patients with sleeping sickness due to *T. b. gambiense.* Eflornithine is an enzyme-activated, selective, irreversible inhibitor of ornithine decarboxylase, which is involved in the first step of polyamine synthesis. Polyamines play an important role in the growth, differentiation and multiplication of trypanosomes.

Eflornithine can be administered intravenously and orally, but large doses, approximately 20 grams a day, are needed to reach therapeutic levels when it is given by mouth.

Compared to the other drugs used for the treatment of African trypanosomiasis, eflornithine is relatively well tolerated. Adverse effects are usually mild and reversible. They include anemia, thrombocytopenia, leukopenia, nausea, vomiting, diarrhea and convulsions. Interestingly, given its mechanism of action, eflornithine is now being produced for topical application as a depilatory.

Unfortunately, the large amount of drug needed and production costs have limited its profitability and availability. As a result, production of eflornithine for the treatment of African trypanosomiasis has ceased, but limited supplies of the drug are available. Even though production may resume in the future with support from government or non-government agencies, an expensive drug such as eflornithine is not likely to be used for the treatment of any disease if residents of impoverished areas are required to pay for it.

Suramin

Suramin has activity against both *T. b. gambiense* and *T. b. rhodesiense*. It does not cross the blood–brain barrier, and it is therefore useful only for the hemolymphatic stage of infection. The precise mechanism of action is not known, but suramin is a polyanionic molecule that complexes with and inhibits a number of enzymes. The anti-trypanosomal effects correlate with inhibition of the parasite's glycerol-3-phosphate oxidase and glycerol-3-phosphate dehydrogenase, two enzymes that are involved in energy metabolism.

Suramin is a toxic drug. Fever and allergic reactions are common. Immediate reactions include nausea, vomiting, shock, loss of consciousness and occasionally death. Other side-effects include renal dysfunction, exfoliative dermatitis, stomatitis, paresthesias, photophobia, peripheral neuropathy, blood dyscrasias, shock and optic atrophy. Untoward effects seem to be more severe in malnourished hosts and in persons with pre-existing kidney and liver disease.

Pentamidine

Pentamidine isethionate, a diamidine, is active against *T. b. gambiense* and *T. b. rhodesiense* in the bloodstream and lymph nodes, but it does not cross the blood–brain barrier. It has also been widely used as an alternative to trimethoprim–sulfamethoxazole for the treatment of *Pneumocystis carinii* pneumonia. Pentamidine must be administered parenterally, either by slow intravenous infusion or by intramuscular injection. An oral preparation is not available.

Its mechanism of action against trypanosomes is uncertain. Pentamidine is known to bind to DNA in a non-intercalative manner. It also poisons trypanosomal topoisomerase II, resulting in linearization of minicircle DNA in the kinetoplast. It has other effects that may be relevant to its anti-parasitic action, including decreased amounts of membrane phospholipids and altered polyamine function through a reduction in the activity of ornithine decarboxylase.

Pentamidine is associated with a number of side-effects, which include tachycardia, nausea, vomiting, dizziness, facial flushing, breathlessness and a metallic taste. Severe hypotension can result if it is given too rapidly by intravenous infusion. Hypoglycemia due to pancreatic beta-cell damage and insulin release can be severe and occasionally fatal. Insulin-dependent diabetes mellitus may follow in patients experiencing beta-cell damage. Reversible azotemia, hypocalcemia and hyperkalemia and bone marrow suppression occur. Fatal cases of pancreatitis have been reported in persons with AIDS who have received pentamidine.

Melarsoprol

Melarsoprol, or Mel B, is a trivalent arsenical. It has the distinction of being one of the most toxic drugs administered to humans. It is used for the treatment of central nervous system trypanosomiasis. Melarsoprol is given intravenously. A small, but adequate concentration of the drug is found in the central nervous system. Melarsoprol is thought to enter the parasite through an adenosine transporter, and like other arsenicals, reacts with sulfhydryl groups to inactivate enzymes. It may act differentially on the parasite's pyruvate kinase.

Reactive encephalopathy is the most feared complication with melarsoprol and can cause death in as many as 4–6% of those treated. The syndrome is characterized by headache, dizziness, mental dullness, confusion, ataxia and grand mal seizures with progression to obtundation and coma. It is thought to be due at least in part to an immunologic reaction. Corticosteroids such as prednisolone or prednisone reduce the risk of encephalopathy and mortality. Other untoward effects are common with melarsoprol and include allergic reactions such as rash and fever, abdominal pain, vomiting, arthralgias, a Guillain–Barré-like syndrome, hepatitis and renal toxicity. Severe hemolysis can occur in patients with G-6-PD deficiency. Erythema nodosum has been reported in patients who were concurrently infected with *Mycobacterium leprae.*

Future considerations

The first priority in African sleeping sickness is to secure adequate supplies of eflornithine for the treatment of *T. b. gambiense* infection. Its production for oral and intravenous administration has ceased, but it is now being used topically as a depilatory, which raises the possibility that the manufacturer could donate supplies in the future. It would be advantageous to have a drug with a similar mechanism of action, but with greater potency and improved pharmacodynamics.

Even with adequate supplies of eflornithine, there would still be substantial problems in the treatment of *T. b. rhodesiense* infections. The organism is resistant to eflornithine, and the drugs available to treat it are highly toxic.

The African trypanosomes evade humoral host defenses by antigenic variation of surface glycoproteins, as discussed in Chapter 5. An attractive target for novel chemotherapeutic interventions would be the parasite's mechanism of antigenic mimicry. If the switch in variant surface glycoproteins were blocked, host defenses would probably be effective in eliminating infection. Other alternatives include targeting the kinetoplast, as discussed in reference to *T. cruzi* above. It is conceivable that new drugs developed for one kinetoplastid will be found to have activity against others.

Leishmania *spp. (cutaneous, mucosal and visceral leishmaniasis)*

Pentavalent antimonials have been used for decades for the treatment of cutaneous, mucosal and visceral leishmaniasis. They are also associated with substantial toxicity, and there has been evidence of increasing resistance among *Leishmania* spp. in widely scattered areas. Amphotericin B deoxycholate and pentamidine isethionate have been effective, but toxic, alternatives. Major advances have been made in the treatment of visceral leishmaniasis in recent years. The most dramatic was the introduction of liposomal amphotericin B for the treatment of visceral leishmaniasis. It recently became the first drug approved by the US Food and Drug Administration for this indication. A number of alternative drugs and therapeutic approaches have been examined over the years and are discussed below.

Pentavalent antimony (stibogluconate sodium and meglumine antimoniate)

Two pentavalent antimony-containing drugs, stibogluconate sodium and meglumine antimoniate, have been used for the treatment of leishmaniasis. They are of comparable efficacy and toxicity when administered on the basis of their pentavalent antimony content. Their efficacy varies by geographic region and the infecting *Leishmania* species. Treatment failures have become increasingly common in patients infected with *L. donovani* in India and with other *Leishmania* species in different areas.

Both stibogluconate sodium and meglumine antimoniate are available only for parenteral use. They are administered on the basis of their pentavalent antimony content, 20 mg of Sb^V per kg body weight per day. The mechanism of action is uncertain. Pentavalent antimony appears to be concentrated in cells of reticuloendothelial origin, the only site where *Leishmania* resides in humans.

The pentavalent antimonial drugs have important untoward effects. Chemical pancreatitis, muscle and joint pain, fatigue, nausea, weakness, leukopenia, thrombocytopenia and rash are common. Severe pancreatitis can occur. It seems to be more likely in persons with chronic renal insufficiency. Most recipients of pentavalent antimony evidence nonspecific ST-T wave changes by ECG. Sudden death, apparently due to cardiac arrhythmia, can occur when doses greater than 20 mg Sb^V per kg body weight per day are administered, or at lower doses in persons with underlying heart disease.

Amphotericin B deoxycholate and liposomal amphotericin B

Amphotericin B deoxycholate has been used as an alternative to pentavalent antimony for the treatment of leishmaniasis for decades. Amphotericin binds to sterols and probably damages the surface membrane of *Leishmania* spp. in the same manner that it affects fungi. The major problem with amphotericin B deoxycholate has been the requirement for parenteral administration over long periods of time, and its toxicity, which includes fever, malaise and, most importantly, dose-dependent renal toxicity and accompanying electrolyte disturbances.

A recent advance has been the use of liposomal amphotericin B for the treatment of visceral leishmaniasis. *Leishmania* resides solely in mononuclear phagocytes, and cells of reticuloendothelial origin which preferentially take up liposomes. In addition, liposomal amphotericin B is substantially less toxic than amphotericin B deoxycholate. Liposomal amphotericin B was the first drug to receive FDA approval for the treatment of visceral leishmaniasis. It is as effective as pentavalent antimonials in the treatment of antimony-susceptible strains of *L. donovani*, and it is active against antimony-resistant strains. Liposomal amphotericin B is less toxic than either pentavalent antimonials or amphotericin B deoxycholate. Although not as extensively studied, other lipid-associated amphotericin B preparations appear to act in the same manner as liposomal amphotericin B and have similar toxicity profiles. The major factors limiting the use of liposomal amphotericin B and lipid-associated preparations is their cost and the requirement for parenteral administration.

Future considerations

Over the years intense interest has focused on new approaches to the treatment of leishmaniasis. Immunotherapy has been of potential interest. *Leishmania* is found only within mononuclear phagocytes in humans and other mammals. It is clear from animal models of leishmaniasis and field studies in humans that spontaneous resolution of infection and prevention of re-infection are associated with activation of macrophages by interferon-γ in concert with other cytokines of the TH1 type.

A logical extension from these findings was the use of recombinant interferon-γ for the treatment of human leishmaniasis. Unfortunately, interferon-γ was not curative when used alone, but the combination of interferon-γ and pentavalent antimony was effective in cases where antimony alone had failed. Interestingly, 'immunotherapy', in which a crude leishmanial promastigote antigen preparation is administered concurrently with BCG, results in slow healing of cutaneous leishmaniasis. It is possible that effective immunological manipulation will emerge as the treatment of choice for leishmaniasis in the future. It is also likely that an effective anti-leishmanial vaccine will eventually be developed.

Over the years a number of case reports and uncontrolled case series have suggested that various antibiotics and other drugs might be effective. Interpretation was difficult because cutaneous leishmaniasis eventually heals spontaneously. Without appropriate controls, data suggesting a positive outcome are difficult to interpret. More recently, well controlled trials indicate that locally applied heat is effective in the treatment of cutaneous leishmaniasis, but it is difficult to administer. Ketoconazole and itraconazole, both azole antifungals, also have some activity against selected *Leishmania* spp. However, none of these drugs or approaches has proven to be sufficiently effective to be recommended for general clinical use.

Three experimental approaches are worthy of further discussion. In the first, elegant studies of pyrazolopyrimidines demonstrated how a detailed understanding of the parasite's biochemistry could lead to rational drug design. In the second, the synthesis of inhibitors selectively active against leishmanial topoisomerase II was shown to be possible. And third, the development of an anti-cancer drug, miltefosine, appears particularly promising in studies of human visceral leishmaniasis in India.

Pyrazolopyrimidines (allopurinol and allopurinol ribonucleoside)

The metabolism of purines in trypanosomes and leishmania differs significantly from that in humans. *Leishmania* and trypanosomes rely on salvage pathways to obtain purine analogs, whereas humans synthesize purines *de novo*. Certain purine analogs are metabolized by parasites to nucleotides and aminated to analogs of adenine nucleotides. These stop protein synthesis and result in breakdown of RNA. Pyrazolo[3,4-d]-pyrimidines, such as allopurinol and allopurinol ribonucleoside, are aminated to adenine nucleotide analogs in this manner. They have been shown to kill *Leishmania* spp. and *T. cruzi in vitro* and in animal models. Activity was also demonstrated in a study of American cutaneous leishmaniasis, but the efficacy was not sufficient for these drugs to be recommended for routine clinical use. Nonetheless, the development of the pyrazolopyrimidines illustrates how differences in metabolic pathways between parasites and humans can potentially be exploited in the development of new approaches to chemotherapy.

Topoisomerase II inhibitors

Building on studies of topoisomerase II in trypanosomes, leishmanial topoisomerase II has been cloned, sequenced and studied. Selective activators and inhibitors that are lethal for *Leishmania* have been synthesized. They appear to act in a manner similar to the fluoroquinolone antibiotics, such as ciprofloxacin, that inhibit topoisomerase II (which is known as DNA gyrase) in susceptible bacteria. The findings suggest that specific topoisomerase

II-directed drugs preferentially active against trypanosomes and *Leishmania* can be synthesized *de novo* or identified from the large number of topoisomerase II inhibitors synthesized as potential anti-neoplastic drugs. Microarray technology may facilitate the process of screening in the future.

Miltefosine

Most recently, miltefosine, a phosphocholine analog that interferes with cell-signaling pathways and membrane synthesis, was shown to be effective in the treatment of visceral leishmaniasis in India at a site where substantial antimony resistance had been documented. Miltefosine is closely related to lecithin (phosphatidylcholine). It was originally developed as an anti-neoplastic drug, but it does not have efficacy against tumors when administered orally. The exact mechanism of its anti-leishmanial effective is not known. In a phase II study miltefosine appeared to be highly effective in the treatment of visceral leishmaniasis. Gastrointestinal side-effects were frequent, but mild to moderate in severity, and not severe enough to require discontinuation of therapy. Reversible transaminase elevations and increased creatinine were also reported. Further research is underway to confirm these preliminary observations and to determine whether miltefosine will have a role in treating other *Leishmania* spp. and other clinical forms of leishmaniasis.

Treatment of luminal protozoa

A number of pathogens such as *Entamoeba histolytica* and *Giardia lamblia* live under anaerobic conditions in the intestinal tract, and *Trichomonas vaginalis* in the vagina. Metronidazole, tinidazole and ornidazole, which are 5-nitroimidazole derivatives, are selectively active against them as well as anaerobic bacteria. Of the three, only metronidazole is approved for use in the United States. It is the treatment of choice for amebic colitis and amebic liver abscess. It is active against amebic trophozoites, but does not invariably kill cysts. A luminal agent such as paromomycin, iodoquinol or diloxanide furoate is administered with metronidazole for invasive amebiasis and used alone for asymptomatic infections to prevent the later development of disease. Metronidazole is also effective for the treatment of enteritis due to *G. lamblia* and vaginitis due to *T. vaginalis*. Furazolidone is an alternative for the treatment of giardiasis. It is often used in children because of its formulation as a liquid.

Metronidazole

Metronidazole is active against a number of protozoa that live under anaerobic conditions, including *E. histolytica, G. lamblia, Balantidium coli* and *T. vaginalis*, as well as many anaerobic bacteria. It is activated in susceptible anaerobes by reduction of the 5-nitro group through a sequence of intermediate steps involving microbial electron transport proteins of low redox potential. Metronidazole serves as an electron sink, depriving anaerobes of reducing equivalents. In addition, the reduced form of metronidazole can affect the helical structure of DNA, resulting in strand breakage and impaired template function.

Metronidazole is well absorbed orally and is also available for parenteral administration. It is metabolized in the liver. The most common side-effects are nausea, vomiting, diarrhea and a metallic taste. Others include headache, dizziness, dry mouth, stomatitis, vaginal or oral candidiasis and reversible neutropenia. Rarely, recipients of metronidazole develop peripheral neuropathy, psychosis or central nervous system abnormalities. Metronidazole is also associated with disulfiram (Antabuse)-like effects, resulting in severe nausea and vomiting in persons who concurrently ingest alcohol. There has been concern about possible carcinogenesis and birth defects with metronidazole, but neither has been documented in humans. Nonetheless, metronidazole is not approved for use during pregnancy.

Paromomycin

Paromomycin is an aminoglycoside antibiotic. It is poorly absorbed after oral administration, resulting in extremely high concentrations in the lumen of the bowel, but not in the body. It is effective because of its restricted area of distribution. Paromomycin acts at the level of the 30S ribosome and inhibits protein synthesis. It kills cysts of *E. histolytica* and is administered with metronidazole in symptomatic amebiasis and alone in persons with asymptomatic infection. It is also active against another enteric pathogen, *Dientamoeba fragilis,* and it has been used topically to treat vaginal trichomoniasis.

Side-effects with paromomycin are primarily gastrointestinal. They include nausea, vomiting, abdominal cramps and diarrhea. A small amount of paromomycin is absorbed and cleared through the kidney. Renal toxicity and ototoxicity can occur, particularly in patients with underlying renal disease. Pancreatitis has been reported in the setting of AIDS.

Iodoquinol

Iodoquinol (diiodohydroxyquin), a halogenated oxyquinoline, is a luminally active drug that kills cysts of *E. histolytica* and is effective in the treatment of *D. fragilis* and *B. coli.* The mechanism of action is unknown. The side-effects include nausea, vomiting, diarrhea, headache, abdominal pain, fever and itching. On occasion iodine-related dermatitis develops. Seizures and encephalopathy are rare. Because of the high iodine content, iodoquinol can interfere with thyroid function tests for months after administration. It is contraindicated in anyone with iodine intolerance. Optic nerve damage or inflammation and peripheral neuropathy can occur after prolonged administration of high doses. The recommended dosage should not be exceeded.

Diloxanide furoate

Diloxanide furoate, a substituted acetanilide, is also effective in eradicating cysts of *E. histolytica* following treatment with metronidazole for invasive amebiasis or when used alone in asymptomatic infection. Its mechanism of action is unknown. The most common side-effect is flatulence and mild gastrointestinal disturbances. Serious untoward effects are rare at the recommended doses. While not approved for use in the United States, the low cost of diloxanide furoate makes it an attractive alternative in developing countries.

Furazolidone

Furazolidone, a nitrofuran derivative, is the only drug for giardiasis that is available in liquid form. The precise mechanism of action is not known, but furazolidone is thought to damage the parasite's DNA. Common side-effects include nausea, vomiting, diarrhea and fever. Rare side-effects include hypotension, urticaria, serum sickness, mood disorders and hypersensitivity reactions. Mild to moderate hemolysis may occur in patients with G-6-PD deficiency. Like metronidazole, furazolidone has disulfiram-like activity. It is also a monoamine oxidase inhibitor. It should not be given to breast-feeding mothers because neonates can develop hemolytic anemia due to glutathione instability.

Future considerations

Treatment of cryptosporidiosis in immunocompromised individuals is the most urgent need related to intestinal protozoal infections. Infection with *Cryptosporidium parvum* in a normal host results in days to occasionally several weeks of non-bloody diarrhea, which is self-limited. In persons with AIDS, cryptosporidium infection may be persistent and require therapy. The most effective approach in patients with AIDS is to reverse the immunodeficiency state with highly active antiretroviral treatment (HAART) of HIV, which has led to a 90% decrease in the incidence of cryptosporidiosis in those treated in the United States. In patients with severe AIDS who are not responding to, or are intolerant of HAART, there is little to offer in the way of treatment for cryptosporidiosis. The one drug thought to be effective was paromomycin, which was recently shown to be no more active than placebo.

An attractive approach for new chemotherapeutic agents would be to interfere with the intracellular replication of the parasite. Upon invasion of epithelial cells, cryptosporidia undergo both the sexual and asexual stages of the life cycle. It has long been appreciated that the parasite resides within a unique intracellular, yet extracytoplasmic compartment that bulges into the lumen of the small bowel. This parasite compartment is separated from the host cytoplasm by an electron-dense material. The nature of this material has recently been shown by immunofluorescence to be filamentous actin. This filamentous actin appears to be intimately associated with the epithelial-cell plasma membrane, and contains the actin-binding protein alpha-actinin. Understanding the mechanism by which the parasite induces host cell actin polymerization, and delineating its importance in intracellular replication of the parasite, are important applications of basic cell biology towards the search for effective therapies for cyptosporidiosis.

Intestinal helminths

The intestinal helminths, or roundworms, are the most prevalent of the human parasites. In areas of the world where sanitation is poor, the majority of residents harbor one or more species. The major intestinal roundworms are *Ascaris lumbricoides*, which is the only roundworm that looks like an earthworm, the hookworms *Ancylostoma duodenale* and *Necator americanus*, and the whipworm, *Trichuris trichiura*. All are susceptible to albendazole and mebendazole.

Of the intestinal worms, *Ascaris* is the most prevalent, with more than 1.5 billion people infected worldwide. Heavy infections can result in intestinal obstruction due to masses of worms in the intestine. On occasion, adult *Ascaris* migrate into the biliary tract, resulting in ascending cholangitis or pancreatitis. The hookworms are the next most prevalent and are a major cause of iron deficiency anemia. *Trichuris trichiura* causes gastrointestinal symptoms including diarrhea, anemia and in heavy infection, rectal prolapse. Recent evidence suggests that pediatric infections with intestinal helminths contribute to growth retardation and adversely effect performance at school and work.

Strongyloides stercoralis is the fourth major intestinal helminth. It is important because of its capacity for autoinfection. *Strongyloides stercoralis* can persist indefinitely and progress to disseminated hyperinfection in persons who are taking corticosteroids and/or become immunocompromised. The treatment of choice for *S. stercoralis* is ivermectin, which is discussed in reference to onchocerciasis (see Filarial Diseases). Unfortunately, ivermectin is not curative in all cases. Albendazole has activity against *S. stercoralis*, but it cures less than half of those infected. Thiabendazole, a well absorbed benzimidazole, was once widely used for *S. stercoralis*, but the majority of recipients experienced adverse effects. Some physicians still consider it the treatment of choice for disseminated strongyloidiasis, but it is no longer manufactured in the United States.

Albendazole

A number of drugs have been used over the years to treat intestinal helminths. The benzimidazole albendazole has emerged as the treatment of choice. It has been used widely in targeted mass treatment programs in developing areas. It is effective as a single dose of 400 mg against all three of the major intestinal helminths, although daily doses for 3 days are recommended in persons with heavy *T. trichiura* infection. Retreatment is recommended at 4-month intervals in highly endemic areas because of reinfection. Mebendazole has a similar spectrum of activity against intestinal helminths, but it has less favorable pharmacodynamics. It is typically administered twice daily for three days.

Albendazole has activity against a number of other intestinal nematodes, including *Capillaria philippinensis, Enterobius vermicularis, Trichostronglus* species, and *Ancylostoma braziliense,* the animal hookworm that causes cutaneous larva migrans. It is also active against some microsporidium species including *Enterocytozoon bieneusi* and *Encephalitozoon intestinalis.* Finally, albendazole is effective in the treatment of neurocysticercosis due to the pork tapeworm, *Taenia solium,* and it is the medical treatment of choice for echinococcosis (see Platyhelminths: Trematodes and Cestodes).

Albendazole, like mebendazole and other benzimidazoles, binds to β-tubulin in susceptible helminths, inhibits its assembly, and impairs the uptake of glucose, leading to depletion of glycogen stores. It also inhibits helminthic fumarate reductase, but the importance of this in its anthelminthic activity is uncertain. Albendazole is practically insoluble in water. Absorption is enhanced when it is taken with a fatty meal. It undergoes extensive first-pass metabolism in the liver. Only albendazole sulfoxide, which is biologically active against helminths, is present in serum. Sulfoxidation of albendazole also occurs in the gut.

Albendazole is generally well tolerated as a single dose, but diarrhea, abdominal pain, rash and migration of *Ascaris* from the mouth or anus can occur. When used at high doses for prolonged periods, albendazole can be associated with liver enzyme elevations, transient bone marrow suppression, and less commonly, alopecia. Albendazole, like other benzimidazoles, is teratogenic and contraindicated during pregnancy.

Pyrantel pamoate

Pyrantel pamoate is available over the counter in the United States for the treatment of the pinworm, *Enterobius vermicularis.* It is also active against *A. lumbricoides*, the hookworms and *Trichostrongylus* spp. It is not effective against *Trichuris trichiura.* Pyrantel acts as an agonist at nicotinic acetylcholine receptors and is a depolarizing neuromuscular blocking agent. It results in a short period of calcium-dependent stimulation of the worm followed by paralysis. Pyrantel also inhibits human acetylcholinesterases, but it is minimally toxic in the doses used to treat intestinal helminths. Side-effects include mild gastrointestinal symptoms, headache, drowsiness, insomnia and dizziness.

Future considerations

In summary albendazole, pyrantel and ivermectin are orally administered, well tolerated and effective when used for the treatment of various intestinal helminths as outlined above. Although development of resistance may occur in time, it has not yet emerged as a problem with these compounds, even with the widespread use of albendazole in mass treatment programs. As good as albendazole is for the treatment of intestinal helminths, it would be advantageous to have a single drug that was highly active against *A. lumbricoides,* the hookworms, *T. trichiura* and *S. stercoralis.* In addition, economic development and improved sanitation should result in reduced prevalences of the intestinal helminths in the future.

Filarial diseases

Diethylcarbamazine is the drug of choice for the treatment of lymphatic filariasis due to *Wuchereria bancrofti, Brugia malayi* and *Brugia timori.* It was previously used for the treatment of onchocerciasis (river blindness) but the inflammatory response elicited by released microfilarial antigens resulted in serious systemic and ocular toxicity. A major advance in the treatment of onchocerciasis was the introduction of ivermectin, which kills microfilariae slowly. While it does not kill adult worms, it inhibits oviposition in *Onchocerca volvulus.* In community-based treatment programs in *O. volvulus* endemic areas, ivermectin has proven effective, safe and well tolerated in comparison to previous treatment with diethylcarbamazine. Successive annual treatments are associated with a sequential reduction in microfilaria. As noted above, ivermectin is also the treatment of choice for *S. stercoralis.* It is active against *A. lumbricoides* and *E. vermicularis,* but it is variably effective against *T. trichiura* and does not kill hookworms. Ivermectin is also effective against arthropods, and is used for the treatment of scabies.

Diethylcarbamazine

Diethylcarbamazine, a piperazine derivative, is the drug of choice for the treatment of lymphatic filariasis due to *Wuchereria bancrofti, Brugia malayi* and *Brugia timori.* It kills microfilariae and has some activity against adult worms. Diethylcarbamazine seems to affect filariae in multiple ways. It causes a decrease in muscle activity, which eventually leads to immobilization. This may be due to the piperazine moiety.

Piperazine acts as a low-potency agonist at extra-synaptic γ-aminobutyric acid-gated chloride channels in nematode muscle. This increases chloride conductance. There is hyperpolarization and suppression of spontaneous action potentials. Worms are paralyzed. It is possible that diethylcarbamazine acts by this mechanism.

Diethylcarbamazine also alters the surface characteristics of the worm, enhancing killing by human host defenses. Finally, it affects arachidonic acid metabolism and disrupts parasite microtubules. Untoward effects due to the release of parasite antigens are common, including fever, pruritus and occasionally hypotension. Dose-related gastrointestinal side-effects also occur. Acute psychotic reactions have been reported, but are rare. Life-threatening reactions, including encephalopathy, have been reported in patients with heavy *Loa loa* infections. They are thought to be due to the release of worm antigens in the central nervous system and concomitant inflammatory responses.

Ivermectin

Ivermectin is the 22,23-dihydro derivative of ivermectin B1, a macrocyclic lactone produced by the actinomycete, *Streptomyces avermitilis.* It was first developed as an insecticide and as a veterinary drug for treatment and prevention of nematode and arthropod infections in cattle and domestic pets. It is well absorbed orally and typically given as a single dose. Ivermectin potentiates the opening of glutamate-gated chloride channels in nematodes and arthropods. The result is an influx of chloride. In nematodes this is associated with paralysis of the pharyngeal pumping motion. Ivermectin is generally well tolerated in humans, but therapy may be complicated by fever, pruritus, arthralgias, myalgias and cutaneous edema, due in part to release of worm antigens.

Future considerations

The availability of ivermectin and diethylcarbamazine allows for the treatment of the major filarial infections. Ivermectin is indicated for onchocerciasis and is effective against *Mansonella streptocerca.* Diethylcarbamazine is the drug of choice for *W. bancrofti, B. malayi, L. loa* and tropical pulmonary eosinophilia. Both are active against *Mansonella streptocerca.* Neither works against *Mansonella perstans*; mebendazole is the drug of choice. Diethylcarbamazine is administered orally over 14 to 21 days. A drug(s) that is active against adult filariae, results in minimal or slow release of antigens, and can be administered as a single oral dose would be preferable.

Platyhelminths: trematodes and cestodes

While trematodes and cestodes have, in general, received the least attention from a biochemical or molecular biological perspective, therapy against them is excellent thanks to large scale screening programs. The development of praziquantel, a heterocyclic prazino-isoquinoline derivative, stands among the great therapeutic accomplishments of this century. Its broad spectrum of activity revolutionized the treatment of fluke and tapeworm infections. Praziquantel is now the drug of choice for schistosomiasis, most other fluke infections, and intestinal tapeworms. It is active against *Schistosoma mansoni, S. hematobium, S. japonicum, S. melongi* and *S. intercalatum.* It replaced several highly toxic drugs previously used for *S. mansoni* and S. *haematobium,* and in the case of *S. japonicum,* was the first drug to be effective. The possibility of praziquantel-tolerant *S. mansoni* has been suggested by reports of decreased cure rates in some areas, but re-infection could not be excluded. Frank resistance has not been confirmed.

Praziquantel is also recommended for the treatment of the liver flukes, *Clonorchis sinensis, Opisthorchis viverinni,* and *Metorchis conjunctus,* the intestinal flukes *Fasciolopsis buski, Heterophes heterophes, Metagonimus yokogawi,* and *Nanophyetus salmincola,* and the lung fluke *Paragonimus westermani.* Praziquantel is well tolerated orally and administered in three doses daily for one or two days depending on the indication. Only *Fasciola hepatica,* the sheep liver fluke, seems to be resistant to praziquantel. The veterinary fasciolide, triclabendazole, was effective, but production recently ceased. Bithionol is now the only drug available for treatment of *F. hepatica.*

All of the adult tapeworms that are found in humans, including *Taenia solium,* the pork tapeworm, *Taenia saginata*, the beef tapeworm, *Diphyllobothrium latum*, the fish tape worm, *Dipylidium caninum*, the dog tape worm, and *Hymenolepis nana*, the dwarf tapeworm, are susceptible to a single dose of praziquantel. A higher dose is needed for *H. nana*. Praziquantel is not effective in the treatment of echinococcosis. Albendazole (see Intestinal Helminths) is the drug with the greatest activity against *E. granulosus* and *E. multilocularis*, but high doses are needed for prolonged periods of time, and it is variably effective.

The treatment of neurocysticercosis, which is caused by the larval form of *T. solium*, is controversial. Both praziquantel and albendazole can kill cysticerci in the central nervous system. Corticosteroids are usually administered with them to reduce inflammatory reactions in the brain. The penetration of praziquantel into the central nervous system is decreased by corticosteroids while that of albendazole is increased. Most physicians prefer albendazole for that reason. It is also administered for a shorter period of time than praziquantel. Both drugs are contraindicated in persons with cysticerci in the eye or spinal cord.

Praziquantel

The precise mechanism of action of praziquantel is uncertain. The drug is rapidly taken up by flukes and tapeworms. In adult schistosomes the tegument of the worm is damaged, resulting in intense vacuole formation and calcium influx (Figure 17.2). Tetanic contraction of the worm's musculature and paralysis follow. Damaged worms are often swept back to the liver. A number of surface antigens, including actin, are exposed. The damaged surface is rendered susceptible to immune attack.

In the case of *Hymenolepis* spp., praziquantel causes calcium release from endogenous stores, leading to massive contraction of the worm. This is followed by expulsion from the gastrointestinal tract. The tegument of the neck region of the adult tapeworm develops vacuoles, while the scolex and proglottids appear to be unaffected.

Praziquantel is well absorbed orally. It is generally administered in three doses daily for 1 or 2 days. Mild abdominal discomfort, diarrhea, malaise, headache and dizziness are common, but seldom require the termination of therapy. Some recipients manifest allergic symptoms with fever, rash, pruritus and eosinophilia in response to released worm antigens.

Future considerations

For now, praziquantel provides excellent efficacy with acceptable toxicity and pharmacodynamics for the treatment of all but a few trematode and cestode infections. The only important human fluke that does not respond to praziquantel is *F. hepatica*. The veterinary drug, trichlabendazole, was effective, but it is no longer in production. The only drug now available is bithionol, which is associated with nausea, vomiting, abdominal pain, diarrhea, leukopenia, hepatitis and photosensitivity reactions. Improved chemotherapy is needed for *F. hepatica*.

In respect to cestodes, praziquantel is excellent when administered as a single dose for the treatment of adult tapeworms in the gastrointestinal tract. The treatment of neurocysticercosis remains controversial. Both praziquantel and albendazole can kill cysticerci in the central nervous system. The role of chemotherapy may vary depending on the number and site of the lesions. Neither drug works well against cysts in the ventricles or subarachnoid space. Studies of the natural history of cysticercosis

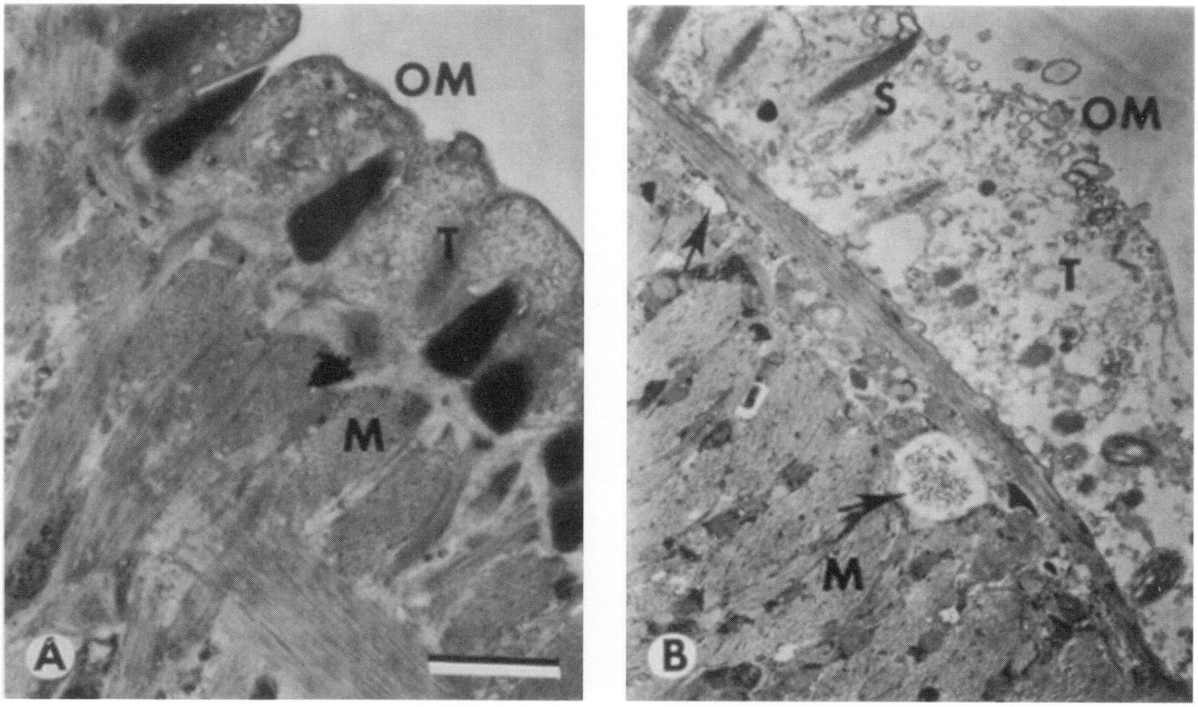

FIGURE 17.2 Effects of praziquantel (1 μM) on adult male schistosome (6–8 week infections) in B (control A) by transmission electron microscopy. Praziquantel resulted in the formation of vacuoles and disruption of the spines. Incubation was for 1 h. OM = outer member; T = tegument; S = spine; M = muscle. (Kindly provided by David P. Thompson, Ph.D., Pharmacia & Upjohn, Inc., Kalamazoo, MI.)

are needed, along with the development of therapeutic approaches that would further minimize the inflammatory response elicited by released cyst antigens.

The treatment of echinococcosis remains poor. Albendazole or mebendazole cure less than half of the cases, and praziquantel is not effective. Surgical resection or percutaneous or laparoscopic aspiration and instillation of a scolacidal agent currently provide the best opportunity for cure. A better understanding of the biochemistry and cell biology of echinococcal cysts is needed for the development of new approaches to chemotherapy.

CONCLUSION

The past two decades have seen unprecedented advances in understanding the molecular biology and biochemistry of the major human parasites. The opportunities for identification of new molecular targets and the synthesis of compounds that have activity against them are unprecedented and offer great hope for the future. As the genomes of the parasites are sequenced new targets will almost certainly be identified, and inhibitors synthesized. Microarray technology offers the potential for mass screening of compounds on a scale once unimaginable.

The excitement engendered by these advances must be tempered by the realities of drug development and production. The first hurdle is drug toxicity. At times predictable, but often idiosyncratic, unexpected untoward effects have prevented the use of many otherwise effective anti-parasitic drugs. Particularly troublesome are serious toxicities that occur at low frequency. They may come to attention only after a drug has been marketed.

To be successful, new drugs must also have favorable pharmacodynamics. Oral administration in a single dose or several doses on one day is particularly important for mass treatment programs in areas where the lack of education, refrigeration and other logistic factors may interfere with compliance.

The economic realities of new drug development must also be considered. Many parasitic diseases are endemic in impoverished areas where the likelihood of profits is too small to engender investment by major pharmaceutical companies. Even if a highly effective anti-parasitic drug is identified, the cost of development may be prohibitive.

The cost of production is also critical as illustrated by eflornithine. It seems tragic that the manufacture of an effective anti-parasitic drugs should cease for economic reasons at a time when the drug is critically needed. The story can be different for drugs that are used in industrialized countries for other human indications or veterinary infections. As illustrated by ivermectin, which is marketed for the treatment of the dog heartworm, *Dirofilaria immitis,* veterinary use and profits have offset production costs. Ivermectin has been donated by the manufacturer for the treatment of onchocerciasis.

There are no simple answers to the economic constraints on basic research, drug development and production. There is increasing realization that health is an issue of international security. The World Health Organization, World Bank, National Institutes of Health, Gates Foundation and other governmental and non-governmental groups have increasingly focused their resources on diseases of the non-industrialized world. Their continued support is critical to the development, production and distribution of new anti-parasitic drugs. The future seems bright, but it will take a concerted effort to bring the fruits of modern science to the bedside of the poor who suffer from parasitic diseases.

FURTHER READING

Anon. (2002). Drugs for parasitic infections. *Med. Lett. Drugs Ther.* 1–12, [online].

Campbell, W.C., Fisher, M.H., Stapley, E.O. *et al.* (1983) Ivermectin: a potent new antiparasitic agent. *Science* **221**, 823–828.

Gillespie, S.I. and Pearson, R.D. (eds) (2001). *Principles and Practice of Clinical Parasitology*, Chichester: John Wiley and Sons.

Guerrant, R.L., Walker, D.H. and Weller, P.F. (eds) (1999). *Tropical Infectious Diseases: Principles, Pathogens and Practice.* New York: Churchill Livingstone.

Guttman, P. and Ehrlich, P. (1891). Ueber die wirkung des methyleneblau bei malaria. *Berlin Klin. Wochenschr.* **28**, 953–956.

Hardman, J.G. and Limbird, L.E. (eds) (2001). *Goodman and Gilman's Pharmacological Basis of Therapeutics*, New York: McGraw-Hill.

Health Information for International Travel, 2001–2002. Atlanta, GA: US Department of Health and Human Services, Public Health Service, Centers for Disease Control and Prevention.

Horton, J. (2000). Albendazole: a review of anthelmintic efficacy and safety in humans. *Parasitology* **121, Suppl.**, S113–S132.

Jha, T.K., Sundar, S., Thakur, C.P. *et al.* (1999) Miltefosine, an oral agent, for the treatment of Indian visceral leishmaniasis. *N. Engl. J. Med.* **341**, 1795–1800.

King, C.H. and Mahmoud, A.A. (1989). Drugs five years later: praziquantel. *Ann. Intern. Med.* **110**, 290–296.

Marr, J.J. (1991). Purine analogs as chemotherapeutic agents in leishmaniasis and American trypanosomiasis. *J. Lab. Clin. Med.* **118**, 111–119.

Meyerhoff, A. (1999). US Food and Drug Administration approval of AmBisome (liposomal amphotericin B) for the treatment of visceral leishmanias. *Clin. Infect. Dis.* **28**, 42–48.

Pearson, R.D. (1999). Agents active against parasites and *Pneumocystis carinii*. In: Mandell, G.L., Bennett, J.E. and Dolin, R. (eds). *Principles and Practice of Infectious Diseases*, 5th edition, New York: Churchill Livingstone, pp. 505–539.

Pearson, R.D. (2001). Atovaquone/proguanil for the treatment and prevention of malaria. *Curr. Infect. Dis. Reps* **3**, 47–49.

Pearson, R.D. (2002). Parasitic diseases: helminths. In: Yamada, T., Alpers, D.H., Laine, L., Owyang, C. and Powell, D.W. (eds). *Textbook of Gastroenterology*, 4th edition, Philadelphia: Lippincott-Raven.

Shapiro, T.A. and Englund, P.T. (1990). Selective cleavage of kinetoplast DNA minicircles promoted by antitrypanosomal drugs. *Proc. Natl. Acad. Sci. USA* **87**, 950–954.

Slunt, K.M., Mann, B.J., Macdonald, T.L. and Pearson, R.D. (1999). The therapeutic potential of DNA topoisomerase II poisons in treating leishmaniasis. *Recent Res. Dev. Antimicrob. Agents Chemother.* **3**, 357–370.

Sousa Estani, S., Segura, E.L., Ruiz, A.M. *et al.* (1998). The efficacy of chemotherapy with benznidazole in children in the indeterminate phase of Chagas' disease. *Am. J. Trop. Med. Hyg.* **59**, 526–529.

Stead, A.M.W., Bray, P.G. and Edwards, I.G. *et al.* (2001). Diamidine components: selective uptake and targeting in *Plasmodium falciparum*. *Mol. Pharmacol.* **59**, 1298–1306.

Wellems, T.E. and Plowe, C.V. (2001). Chloroquine-resistant malaria. *J. Infect. Dis.* **184**, 770–776.

Index